COMPACTION OF SOILS AND GRANULAR MATERIALS:

A review of research performed at the Transport Research Laboratory

DEPARTMENT OF TRANSPORT

Transport Research Laboratory

COMPACTION OF SOILS AND GRANULAR MATERIALS:

A review of research performed at the Transport Research Laboratory

A W Parsons FIHT

London: H M S O

First published 1992

ISBN 0 11 551091 5

PREFACE

This book provides a complete review and compilation of the results of the programme of research on the compaction of soil and granular base material at the Transport Research Laboratory (formerly the Road Research Laboratory and Transport and Road Research Laboratory) from the inception of the studies in the mid 1940s until the end of 1990.

The objectives of the book have been to:–

(i) Provide a permanent record, under one cover, of all valid research data compiled at the Laboratory relating to investigations into the compaction of soil and unbound granular materials;

(ii) To make comparisons of the performances of a wider range of compaction machines than has ever been achieved before; and

(iii) To interpret the data so as to provide the practising engineer with up-to-date guidance on the performance of plant and the preparation and control of compaction specifications.

It is hoped that this single volume will comprise a unique reference book on all aspects of soil compaction.

Results of tests with compaction plant are generally portrayed in this book in terms of relative compaction (dry density expressed as a percentage of the maximum dry density achieved in the 2.5 kg rammer compaction test at the relevant period) so as to eliminate the effects of minor variations in the properties of the test soils over the 45 years in which the investigations took place. There is clear evidence in the data that early results obtained in the programme of research compare favourably with those obtained in more recent work and such early results are shown to be perfectly valid in assessing the performance of modern plant.

The author is indebted to the former Director of the Transport and Road Research Laboratory (Mr D F Cornelius) for arranging that the time and facilities necessary for the writing of the book were made available. The author also acknowledges the role played by the Highways Engineering Division of the Department of Transport in recognising the need for this book and authorising the funding of the project. It must be recognised, however, that this work could not have been undertaken without the initiative, support, guidance and helpful comments of the former Head of the Ground Engineering Division, Dr M P O'Reilly. The author is also grateful to Mrs V Atkinson for the preparation of many of the figures and her assistance in the analytical work.

FOREWORD

This book provides, under a single cover, all the results of the compaction research on soils and granular materials performed since 1946 as part of the Soil Mechanics / Ground Engineering research programme at the Transport Research Laboratory. Its statistical analyses of the results provide, for the first time, general findings on the performance of plant based on all available data. Comparisons have been made of the performances of a wide range of compaction machines, providing the practising engineer with up-to-date guidance on the performance of plant and the preparation and control of compaction specifications. The quantity of data on the performance of compaction plant makes this a unique book for all concerned with the manufacture and operation of such equipment.

The author, Tony Parsons, has been involved in the work on soil compaction since 1952, and has carried out or supervised the various TRL studies and written many of the individual reports produced since then. Research results were initially recorded as comprehensive hand-written notes, tables and graphs in a series of 'Record Books', and these have been used extensively in the preparation of this book. In addition, minor anomalies in published data have been corrected and additional information, omitted from the individual reports, has been included.

Much of this book is factual, but of necessity the interpretations of the data involve the judgements of the author based on his wide experience of compaction processes; such judgements are not necessarily those of TRL or the Department of Transport.

John Wootton
Chief Executive
Transport Research Laboratory

CONTENTS

Paragraph

Paragraph

CHAPTER 1 INTRODUCTION

1.1 The need for compaction has long been recognised in the construction of earth and rock fill structures. The effects of stresses imposed by superimposed structures, water impounded by dams, traffic on roads, and the self weight of the fill, all can introduce changes in the orientation and packing of the solid particles if they are not already formed into a dense pattern by the process of compaction. In addition, the effects of weathering caused by ingress of water, temperature variation, and, on occasion, chemical action, can all be reduced, if not nullified, by the proper application of the compaction process to produce a fill with a sufficiently high value of density.

1.2 The achievement of these beneficial effects is brought about in the compaction process by the reduction of the void content in a material, which in turn reduces the permeability to water, increases the shear strength of the soil or rock, and increases resistance to settlement and other deformations.

1.3 The compaction process was first studied in detail by Proctor (1933) when he demonstrated the effects of moisture content on the state of compaction achieved, using a hand operated rammer in the laboratory. As a tribute to this early, revealing work the relation between dry density and moisture content is often referred to as a 'Proctor curve', even when relating to compaction methods far removed from the method used originally by Proctor.

1.4 Since Proctor's early work, many other researchers have studied the compaction process, but on a laboratory scale only prior to 1940.

1.5 Shortly after the end of the 1939–45 war, when the then Road Research Laboratory was resuming its interest in road construction and road traffic, studies were begun in 1946 of the operation and performance of full-scale compaction plant on a range of soils which were regarded as representative of those encountered in earthwork construction in the United Kingdom. At that time the work was unique and a world-wide reputation was built up as the sole source of information on the performance of various types of compactor. Included in the early work were examinations of the newly developed application of vibration in the form of vibrating rollers and vibrating-plate compactors.

1.6 In parallel with the studies of the performance of various types of compactor, ancillary studies of methods of measuring the in-situ density of compacted soil were also carried out. Thus, in addition to the general application of the British Standard sand-replacement test (British Standards Institution, 1990b) in the studies of the performance of the compactors, rubber balloon methods and nuclear methods for determining in-situ density were also studied at various times.

1.7 Laboratory compaction tests also came under scrutiny from time to time, and when the necessity arose a new laboratory vibratory compaction test, the BS vibrating hammer compaction test (British Standards Institution, 1990a) was developed. The need for this came about when it was found that, where unbound granular pavement materials were concerned, neither the 2.5 kg nor the 4.5 kg rammer tests traditionally used in the laboratory represented the performance of full-scale compaction plant, or provided guidance on the optimum moisture contents for effective full-scale compaction.

1.8 Additional studies were carried out on construction sites as part of the general programme of compaction research. This work highlighted the variability of the measured states of compaction and the difficulty of assessing whether specifications for compaction were complied with. The justification for a change in specifications for the compaction of earthworks was established in this work, and the change to a 'method specification' for compaction was made easy by the availability of the performance data gathered over the previous years.

1.9 In more recent years performance testing of plant has continued in an intermittent fashion, mainly to provide additional information needed to keep the Department of Transport's 'method specification' as up-to-date as possible.

Additionally, considerable attention has been paid to the problems of moisture content control for fill materials in highway embankments, and this has culminated in the development of the 'moisture condition test' which has been introduced into the 6th edition of the 'Specification for Highway Works' (Department of Transport, 1986) and the latest edition of BS 1377 (British Standards Institution, 1990a).

1.10 Research into various aspects of the compaction process and its control has, therefore, taken place over the considerable period of about 45 years. The work has been published in various documents over the years, although much detail was contained in the unpublished reports which were produced by the Laboratory in the period prior to 1966.

1.11 The objective of this book is to bring together under one cover the more important information produced over these many years, and which is still valid today. It is intended to provide information for the practising engineer engaged in earthwork and pavement construction (excluding bituminous materials), and as a reference book on the many facets of compaction.

References

British Standards Institution (1990a). British Standard methods of test for soils for civil engineering purposes: BS1377: Part 4 Compaction related tests. BSI, London.

British Standards Institution (1990b). British Standard methods of test for soils for civil engineering purposes: BS1377: Part 9 In situ tests. BSI, London.

Department of Transport (1986). Specification for highway works. 6th edition. HM Stationery Office, London.

Proctor R R (1933). The design and construction of rolled earth dams. Engineering News Record, 111 (9) 245–8, (10) 216–9, (12) 348–51, (13) 372–6.

CHAPTER 2 BASIC PRINCIPLES OF COMPACTION

2.1 Soil compaction is the process whereby soil particles are constrained to pack more closely together through a reduction in air voids, generally by mechanical means (Road Research Laboratory, 1952).

2.2 When air voids are reduced to a low value in a fill material at or near its equilibrium moisture content for its particular location in an embankment, there will be little potential for variations in moisture content to occur, even in the long term. The settlement in the fill will also be negligible, although settlement may still arise from consolidation of the sub-soil foundation.

2.3 Compaction is measured quantitatively in terms of the dry density of the soil, ie the mass of solids per unit volume of the compacted material. A diagrammatic representation of the solid, water and air phases of compacted soil is given in Figure 2.1. The inter-relations between the various terms, using consistent units throughout, are as follows:-

$$\text{Bulk density } \rho = \frac{M}{V}$$

where M=mass of a given volume of soil,
and V=the given volume of soil.

$$\text{Dry density } \rho_D = \frac{M_s}{V} = \frac{100\,\rho}{(100 + w)}$$

where M_s=mass of solid particles in the given volume of soil,
w =moisture content (per cent)
$=\frac{100M_w}{M_s}$ (per cent),
and M_w=mass of water in the given volume of soil.

Void ratio is defined as $\frac{\text{volume of voids}}{\text{volume of solids}}$

$$\text{Void ratio } e = \frac{V_v}{V_s} = \frac{\rho_s}{\rho_D} - 1$$

where V_v=volume of voids in the given volume of soil,
V_s=volume of solid particles in the given volume of soil,
and ρ_s=particle density.

Air voids or air content is the volume of air expressed as a percentage of the total volume of compacted material:–

$$\text{Air voids } V_a = \frac{100\,V_o}{V} \text{ (per cent)}$$

$$= 100\left\{1 - \rho_D\left(\frac{1}{\rho_s} + \frac{w}{100\,\rho_w}\right)\right\} \text{ (per cent)}$$

where V_o=volume of air in the given volume of soil,
and ρ_w=density of water.

At a given value of moisture content, the maximum possible dry density, at zero air voids, is given by:–

$$\rho_D = \frac{1}{\frac{1}{\rho_s} + \frac{w}{100\rho_w}}$$

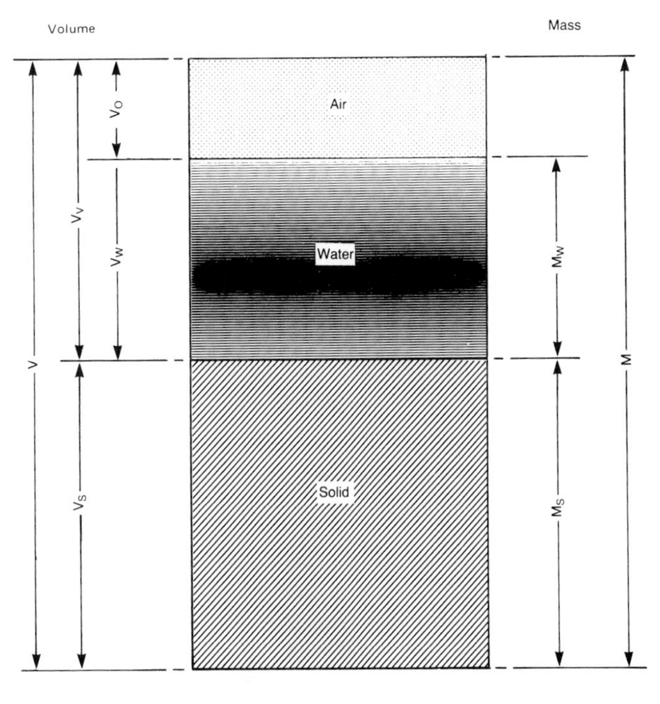

Figure 2.1 Composition of soil

2.4 The principal factors which influence the dry density achieved are:–

1. Moisture content
2. Energy input
3. Volume compacted and
4. Soil characteristics.

Effect of moisture content

2.5 A simplified concept to explain the effect of moisture on the compaction process is that the water acts as a lubricant, with increasing lubrication giving rise to easier compaction and the achievement of increased levels of dry density. This process can continue until a near zero-air-void condition is achieved, beyond which the addition of further water can only be brought about by the displacement of solids, and decreases in dry density are produced with further increases in moisture content. A typical relation between dry density and moisture content is shown in Figure 2.2.

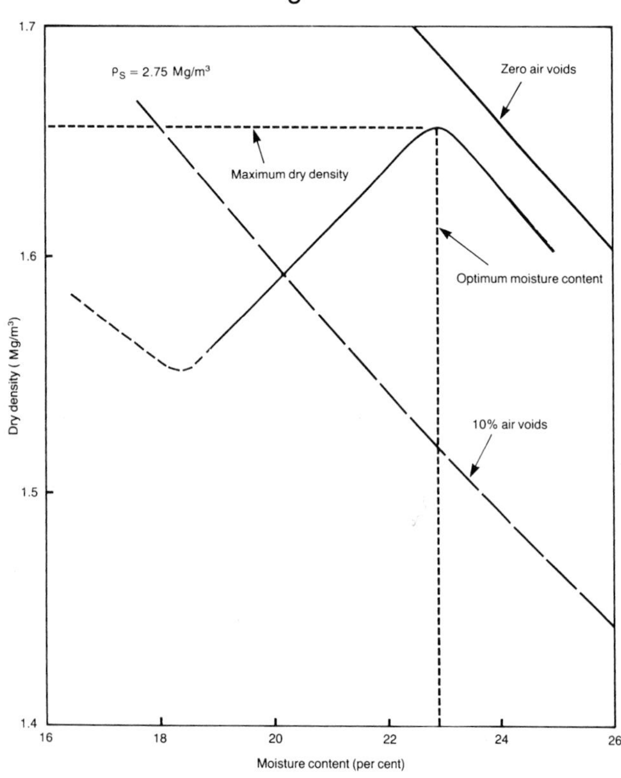

Figure 2.2 Typical relation between the dry density of a clay soil and its moisture content for a constant amount of compaction

2.6 Although the concept of lubrication explains the mechanism of the changes in dry density with variation in moisture content a more accurate explanation can be propounded based on the inter-relation between dry density, moisture content and shear strength. Typical

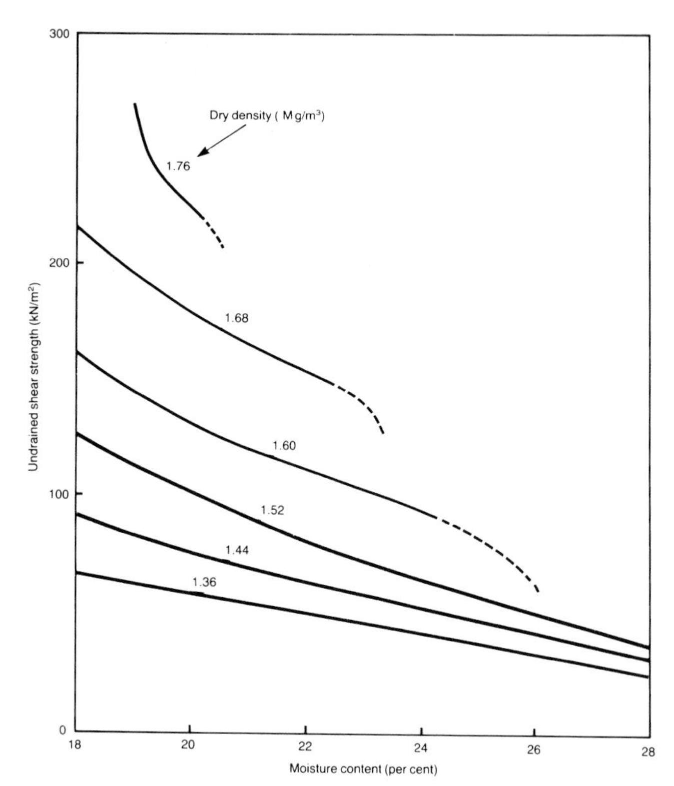

Figure 2.3 Relations between undrained shear strength (determined using the unconfined compression test) and moisture content at various dry densities in a heavy clay soil

relations between undrained shear strength and moisture content of a heavy clay soil, for different values of dry density (Lewis, 1959) are given in Figure 2.3. If the shear stresses exerted by the particular method of compaction exceed the existing shear strength of the soil, further densification will take place unless the material is already in a near-saturated condition, ie near to zero air voids. From Figure 2.3 it can be seen that the effect of adding moisture to a heavy clay soil at a constant dry density is to decrease the strength of the soil and thus an increase in dry density would be required to sustain a shear strength equivalent to a constant compactive effort. As mentioned previously, the dry density increases with increase in moisture content until such time as the zero air voids line is approached. At this stage the water in the soil will take up some of the stress exerted by the compactor and excess pore water pressures will be generated.

2.7 Returning to Figure 2.2, where a typical relation between dry density and moisture content is illustrated, the peak of the curve is defined in terms of the 'maximum dry density' and the 'optimum moisture content'. At moisture contents considerably less than the optimum moisture content the relation between dry density and moisture content often shows a reversal in that dry density increases with decrease in moisture content, as shown in Figure 2.2. This effect is particularly marked in granular soils when vibratory compaction is used.

2.8 With uniformly graded sands and gravels the achievement of a low air content is not possible. The uniform size of the particles prevents the dense packing necessary to produce high values of dry density, and the free-draining nature of such material prevents the voids from retaining water. The effect is that at maximum dry density and optimum moisture content the air content can be as high as 10 to 15 per cent, as shown in Figure 2.4.

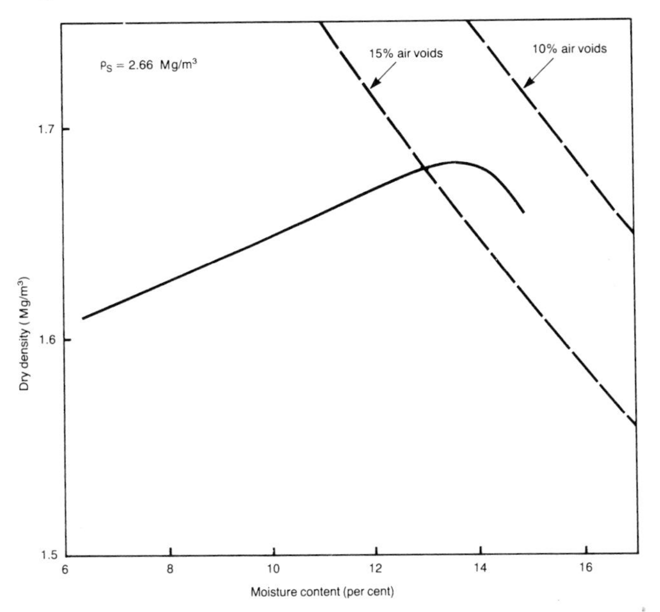

Figure 2.4 Typical relation between dry density and moisture content for a uniformly graded sand (constant compactive effort)

2.9 The relation between dry density and moisture content is not unique for a given sample of soil, but varies depending on the compactive effort employed. An increase in compactive effort will move the relation in Figure 2.2 upwards and to the left of the position shown with an increase in maximum dry density and a reduction in optimum moisture content. Potentially, for any given sample of soil, an infinite number of dry density-moisture content relations can exist, each with its individual values of maximum dry density and optimum moisture content.

Effect of energy input

2.10 The total energy used in the compaction process is dependent on a number of factors. In the laboratory it may be combinations of mass, height of drop and number of blows of a rammer, or of power input, mass and time of vibration of a vibrating hammer. In full-scale compaction the mass, width, vibrating force (if any) and number of passes of the compactor are the main factors involved.

2.11 It is usual, when determining the effect of energy input, to maintain the soil at a constant moisture content and determine the relation between dry density and compactive energy as, for example, with the dry density – passes relation for full-scale compaction plant. It is a common practice to plot graphs of this sort with the energy factor on a logarithmic scale (Morel and Machet, 1980), allowing a straight line relation to be plotted up to the point at which the near-zero air void condition is approached. An example of the relation between dry density and number of passes, with passes plotted on both linear and logarithmic scales, is given in Figure 2.5. The advantage of the logarithmic scale is that it facilitates regression analysis to determine the best-fit relation where results are somewhat

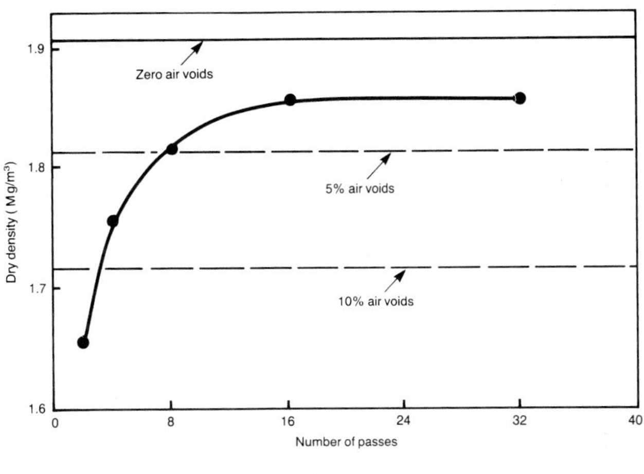

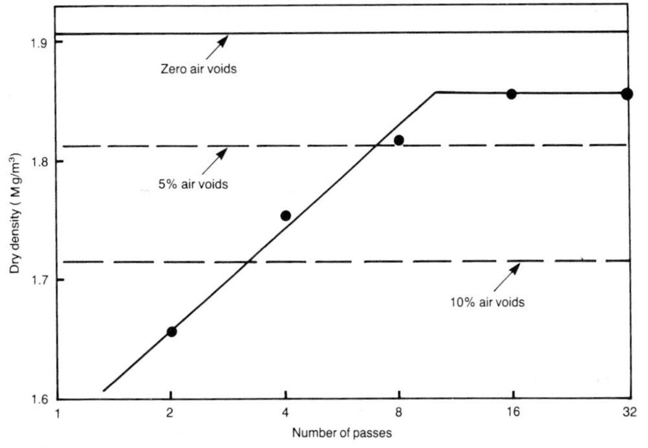

Figure 2.5 The relation between dry density and number of passes with (*top*) passes on a linear scale and (*bottom*) passes on a logarithmic scale. Results obtained on sandy clay at a moisture content of 15.5 per cent using an 8-tonne smooth-wheeled roller

scattered. However, where regression analyses are carried out it must be remembered that the points close to the zero air voids line, where further increases in dry density cannot take place, should not be included in the determination.

2.12 To illustrate the inter-relation between compactive energy, dry density and moisture content, Figure 2.6 gives the results of using four compaction energies, each at four different moisture contents.

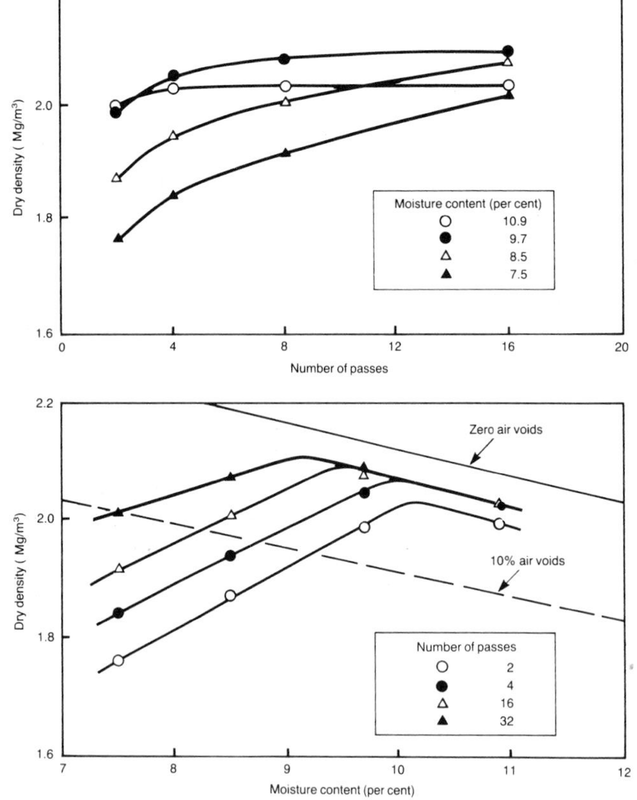

Figure 2.6 Relations between dry density, number of passes and moisture content. These results were obtained on a 200 mm compacted layer of well-graded sand, using a 145 kg vibrating-plate compactor

2.13 The increase in dry density with successive increments of energy, as for example with successive passes of a roller, can be explained on the basis of the shear stress – shear strength concept discussed in Paragraph 2.6. Referring to Figure 2.7, sinkage of the roll on Pass No 1 will increase, with associated increase in dry density, until a shear strength is mobilised which equalises the shear stress associated with the resulting contact area. On Pass No 2 and successive passes, the increased dry density following the previous pass results in a reduced sinkage and contact area; shear stresses are, therefore, increased and dry density is further increased.

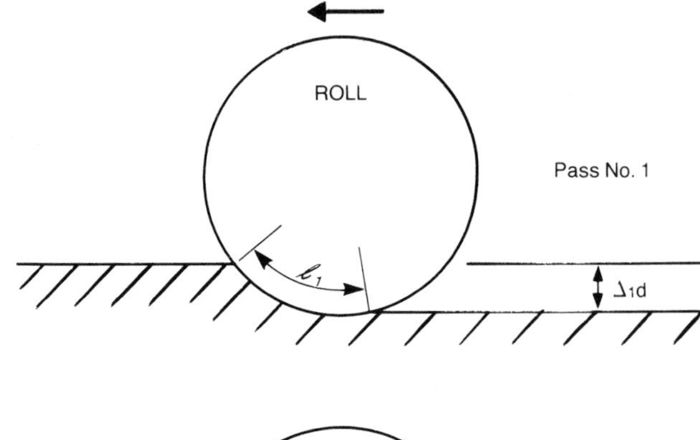

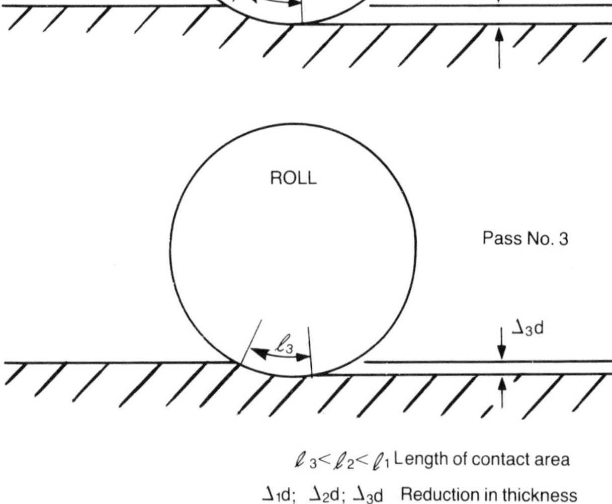

Figure 2.7 Reductions in contact area of a compaction roll on successive passes

2.14 With a rammer, the increased dry density with each blow increases the stiffness of the soil, resulting in greater rates of deceleration during impact and hence increased dynamic stresses with successive blows. It can be seen from this explanation that the shear stresses in the soil during a process of compaction involving multiple passes of a machine, or multiple blows of a rammer, are not necessarily constant but can increase as the dry density increases. Eventually, the increase per pass becomes so small as to be negligible, in which case a state of 'compaction to refusal' has been reached, or the zero air void condition is approached and transfer of stress to the moisture takes place to create excess pore water pressures. When these latter conditions are reached, no further increase in dry density occurs unless the soil is sufficiently permeable to allow the expulsion of water.

Effect of volume compacted

2.15 Shear stresses exerted by the compactor reduce with increasing depth in the layer. Thus the shear strengths mobilised to equalise the shear stresses will also decrease with increasing depth and a gradient of dry density is produced. The general form of the relation between dry density and depth in a compacted layer is given in Figure 2.8. The effect of increasing the thickness of compacted layer, and hence of increasing the effective volume to be compacted, will, therefore, generally be to reduce the average state of compaction in the layer.

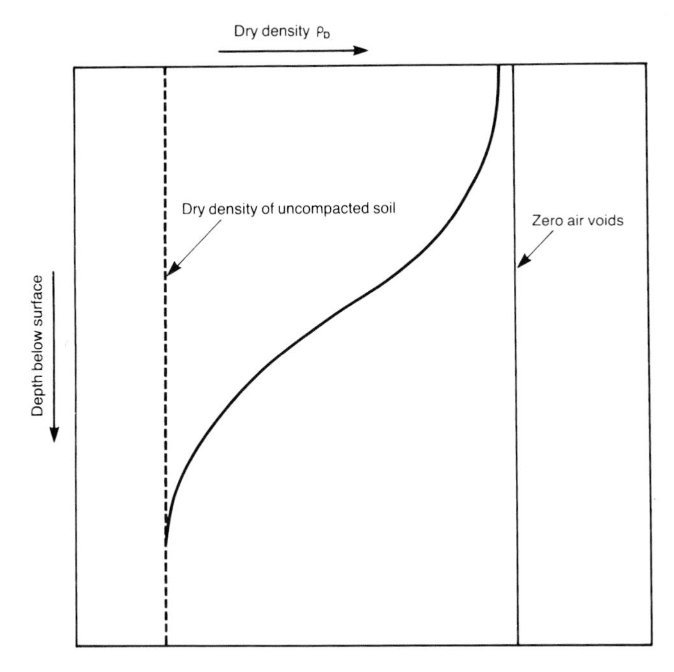

Figure 2.8 General form of relation between dry density and depth in a compacted layer of soil

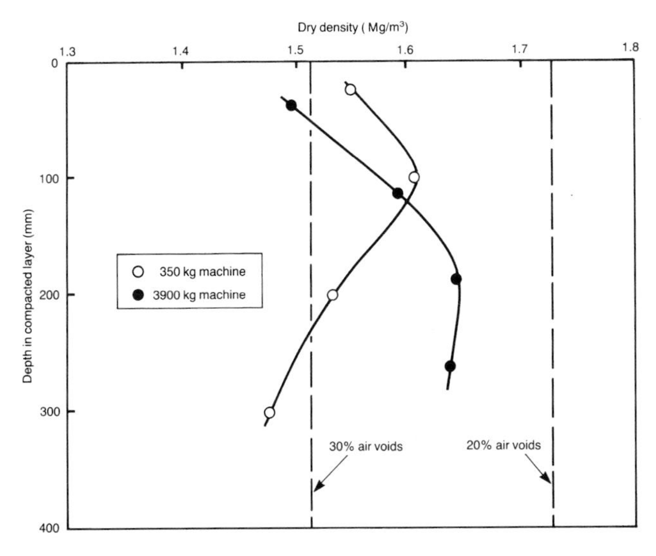

Figure 2.9 Relations between dry density and depth in the compacted layer after 32 passes of two sizes of vibrating roller. The soil is a uniformly graded fine sand with a moisture content of 8.5 per cent

2.16 Where shear stresses near the surface of the layer exceed the maximum possible shear strength, for instance with non-cohesive soils such as sands, over-stressing can occur in the near-surface material, yielding lower values of dry density than occur deeper in the layer. Examples of this are given in Figure 2.9. As the Figure shows, this effect is most marked with heavy vibrating rollers, and is particularly prevalent on sands with low coefficients of uniformity, ie of a single-sized nature.

2.17 Increase in compacted volume, either by thicker layers in the field or by larger moulds in the laboratory, results in movements of the relation between dry density and moisture content downwards and to the right, giving rise to reduced values of maximum dry density and increased values of optimum moisture content. The inter-relation between depth in the compacted layer, dry density and moisture content is illustrated in Figure 2.10, where dry densities were determined at four increments of depth in the layer, each at three values of moisture content (Lewis, 1954). As only three points are available for each relation between dry density and moisture content in this Figure, the most likely curves have been drawn.

Effect of soil characteristics

2.18 Soils of various types have different capacities for absorbing water. Thus a high plasticity clay may have a moisture content in excess of 30 per cent and have a high strength, whereas similar strengths may be attained at moisture contents of only 15 to 20 per cent with a low plasticity clay and at moisture contents less than 10 per cent with sands. The surface area of the coarser particles and the plasticity of the fines determine the moisture capacity and this, in turn, for a well-graded soil, determines the potential dry density that can be achieved with a given compactive effort. Figure 2.11 shows typical laboratory results using the British Standard 2.5 kg rammer compaction test (British Standards Institution, 1990) for various types of soil and a crushed limestone aggregate. As the soil becomes less plastic and more granular the relation between dry density and moisture content generally moves upwards and to the left, with increased values of maximum dry density.

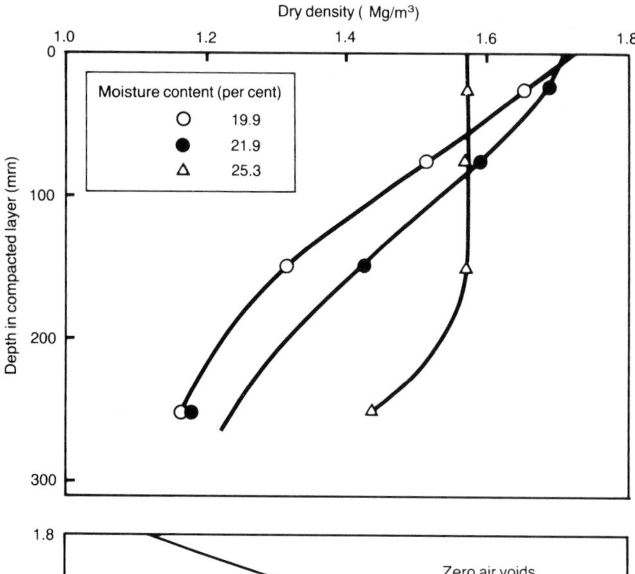

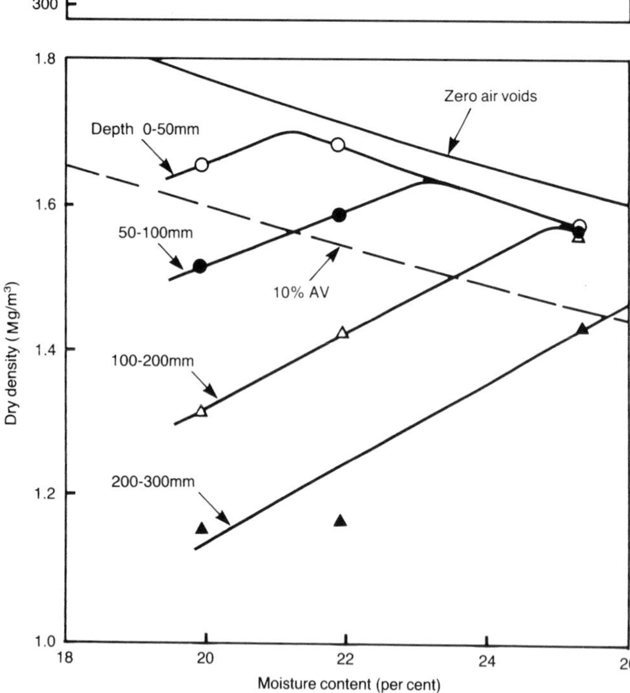

Figure 2.10 Relations between dry density, depth in the compacted layer and moisture content. These results were obtained with 32 passes of an 8-tonne smooth-wheeled roller on heavy clay

Figure 2.11 Relations between dry density and moisture content for various soils and aggregates when compacted in the BS 2.5 kg rammer test (Part 4, BS 1377 : 1990)

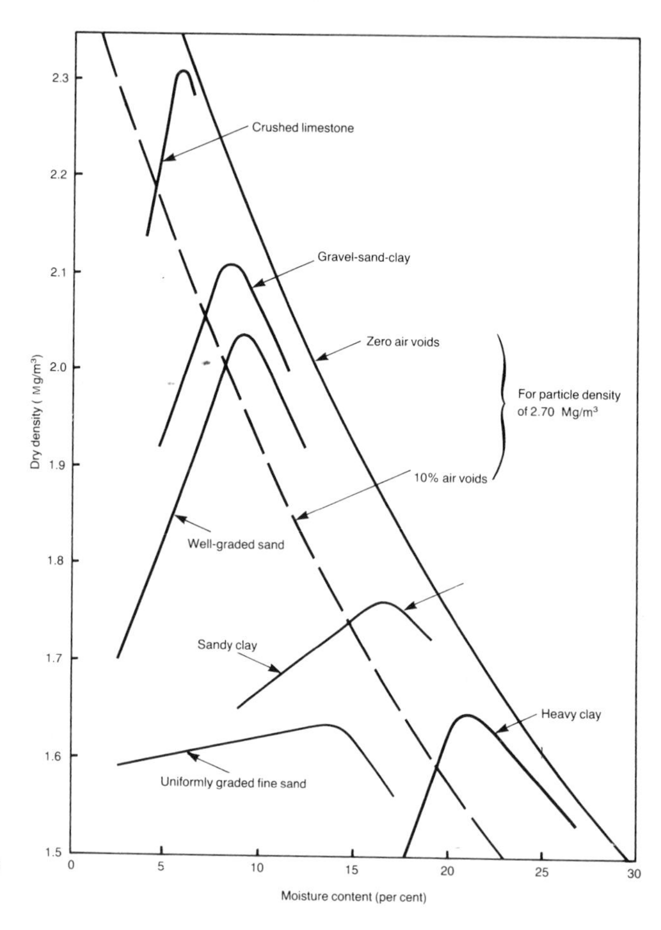

2.19 The exception to this trend is the uniformly graded fine sand (Figure 2.11); with this soil and other uniformly graded materials large voids are left between the single sized particles. The increase in dry density with increase in moisture content is relatively small and an optimum moisture content is difficult to determine because of the free-draining property of the soil.

Secondary factors influencing levels of compaction

2.20 Other factors than those discussed above may also have an influence on the dry density achieved. For example, the stiffness of the layer immediately underlying that being compacted has an effect on the state of compaction achieved, as shown in Figure 2.12. Here the underlying materials were a cement stabilised gravel (E = 250 MPa), a natural sandy gravel (E = 130 MPa) and a low plasticity silt (E = 10 MPa) (Valeux and Morel, 1980). Significant reductions in dry density are shown to occur with the

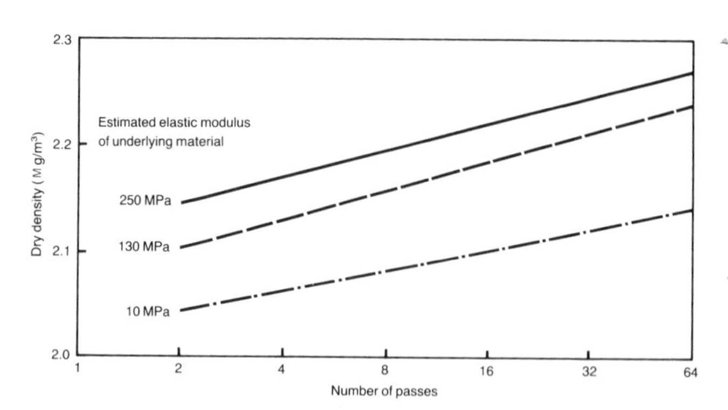

Figure 2.12 Relations between dry density and number of passes obtained with a heavy vibrating roller on crushed gravel (Valeux and Morel, 1980)

weaker underlying material. This can be particularly important in the compaction of road pavement layers, where the achievement of the highest potential values of dry density can benefit the long-term performance of the road.

2.21 With cohesive soils the aggregations of particles to form lumps can influence the level of compaction attained. For example, clay soils excavated from trenches may prove difficult to recompact if large lumps are replaced for compaction by typical small trench compactors. Figure 2.13 compares results obtained when replacing material for compaction in a trench in an 'as excavated' condition with those obtained after treatment of the soil with a rotary cultivator to produce a finer tilth. With sandy clay a reduction of about 1 per cent in dry density occurred when the lumps were not broken down, whereas with heavy clay the loss in dry density was a maximum of about 2.5 per cent. The lower part of Figure 2.13 shows the relation between the proportion of the clay with a lump size in excess of about two-thirds of the thickness of the compacted layer and the moisture content for the tests. This indicates that the main trend was for the proportion of larger lumps to increase with increase in moisture content; this also applies, but to a lesser extent, with material which had been treated with the rotary cultivator. However, at the higher moisture contents such lumps will be weaker and more easily deformed by the compactor. The increase in the proportion of large lumps at low values of moisture content, as shown for the untreated sandy clay, might also be expected as the clay becomes relatively stronger.

2.22 On a larger scale, over-consolidated clay from a deep cutting or borrow pit, excavated by large earthmoving equipment, can also be deposited in fill areas in very large, 'boulder-size' lumps which are difficult to negotiate and compact by even the largest compactors. It is generally considered that the achievement of a relatively fine tilth with no particle aggregation exceeding about two-thirds of the final compacted thickness of the layer, is the best means of ensuring uniformly well compacted clay soils.

Judgement of the quality of compaction

2.23 In Paragraph 2.3 it is stated that compaction is measured quantitatively in terms of the dry density of the soil. However, as can be concluded from the foregoing text, a stark value

Figure 2.13 Effect of 'lumps' on the relations between dry density and moisture content of sandy clay and heavy clay. Results for 100 mm thick layers compacted with 6 passes of a 55 kg vibro-tamper

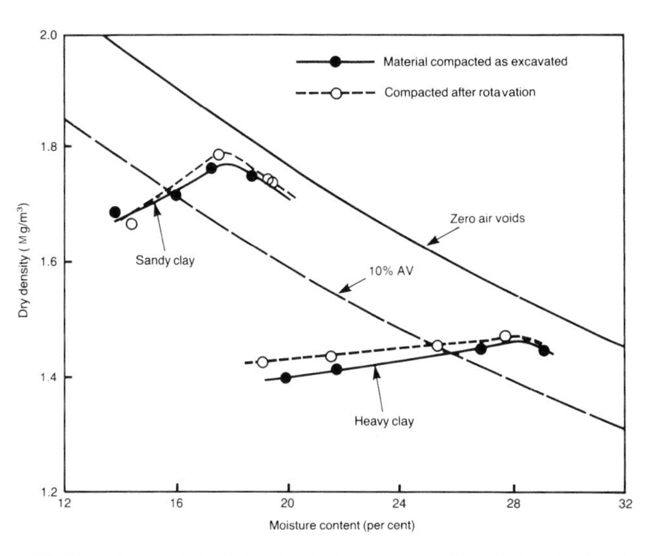

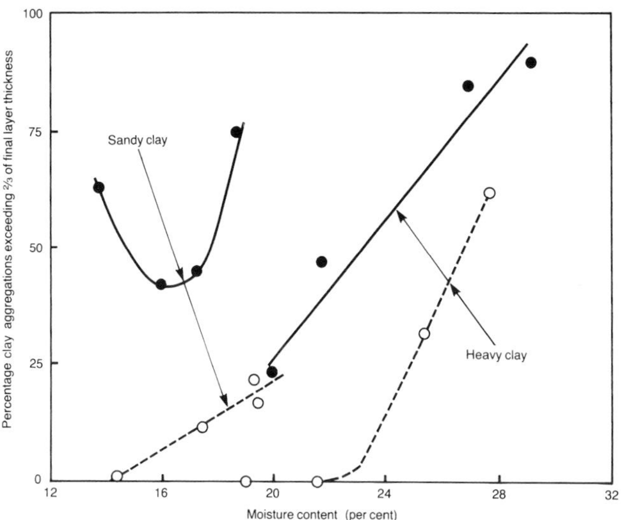

of dry density can be meaningless unless some other information is available on which to judge that value. Figure 2.11 indicates the wide range of values of dry density that can be achieved as the same method of compaction is applied to a range of soil types. The difficulty associated with judging the quality of compaction gives rise to various forms of 'end-product' compaction specifications for interpreting measured values of in-situ dry density.

2.24 A common method of judging compaction is to express the in-situ dry density as a percentage of the maximum dry density determined in a laboratory compaction test on the same soil (Road Research Laboratory, 1952; Reichert, 1980); the resulting percentage is known as the 'relative compaction'. Figure 2.11 illustrates the results for the British Standard

2.5 kg rammer test (British Standards Institution, 1990) for various soils, and the maximum dry density at the peak of each relation between dry density and moisture content could be used as the 'standard' for its respective soil. Thus, for heavy clay, the maximum dry density is 1.65 Mg/m^3 (Figure 2.11). For cohesive materials a common standard is a relative compaction of 90 or 95 per cent of the 2.5 kg rammer test, but with more granular soils, where the rammer method of compaction becomes less efficient (as judged by comparison with the state of compaction achieved by full-scale compaction plant), relative compaction levels of 90 or 95 per cent of the British Standard 4.5 kg rammer test (British Standards Institution, 1990) might be applied. For crushed aggregates a vibratory compaction standard such as the British Standard vibrating hammer compaction test (British Standards Institution, 1990) would be more appropriate.

2.25 An alternative method of judging the quality of compaction is to measure the air voids remaining in the compacted soil. The formula in Paragraph 2.3 is used, and in addition to the in-situ dry density, the moisture content and the particle density must be known. It has been suggested that this method is more appropriate where variations in soil type occur within small distances or limited areas of excavation (Lewis, 1962), as the change in particle density is small in comparison with the potential change in the maximum dry density achieved in a laboratory compaction test. The disadvantage of judging compaction by determining air voids is that air voids can be reduced to a low value simply by increasing the moisture content; strict control of moisture content is necessary, therefore, to ensure that a low air content is not achieved by using a very wet, and hence weak, soil. Common maximum levels of air content for well-graded soils (ie excluding single-sized soils) are 10 or 5 per cent, depending on the location in the fill and the type of material.

2.26 Figure 2.14 gives an example where the two methods discussed in Paragraphs 2.24 and 2.25 are compared. The figure illustrates:–

(i) A laboratory test result, used for establishing the maximum dry density for determining the relative compaction, and the dry density equivalent to 95 per cent relative compaction (assumed to be the specified requirement for the purposes of this example);

(ii) The 10 per cent air voids line (assumed to be an alternative specified state of compaction for the purposes of this example);

(iii) A relation between dry density and moisture content for full-scale compaction. The location of this curve will depend principally on the type of compactor, the number of passes and the thickness of layer, all of which can be varied to suit the conditions and the specified compaction requirements.

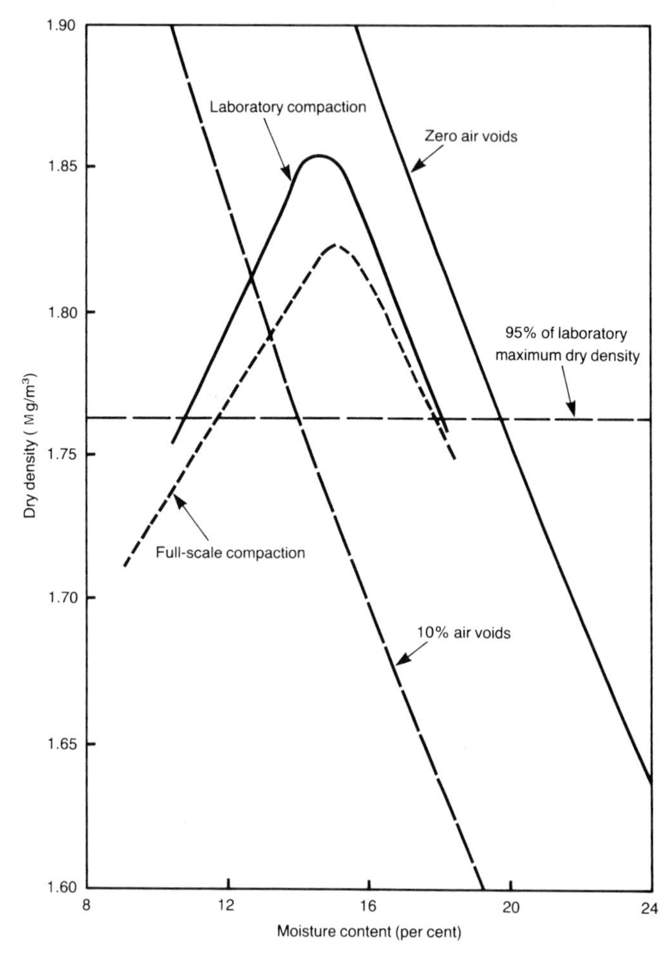

Figure 2.14 Concepts of relative compaction and air voids methods of judging quality of compaction (see Paragraphs 2.26 and 2.27)

2.27 As drawn, the results in Figure 2.14 show that the compactor will be capable of achieving 95 per cent relative compaction within the range of moisture contents where the dry density-moisture content relation lies above the dry density equivalent to 95 per cent relative compaction, ie between about 12 and 18 per cent. Note that an upper limit of moisture content for the soil is automatically imposed; the lower limit of moisture content is determined by the location of the full-scale compaction curve (see Paragraph 2.26 (iii) above). If a maximum of 10 per cent air voids is specified in the example shown, then the compactor would achieve this at all values of moisture content in excess of about 13 per cent, ie the moisture content at which the dry density – moisture content relation for full-scale compaction crosses the 10 per cent air voids line. Thus, the air void method of specification does not incorporate an upper limit of moisture content and it is essential that an additional requirement limiting the moisture content to a maximum value is imposed.

2.28 It is the intention, in subsequent chapters of this book, to make use of both methods of judging compaction to compare the performances of various types of compactor. The criteria set will vary depending on the soil type, as discussed in Chapter 3.

References

British Standards Institution (1990). British Standard methods of test for soils for civil engineering purposes: BS 1377: Part 4 Compaction related tests. BSI, London.

Lewis, W A (1954). Further studies in the compaction of soil and the performance of compaction plant. *Road Research Technical Paper* No 33. HM Stationery Office, London.

Lewis, W A (1959). Investigation of the performance of pneumatic-tyred rollers in the compaction of soil. *Road Research Technical Paper* No 45. HM Stationery Office, London.

Lewis, W A (1962). Compaction of soils and road bases. J. Instn Highway Engrs, *9* (3) 181–202.

Morel, G and Machet, J M (1980). Compaction of pavement layers by vibration (in French). *International Conference on Compaction*, Editions Anciens ENPC, Paris, Vol II, 437–44.

Reichert, J (1980). Report : Various national specifications in control of compaction. *International Conference on Compaction*, Editions Anciens ENPC, Paris, Vol III, 181–203.

Road Research Laboratory (1952). Soil mechanics for road engineers. HM Stationery Office, London, Chapter 9.

Valeux, J C and Morel, G (1980). Influence of bearing capacity of underlying materials on the compaction of pavement layers (in French). *International Conference on Compaction*, Editions Anciens ENPC, Paris, Vol II, 475–80.

CHAPTER 3 TEST FACILITIES FOR INVESTIGATING THE PERFORMANCE OF COMPACTION PLANT

3.1 Investigations of the performance of compaction plant, which have been pursued at TRL (formerly RRL and TRRL) since 1946, have required purpose-built test facilities in which soil conditions can be rigorously controlled. Studies have been made using soils of various types which are representative of those frequently encountered in earthwork construction in the United Kingdom. Altogether, three different pilot-scale facilities have been used for the investigations of compaction plant, and the soils used have been perpetuated to a large extent over the five decades in which the work was carried out. In this Chapter descriptions are given of the three facilities, the soils involved, the methods of their preparation and general test procedures.

Plate 3.2 Use of 'Road Machine No 3' to test a pneumatic-tyred roller in 1946

Early tests, 1946 to 1953

3.2 Early investigations of the performance of compaction plant were made in a covered circular track, originally built to study the wear of bituminous road surfacings, at the, then, Road Research Laboratory's site at Harmondsworth; this track was known as 'Road Machine No 3'. Views of the road machine are shown in Plates 3.1 and 3.2 and a plan and cross-section of the track, showing the position of the test areas of soil, are shown in Figure 3.1. The soils were contained in bays with concrete sides and floor, each 600 mm deep, 3.5 m wide and 12 to 15 m long, arranged around the circular track.

Plate 3.1 View of 'Road Machine No 3' as used in compaction investigations from 1946 to 1953

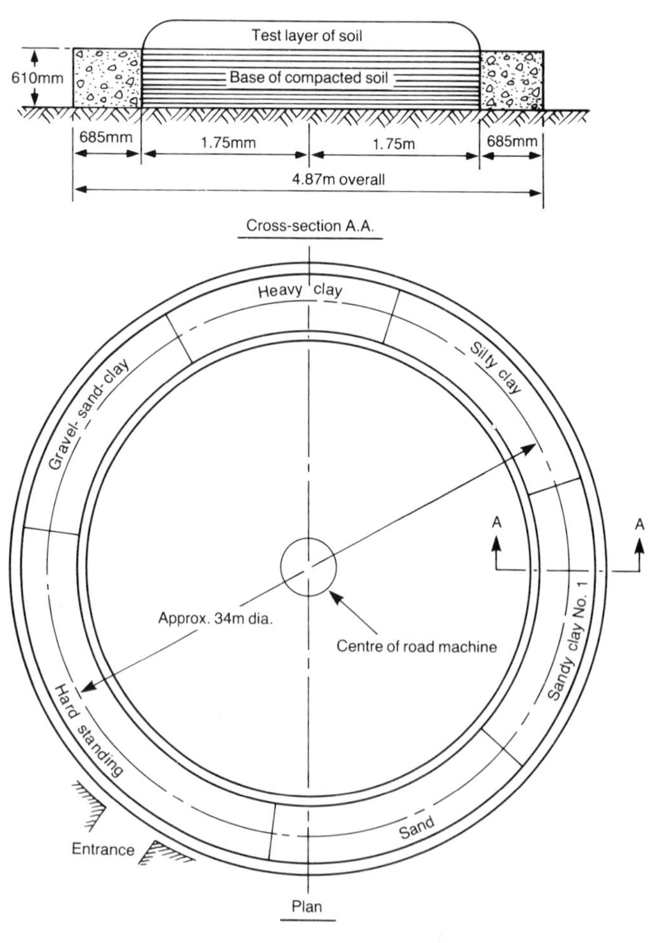

Figure 3.1 Details of 'Road Machine No 3' and layout of soils as used in compaction investigations from 1946 to 1953

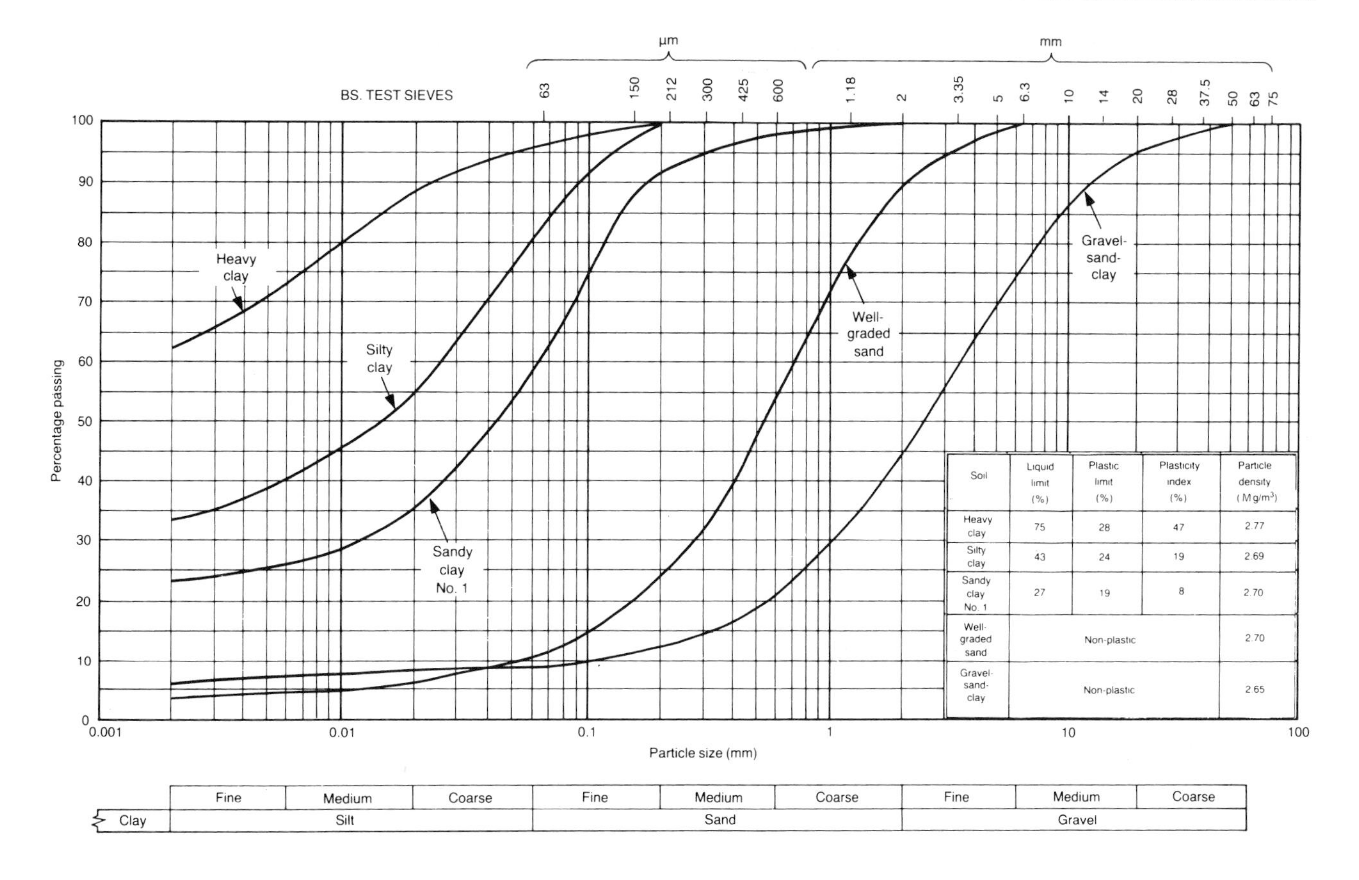

Soil	Liquid limit (%)	Plastic limit (%)	Plasticity index (%)	Particle density (Mg/m^3)
Heavy clay	75	28	47	2.77
Silty clay	43	24	19	2.69
Sandy clay No. 1	27	19	8	2.70
Well-graded sand	Non-plastic			2.70
Gravel-sand-clay	Non-plastic			2.65

Figure 3.2 Particle-size distributions and results of plasticity and particle density tests for the soils used in early compaction investigations from 1946 to 1953

3.3 Items of plant which were not self-propelled were towed by a remotely controlled, electrically powered lorry running on the concrete walls of the soil bays (Plate 3.2). In this way the possible compacting effect of the prime mover was avoided. The soils used (Figure 3.1) were a heavy clay, silty clay, sandy clay No 1 (so numbered to differentiate from a slightly different sandy clay used in subsequent test work), well-graded sand and a gravel-sand-clay. Details of typical particle-size distributions, plasticity properties and particle densities of the soils are given in Figure 3.2.

Plate 3.3 Rotary cultivator used in early compaction investigations from 1946 to 1953

3.4 The test procedure with any item of compaction plant was to break up the soil to the desired depth by means of a rotary cultivator to form a loose tilth; the machine used in the early tests is shown in Plate 3.3. Any adjustment of moisture content was achieved by spraying water on to the soil followed by further mixing with the rotary cultivator, or by aeration with intermittent mixing.

3.5 The state of compaction of the soil after operation of the compactor under test was determined by measuring the dry density using the sand-replacement (small pouring cylinder) method (British Standards Institution, 1990b), see Plate 3.4; usually a minimum of 10 determinations of dry density were made over a compacted area to obtain a mean value of the state of compaction of the soil.

3.6 Various types of compactor were tested in this first pilot-scale facility and the results have been reported by Williams and Maclean (1950) and by Lewis (1954). The machines tested are listed in Table 3.1, which also contains references to the chapters of this book containing the test results.

Plate 3.4 Determination of dry density by the sand-replacement test during the early compaction investigations from 1946 to 1953

Table 3.1 Compaction plant investigated in the early tests from 1946 to 1953

Type of machine	Further details	Chapter containing results
Smooth-wheeled roller	2.8 t 3-wheel	4
	8.6 t 3-wheel	4
Pneumatic-tyred roller	12 t 9-wheel	5
Sheepsfoot roller	4.5 t taper-foot	6
	5.0 t club-foot	6
Vibrating roller (smooth wheel)	220 kg, single roll, pedestrian operated	7
	2.4 t tandem	7
Vibrating-plate compactor	240 kg	9
	1.5 t	9
	2.0 t	9
Rammer	100 kg power rammer	11
	600 kg 'frog' rammer	11
Tracklaying tractor	6 t	13
	11 t	13
Agricultural roller	840 kg roller	14

Middle period of investigations of the performance of compaction plant, 1954 to 1966

3.7 Although the circular test track used in the early tests was perfectly adequate for relatively small types of plant, it was not suitable for testing the largest types of compaction equipment that were being used in earthwork construction, such as 20 to 50 tonne pneumatic-tyred rollers. These heavier rollers required a greater drawbar pull than could be provided by the electrically operated towing lorry of the circular track, and the 600 mm depth of test soil was also considered insufficient for the larger types of compaction plant.

3.8 In view of these limitations, a special building was constructed in 1954 in the Laboratory's grounds at Harmondsworth. Details of the building and layout of the test soils are shown in Figure 3.3 and a general view is given in Plate 3.5. It should be noted that, although the heavy clay, well-graded sand and gravel-sand-clay were retained from the earlier test facility, a sandy clay No 2 (intermediate between the silty clay and sandy clay No 1) and a uniformly graded fine sand were introduced. Additionally, two extra test pits were included for tests with granular base materials, a limestone wet-mix macadam and a slag wet-mix macadam. Typical particle size distributions and values of plasticity and particle density are given in Figure 3.4.

Plate 3.5 Interior view of the pilot-scale facility used for compaction investigations from 1954 to 1966

3.9 The new building was about 30 m long and 27 m wide, with about 5 m of head-room beneath the roof trusses. Each of the five soil pits was about 10.5 m long, 4.6 m wide and 0.9 m deep. Each of the two pits containing granular base material was about 10.5 m long, 3.5 m wide and 0.5 m deep. The floor of the test pits was left as the natural foundation soil (sandy clay), see Figure 3.3, as it was considered that this would have similar elastic properties to those of the compacted test soils and the conditions, therefore, would approximate more closely to those found on construction sites than the concrete floor used in the earlier facility (see Paragraph 3.2).

3.10 Methods of preparation of the soils and the subsequent determination of the state of

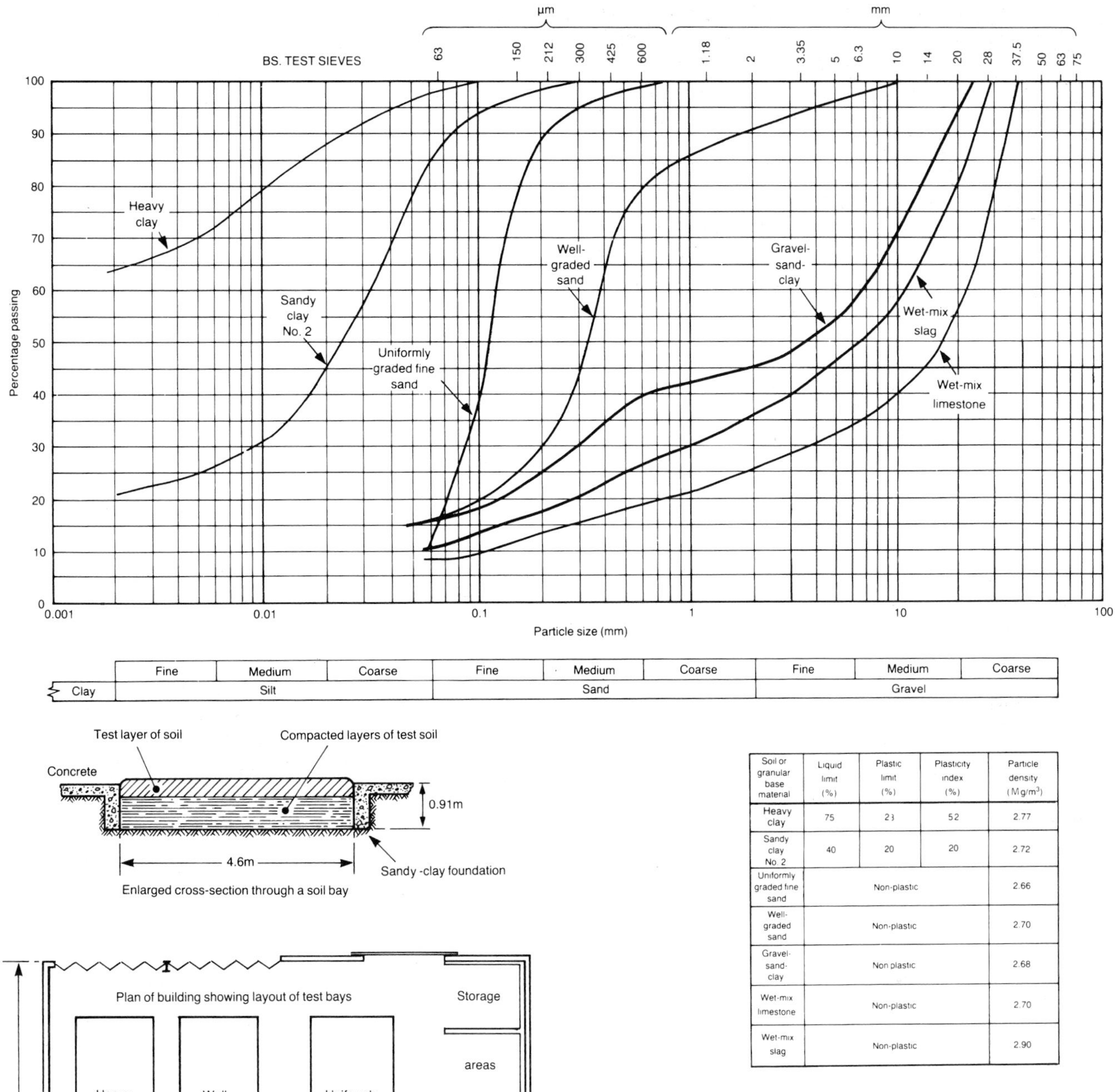

Figure 3.3 Layout of soils in the pilot-scale facility used for compaction investigations from 1954 to 1966

Soil or granular base material	Liquid limit (%)	Plastic limit (%)	Plasticity index (%)	Particle density (Mg/m³)
Heavy clay	75	23	52	2.77
Sandy clay No. 2	40	20	20	2.72
Uniformly graded fine sand	Non-plastic			2.66
Well-graded sand	Non-plastic			2.70
Gravel-sand-clay	Non plastic			2.68
Wet-mix limestone	Non-plastic			2.70
Wet-mix slag	Non-plastic			2.90

Figure 3.4 Particle-size distributions and results of plasticity and particle density tests for the soils and granular base materials used in compaction investigations from 1954 to 1966

compaction involved similar procedures to those used in the early tests (described in Paragraphs 3.4 and 3.5) although more modern rotary cultivators, mounted respectively on a 22 kW crawler tractor (Plate 3.6) and a 34 kW wheeled tractor (Plate 3.7), were used to prepare loose layers with a maximum thickness of 360 mm. In addition, a rotary mixer capable of preparing extra-deep loose layers, up to 600 mm thickness, mounted on a 48 kW crawler tractor, was introduced (Plate 3.8). The determination of dry density in uniformly graded fine sand, using the sand-replacement (small pouring cylinder) method, is shown in Plate 3.9.

Plate 3.6 Crawler tractor mounted rotary cultivator used for preparing the clay soils and the uniformly graded fine sand in compaction investigations during the period from 1954 to 1966

Plate 3.7 Wheeled tractor mounted rotary cultivator used for preparing well-graded granular soils in compaction investigations during the period from 1954 to 1966. A two-furrow plough could also be mounted on this tractor for the preparation of the wet-mix granular base materials

Plate 3.8 Rotary cultivator mounted on a crawler tractor for preparation of loose layers of soil up to 600 mm thick for compaction investigations from 1954 onwards

Plate 3.9 Determination of dry density by the sand-replacement test during the middle period of compaction investigations from 1954 to 1966. The soil being tested is the uniformly graded fine sand

3.11 With the granular base materials it was found that rotary cultivation caused segregation of the coarser particles. The problem was solved by breaking up the compacted material, and mixing in moisture and mixing for aeration, by use of a ploughing process; a two-furrow agricultural plough mounted as an alternative to the rotary cultivator on the 34 kW wheeled tractor was introduced for this purpose. The method of test for determining the state of compaction of the granular base materials was the sand replacement (large pouring cylinder) method (British Standards Institution, 1990b).

3.12 The various types of compactor tested on soils in the middle period of pilot-scale compaction investigations are given in Table 3.2, with references to the chapters of this book containing the test results. The machines tested on the wet-mix macadams are given in Table 3.3; all results for the granular base materials are given in Chapter 16. Results of the work in this middle period have been reported by Lewis (1959, 1961, 1966) and by Lewis and Parsons (1961).

Most recent phase of investigations of the performance of compaction plant since 1967

3.13 With the move of the Laboratory to its present site at Crowthorne, Berkshire, the decision was made to provide a facility capable of carrying out similar investigations to those in the middle period, but excluding the granular base materials. A specially designed building was constructed with larger soil pits than those

Table 3.2 Compaction plant investigated using the natural soils in the second pilot-scale facility in the period from 1954 to 1966

Type of machine	Further details	Chapter containing results
Smooth-wheeled roller	Test rig, towed (study of variation in roll diameter)	4
	8.6 t, 3-wheel	4
	8.9 t, tandem	4
Pneumatic-tyred roller	12 t, 9-wheel, towed	5
	18.5t, 9-wheel, self-propelled	5
	20 t, 9-wheel, towed	5
	45 t, 4-wheel, towed	5
Tamping roller	11 t, self-propelled	6
Grid roller	13.6t, towed	6
Vibrating roller	350 kg, single roll, pedestrian operated	7
	990 kg, tandem	7
	2.2 t, tandem	7
	3.9 t, towed	7
	3.9 t, tandem	7
	4.9 t, towed	7
	8.6 t, towed	7
Vibrating sheepsfoot roller	4.3 t, towed	8
	5.1 t, towed	8
Vibrating-plate compactor	180 kg (Mass per unit area 820 kg/m^2)	9
	330 kg	9
	450 kg	9
	670 kg	9
	710 kg	9
Vibro-tamper	56 kg	10
	74 kg	10
	100 kg	10
Dropping-weight compactor	Multi-weight machine	12
	Experimental rig	12

Table 3.3 Compaction plant investigated using the granular base materials in the second pilot-scale facility in the period from 1954 to 1966. Results are given in Chapter 16

Type of machine	Further details
Smooth-wheeled roller	2.7 t, 3-wheel 7.4 t, 3-wheel
Pneumatic-tyred roller	13 t, 8-wheel, self-propelled
Vibrating roller	350 kg, single roll, pedestrian operated 990 kg, tandem
Vibrating-plate compactor	670 kg 710 kg

Plate 3.10 Interior view of the pilot-scale facility used for compaction investigations since 1967

used earlier. A plan of the building with a typical cross-section of a soil pit is given in Figure 3.5 and a general view of the interior is given in Plate 3.10. The building was about 54 m long and 27 m wide, and the soil pits were about 17 m long, 5.5 m wide and 1 m deep. Each pit was formed with concrete-lined sides and with a base of the natural foundation soil as was the case in the earlier facility. As can be seen by comparing Figures 3.3 and 3.5, the latest facility had the advantage of enabling compaction plant to traverse a test soil without having to cross a second soil. Thus mutual contamination of the test soils was considerably reduced. An area set aside for additional soils was eventually used for research on pavement design problems (Figure 3.5).

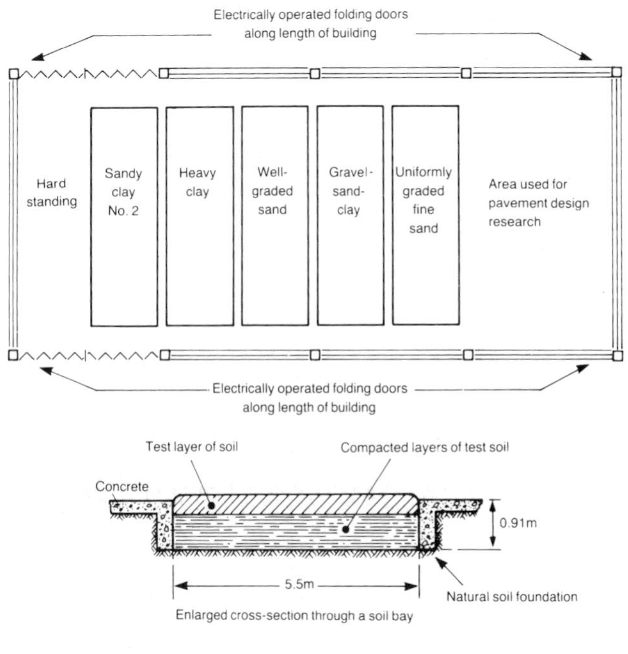

Figure 3.5 Layout of soils in the pilot-scale facility used for compaction investigations since 1967

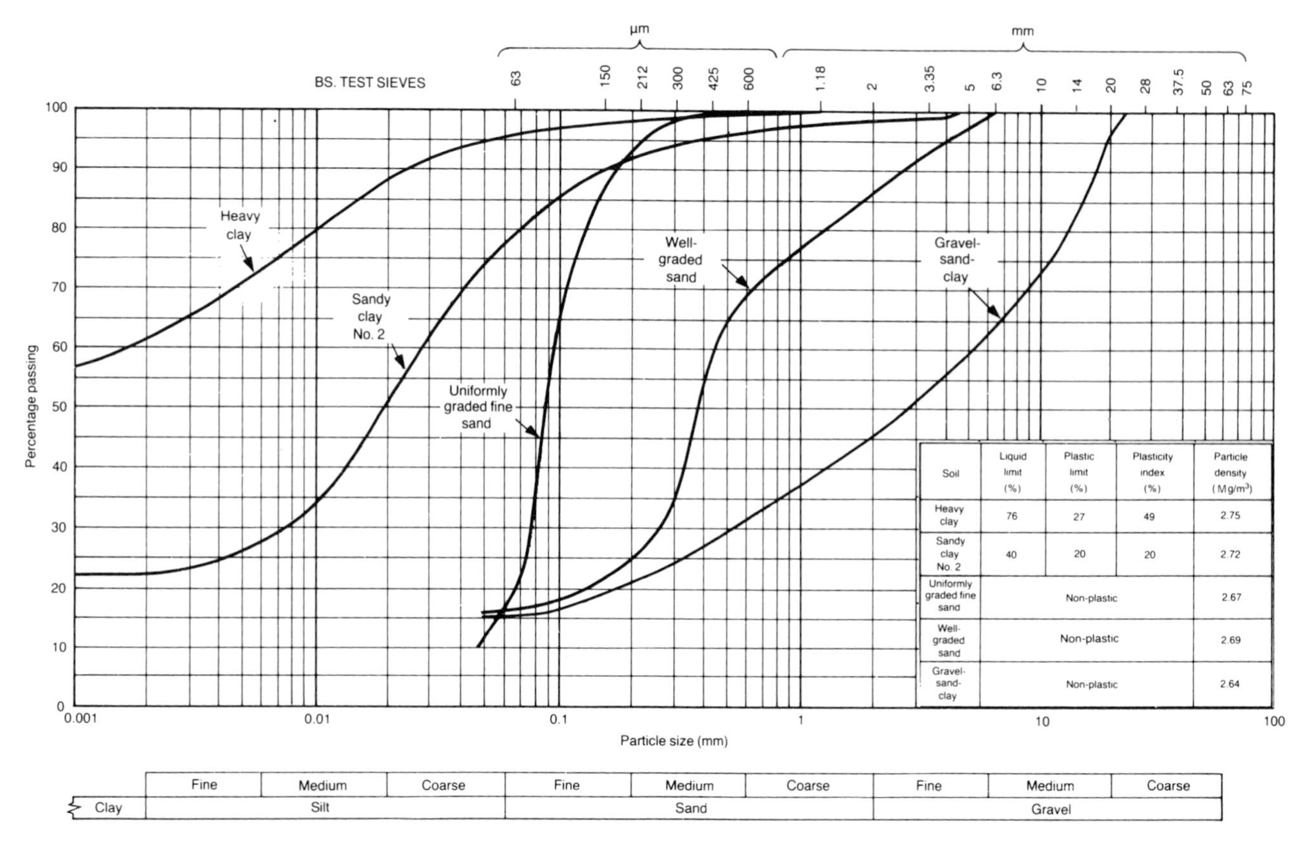

Soil	Liquid limit (%)	Plastic limit (%)	Plasticity index (%)	Particle density (Mg/m³)
Heavy clay	76	27	49	2.75
Sandy clay No. 2	40	20	20	2.72
Uniformly graded fine sand	Non-plastic			2.67
Well-graded sand	Non-plastic			2.69
Gravel-sand-clay	Non-plastic			2.64

Figure 3.6 Particle-size distributions and results of plasticity and particle density tests for the soils used in compaction investigations since 1967

3.14 By careful selection and, on occasion, by reconstitution, the test soils used in the new facility were intended to be identical to the five soils used previously. Typical particle-size distributions and values of plasticity and particle density are given in Figure 3.6.

3.15 The rotary cultivators combined with the 34 kW wheeled tractor (Plate 3.7) and the 48 kW crawler tractor (for deeper mixing) (Plate 3.8) were retained for the preparation of the soils in the facility at Crowthorne. The rotary cultivator with the 22 kW crawler tractor (Plate 3.6), however, was replaced by a new 30 kW crawler tractor with a mounted rotary cultivator, as shown in Plate 3.11. The methods of determining the states of compaction were also similar, although latterly use has been made of nuclear moisture-density gauges (see Chapter 18).

Plate 3.11 Rotary cultivator mounted on a 30 kW crawler tractor for preparation of clay soils and uniformly graded fine sand in compaction investigations since 1967

3.16 The types of compactor tested in the facility at Crowthorne in the third phase of investigations are given in Table 3.4, with references to the chapters of this book containing the test results. No general reviews of this later work have been published, but reports containing the results for individual machines have been published since 1966 in the TRRL Laboratory Report and, latterly, Research Report series (see individual references in the relevant chapters of this book).

Variation in soil characteristics

3.17 Although attempts were made to perpetuate the soils throughout two, or even all three, phases of the studies of compaction plant, some variation in characteristics occurred, as

Table 3.4 Compaction plant investigated in the third pilot-scale facility, built at Crowthorne, since 1967

Type of machine	Further details	Chapter containing results
Smooth-wheeled roller	8.6 t, 3-wheel	4
Tamping roller	17 t, self-propelled	6
Vibrating roller	1.0 t, pedestrian operated, double vibrating rolls	7
	1.5 t, pedestrian operated, double vibrating rolls	7
	1.7 t, pedestrian operated, double vibrating rolls	7
	6.9 t, double vibrating rolls	7
	7.3 t, double vibrating rolls	7
	7.7 t, single roll, self-propelled	7
	12 t, towed	7
Vibrating-plate compactor	80 kg (Mass per unit area 410 kg/m^2)	9
	80 kg (Mass per unit area 450 kg/m^2)	9
	140 kg	9
	150 kg	9
	180 kg (Mass per unit area 470 kg/m^2)	9
	180 kg (Mass per unit area 710 kg/m^2)	9
Vibro-tamper	59 kg	10
Dropping-weight compactor	Mobile machine, 590 kg rammer	12

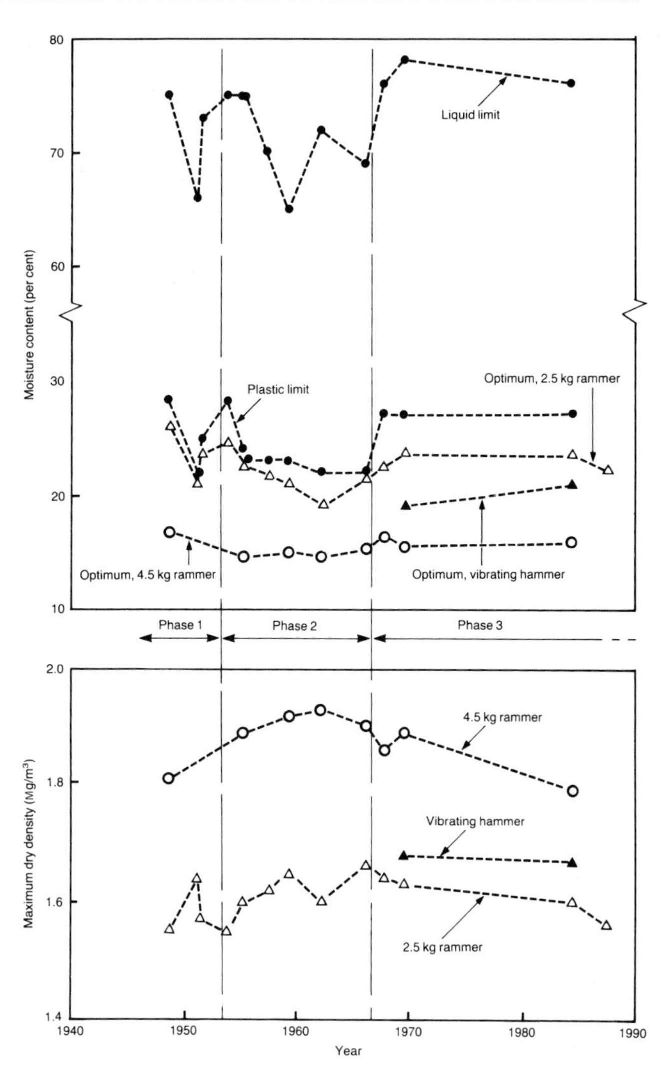

Figure 3.7 Variation with time of British Standard plasticity and laboratory compaction test results obtained on the heavy clay soil

might be expected, particularly as a result of contamination due to repeated testing and, in the case of the granular soils, some degradation of the coarser particles. Replacement by alternative soils or by reconstituted materials which were not perfectly identical also contributed to variations. For the clay soils, the variations over the years in plasticity and laboratory compaction test results are given in Figures 3.7, 3.8 and 3.9; for the granular soils variations in laboratory compaction test results and particle-size distributions are given in Figures 3.10, 3.11 and 3.12. The granular base materials used in investigations during the second phase of pilot-scale studies were employed for a relatively short period of time and little variation occurred about the values given in Figure 3.4. For completeness the laboratory compaction test results for these materials are given in Table 3.5 (Lewis and Parsons, 1961).

Table 3.5 Results of British Standard laboratory compaction tests on the granular base materials used in the second phase of pilot-scale studies

Material:–	Wet-mix limestone		Wet-mix slag	
Test	Maximum dry density (Mg/m^3)	Optimum moisture content (%)	Maximum dry density (Mg/m^3)	Optimum moisture content (%)
2.5 kg rammer	2.31	5.5	2.31	7.3
4.5 kg rammer	2.38	4.6	2.33	7.1
Vibrating hammer	2.46	3.3	2.41	6.6

3.18 It is interesting to note the variations in soil properties that have occurred over the years since 1946, as given in Figures 3.7 to 3.12. Rapid fluctuations can be ignored as they were probably caused by sampling or experimental errors, but continuing trends upwards or

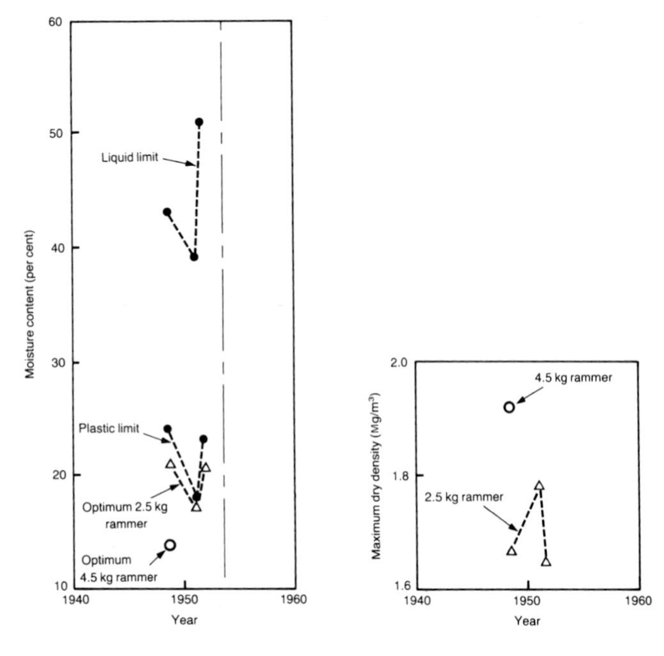

Figure 3.8 Variation with time of British Standard plasticity and laboratory compaction test results obtained on the silty clay soil during the first phase of pilot-scale investigations

downwards, especially when supported by the results of other tests, indicate a change in material properties which should be taken into account when interpreting and comparing results produced by different types of compaction plant at widely spaced intervals of time.

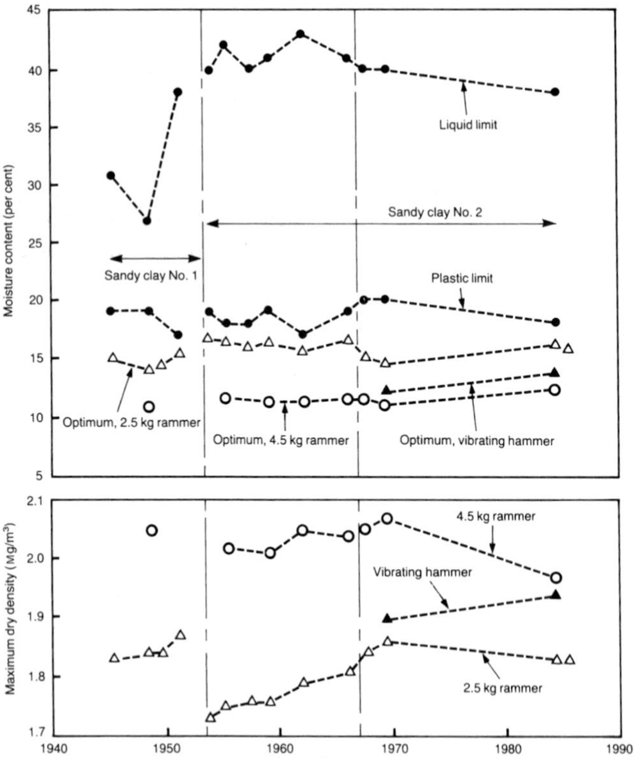

Figure 3.9 Variation with time of British Standard plasticity and laboratory compaction test results obtained on the sandy clay soils

3.19 As a normalising procedure to take account of these effects, therefore, the results given in the following Chapters of this book are expressed in terms of relative compaction (see Paragraph 2.24). For compaction of earthwork materials the 2.5 kg rammer test (British Standards Institution, 1990a) is generally regarded as the appropriate test by which to set standards, although with some types of plant at low values of moisture content states of compaction well in excess of the maximum dry density of that test may well be produced, ie values of relative compaction in excess of 100 per cent will be obtained. However, as can be seen from Figures 3.7 to 3.12, more results are available for the 2.5 kg rammer test than for the 4.5 kg rammer test and so the values of the maximum dry density at any given point in time will be more easily determined.

3.20 Using Figures 3.7 to 3.12, Figure 3.13 gives the general trend of maximum dry density and optimum moisture content in the 2.5 kg

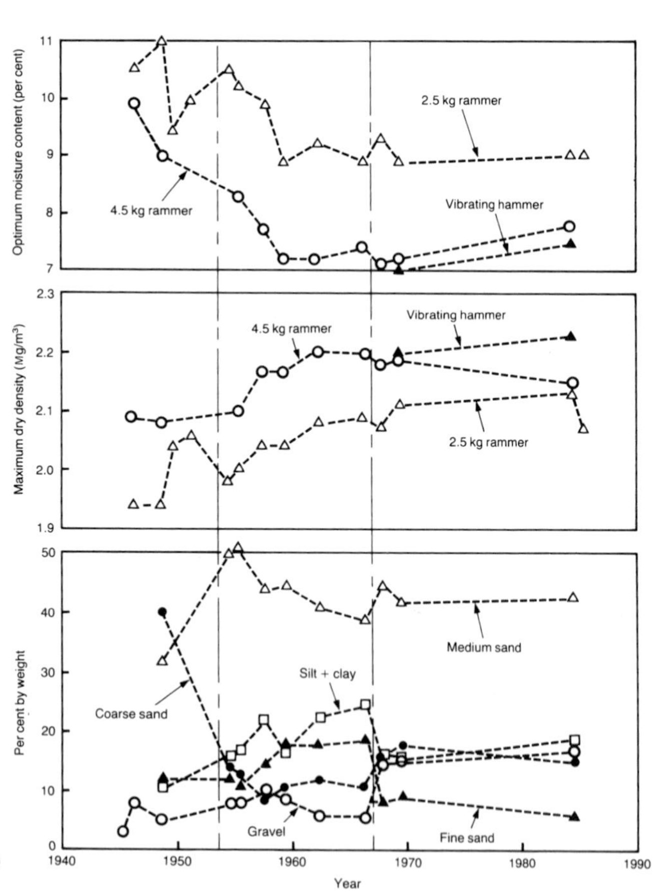

Figure 3.10 Variation with time of British Standard laboratory compaction and particle-size distribution test results on the well-graded sand soil

rammer compaction test for each of the soils used in the pilot-scale investigations of the performance of compaction plant. This Figure will be used as a basis for assessing the results produced by compaction plant at any given period, as discussed in the following Chapters.

Figure 3.11 Variation with time of British Standard laboratory compaction and particle-size distribution test results on the gravel-sand-clay soil

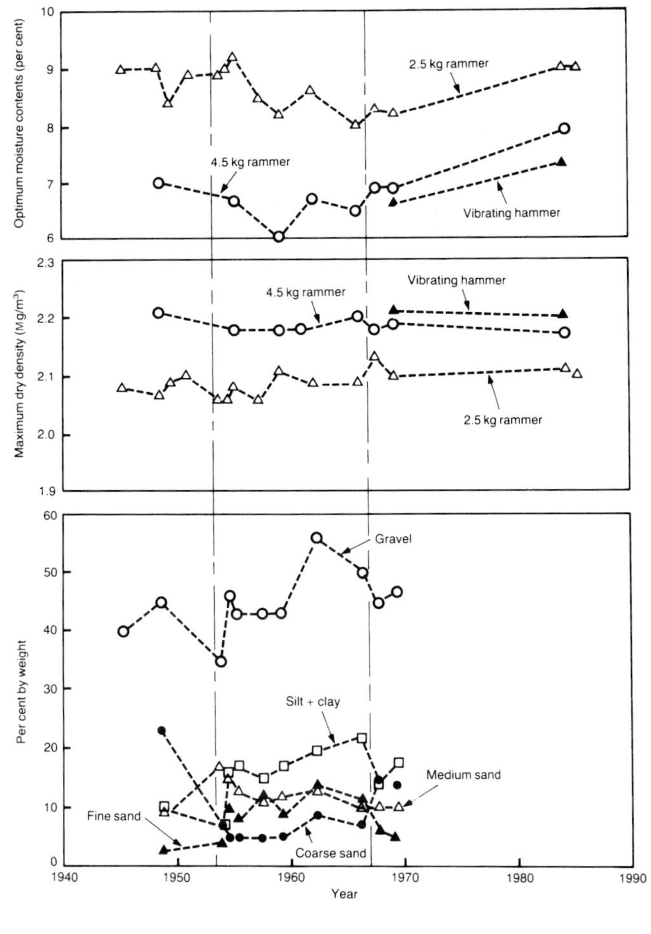

Figure 3.12 Variation with time of British Standard laboratory compaction and particle-size distribution test results on the uniformly graded fine sand soil

Figure 3.13 Variation with time of the maximum dry densities and optimum moisture contents using the 2.5 kg rammer compaction test with the various soils used in pilot-scale studies of compaction plant

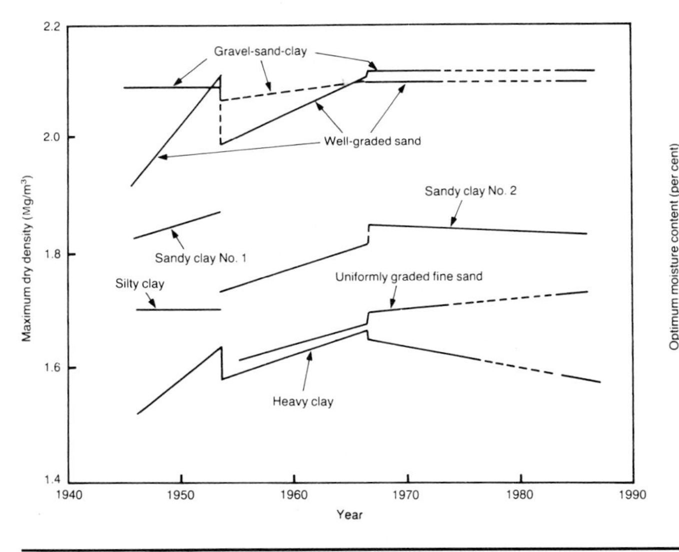

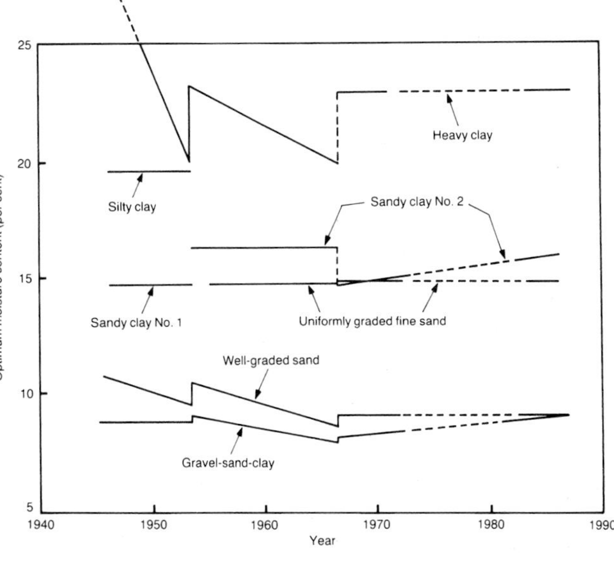

References

British Standards Institution (1990a). British Standard methods of test for soils for civil engineering purposes: BS 1377: Part 4 Compaction related tests. BSI, London.

British Standards Institution (1990b). British Standard methods of test for soils for civil engineering purposes: BS 1377: Part 9 In situ tests. BSI, London.

Lewis, W A (1954). Further studies in the compaction of soil and the performance of compaction plant. *Road Research Technical Paper* No 33. HM Stationery Office, London.

Lewis, W A (1959). Investigation of the performance of pneumatic-tyred rollers in the compaction of soil. *Road Research Technical Paper* No 45. HM Stationery Office, London.

Lewis, W A (1961). Recent research into the compaction of soil by vibratory compaction equipment. *Proc. 5th Internat. Conf. Soil Mech*, Vol II, Dunod, Paris.

Lewis, W A (1966). Full-scale studies of the performance of plant in the compaction of soils and granular base materials. *Proc. Instn. Mech. Engrs*, 181, Pt 2A, No 3, 79–90.

Lewis, W A and Parsons, A W (1961). The performance of compaction plant in the compaction of two types of granular base material. *Road Research Technical Paper* No 53. HM Stationery Office, London.

Williams, F H P and Maclean, D J (1950). The compaction of soil: a study of the performance of plant. *Road Research Technical Paper* No 17. HM Stationery Office, London.

CHAPTER 4 COMPACTION OF SOIL BY SMOOTH-WHEELED ROLLERS

Description of machines

4.1 Smooth-wheeled rollers are characterised by having smooth steel rolls and rely solely on their dead-weight to generate the shear stresses necessary for compaction. They can be self-propelled, or towed by a tractor which may be of the wheeled or track-laying (crawler) type. By the above definition, vibrating rollers with smooth steel rolls, when employed with the vibrating mechanism inoperative, can also be treated as smooth-wheeled rollers, and results obtained in this way have been used to enhance the information in this Chapter. In the course of investigations of the performance of compaction plant at the Laboratory a total of eleven different machines were used as smooth-wheeled rollers, although some only to a very limited extent. The relevant details of the machines are given in Table 4.1 and they are illustrated in Plates 4.1 to 4.11.

4.2 The 2.8 t and 8.6 t smooth-wheeled rollers (Plates 4.1 and 4.2) are traditional self-propelled three-wheel (tricycle configuration) rollers. As can be seen from the photographs, these machines would have long been treated as obsolete, although the results produced by them in the first phase of the investigations, from 1946 to 1953 (see Table 3.1) will remain valid for modern smooth-wheeled rollers with the same loading on the rolls. Plate 4.3 illustrates an 8.9 t tandem smooth-wheeled roller which was used in a limited series of tests in 1962.

Plate 4.1 2.8 t smooth-wheeled roller investigated in 1947 and 1951

Table 4.1 Details of the various machines contributing data on the performance of smooth-wheeled rollers. The rollers are illustrated in Plates 4.1 to 4.11

Smooth-wheeled roller or vibrating roller used without vibration	Total mass	Front roll				Rear roll(s)			
		Width	Diameter	Mass on roll	Mass per unit width	Width	Diameter	Mass on roll(s)	Mass per unit width
	(kg)	(m)	(m)	(kg)	(kg/m)	(m)	(m)	(kg)	(kg/m)
2.8 t	2790	0.61	0.86	860	1410	0.38x2	0.91	1930	2540
8.6 t	8620	1.07	1.07	3540	3320	0.46x2	1.37	5080	5550
8.9 t tandem	8890	1.32	1.30	5520	4180	1.30	1.02	3370	2590
Test rig	1550	0.61	0.61 + 0.91	1550	2540	–	–	–	–
220 kg pedestrian-operated	220	0.61	0.53	220	360	–	–	–	–
350 kg pedestrian-operated	350	0.71	0.57	350	490	–	–	–	–
2.4 t tandem	2440	0.81	0.76	990	1220	0.81	0.76	1450	1790
3.9 t tandem	3880	1.00	0.89	2790	2790	0.70	0.76	1090	1560
3.9 t towed	3910	1.83	1.22	3880	2120	–	–	–	–
7.7 t	7670	1.83	1.37	4140	2260	Pneumatic-tyred wheels		3530	–
8.6 t towed	8600	1.91	1.60	8580	4490	–	–	–	–

Plate 4.2 8.6 t smooth-wheeled roller investigated in 1948–51 with further tests in 1953, 1958 and 1967

Plate 4.4 Test rig used for investigations of the effect of diameter of smooth-wheeled rollers, carried out in 1954

Plate 4.3 8.9 t tandem smooth-wheeled roller investigated in 1962

Plate 4.5 220 kg pedestrian-operated vibrating roller which, when used without vibration, contributed data on the performance of smooth-wheeled rollers; investigated in 1949

4.3 A specific study of the effect of varying the diameter of a smooth-wheeled roll was carried out in 1954 and the test rig used for this purpose is shown in Plate 4.4. This was a towed roller which could be ballasted to the required loading condition and fitted with either of three rolls of different diameters.

4.4 Plates 4.5 to 4.11 illustrate various types of vibrating roller tested without vibration in limited studies at various periods. The machine in Plate 4.5 was pedestrian propelled with a mass of only 220 kg, while Plate 4.6 illustrates a pedestrian-operated self-propelled roller of 350 kg. These machines, when operated without vibration, cannot be expected to produce an effective compactive effort, but the results obtained have been included for completeness. Plates 4.7 and 4.8 show self-propelled tandem vibrating rollers whilst Plates 4.9 to 4.11 show single-roll vibrating rollers with masses on the roll (Table 4.1) ranging from 3.8 t to 8.6 t. Two of these machines are of the towed variety, normally utilising a crawler tractor for propulsion, whilst the machine shown in Plate 4.10 is self-propelled with traction provided by pneumatic-tyred wheels at the rear.

Plate 4.6 350 kg pedestrian-operated vibrating roller which, when used without vibration, contributed data on the performance of smooth-wheeled rollers; investigated in 1956

4.5 The range of machines illustrated in Plates 4.1 to 4.11 are considered to include all common configurations of smooth-wheeled rollers likely to be encountered. The principal characteristics of smooth-wheeled rollers affecting their performance in compacting soil is the load on the rolls and the width and diameter of the rolls. These factors, taken together, control the stresses exerted in the surface layer of the soil, while the dimensions of the rolls affect the distribution of the stresses throughout the layer being compacted. For specification purposes it is usual to categorise smooth-wheeled rollers by their mass per unit width of roll (Department of Transport, 1986) as the principal indicator of their potential performance. The values of mass per unit width of roll for the various smooth-wheeled rollers investigated at the Laboratory are included in Table 4.1.

Plate 4.7 2.4 t tandem vibrating roller, used without vibration during investigations in 1949–50

Plate 4.8 3.9 t tandem vibrating roller, used without vibration during investigations in 1960

Relations between state of compaction and moisture content

4.6 Relations between the state of compaction produced by smooth-wheeled rollers and the moisture content of the soil are given in Figures 4.1 to 4.6. In all cases the dry density has been expressed as a percentage of the maximum dry density obtained in the 2.5 kg rammer compaction test (British Standards Institution, 1990) and termed the relative compaction value.

Plate 4.9 3.9 t towed vibrating roller, used without vibration during investigations in 1956–57

Plate 4.10 7.7 t vibrating roller, used without vibration during investigations in 1968–69

Plate 4.11 8.6 t towed vibrating roller, used without vibration during investigations in 1960–61

Figure 4.1 Relations between dry density and moisture content for smooth-wheeled rollers on heavy clay, expressed in terms of relative compaction with the 2.5 kg rammer test. Layer thickness 150 mm

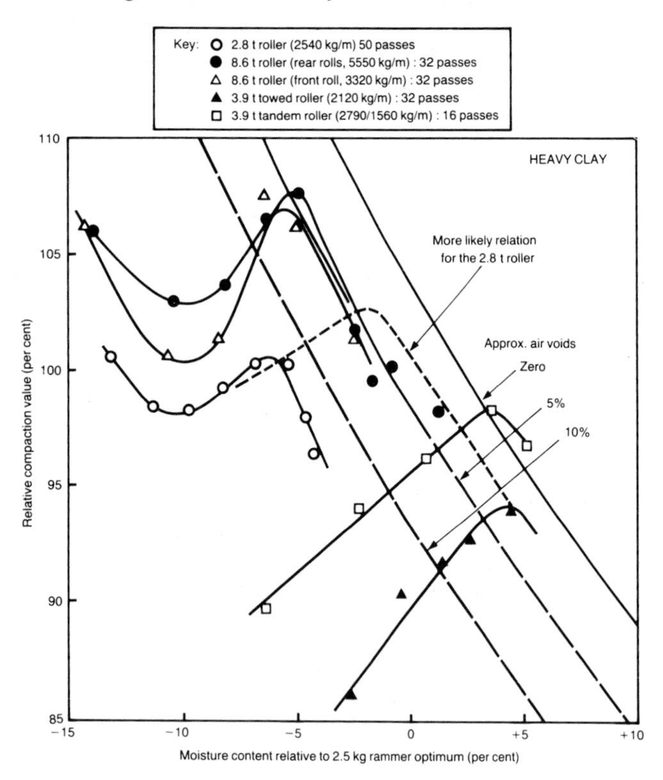

Figure 4.2 Relations between dry density and moisture content for smooth-wheeled rollers on silty clay, expressed in terms of relative compaction with the 2.5 kg rammer test. Layer thickness 150 mm

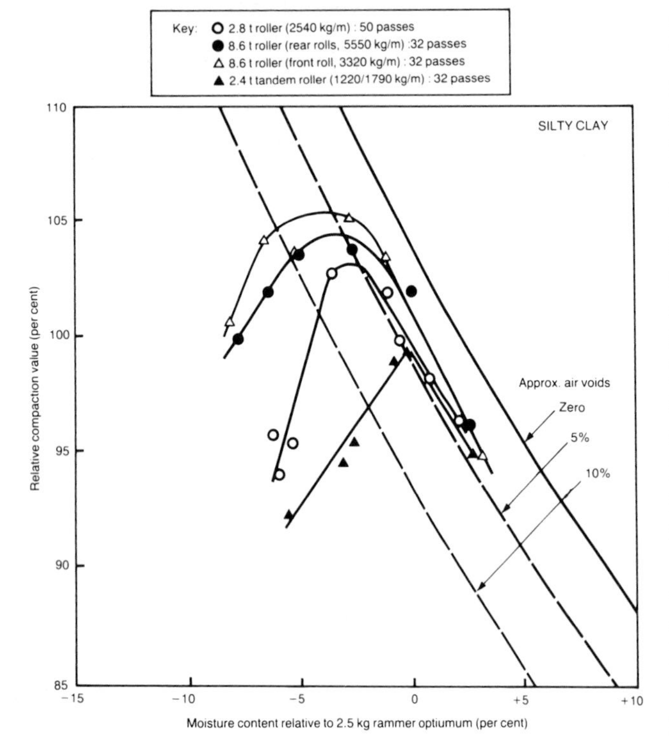

Figure 4.3 Relations between dry density and moisture content for smooth-wheeled rollers on sandy clay No 2, expressed in terms of relative compaction with the 2.5 kg rammer test. Layer thickness 150 mm

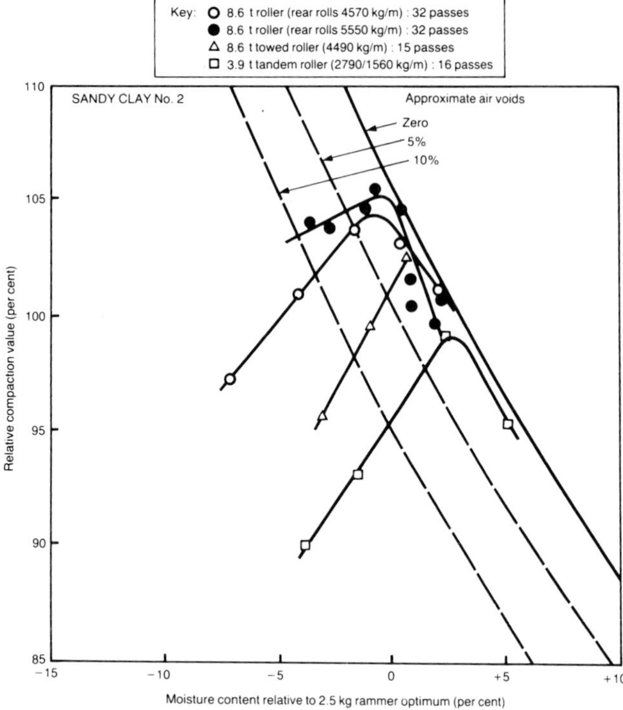

The moisture content has been expressed as the difference from the optimum moisture content obtained in the same compaction test. In this way the effects of variations in soil characteristics which occurred over long intervals of time, as discussed in Paragraph 3.17, have been minimised. The period in which each machine was tested to produce the results in Figures 4.1 to 4.6 is given in the caption to the relevant Plate illustrating the machine; for example, the 2.8 t roller providing results in Figures 4.1, 4.2 and 4.4 to 4.6 is illustrated in Plate 4.1 and the caption states that the machine was tested in 1947 and 1951. The most likely maximum dry density and optimum moisture content obtained in the 2.5 kg rammer compaction test was determined from Figure 3.13. Also included in Figures 4.1 to 4.6 are lines indicating the zero, 5 per cent and 10 per cent air void condition. These are necessarily approximate as they are based on the average maximum dry density and optimum moisture content of the 2.5 kg rammer test for the test results given in the individual figure.

4.7 The compactive effort used in producing each of the relations given in Figures 4.1 to 4.6 is given in the key to each figure in terms of the mass per unit width of roll and the number of passes of the machine. In all cases the number

Figure 4.4 Relations between dry density and moisture content for smooth-wheeled rollers on sandy clay No 1, expressed in terms of relative compaction with the 2.5 kg rammer test. Layer thickness 150 mm

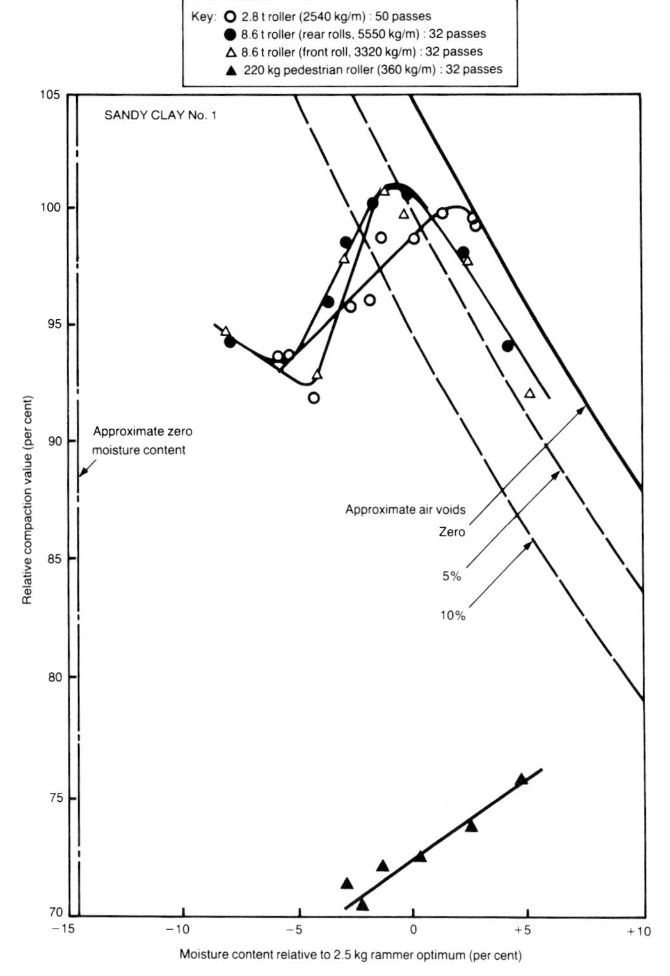

of passes was relatively high and the relations can be regarded as representing the maximum states of compaction to be achieved, ie compaction to refusal, for the 150 mm thick compacted layers used.

4.8 With the 8.6 t roller, results were obtained for compaction by the front roll and the rear rolls separately. As the machine had a tricycle configuration such tests were perfectly feasible and the number of passes of the roll over the soil was equal to the number of machine passes. Figure 4.3 additionally contains results for this machine (rear rolls only) when operated with a reduced mass per unit width of roll, achieved by a reduction in the quantity of ballast carried. For the 2.8 t roller, also with a tricycle configuration, no information is on record as to the location of the soil tested, but it is a fairly safe assumption that the rear rolls, ie the most heavily loaded rolls (Table 4.1), contributed at least in part if not entirely, to the results given. Where tandem rollers are concerned, it must be remembered that the number of passes of a roll over the surface of the compacted layer is double the number of machine passes given in the key to each figure. For purposes of determining the effect of variations in roll loading on the results produced, the effective mass per unit width of roll of tandem rollers, 'm', has been taken as:–

$$m = \sqrt{(m_1 m_2)}$$

where m_1 is the mass per unit width of the front roll
and m_2 is the mass per unit width of the rear roll.

4.9 In general the relations given in Figures 4.1 to 4.6 follow the general trends illustrated in

Figure 4.5 Relations between dry density and moisture content for smooth-wheeled rollers on well-graded sand, expressed in terms of relative compaction with the 2.5 kg rammer test. Layer thickness 150 mm

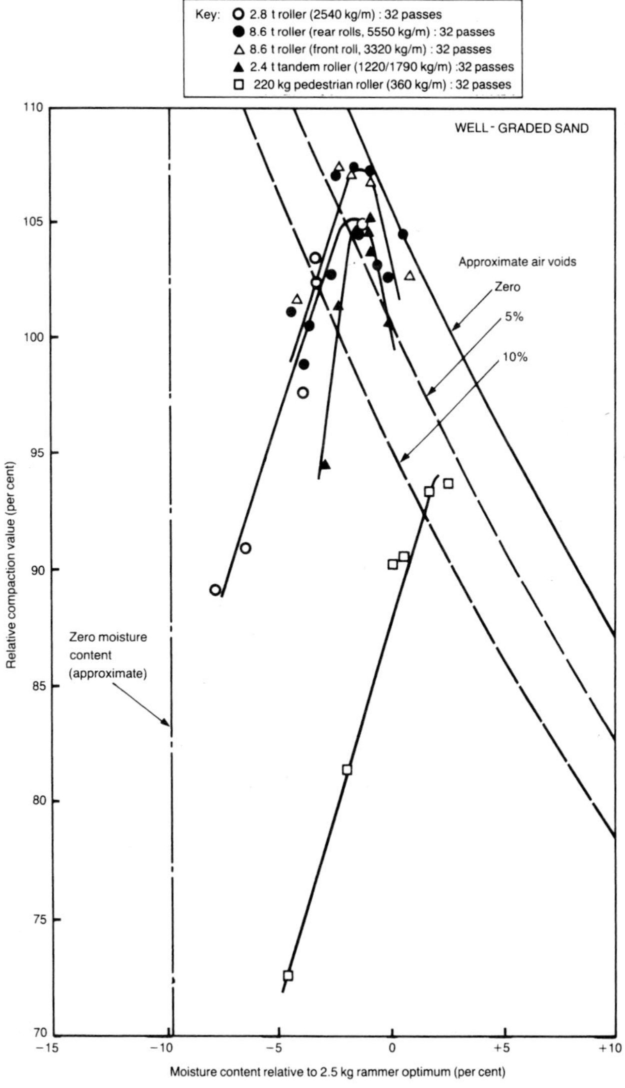

Figure 4.6 Relations between dry density and moisture content for smooth-wheeled rollers on gravel-sand-clay, expressed in terms of relative compaction with the 2.5 kg rammer test. Layer thickness 150 mm

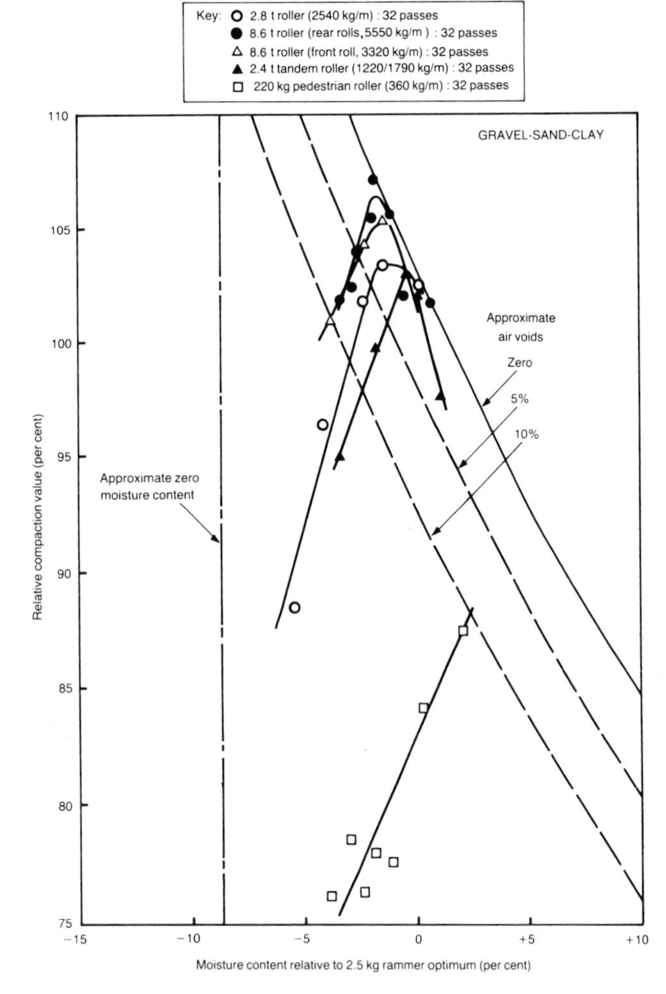

Figure 2.2 and discussed in Paragraphs 2.5 to 2.9. Thus the maximum relative compaction value decreased, and the optimum relative moisture content increased as the mass per unit width of roll decreased. The air content reduced to a low value at the higher moisture contents, as is to be expected, except with the 2.8 t roller on heavy clay (Figure 4.1). The apparent inability to reduce the air content below 10 per cent in the results shown, reproduced from published data (Lewis, 1954) is inexplicable. In fact, results are on record and are given later in Figure 4.8(c) indicating that at high values of moisture content, equivalent to the 2.5 kg rammer optimum moisture content plus 5 per cent, a low air content was achieved. For this reason, a more likely relation between the state of compaction and moisture content for the 2.8 t roller has been added to Figure 4.1 with a maximum relative compaction value at the more usual level of about 4 to 5 per cent air voids.

4.10 The effect of mass per unit width of roll on the maximum state of compaction and optimum moisture content is illustrated in Figure 4.7. Regression lines have been drawn through the points, taking together the pairs of heavy clay and silty clay soils, the two sandy clays and the granular soils to produce three separate relations. The result for the 2.8 t roller on heavy clay has been omitted for the reason given above. The equations of the regression lines and the correlation coefficients are given in Table 4.2. The results summarised in Figure 4.7 and Table 4.2 are for compaction to refusal by the smooth-wheeled rollers (in general about 32 roll passes) with a compacted layer thickness of 150 mm. Any reduction in number of passes or increase in thickness of layer will result in a reduction in the maximum relative compaction value and an increase in optimum moisture content as discussed in Chapter 2.

Table 4.2 Equations and correlation coefficients for the relations given in Figure 4.7

Soil	Regression equation*	Number of points	Correlation coefficient
Heavy clay & silty clay	$c = 56.0\ m^{.075}$	9	0.75
	$w = 1620\ m^{-.67} - 10$	9	0.73
Sandy clays	$c = 48.6\ m^{.089}$	8	0.92
	$w = 143\ m^{-.32} - 10$	8	0.96
Well-graded sand & gravel-sand-clay	$c = 69.7\ m^{.051}$	10	0.93
	$w = 34.1\ m^{-.17} - 10$	10	0.94

*c = maximum dry density for the machine as percentage of the maximum dry density obtained in the 2.5 kg rammer compaction test

w = difference between optimum moisture content for the machine and optimum moisture content obtained in the 2.5 kg rammer compaction test (per cent)

m = mass per unit width of smooth-wheeled roller (kg/m)
(For tandem rollers $m = \sqrt{(m_1 m_2)}$, see Paragraph 4.8)

The above results apply in general to 32 roll passes on 150 mm compacted layers

Relations between state of compaction and number of passes

4.11 Results of investigations to determine the effect on compaction of variations in the number of passes of smooth-wheeled rollers are given in Figures 4.8 to 4.13. Results for each soil used are illustrated under a separate Figure number, whilst different values of moisture content are given in separate graphs within each figure. Thus Figure 4.8, for heavy clay, contains results at three values of moisture content; Figure 4.8(a) has results at a moisture content of optimum − 6.5 per cent, Figure 4.8(b) at optimum −0.5 per cent and Figure 4.8(c) at optimum +5 per cent.

Figure 4.7 Maximum dry densities and optimum moisture contents related to the mass per unit width of smooth-wheeled rollers. Results from Figures 4.1 to 4.6

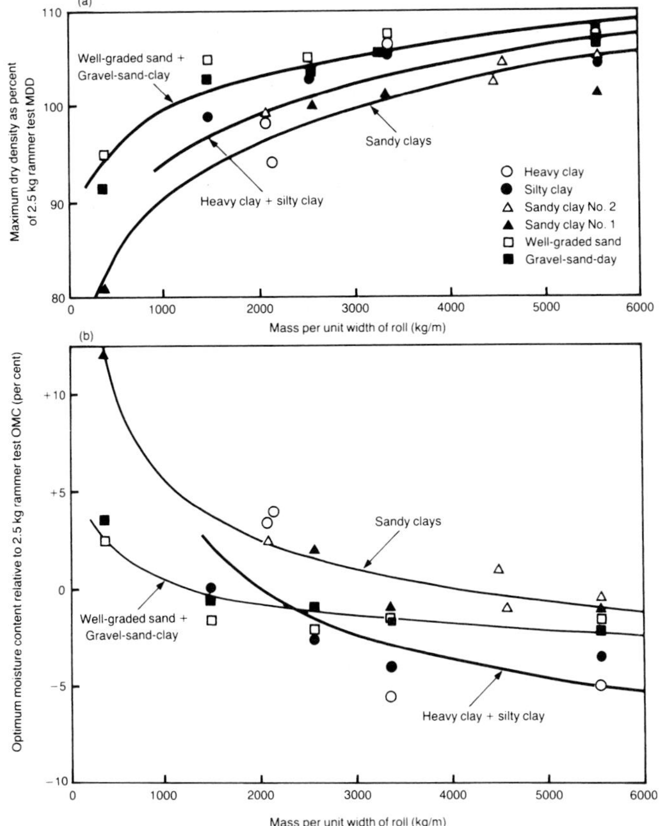

In all cases optimum refers to the optimum moisture content for the 2.5 kg rammer compaction test and the relative compaction value is the dry density produced by the roller expressed as a percentage of the maximum dry density obtained in the same test. The results of the 2.5 kg rammer compaction test pertaining at the relevant times of the tests were obtained from Figure 3.13. Approximate air void lines for the average condition for the results are also included in each graph. The horizontal axis of each graph gives the number of passes on a logarithmic scale to allow maximum use of straight line relations (see Paragraph 2.11 and Figure 2.5). All the results apply to a layer thickness of 150 mm.

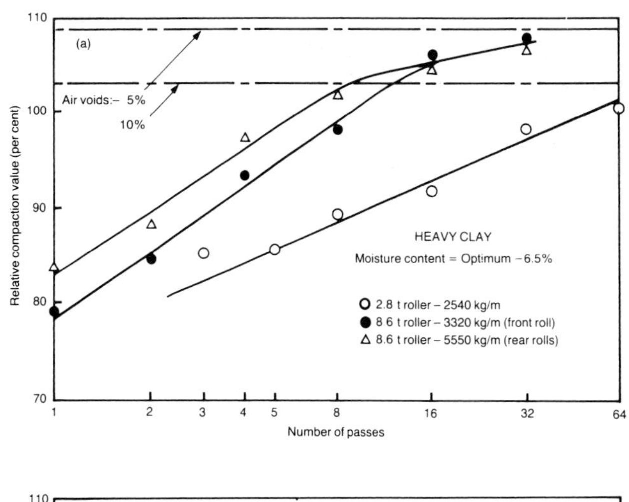

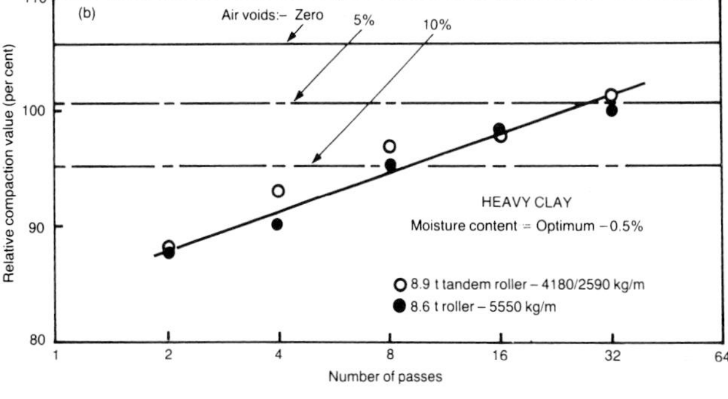

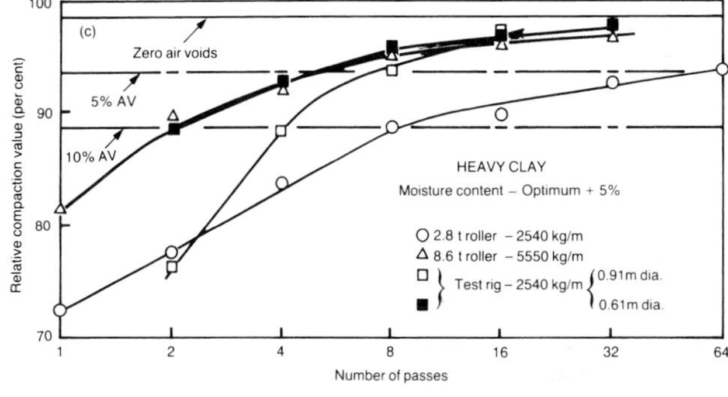

Figure 4.8 Relations between relative compaction and number of passes of smooth-wheeled rollers on heavy clay at three different values of moisture content and a layer thickness of 150 mm. All values are related to the maximum dry density and optimum moisture content obtained in the 2.5 kg rammer compaction test (Figure 3.13)

4.12 A study of Figures 4.8 to 4.13 reveals that the 8.6 t roller contributed the majority of the data, with the 2.8 t roller also contributing to the data on most soils. The occasional results produced by the vibrating rollers (when used with the vibrating mechanism inoperative) provide useful comparisons. In general, the figures confirm that as the moisture content of a soil is increased the number of passes required to achieve a given level of compaction is reduced. Additionally, the level of compaction at any given number of passes increases with increase in mass per unit width of roll. It should be remembered that the number of passes plotted in Figures 4.8 to 4.13 are those of the machine, so that for tandem rollers the number of roll passes will be double the plotted number of passes.

4.13 All results have been included for completeness, but there are two inconsistencies in the Figures that have to be taken into account when reaching general conclusions.

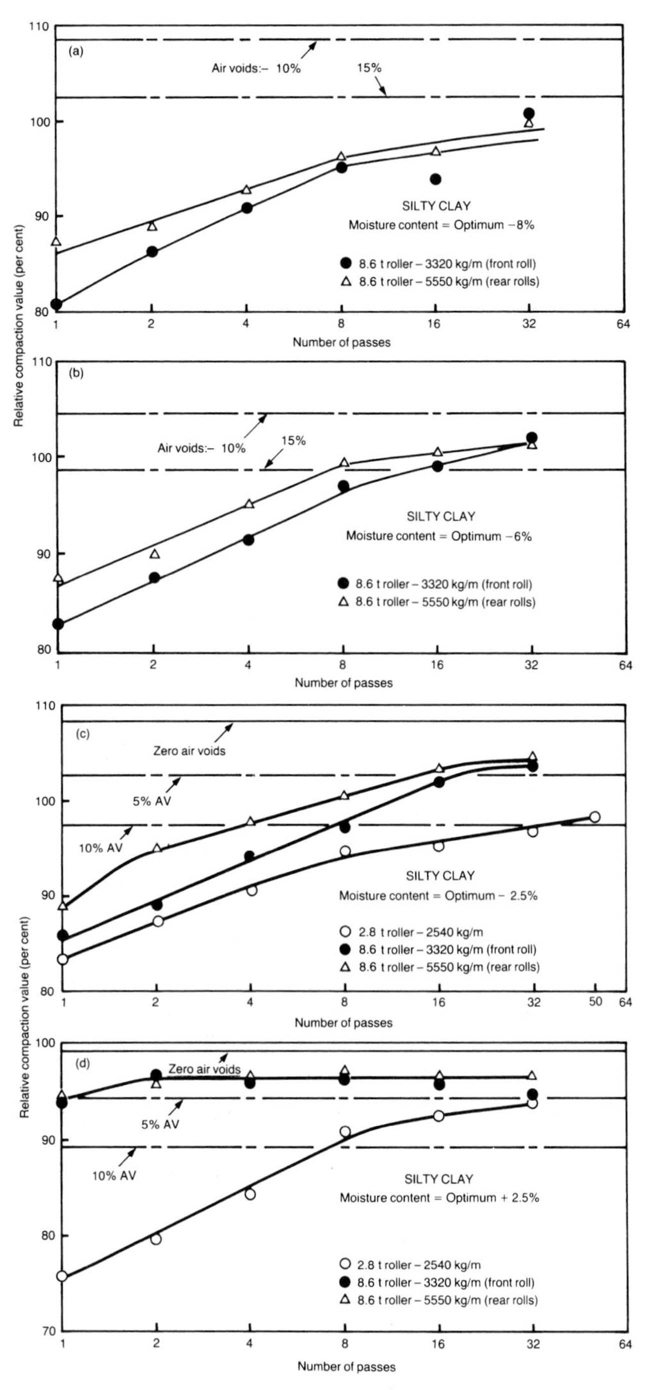

Figure 4.9 Relations between relative compaction and number of passes of smooth-wheeled rollers on silty clay at four different values of moisture content and a layer thickness of 150 mm. All values are related to the maximum dry density and optimum moisture content obtained in the 2.5 kg rammer compaction test (Figure 3.13)

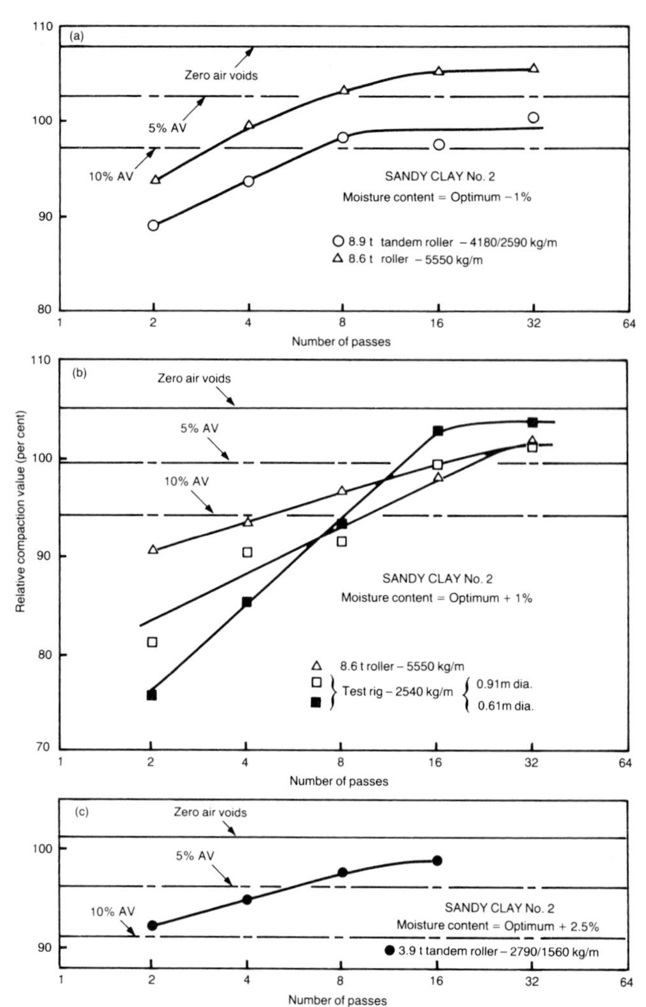

Figure 4.10 Relations between relative compaction and number of passes of smooth-wheeled rollers on sandy clay No 2 at three different values of moisture content and a layer thickness of 150 mm. All values are related to the maximum dry density and optimum moisture content obtained in the 2.5 kg rammer compaction test (Figure 3.13)

4.14 Firstly, in Figure 4.8(c) the results obtained with the 2.8 t roller do not compare well with those produced by the test rig with the same loading per unit width of roll of 2540 kg/m. At this particularly high value of moisture content the similarity between the results produced by the 8.6 t roller and the test rig at the higher numbers of passes is as would be expected. Despite the discrepancy in the results obtained with the 2.8 t roller, a low air content was achieved with a high number of passes; the termination of the relation given in Figure 4.1, taken from Lewis (1954), cannot, therefore, be justified (see Paragraph 4.9).

4.15 Secondly, in Figure 4.12(a), the states of compaction produced by the 2.8 t roller are shown as higher than those for the 8.6 t roller.

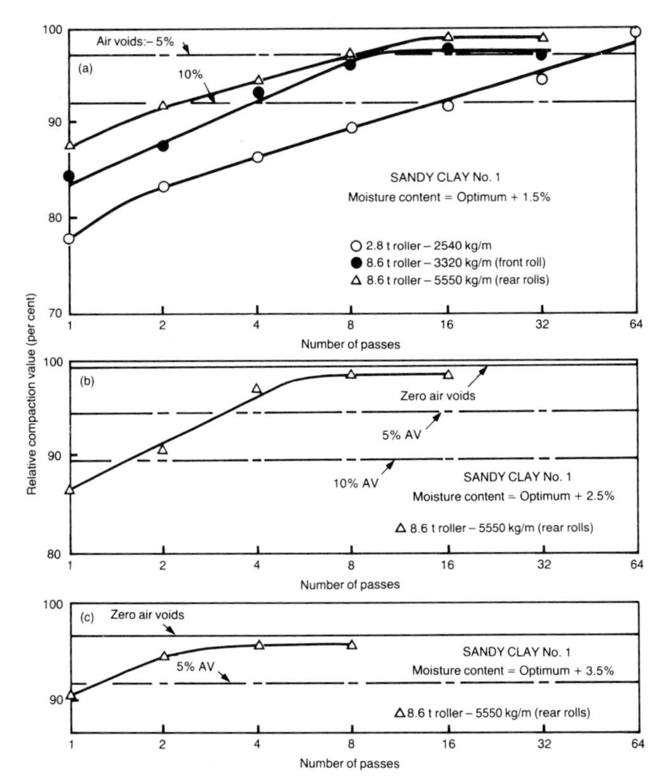

Figure 4.11 Relations between relative compaction and number of passes of smooth-wheeled rollers on sandy clay No 1 at three different values of moisture content and a layer thickness of 150 mm. All values are related to the maximum dry density and optimum moisture content obtained in the 2.5 kg rammer compaction test (Figure 3.13)

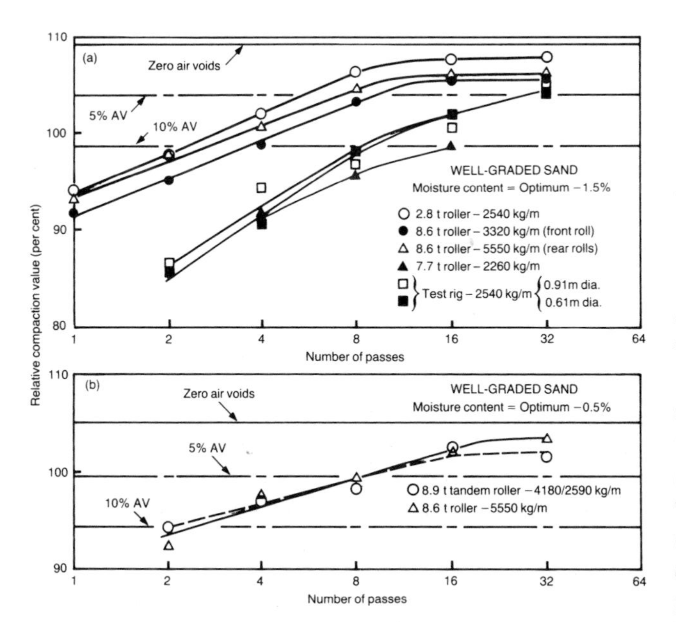

Figure 4.12 Relations between relative compaction and number of passes of smooth-wheeled rollers on well-graded sand at two different values of moisture content and a layer thickness of 150 mm. All values are related to the maximum dry density and optimum moisture content obtained in the 2.5 kg rammer compaction test (Figure 3.13)

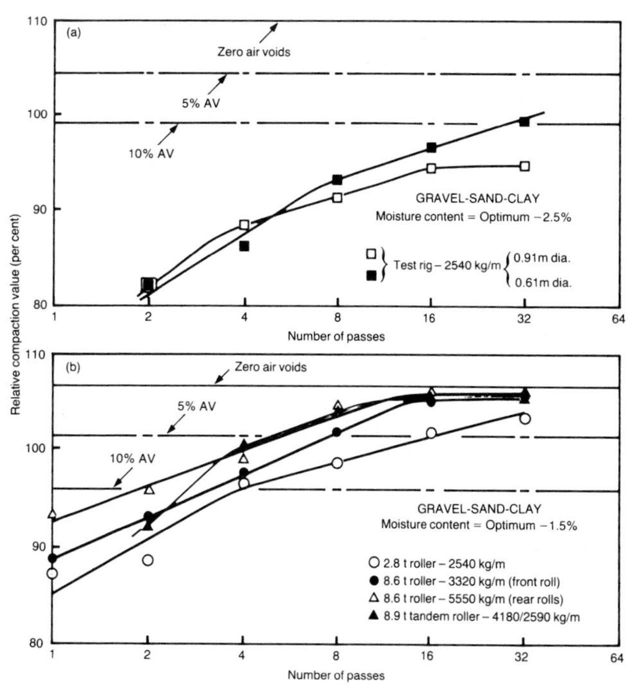

Figure 4.13 Relations between relative compaction and number of passes of smooth-wheeled rollers on gravel-sand-clay at two different values of moisture content and a layer thickness of 150 mm. All values are related to the maximum dry density and optimum moisture content obtained in the 2.5 kg rammer compaction test (Figure 3.13)

The results achieved with the test rig for the same mass per unit width of roll (2540 kg/m) appear to be more realistic and are comparable with the results for the 8.6 t and 7.7 t rollers included in Figure 4.12(a).

4.16 Taking the results given in Figures 4.1 to 4.6 and 4.8 to 4.13, general conclusions can be reached with regard to the number of passes needed to achieve given levels of compaction in 150 mm thick layers over a range of moisture contents. The levels of compaction chosen were those commonly used in earthwork construction, namely 95 per cent relative compaction (a dry density equivalent to 95 per cent of the maximum dry density obtained in the 2.5 kg rammer compaction test) and 10 per cent air voids. Regression analyses have been carried out for each soil individually and for various combinations of soils, and the most significant relations overall were achieved by taking the heavy clay and silty clay together (Figure 4.14), the two sandy clays together (Figure 4.15) and the well-graded sand and the gravel-sand-clay together (Figure 4.16). The Figures illustrate the results of the regression analyses in the form of relations between the number of roll passes necessary to achieve the required state of compaction and the mass per unit width of roll for various relative moisture contents (moisture content of soil minus the optimum moisture content obtained in the 2.5 kg rammer compaction test).

4.17 The regression equations on which the graphs in Figures 4.14 to 4.16 are based, and their correlation coefficients, are given in Table 4.3. The use of regression lines compensates in an objective way for the anomalies between results such as those discussed in Paragraphs 4.14 and 4.15; these anomalies have clearly contributed to the somewhat reduced values of some of the correlation coefficients given in Table 4.3. However, Figures 4.14 to 4.16 give a general summary of the results of the investigations of smooth-wheeled rollers on 150 mm thick compacted layers, and they can be taken as a guide to the performance of this type of compactor over a fairly wide range of soil types.

Table 4.3 Equations and correlation coefficients for the relations given in Figures 4.14 to 4.16

Soil	Level*	Regression equation†	No. of data points	Correlation coefficient
Heavy clay & silty clay	(a)	P=7.24–1.80 M–.020 w	20	0.75
	(b)	P=7.09–1.74 M–.098 w	22	0.85
Sandy clays	(a)	P=4.83–1.08 M–.116 w	16	0.84
	(b)	P=5.40–1.25 M–.209 w	18	0.91
Well-graded sand & gravel-sand clay	(a)	P=5.71–1.53 M–.213 w	17	0.89
	(b)	P=6.40–1.73 M–.327 w	22	0.87

*Level (a)=95 per cent of the maximum dry density obtained in the 2.5 kg rammer compaction test
Level (b)=10 per cent air voids
†P =Log_{10} (Number of roll passes required)
M=Log_{10} (Mass per unit width of roll, kg/m)
w=Moisture content of soil, per cent, minus the optimum moisture content, per cent, obtained in the 2.5 kg rammer compaction test

Figure 4.14 Smooth-wheeled rollers on heavy- and silty-clay soils. Relations between number of roll passes required and mass per unit width of roll to achieve (a) 95 per cent of the maximum dry density obtained in the 2.5 kg rammer compaction test and (b) 10 per cent air voids. These graphs were derived by regression analyses of the results given in Figures 4.1, 4.2, 4.8 and 4.9 and apply to 150 mm compacted layers

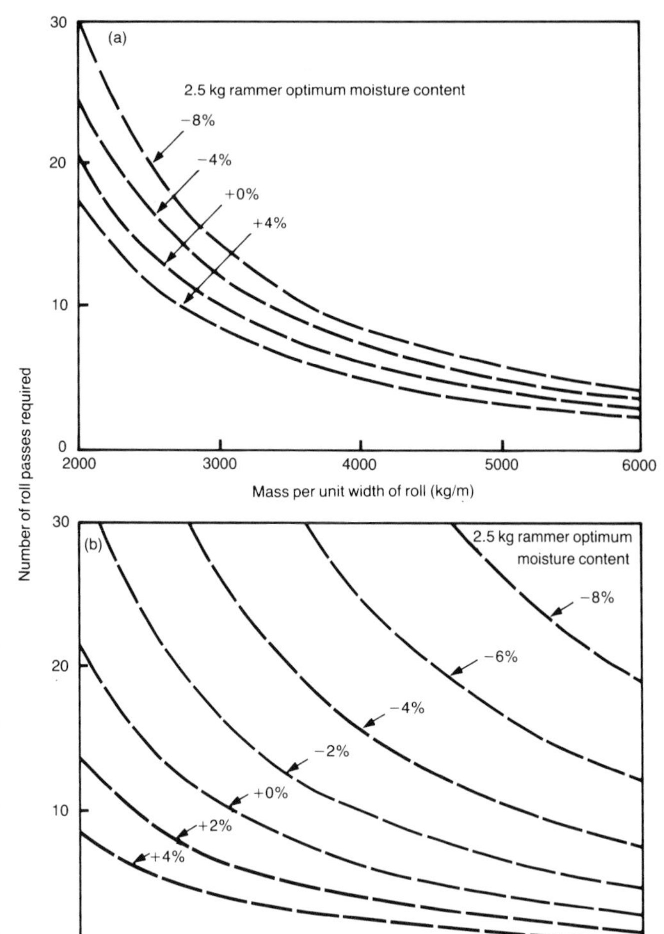

Relations between state of compaction and depth within the compacted layer

4.18 The variation of the state of compaction within the compacted layer, as determined in the investigations of the performance of plant at the Laboratory, provides a guide to the potential maximum thickness which may be compacted, given that the number of passes used is taken into account. The results were obtained by compacting fairly thick layers, 300 to 400 mm deep in general, and determining the dry density of the soil at various depths in the layer after successive removal, by careful excavation, of overlying material. Because of this necessary procedure the results are likely to be less accurate than those obtained when investigating 150mm thick surface layers.

4.19 The number of results available for smooth-wheeled rollers is relatively small, and as can be seen from a study of the relations given in Figures 4.17 to 4.22, almost all the tests were conducted with the 8.6 t roller. However, with the well-graded sand (Figure 4.21) investigations were also carried out with the 2.4 t tandem roller (Plate 4.7) and the 350 kg pedestrian roller (Plate 4.6).

Figure 4.15 Smooth-wheeled rollers on sandy clay soils. Relations between number of roll passes required and mass per unit width of roll to achieve (a) 95 per cent of the maximum dry density obtained in the 2.5 kg rammer compaction test and (b) 10 per cent air voids. These graphs were derived by regression analyses of the results given in Figures 4.3, 4.4, 4.10 and 4.11 and apply to 150 mm compacted layers

Figure 4.16 Smooth-wheeled rollers on well-graded sand and gravel-sand-clay. Relations between number of roll passes required and mass per unit width of roll to achieve (a) 95 per cent of the maximum dry density obtained in the 2.5 kg rammer compaction test and (b) 10 per cent air voids. These graphs were derived by regression analyses of the results given in Figures 4.5, 4.6, 4.12 and 4.13 and apply to 150 mm compacted layers

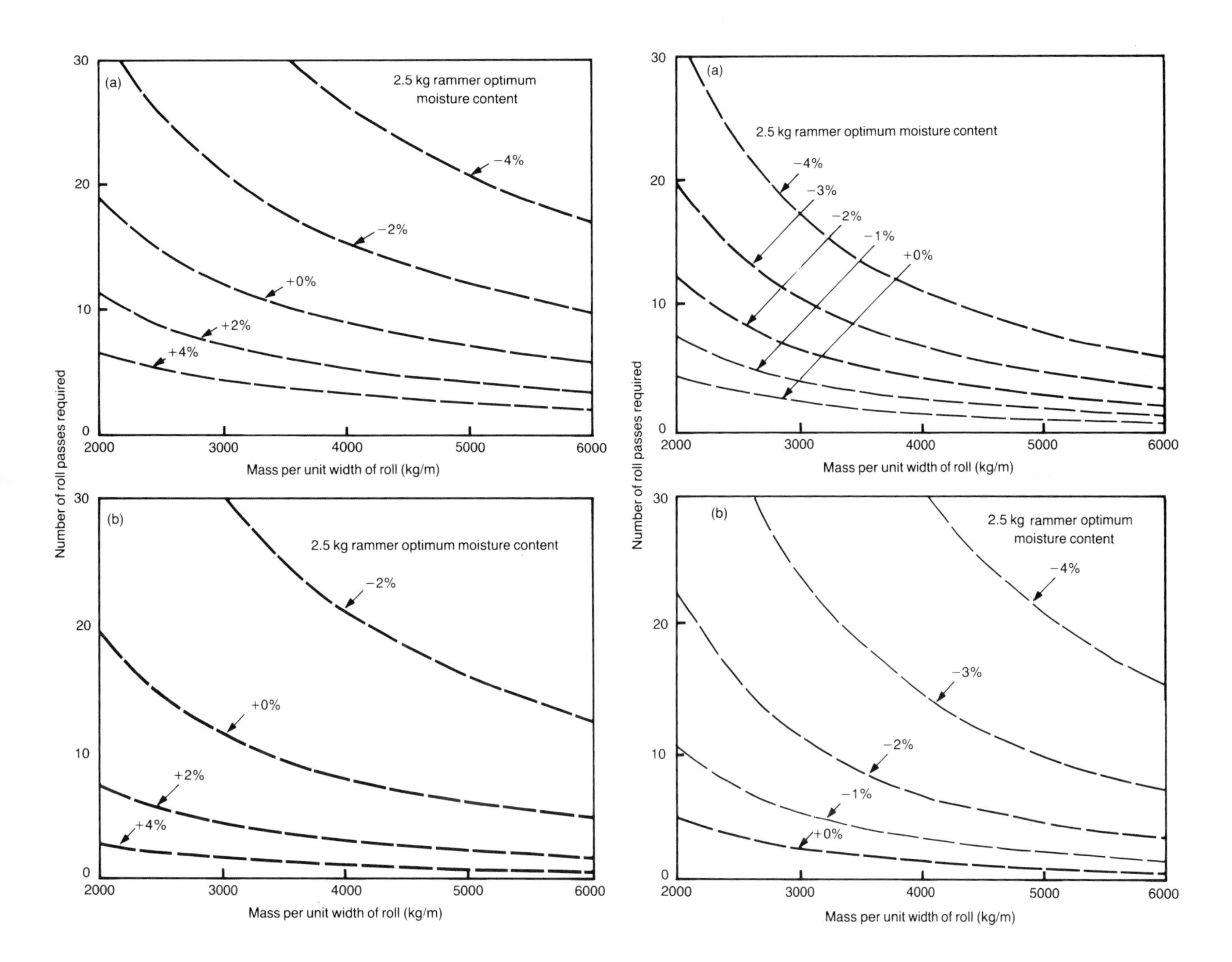

4.20 Figures 4.17 to 4.19, 4.21 and 4.22 contain results produced by 32 passes of the front roll of the 8.6 t roller, with tests at three different moisture contents on each soil. These relations clearly show the effect of variations in moisture content, as discussed in Paragraph 2.17, with the density gradient (rate of decrease in dry density with increasing depth below the surface of the layer) reducing as the moisture content increased. The relation given in Figure 4.20 for 32 passes of the rear rolls of the 8.6 t roller on sandy clay No 2 is the result of a solitary test carried out in the second phase of the compaction investigations (see Table 3.2).

4.21 In summarising the results relating states of compaction to moisture content and number of roll passes earlier in this Chapter, the criterion used was the average state of compaction in the 150 mm thick compacted layer. In considering the relation between state of compaction and depth in the compacted layer the state of compaction at the bottom of the layer is the best criterion on which to base estimates of the maximum thicknesses of layer to be compacted. Precedents for this approach are provided by Valeux and Morel (1980) and Morel and Machet (1980). Thus, for an average state of compaction equivalent to a relative compaction of 95 per

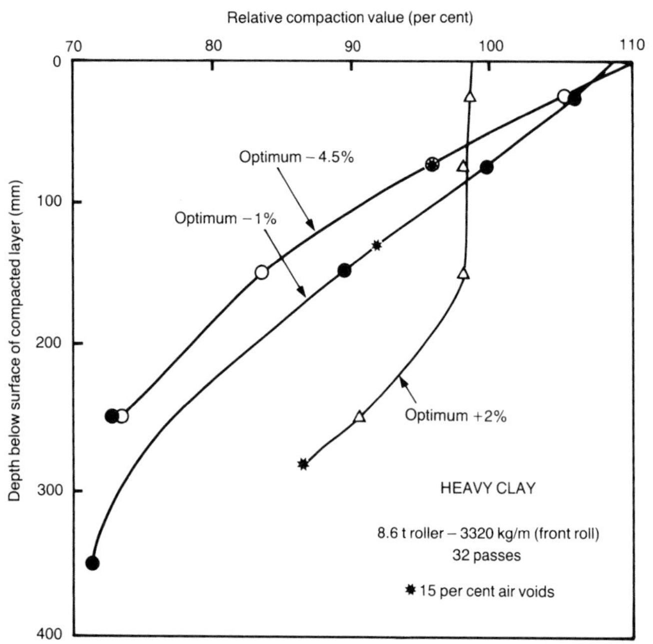

Figure 4.17 Relations between relative compaction and depth in the compacted layer of heavy clay at three different values of moisture content obtained with the 8.6 t smooth-wheeled roller. All values are related to the maximum dry density and optimum moisture content obtained in the 2.5 kg rammer compaction test (Figure 3.13)

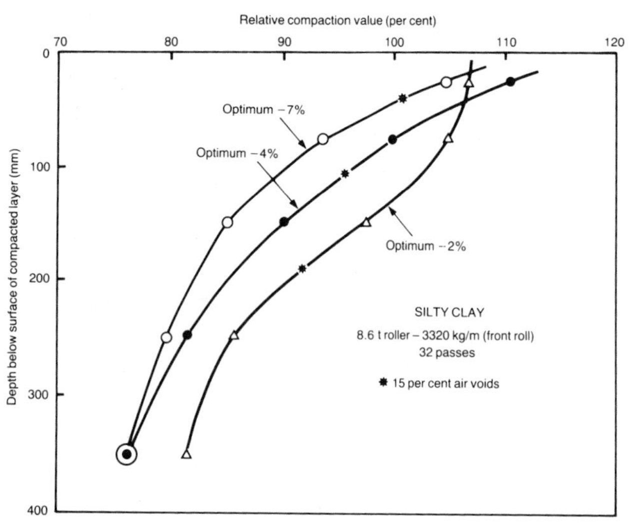

Figure 4.18 Relations between relative compaction and depth in the compacted layer of silty clay at three different values of moisture content obtained with the 8.6 t smooth-wheeled roller. All values are related to the maximum dry density and optimum moisture content obtained in the 2.5 kg rammer compaction test (Figure 3.13)

cent (of the maximum dry density obtained in the 2.5 kg rammer compaction test) a criterion for the required state of compaction at the bottom of the layer would be a relative compaction value of 90 per cent. By similar reasoning an average of 10 per cent air voids might be achieved if the air

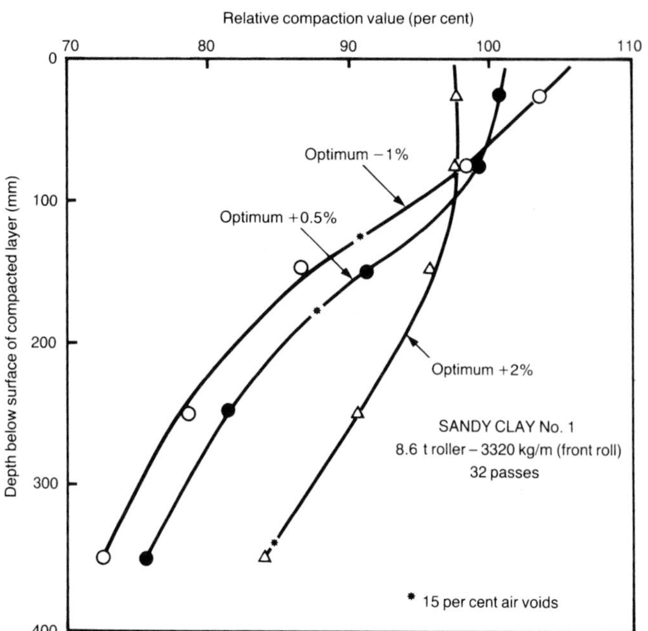

Figure 4.19 Relations between relative compaction and depth in the compacted layer of sandy clay No 1 at three different values of moisture content obtained with the 8.6 t smooth-wheeled roller. All values are related to the maximum dry density and optimum moisture content obtained in the 2.5 kg rammer compaction test (Figure 3.13)

Figure 4.20 Relation between relative compaction and depth in the compacted layer of sandy clay No 2 obtained with the 8.6 t smooth-wheeled roller. All values are related to the maximum dry density and optimum moisture content obtained in the 2.5 kg rammer compaction test (Figure 3.13)

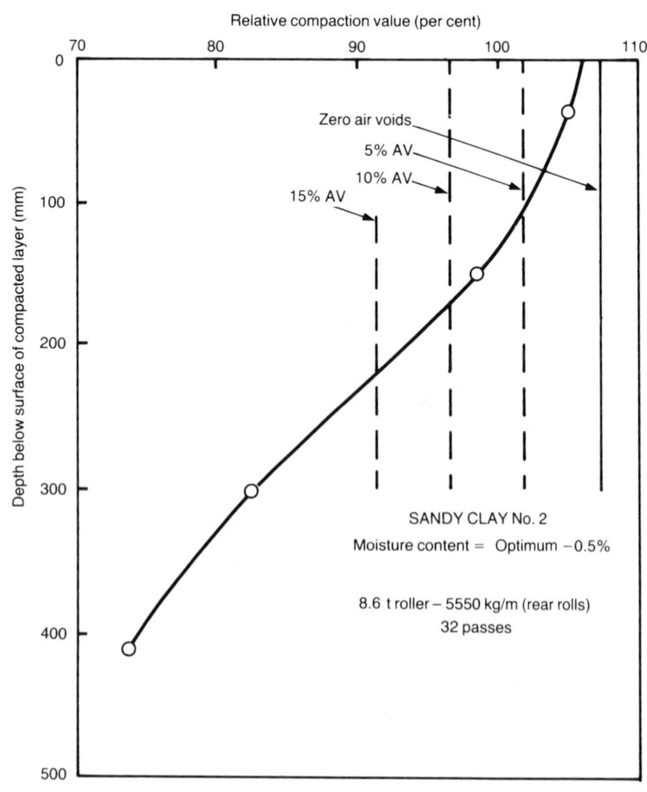

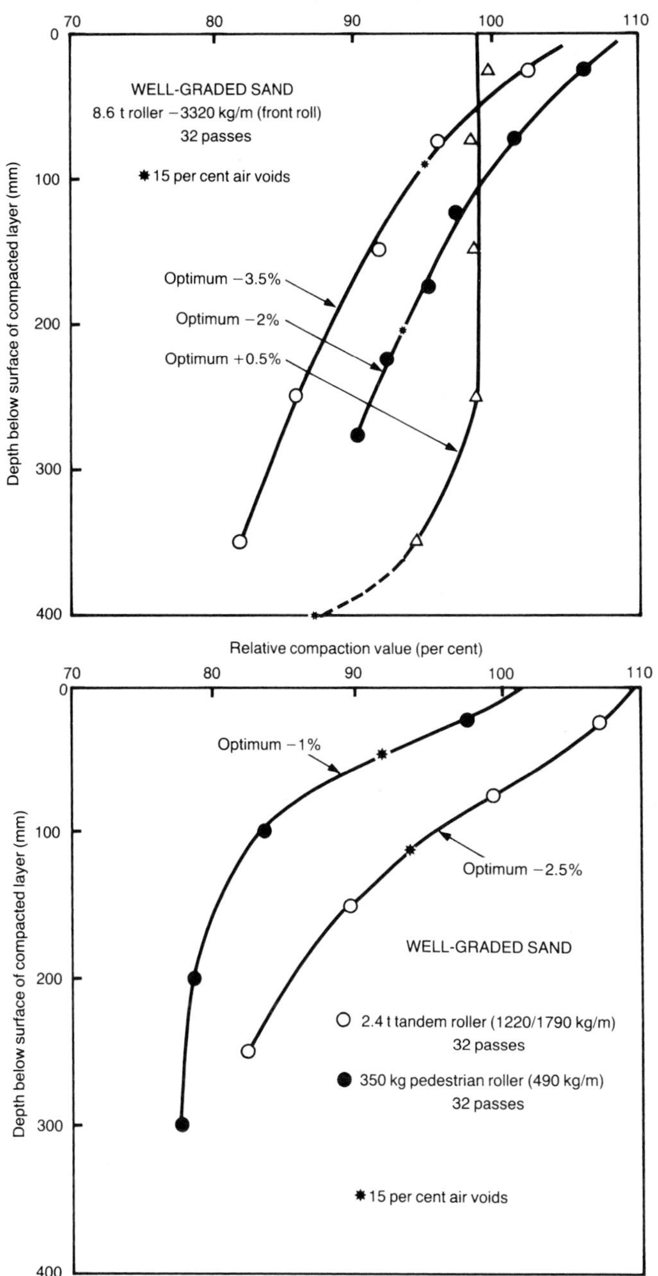

Figure 4.21 Relations between relative compaction and depth in the compacted layer of well-graded sand obtained with smooth-wheeled rollers. All values are related to the maximum dry density and optimum moisture content obtained in the 2.5 kg rammer compaction test (Figure 3.13)

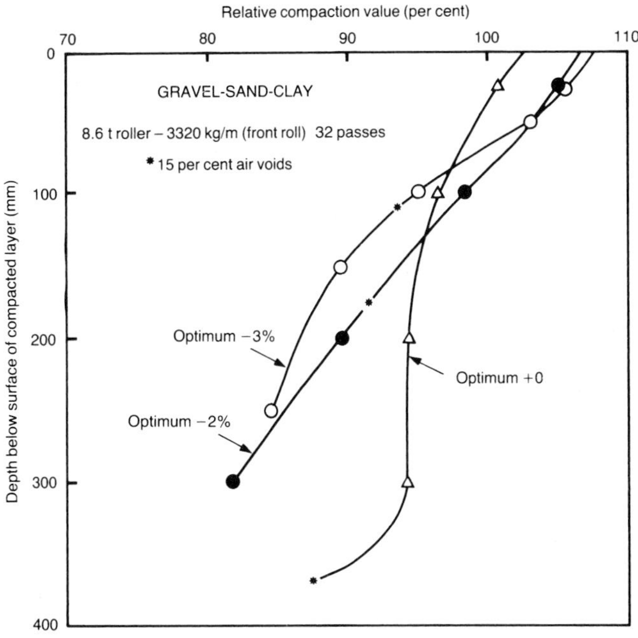

Figure 4.22 Relations between relative compaction and depth in the compacted layer of gravel-sand-clay at three different values of moisture content obtained with the 8.6 t smooth-wheeled roller. All values are related to the maximum dry density and optimum moisture content obtained in the 2.5 kg rammer compaction test (Figure 3.13)

content at the bottom of the layer did not exceed 15 per cent. These two criteria for the minimum state of compaction at the bottom of the layer have been used to summarise the results given in Figures 4.17 to 4.22.

4.22 The available results only allow the effect of moisture content on the maximum depth of layer which can be compacted to be determined for 32 passes of the 8.6 t roller. Relations between depth of compacted layer and relative moisture content for this compactor are given in Figure 4.23. Figure 4.23(a) applies to a minimum state of compaction (at the bottom of the layer) of 90 per cent of the 2.5 kg rammer maximum dry density and Figure 4.23(b) to a maximum air void content of 15 per cent. The depths of layer were taken directly from the relations given in Figures 4.17 to 4.22, 90 per cent relative compaction being read on the horizontal axis; the relative compaction value equivalent to 15 per cent air voids is indicated by an asterisk on each relation. The result given in Figure 4.20 has also been included, although strictly it applies to the rear rolls of the machine, whereas the remainder of the results applies to the front roll. A single relation has been drawn 'by eye' through the points for all the clay soils, and a second relation through the points for the well-graded sand and the gravel-sand-clay. Thus it has not been necessary to divide the clay soils into two sets as was the case for the relations between state of compaction and moisture content and between state of compaction and number of passes.

4.23 From Figure 4.23(a) it can be seen that with clay soils the maximum thickness of compacted layer which may be compacted by 32 passes of the machine, given that a minimum of 90 per cent relative compaction must be

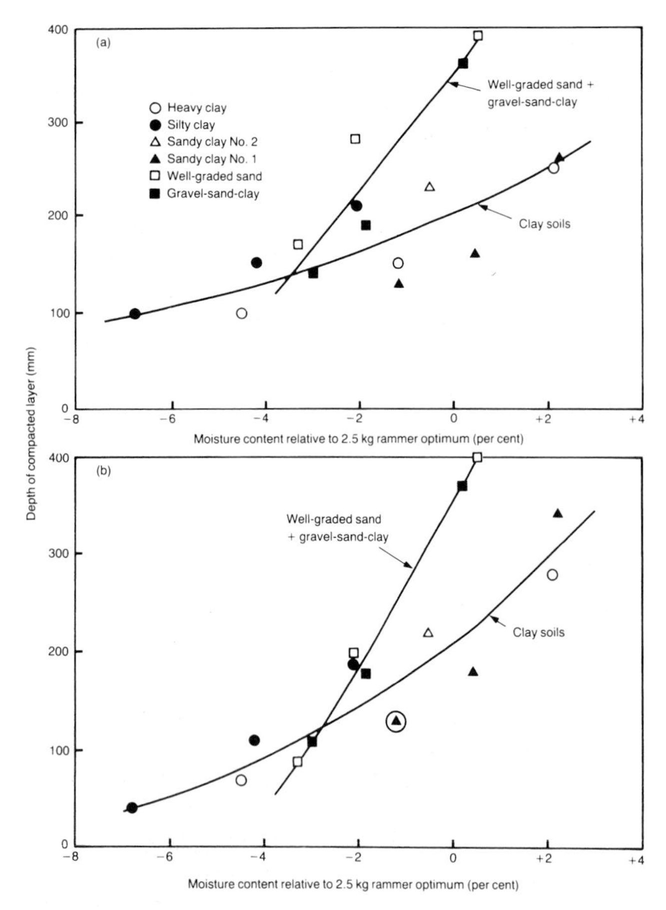

Figure 4.23 Maximum depth of compacted layer, to achieve a dry density at the bottom of the layer equivalent to (a) 90 per cent of the maximum dry density obtained in the 2.5 kg rammer compaction test and (b) 15 per cent air voids, related to moisture content. These graphs are derived from Figures 4.17 to 4.22 and apply only to 32 passes of the 8.6 t smooth-wheeled roller

achieved, varied from about 100 mm at 2.5 kg rammer optimum moisture content minus 6 per cent, to about 250 mm at optimum moisture content plus 2 per cent. If the minimum state of compaction required is equivalent to 15 per cent air voids (Figure 4.23(b)), the maximum depth of layer varies from 50 mm at optimum minus 6 per cent to 300 mm at optimum moisture content plus 2 per cent. With the granular soils the effect of moisture content is much more marked; for the 90 per cent relative compaction criterion, the depth of layer varies from about 150 mm at optimum moisture content minus 3 per cent to 350 mm at a moisture content equal to optimum of the 2.5 kg rammer compaction test. For the same moisture content range, using the 15 per cent air void criterion, depth of layer varies from 100 mm to 350 mm. It is worth repeating here that these results apply to 32 passes of the 8.6 t roller, ie a condition equivalent to compaction to refusal. *Any reduction in the number of passes will result in a reduction in the thickness of layer which may be compacted to the same criterion.*

4.24 Only a very limited amount of information is available on the influence of the size of roller, in terms of mass per unit width of roll, on the maximum thickness of layer which may be compacted. Taking the results from Figure 4.21 which were obtained at roughly equal values of relative moisture content, optimum moisture content minus 2.5 per cent to optimum minus 1 per cent, results for 32 passes of three different machines can be compared. Relations between depth of layer and mass per unit width of roll are given in Figure 4.24, using the same two criteria as discussed in Paragraph 4.21 above. The

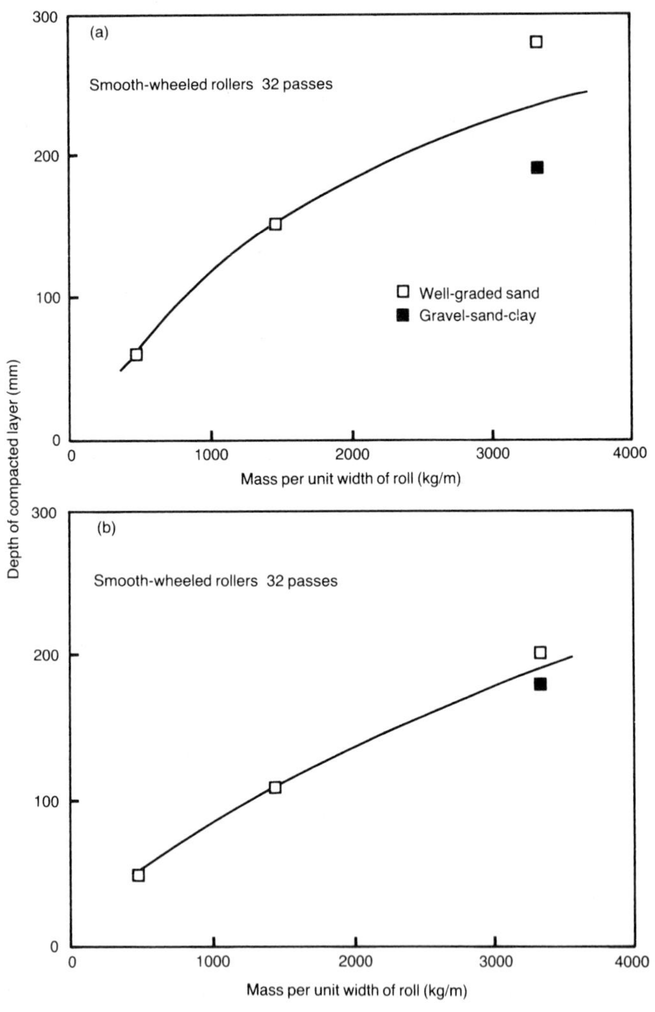

Figure 4.24 Maximum depth of compacted layer in the granular soils, to achieve a dry density at the bottom of the layer equivalent to (a) 90 per cent of the maximum dry density obtained in the 2.5 kg rammer compaction test and (b) 15 per cent air voids, related to the mass per unit width of roll of smooth-wheeled rollers. These data are taken from Figures 4.21 and 4.22 for moisture contents, relative to the 2.5 kg rammer optimum, of -2.5 to -1 per cent

equivalent result for the 8.6 t roller on gravel-sand-clay (Figure 4.22) has also been included. Thus, with the granular soils, it appears that at moisture contents in the range of 1 to 2.5 per cent below the optimum moisture content obtained in the 2.5 kg rammer compaction test, the maximum thickness of layer which may be compacted with 32 roll passes varies from 50 mm for a loading of about 500 kg/m up to 200 to 250 mm (depending on the criterion used) at a loading of 3500 kg/m.

4.25 The results presented in Figures 4.23 and 4.24 provide some general guidance on the potential maximum thickness of layer which may be compacted; the depths of layer obtained in this way were determined from results of tests of 300 to 400 mm thick layers and the effect of the low stiffness of the low-density material in the lower levels of the layer would have caused some loss of density at the intermediate levels. For this reason the assessment of the maximum thickness of layer from these data will be conservative and there is some discrepancy between the summary results given in Figures 4.14 to 4.16 for 150 mm thick layers and those in Figures 4.23 and 4.24. The latter figures are most useful when considering whether an increase in thickness beyond 150 mm would be feasible.

4.26 As an example of this approach, take well-graded sand and gravel-sand-clay soils at a moisture content of about 2 per cent below the 2.5 kg rammer optimum. Figure 4.16(b) indicates that with a roll loading of 3000 kg/m, an average state of compaction equivalent to 10 per cent air voids in a 150 mm layer can be achieved with about 11 passes. Figure 4.24(b) indicates that for the same roll loading and at the same moisture content a depth of 180 mm could possibly be compacted with an air content of 15 per cent at the bottom of the layer, provided 32 roll passes are applied. An obvious conclusion from this is that the use of a 150 mm thick layer with 11 passes is by far the more economical method of compaction. The economics of increasing the thickness of compacted layer by raising the moisture content can also be considered by reference to Figure 4.23. However, with the dearth of results relating the state of compaction achieved to depth in the layer for machines other than the 8.6 t roller, such comparisons as that given above cannot be widely applied.

Effect of diameter of roll on the performance of smooth-wheeled rollers

4.27 Details of the machines investigated as smooth-wheeled rollers are given in Table 4.1. The diameter of rolls used in the various machines varied considerably, from 0.53 m for the 220 kg pedestrian-operated roller to 1.60 m for the 8.6 t roller. However, as can be seen in Figure 4.25, the diameter is related to the roll loading, with a correlation coefficient of 0.84, for the 15 sets of data applying to the standard machines, ie with the exception of the test rig.

4.28 The experimental test rig (Plates 4.4 and 4.12) was manufactured in 1954 with the intention of specifically studying the effects of varying the roll diameter whilst maintaining a constant roll loading. The three different rolls are shown in Plate 4.12, but in the event it was found that the smallest of the rolls, with a diameter of 0.30 m, would not rotate on loose soil at the roll loading chosen for the tests

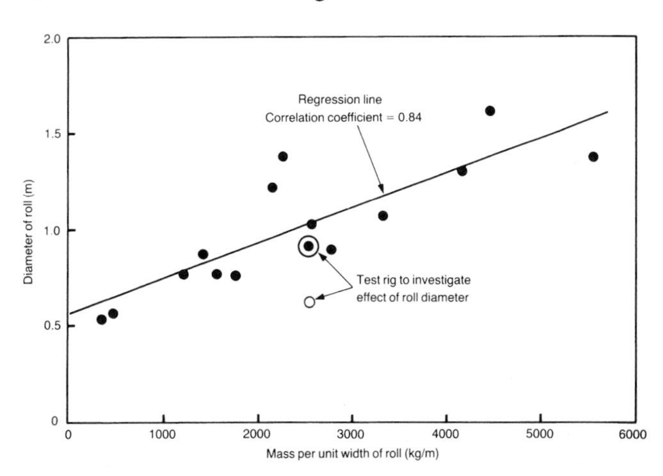

Figure 4.25 Relation between roll diameter and roll loading for the smooth-wheeled rollers included in the investigations. Data from Table 4.1

Plate 4.12 The three sizes of roller used in the test rig to investigate the effect of variations in diameter of a smooth-wheeled roll

(2540 kg/m), the same loading as applied to the rear rolls of the 2.8 t roller (Table 4.1). Thus only two sizes of roll were investigated and the results were reported by Parsons and Smith (1955). The results have been included in Figures 4.8(c), 4.10(b), 4.12(a) and 4.13(a).

4.29 The conclusions reached at the time were that there was little significant difference between the results produced by the two sizes of roll, but that the handling difficulties experienced with the roll having a diameter of 0.61 m indicated that the diameter of about 0.9 m was the minimum that should be used for the particular roll loading of 2540 kg/m. The relation given in Figure 4.25 indicates that generally the diameters of rolls increase with roll loading, and at the loading of about 2500 kg/m, the smaller roll used in the test rig was well outside the range to be expected.

4.30 From the data given it can be concluded that the performance of smooth-wheeled rollers can be directly related to the mass per unit width of roll; the diameter of the roll, by design, is a function of the mass per unit width as indicated by Figure 4.25.

References

British Standards Institution (1990). British Standard methods of test for soils for civil engineering purposes: BS 1377: Part 4 Compaction related tests. BSI, London.

Department of Transport (1986). Specification for highway works. 6th edition. HM Stationery Office, London.

Lewis, W A (1954). Further studies in the compaction of soil and the performance of compaction plant. *Road Research Technical Paper* No 33. HM Stationery Office, London.

Morel, G and Machet, J M (1980). Compaction of pavement layers by vibration (in French). *International Conference on Compaction*, Editions Anciens ENPC, Paris, Vol II, 437–44.

Parsons, A W and Smith, P (1955). An investigation of the effect on the compaction of soil of varying the diameter of a smooth-wheeled roller. *D.S.I.R, RRL Research Note* No RN/2469/AWP.PS (Unpublished).

Valeux, J C and Morel, G (1980). Influence of bearing capacity of underlying materials on the compaction of pavement layers (in French). *International Conference on Compaction*, Editions Anciens ENPC, Paris, Vol II, 475–80.

CHAPTER 5 COMPACTION OF SOIL BY PNEUMATIC-TYRED ROLLERS

Description of machines

5.1 Pneumatic-tyred rollers rely upon the weight imposed on a number of pneumatic-tyred wheels to provide the stresses necessary for the effective compaction of soil. They can have two possible configurations:–

1. Two axle machines – self-propelled or towed by tractor, with two axle sets, the rear axle set arranged so that its wheels track in the gaps in the wheel tracks of the front axle set; the total number of wheels usually varies between 7 and 13.

2. Single axle machines – towed by tractor with a single axle set of wheels, usually three or four.

5.2 Wheel loads can vary from about 1 tonne (t) to a maximum for large single-axle rollers of about 25 t. With any one machine a significant range of wheel loads can be obtained by the addition or removal of ballast. Pneumatic-tyred rollers are usually designed with a form of suspension system that tends to equalise the load on each wheel, either by arranging the wheels in pairs on oscillating axles or by using a hydraulic pressure equalising system.

5.3 In the course of investigations of the performance of compaction plant at the Laboratory four different pneumatic-tyred rollers were tested, but various combinations of wheel load, by appropriate adjustments to the quantity of ballast carried, and tyre inflation pressure were achieved. The relevant details of the machines are given in Table 5.1 and they are illustrated in Plates 5.1 to 5.5.

5.4 Table 5.1 includes the various combinations of wheel load and tyre inflation pressure used to produce the results given in this Chapter. The resulting contact pressure, determined from the wheel load and tyre contact area, is also given for each test condition.

5.5 The 12 t and 20 t towed rollers (Plates 5.1 and 5.2) were of a very similar general design and consisted of a load box to which were fitted four wheels in front and five at the rear. The arrangement of wheels was such that the rear

Table 5.1 Details of the pneumatic-tyred rollers used in the investigations

Machine (maximum laden mass)	Number of wheels	Size of tyres	Rolling width (between outer edges of outer wheels) (m)	Test configuration			
				Wheel load (t)	Tyre inflation pressure (kN/m²)	(bar)	Tyre contact pressure (kN/m²)
12 t towed	9	11.0 in× 12 in× 8 ply	2.08	0.7	250	2.5	240
				1.4	250	2.5	260
18.5 t self-propelled	9	9.0 in× 20 in× 12 ply	1.98	2.1	550	5.5	520
20 t towed	9	9.0 in× 20 in× 14 ply	2.13	2.3	550	5.5	520
45 t towed	4	16.0 in× 21 in× 36 ply	2.36	5.1	620	6.2	460
				5.1	970	9.7	540
				10.2	620	6.2	610
				10.2	970	9.7	770

Plate 5.1 12 t pneumatic-tyred roller investigated at various times between 1946 and 1957

Plate 5.2 20 t pneumatic-tyred roller investigated in 1953 to 1956

wheels ran between the tracks of those in front when the roller was travelling in a straight line. Except for the rear central wheel, which was mounted on a fixed axle, the wheels were arranged in pairs on axles free to oscillate so as to equalise the load on each wheel of the pair when the roller was operated on uneven ground. The frame carrying the front four wheels was attached to the load box by means of a spherical-ball mounting to enable the roller to be steered, even on rough ground.

5.6 The 45 t towed roller (Plate 5.3) also had a load box, but with only one transverse row of four wheels, arranged in pairs, across the centre of the load box. Each pair of wheels was mounted on an axle pivoted at the centre to allow oscillation over rough ground. The wheel arrangement across the width of the roller is shown in Plate 5.4.

5.7 A limited number of tests were carried out with the 18.5 t self-propelled pneumatic-tyred roller (Plate 5.5). This also had nine wheels, with four at the front and five at the rear and as can be seen from Table 5.1, the test conditions were very similar to those for the 20 t towed roller. The 18.5 t roller, which had a 55 kW diesel engine driving the wheels of the front axle, had an on-board air compressor to allow adjustment of tyre inflation pressure 'on the run'. Steering was achieved by articulation of the chassis at a central pivot point. Variation of wheel load could be achieved by adjusting the amount of ballast in a load box over the rear axle and by varying the contents of a large water tank along each side of the front section.

Investigations prior to 1954

5.8 The 12 t towed roller (Plate 5.1) was extensively tested in the early series of investigations from 1946 to 1953 (see Table 3.1). These early tests were carried out on the circular track (Figure 3.1) which, for this particular machine, resulted in the rear wheels running in the tracks made by the front wheels instead of between them as occurs when the machines travel in a straight line. This, combined with the towing lorry running in a fixed radius track (see Paragraph 3.3), resulted in a series of wheel ruts with only slightly compacted soil between them, and the states of compaction measured in such tests were found to be lower than those achieved when the same machine was tested in the second pilot-scale facility (Table 3.2 and Figure 3.3). The differences in results were especially marked when low numbers of passes

Plate 5.3 45 t pneumatic-tyred roller investigated in 1955–56

Plate 5.4 Wheel configuration of the 45 t pneumatic-tyred roller

Plate 5.5 18.5 t self-propelled pneumatic-tyred roller investigated in 1961

of the roller were employed. With the exception of the tests on silty clay (as discussed later) all results obtained with the 12 t pneumatic-tyred roller in the early investigations and reported by Williams and Maclean (1950) and by Lewis (1954) have been excluded from this review.

Figure 5.1 Relations between dry density and moisture content for pneumatic-tyred rollers on heavy clay, expressed in terms of relative compaction with the 2.5 kg rammer test. Layer thickness = 150 mm; number of machine passes = 32

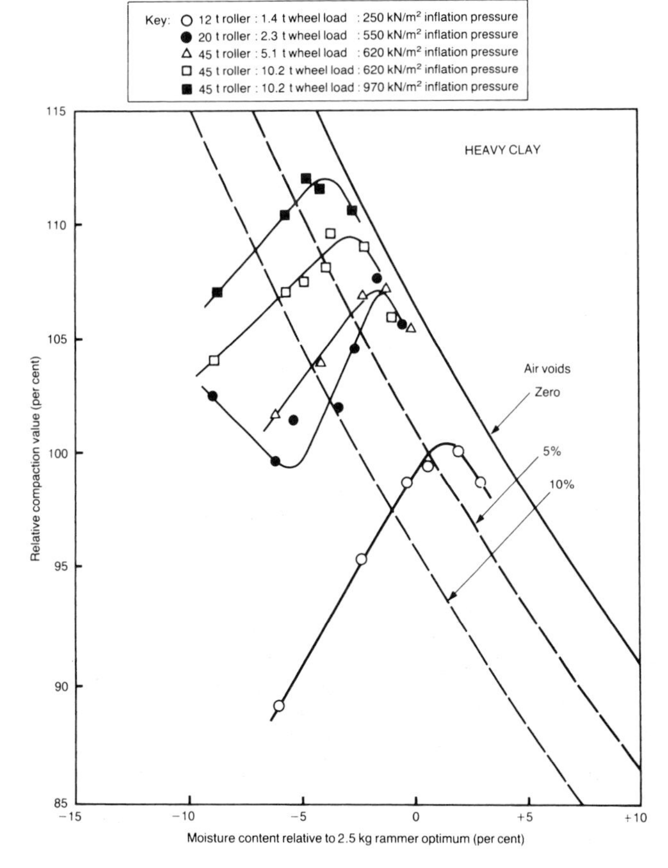

Relations between state of compaction and moisture content

5.9 Relations between the state of compaction produced by pneumatic-tyred rollers on 150 mm compacted layers and the moisture content of the soil are given in Figures 5.1 to 5.6. As for the tests described in Chapter 4, the dry density has been expressed as a percentage of the maximum dry density obtained in the 2.5 kg rammer compaction test (British Standards Institution, 1990) and termed the relative compaction value. The moisture content has been expressed as the difference from the optimum moisture content obtained in the same compaction test. The appropriate values of maximum dry density and optimum moisture content for the 2.5 kg rammer compaction test at the time of the tests were obtained from Figure 3.13. The air void lines in the Figures are based on the average 2.5 kg rammer compaction test values used in the results given in each Figure.

5.10 The results for 32 passes of the rollers on heavy clay (Figure 5.1), sandy clay No 2 (Figure 5.3), well-graded sand (Figure 5.4) and gravel-sand-clay (Figure 5.5) are those reported by Lewis (1959). The single relation obtained on

Figure 5.2 Relation between dry density and moisture content for the 12 t pneumatic-tyred roller on silty clay, expressed in terms of relative compaction with the 2.5 kg rammer test. Layer thickness = 150 mm; number of machine passes = 32; wheel load = 1.4 t; tyre inflation pressure = 250 kN/m²

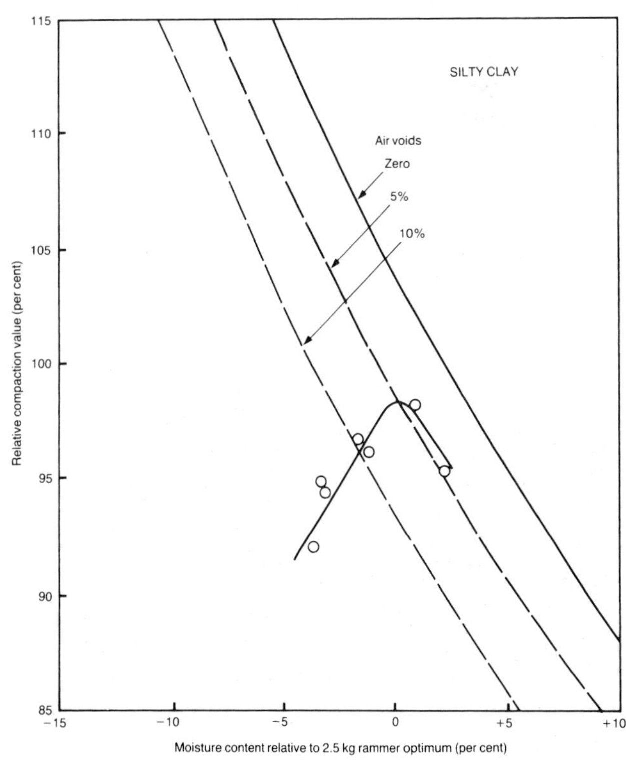

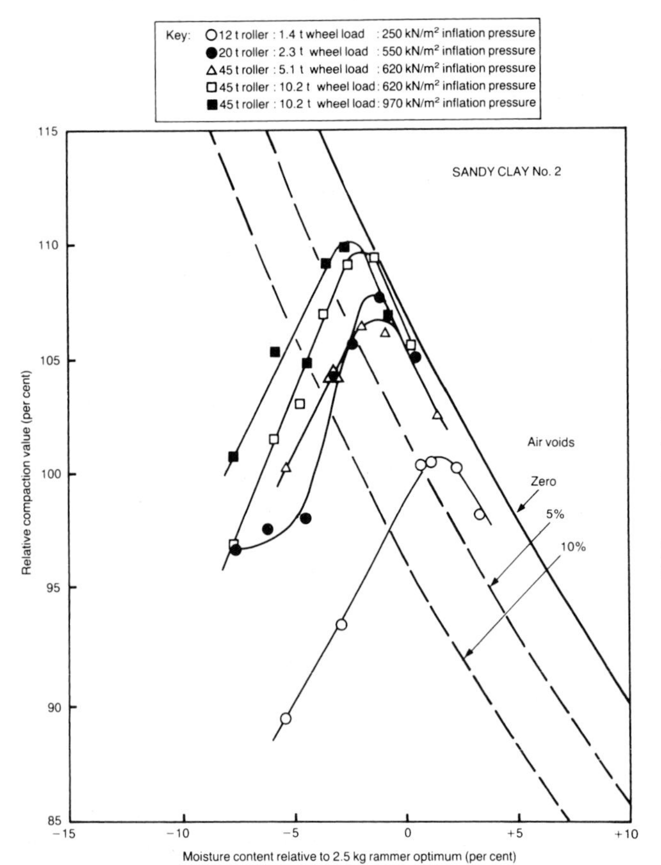

Figure 5.3 Relations between dry density and moisture content for pneumatic-tyred rollers on sandy clay No 2, expressed in terms of relative compaction with the 2.5 kg rammer test. Layer thickness = 150 mm; number of machine passes = 32

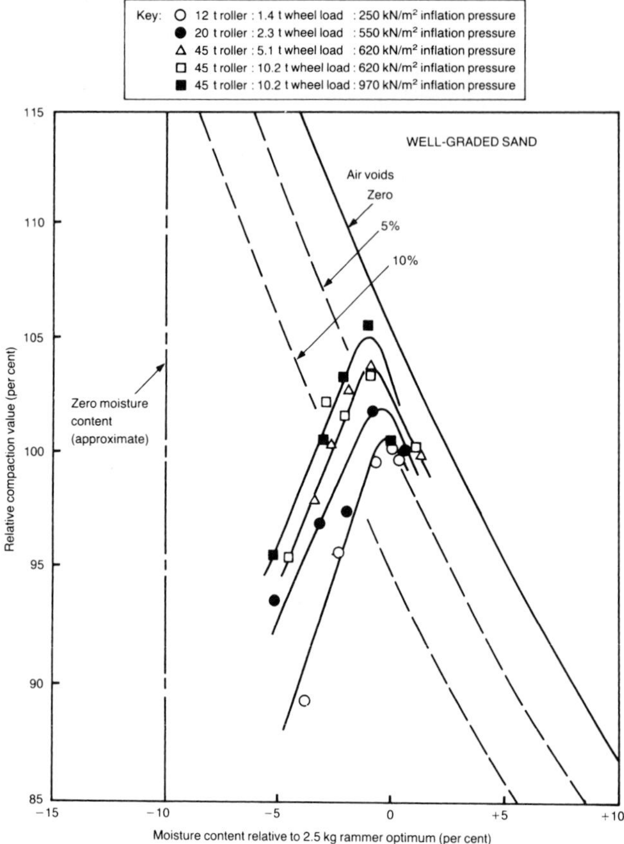

Figure 5.4 Relations between dry density and moisture content for pneumatic-tyred rollers on well-graded sand, expressed in terms of relative compaction with the 2.5 kg rammer test. Layer thickness = 150 mm; number of machine passes = 32

silty clay and given in Figure 5.2 was determined in the early investigations reported by Williams and Maclean (1950) and Lewis (1954). It is considered that the high number of passes (32) involved make these results valid despite the reservations expressed in Paragraph 5.8 above, and the result has been included here to increase the number of soil types on which information is available. In all the above cases, with the use of 32 passes of the rollers, the results can be regarded as representing the maximum states of compaction to be achieved in 150 mm thick compacted layers. Because the rear wheels of the two-axle machines always run between the tracks of the front wheels, the number of wheel passes over any part of the compacted area is equal to the number of roller passes. All references to numbers of passes in this Chapter can equally apply, therefore, to wheel passes or to roller passes.

5.11 The results of tests on uniformly graded fine sand are given in Figure 5.6. Here a low wheel load (0.7 t) and tyre inflation pressure (210 kN/m^2) and only 8 passes were used, but the interest in the results is the absence of any effect of variations in moisture content on the relative compaction value achieved. This property of uniformly graded sand, together with the large proportion of air voids left in the compacted material, has been discussed in Paragraph 2.19.

5.12 In general, the relations given in Figures 5.1 to 5.5 follow the general trend illustrated in Figure 2.2 and discussed in Paragraphs 2.5 to 2.9. Thus the maximum relative compaction value increased and the optimum relative moisture content decreased with increase in wheel load and/or tyre inflation pressure. As will be discussed later, the main factors influencing the performance of pneumatic-tyred rollers are the wheel load and the tyre contact pressure (resulting from the combination of wheel load, tyre inflation pressure and tyre characteristics). The values of both these factors used in the investigations are given in Table 5.1. The relative effects of these factors, taken separately and in combination, on the maximum relative compaction value and the optimum relative

Figure 5.5 Relations between dry density and moisture content for pneumatic-tyred rollers on gravel-sand-clay, expressed in terms of relative compaction with the 2.5 kg rammer test. Layer thickness = 150 mm; number of machine passes = 32

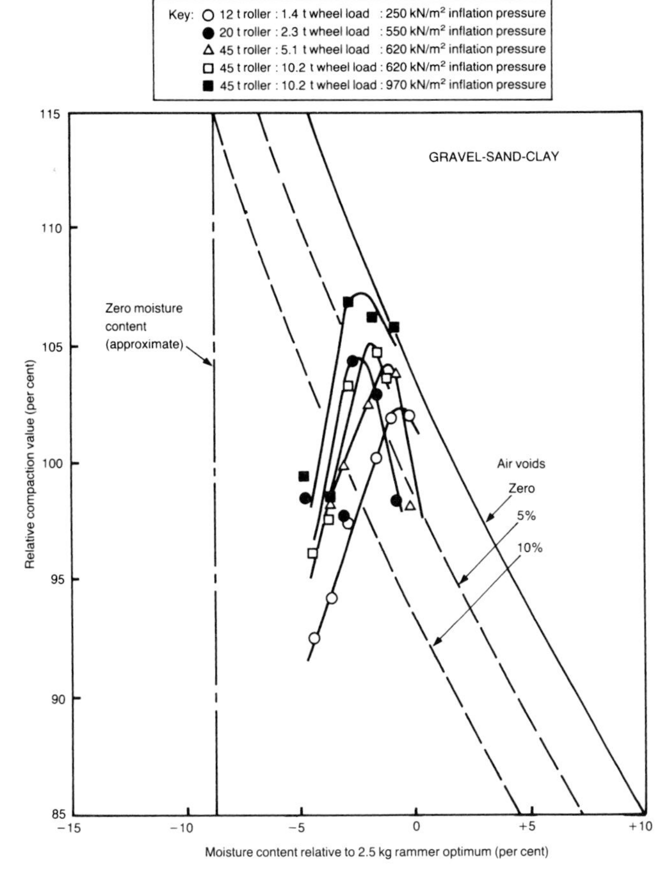

moisture content were studied by regression analyses. It was found that good correlations were obtained with the use of tyre contact pressure alone, and the relations obtained are

Figure 5.6 Relation between dry density and moisture content for the 12 t pneumatic-tyred roller on uniformly graded fine sand, expressed in terms of relative compaction with the 2.5 kg rammer test. Layer thickness = 150 mm; number of machine passes = 8; wheel load = 0.7 t; tyre inflation pressure = 210 kN/m²

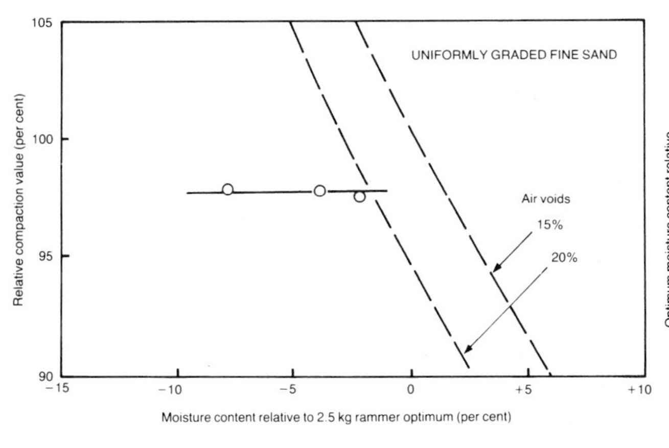

shown in Figure 5.7 and their equations and correlation coefficients are given in Table 5.2.

5.13 The fact that tyre contact pressure is the most significant factor governing the performance of the rollers in Figures 5.1 to 5.5 is likely to be associated with the relatively shallow thickness of layer involved. The stress distribution through the entirety of the 150 mm layer will be little affected by differences in the contact areas of the tyres. (In thicker layers the wheel load has an overriding influence, as discussed later.) As illustrated in Figure 5.7, a single relation can be drawn through the results for all the clay soils, and again for the two granular soils. However, in the latter case, relatively poor correlation coefficients were obtained (Table 5.2) and it can be seen, by examination of the individual results in Figure 5.7, that the well-graded sand consistently produced lower maximum dry densities and higher optimum moisture contents than were obtained on the gravel-sand-clay. It is likely that this is a result of the over-stressing of the surface of the compacted layer of the sand and this matter will be discussed later in this Chapter. A notable feature of these results is the much smaller influence of tyre contact pressure on the maximum relative compaction value (and hence

Figure 5.7 Maximum dry densities and optimum moisture contents related to the contact pressures of pneumatic-tyred rollers. Results from Figures 5.1 to 5.5

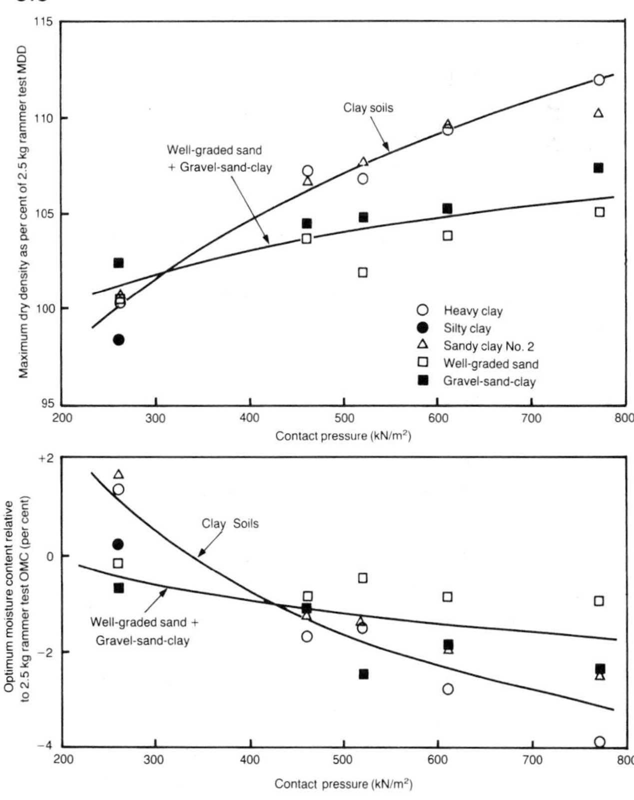

on optimum moisture content) with the granular soils than with the clay soils.

5.14 The results summarised in Figure 5.7 and Table 5.2 are for compaction to refusal (32 passes) by the pneumatic-tyred rollers with a compacted layer thickness of 150 mm. Any reduction in the number of passes or increase in thickness of layer will result in a reduction in the maximum relative compaction value and an increase in optimum moisture content as already discussed in Chapter 2.

Table 5.2 Equations and correlation coefficients for the relations given in Figure 5.7

Soil	Regression equation*	Number of points	Correlation coefficient
Clay soils	$c = 56.8\ p^{.102}$	11	0.98
	$w = 126\ p^{-.437} - 10$	11	0.94
Well-graded sand & gravel-sand-clay	$c = 81.3\ p^{.0395}$	10	0.80
	$w = 20.0\ p^{-.133} - 10$	10	0.55

*c = maximum dry density for the roller as percentage of the maximum dry density obtained in the 2.5 kg rammer compaction test
w = difference between optimum moisture content for the roller and optimum moisture content obtained in the 2.5 kg rammer compaction test (per cent)
p = tyre contact pressure of the roller (kN/m²)
The above results apply in general to 32 passes of the rollers on 150 mm compacted layers

Relations between state of compaction produced and number of passes

5.15 Results of investigations to determine the effect on compaction of variations in the number of passes of pneumatic-tyred rollers are given in Figures 5.8 to 5.11. Results for each soil used are illustrated under a separate Figure number, whilst different values of moisture content are given in separate graphs within each Figure. Thus Figure 5.8, for heavy clay, contains results at four values of moisture content; Figure 5.8(a) has results at a moisture content of optimum – 3.5 per cent, Figure 5.8(b) at optimum –2.5 per cent, Figure 5.8(c) at optimum + 1.5 per cent and Figure 5.8(d) at optimum + 4 per cent. In all cases optimum refers to the optimum moisture content for the 2.5 kg rammer compaction test and the relative compaction value is the dry density produced by the roller expressed as a percentage of the maximum dry density obtained in the same test. As before, the results for the 2.5 kg rammer test relevant to the times at which the investigations were carried out were obtained from Figure 3.13. Approximate air void lines for

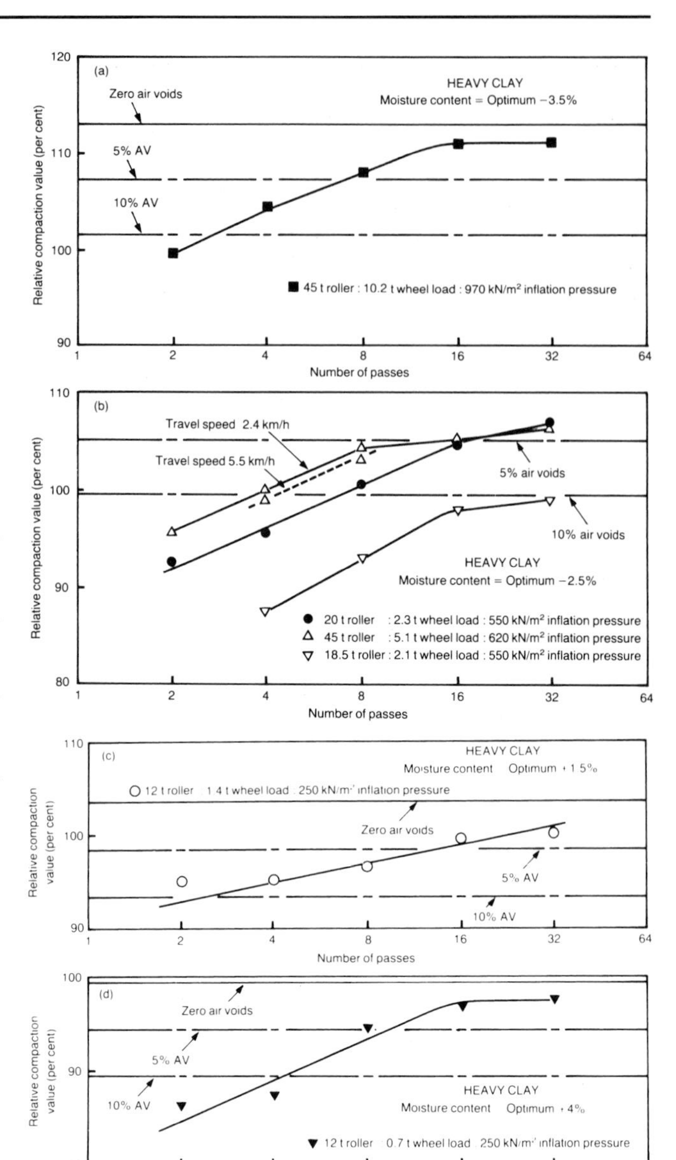

Figure 5.8 Relations between relative compaction and number of passes of pneumatic-tyred rollers on heavy clay at four different values of moisture content and with a layer thickness of 150 mm. All values are related to the maximum dry density and optimum moisture content obtained in the 2.5 kg rammer compaction test (Figure 3.13)

the average condition for the results are also included in each graph. The horizontal axis of each graph gives the number of passes on a logarithmic scale to allow maximum use of straight line relations (see Paragraph 2.11 and Figure 2.5). All the results apply to a layer thickness of 150 mm.

5.16 The combinations of wheel load and tyre inflation pressure used to determine the relations given in Figures 5.8 to 5.11 were similar on each of the four soils used, but with the addition of results for the 18.5 t self-propelled pneumatic-tyred roller on heavy clay (Figure 5.8(b)) and sandy clay No 2 (Figure 5.9(a)). The results for this machine on heavy clay do not compare well with those for the 20 t towed roller, also included in Figure 5.8(b). The contact pressures for these two machines were identical (Table 5.1) and

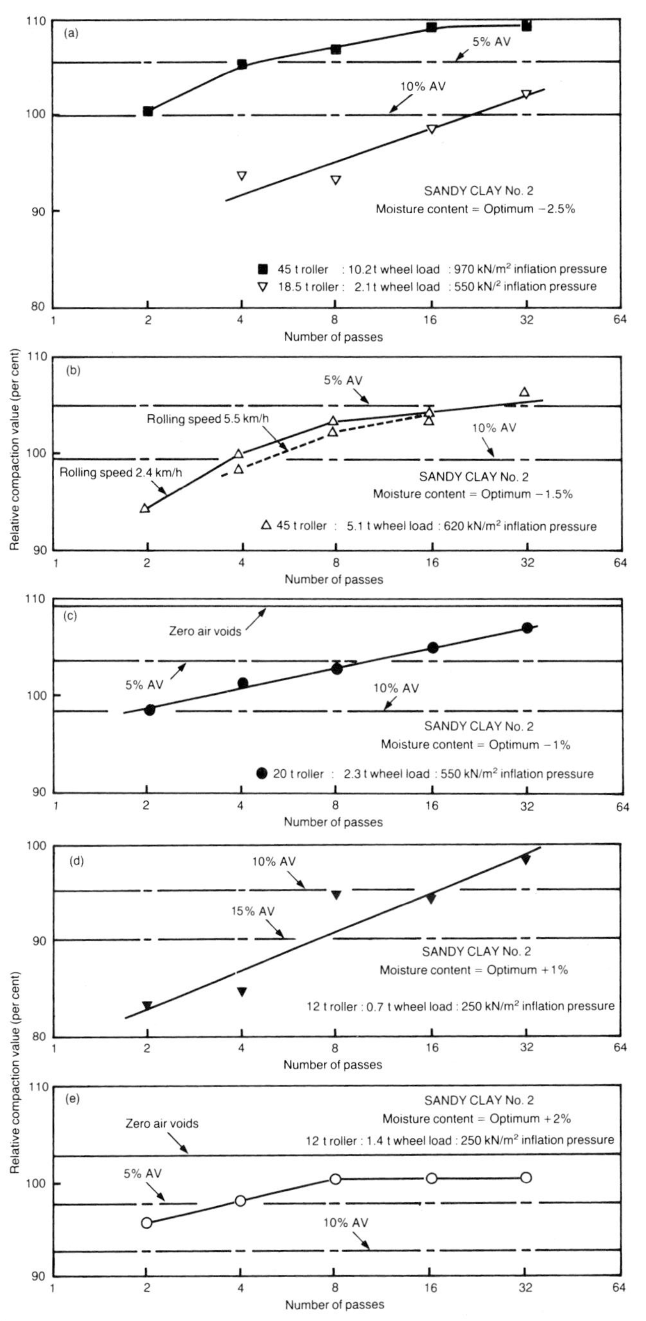

Figure 5.9 Relations between relative compaction and number of passes of pneumatic-tyred rollers on sandy clay No 2 at five different values of moisture content and with a layer thickness of 150 mm. All values are related to the maximum dry density and optimum moisture content obtained in the 2.5 kg rammer compaction test (Figure 3.13)

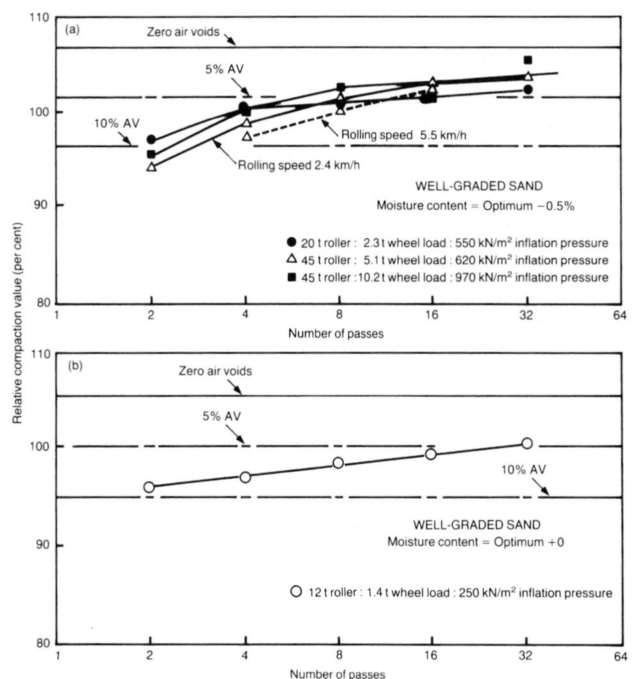

Figure 5.10 Relations between relative compaction and number of passes of pneumatic-tyred rollers on well-graded sand at two different values of moisture content and with a layer thickness of 150 mm. All values are related to the maximum dry density and optimum moisture content obtained in the 2.5 kg rammer compaction test (Figure 3.13)

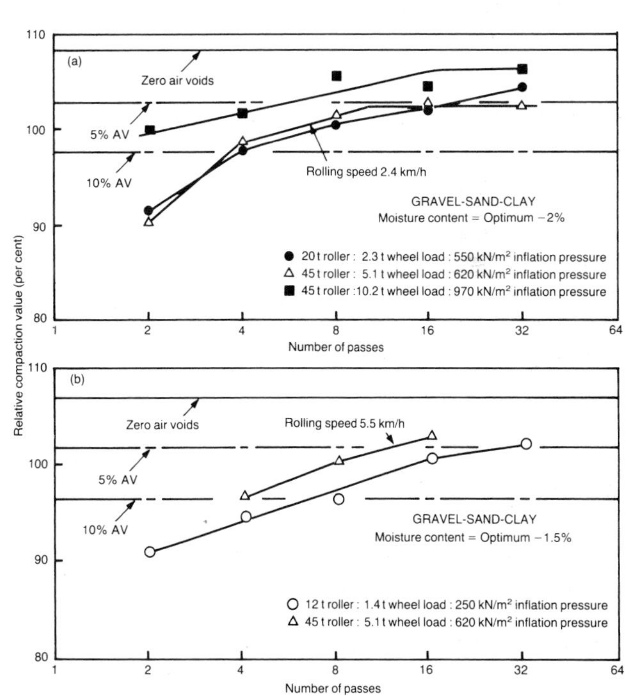

Figure 5.11 Relations between relative compaction and number of passes of pneumatic-tyred rollers on gravel-sand-clay at two different values of moisture content and with a layer thickness of 150 mm. All values are related to the maximum dry density and optimum moisture content obtained in the 2.5 kg rammer compaction test (Figure 3.13)

such differences would not normally be expected. These two machines were also tested on the sandy clay No 2, but at different moisture contents, see results in Figures 5.9(a) and 5.9(c). Allowing for the difference in moisture contents the results appear to be more comparable in this instance.

5.17 Tests with the 45 t roller ballasted to 5.1 t per wheel were carried out at two rolling speeds, 2.4 km/h and 5.5 km/h, see Figures 5.8(b), 5.9(b), 5.10(a) and 5.11(a) and (b). All other tests were carried out at the slower speed of about 2.4 km/h. The effect of speed of rolling on the results will be discussed later.

5.18 Taking the results given in Figures 5.1, 5.3 to 5.5 and 5.8 to 5.11, general conclusions can be reached with regard to the number of passes needed to achieve given levels of compaction in 150 mm thick layers over a range of moisture contents. The levels of compaction chosen were the same as those used in the previous Chapter for smooth-wheeled rollers, ie 95 per cent relative compaction and 10 per cent air voids, see Paragraph 4.16. Regression analyses have been carried out for each soil individually and for various combinations of soils, and the most significant relations were achieved by taking the heavy clay and sandy clay No 2 separately, as shown in Figures 5.12 and 5.13 respectively, but combining the well-graded sand and gravel-sand-clay (Figure 5.14). The Figures illustrate the results of regression analyses in the form of relations between the number of passes necessary to achieve the required state of compaction and the tyre contact pressure for various relative moisture contents (moisture content of soil minus the optimum moisture content obtained in the 2.5 kg rammer compaction test).

5.19 The regression equations on which the graphs in Figures 5.12 to 5.14 are based, and their correlation coefficients, are given in Table 5.3. Figures 5.12 to 5.14 present a general

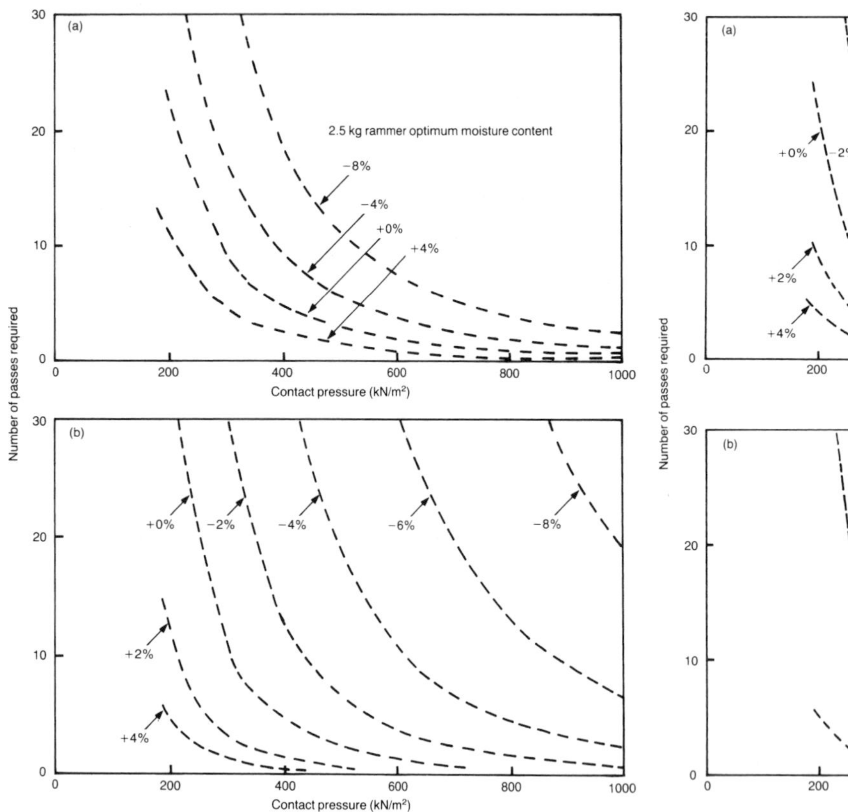

Figure 5.12 Pneumatic-tyred rollers on heavy clay soil. Relations between number of passes required and contact pressure to achieve (a) 95 per cent of the maximum dry density obtained in the 2.5 kg rammer compaction test and (b) 10 per cent air voids. These graphs were derived by regression analyses of the results given in Figures 5.1 and 5.8 and apply to 150 mm compacted layers

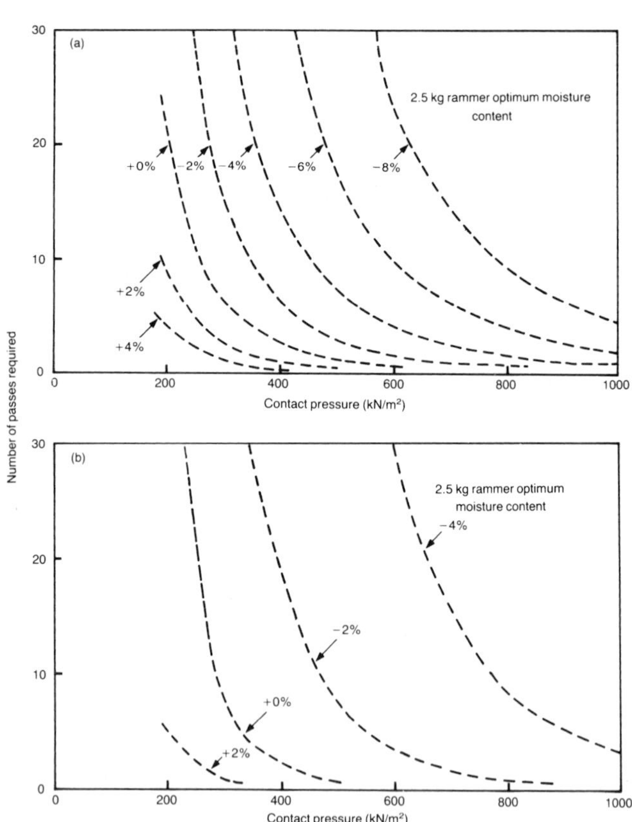

Figure 5.13 Pneumatic-tyred rollers on sandy clay soil. Relations between number of passes required and contact pressure to achieve (a) 95 per cent of the maximum dry density obtained in the 2.5 kg rammer compaction test and (b) 10 per cent air voids. These graphs were derived by regression analyses of the results given in Figures 5.3 and 5.9 and apply to 150 mm compacted layers

summary of the results of the investigations of pneumatic-tyred rollers on 150 mm thick compacted layers, and can be taken as a guide to the performance of this type of compactor over a fairly wide range of soil types.

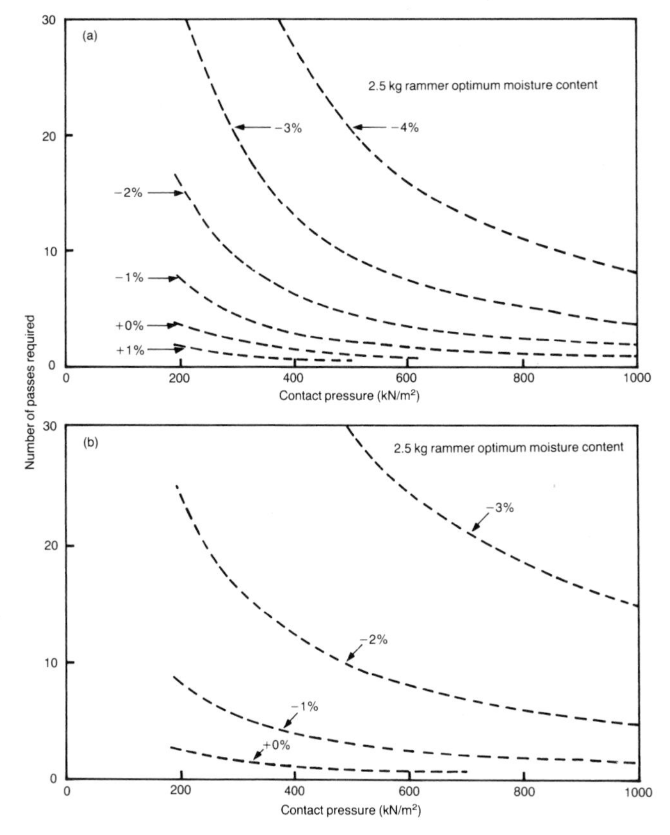

Figure 5.14 Pneumatic-tyred rollers on well-graded sand and gravel-sand-clay soils. Relations between number of passes required and contact pressure to achieve (a) 95 per cent of the maximum dry density obtained in the 2.5 kg rammer compaction test and (b) 10 per cent air voids. These graphs were derived by regression analyses of the results given in Figures 5.4, 5.5, 5.10 and 5.11 and apply to 150 mm compacted layers

Relations between state of compaction and depth within the compacted layer

5.20 Results were obtained by using 32 passes of the roller under test to compact fairly deep layers of soil, 300 to 400 mm deep in general, and determining the dry density of the soil at various depths in the layer. For each soil (Figures 5.15 to 5.18) five combinations of wheel load and tyre inflation pressure, involving the 20 t and 45 t towed rollers, were used. In addition, on sandy clay No 2 and well-graded sand (Figures 5.16(b) and 5.17(b)) tests were carried out with the 18.5 t self-propelled pneumatic-tyred roller.

Table 5.3 Equations and correlation coefficients for the relations given in Figures 5.12 to 5.14.

Soil	Level*	Regression equation†	No of data points	Correlation coefficient
Heavy clay	(a)	P=6.38–2.19 p–.072 w	10	0.86
	(b)	P=8.38–2.97 p–.227 w	10	0.84
Sandy clay No 2	(a)	P=8.46–3.09 p–.186 w	11	0.94
	(b)	P=11.34–4.22 p–.458 w	11	0.93
Well-graded sand & gravel-sand-clay	(a)	P=3.62–1.33 p–.319 w	19	0.94
	(b)	P=2.71–.993 p–.479 w	19	0.84

*Level (a)=95 per cent of the maximum dry density obtained in the 2.5 kg rammer compaction test
Level (b)=10 per cent air voids
†P=Log_{10} (Number of roller passes required)
p=Log_{10} (Tyre contact pressure, kN/m²)
w=Moisture content of soil, per cent, minus the optimum moisture content, per cent, obtained in the 2.5 kg rammer compaction test

5.21 The relations between relative compaction value and depth below the surface of the compacted layer for heavy clay (Figure 5.15), sandy clay No 2 (Figure 5.16) and gravel-sand-clay (Figure 5.18) all follow the expected trend of decreasing state of compaction with increasing

Figure 5.15 Relations between relative compaction and depth in the compacted layer of heavy clay obtained with 32 passes of pneumatic-tyred rollers. All values are related to the maximum dry density and optimum moisture content obtained in the 2.5 kg rammer compaction test (Figure 3.13)

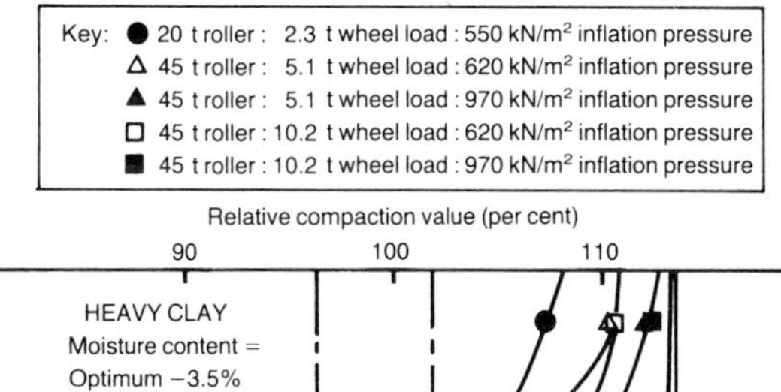

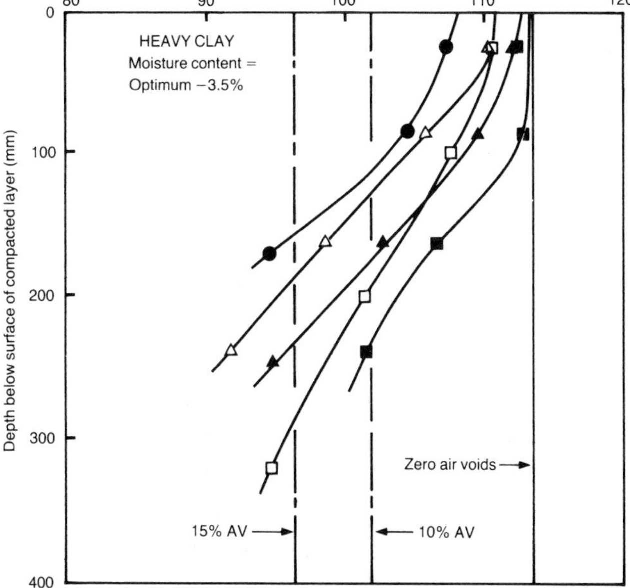

depth in the layer (see Paragraph 2.15 and Figure 2.8). With the well-graded sand (Figure 5.17), however, evidence is apparent of over-stressing of the near-surface soil, with increasing relative compaction values with increasing depth in the top 100 mm of the layer (see Paragraph 2.16 and Figure 2.9). The one result in Figure 5.17 which does not exhibit this over-stressing effect is that for the 18.5 t roller, which was tested about 5 years later than the other rollers. It is possible that small changes in the particle-size distribution of the well-graded sand in the period concerned, 1956 to 1961, as illustrated in Figure 3.10, may have reduced the susceptibility of this soil to over-stressing at the surface.

5.22 In summarising the results relating states of compaction and depth in the compacted layer, the same approach has been adopted as was used for smooth-wheeled rollers in Chapter 4, see Paragraph 4.21. Thus, the thickness of layer capable of being compacted by 32 passes such

Figure 5.16 Relations between relative compaction and depth in the compacted layer of sandy clay No 2 obtained with 32 passes of pneumatic-tyred rollers. All values are related to the maximum dry density and optimum moisture content obtained in the 2.5 kg rammer compaction test (Figure 3.13)

Figure 5.17 Relations between relative compaction and depth in the compacted layer of well-graded sand obtained with 32 passes of pneumatic-tyred rollers. All values are related to the maximum dry density and optimum moisture content obtained in the 2.5 kg rammer compaction test (Figure 3.13)

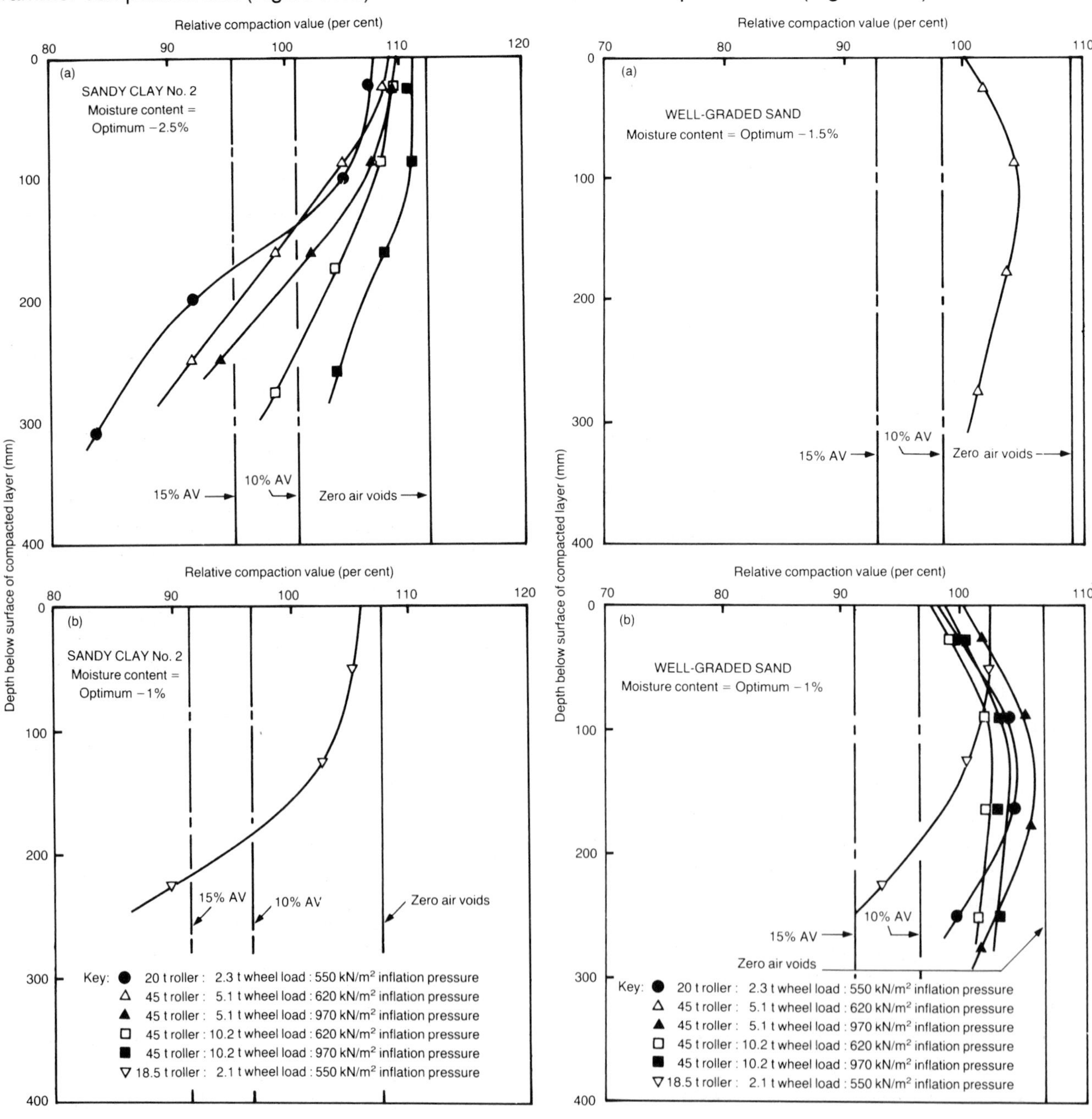

Figure 5.18 Relations between relative compaction and depth in the compacted layer of gravel-sand-clay obtained with 32 passes of pneumatic-tyred rollers. All values are related to the maximum dry density and optimum moisture content obtained in the 2.5 kg rammer compaction test (Figure 3.13)

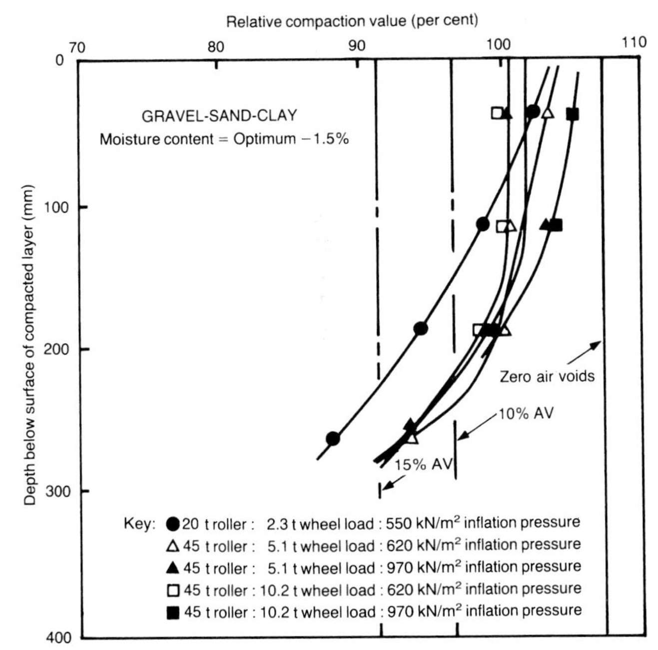

that the state of compaction at the bottom of the layer is no worse than 90 per cent relative compaction or 15 per cent air voids has been determined by regression analyses. Results for the well-graded sand were omitted from the analyses because of the limited depth of layer tested compared with the depth at which either 90 per cent relative compaction or 15 per cent air voids could be judged to occur. Because of the virtual limitation to one value of moisture content only with each of the remaining three soils, the effect of variations in moisture content on the depth capable of being compacted cannot be explored. Regression analyses have been carried out with wheel load and tyre contact pressure separately and in combination and the most influential factor was found to be the wheel load.

5.23 The relations between depth of layer capable of being compacted to the stated levels and the wheel load of the pneumatic-tyred roller are given in Figure 5.19 and their equations and correlation coefficients are given in Table 5.4. It must be remembered that these relations are for 32 passes of the rollers and apply only to the particular moisture contents at which the soils were tested. However, it can be seen that under these conditions layer thicknesses of 150 to 250 mm may be compacted with wheel loads of 2 t and thicknesses of 300 to 450 mm with wheel loads of 10 t. Given that the relation for gravel-sand-clay applies to a higher relative moisture content than those for the clay soils (−1.5 per cent compared with −2.5 to −3.5 per cent) it is clear that for the higher wheel loads pneumatic-tyred rollers are most effective on the cohesive soils, although this conclusion could well be modified if the results on the well-graded sand were capable of interpretation. In addition, the effect of variation in wheel load on the compacted depth is considerably smaller with the gravel-sand-clay.

Table 5.4 Equations and correlation coefficients for the relations given in Figure 5.19

Soil	Level*	Regression equation†	No of data points	Correlation coefficient
Heavy clay	(a)	$D=128\ L^{.486}$	5	0.98
	(b)	$D=107\ L^{.423}$	5	0.93
Sandy clay No 2	(a)	$D=121\ L^{.590}$	5	0.93
	(b)	$D=92.9\ L^{.597}$	5	0.92
Gravel-sand-clay	(a)	$D=206\ L^{.200}$	5	0.97
	(b)	$D=200\ L^{.184}$	5	0.98

*Level (a)=Achievement at the bottom of the layer of 90 per cent of the maximum dry density obtained in the 2.5 kg rammer test
Level (b)=Achievement at the bottom of the layer of 15 per cent air voids
†D=Maximum depth of layer to achieve stated level of compaction (mm)
L =Wheel load (t)

The above results apply only to the soils at the moisture contents given in Figure 5.19 and for 32 passes of the rollers

5.24 From the foregoing analyses it can be concluded that the state of compaction near the surface of the layer, say the upper 150 mm, is mainly influenced by the tyre contact pressure. Tyre contact pressure can be adjusted by varying either wheel load or tyre inflation pressure, as indicated by the data for the 45 t towed roller in Table 5.1. Compaction at the bottom of the compacted layer, say below the depth of 150 mm, is more likely to be affected by variations in wheel load alone.

Effect of rolling speed on the performance of pneumatic-tyred rollers

5.25 The effect of the speed of rolling was investigated with the 45 t towed roller, ballasted to a wheel load of 5.1 t and with a tyre inflation pressure of 620 kN/m² (Figures 5.8(b), 5.9(b), 5.10(a) and 5.11(a) and (b)). In all cases the state of compaction obtained after a given

Figure 5.19 Maximum depth of compacted layer, to achieve a dry density at the bottom of the layer equivalent to (a) 90 per cent of the maximum dry density obtained in the 2.5 kg rammer compaction test and (b) 15 per cent air voids, related to wheel load of pneumatic-tyred rollers. These graphs are derived from Figures 5.15, 5.16 and 5.18 and apply to 32 passes of the rollers

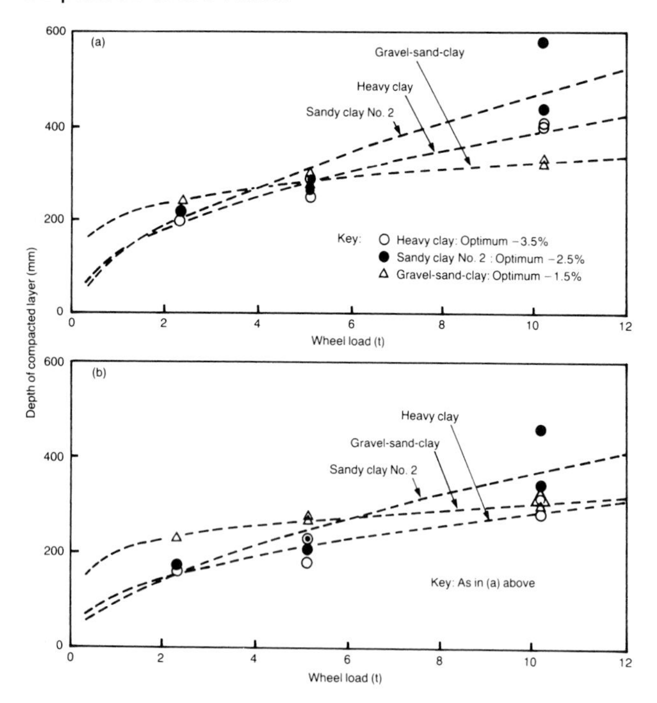

number of passes was lower at a speed of 5.5 km/h than at 2.4 km/h. The comparative numbers of passes to achieve 95 per cent relative compaction and 10 per cent air voids in the 150 mm compacted layers at both speeds are given in Table 5.5. The results can be expected to be affected to some extent by moisture content and the soil type involved, but for the results portrayed in Table 5.5 an average increase in output of compacted soil of about 100 per cent was achieved by an increase in rolling speed of about 130 per cent.

Table 5.5 Results of tests at two rolling speeds with the 45 t pneumatic-tyred roller with a wheel load of 5.1 t and a tyre inflation pressure of 620 kN/m^2

Soil	Moisture content	No of passes required to achieve:–			
		95% relative compaction		10% air voids	
		2.4 km/h	5.5 km/h	2.4 km/h	5.5 km/h
Heavy clay	Optimum – 2.5%	2	2	4	5
Sandy clay No 2	Optimum – 1.5%	3	3	4	5
Well-graded sand	Optimum – 0.5%	3	3	3	4
Gravel-sand-clay	Optimum – 2% to optimum – 1.5%	4	4	4	4

5.26 These results indicate that little adjustment in numbers of passes is required for variations in speed of travel of pneumatic-tyred rollers; the results plotted in Figures 5.8(b), 5.9(b), 5.10(a) and 5.11(a) and (b) indicate that the number of passes should be increased by 10 per cent for a doubling of the speed of travel. The number of passes determined by reference to Figures 5.12 to 5.14, which apply to a rolling speed of about 2.4 km/h, will be only slightly affected, therefore, by variations in rolling speed within the practical range available on construction sites.

Calculation of the states of compaction produced by pneumatic-tyred rollers

5.27 Analyses in earlier parts of this Chapter have provided empirical relations, based on the results obtained in investigations under controlled conditions, from which numbers of passes and thicknesses of layers may be determined for certain required states of compaction. Lewis (1959) approached this problem theoretically in formulating a method for determining the states of compaction produced when soil has been 'compacted to refusal' by pneumatic-tyred rollers.

Cohesive soils

5.28 With cohesive soils the basic assumption made was that on compaction to refusal the soil is compressed to a state in which its shear strength is equal to the shear stresses developed in the compacted layer by the roller. The limiting condition will be when the compacted soil approaches the saturated state. If, in this condition, the shear strength of the soil is insufficient to sustain the shear stresses developed by the roller, serious plastic deformations will occur. Using this approach, if the relations between shear strength, dry density and moisture content are known for the soil and the shear stresses developed by the roller can be calculated, the likely maximum state of compaction that can be produced at any moisture content can also be estimated.

5.29 To determine the dry density produced by a given roller, the maximum shear stresses likely to be developed in the layer were first calculated from a knowledge of the contact areas and pressures of the tyres of the roller (Table 5.1). The calculated results for the rollers investigated are given in Figure 5.20. Although the contact area of a tyre in general approaches an ellipse in shape, an equivalent circular area was employed in order to make use of existing tables giving the shear stress distribution in a homogeneous elastic medium beneath a circular loaded area (Jurgenson, 1934). To determine the average dry density over a given depth of soil, the average shear stress over a similar depth was used. The average shear stresses in the top 150 mm of soil for the rollers used are given in Table 5.6.

Table 5.6 Average shear stress in top 150 mm of soil for various combinations of wheel load and tyre inflation pressure

Roller	Wheel load (t)	Tyre inflation pressure (kN/m²)	Average shear stress in top 150 mm of soil (kN/m²)
12 t towed	1.4	250	74
20 t towed	2.3	550	152
45 t towed	5.1	620	138
	5.1	970	161
	10.2	620	188
	10.2	970	232

5.30 Relations between shear strength and moisture content for the heavy clay and sandy clay No 2 at various relative compaction values are given in Figures 5.21 and 5.22. These relations were based on unconfined compression tests on undisturbed samples taken at the same time as the measurements of dry density and moisture content during the investigations. The values of relative compaction used are direct conversions of the dry density values for which

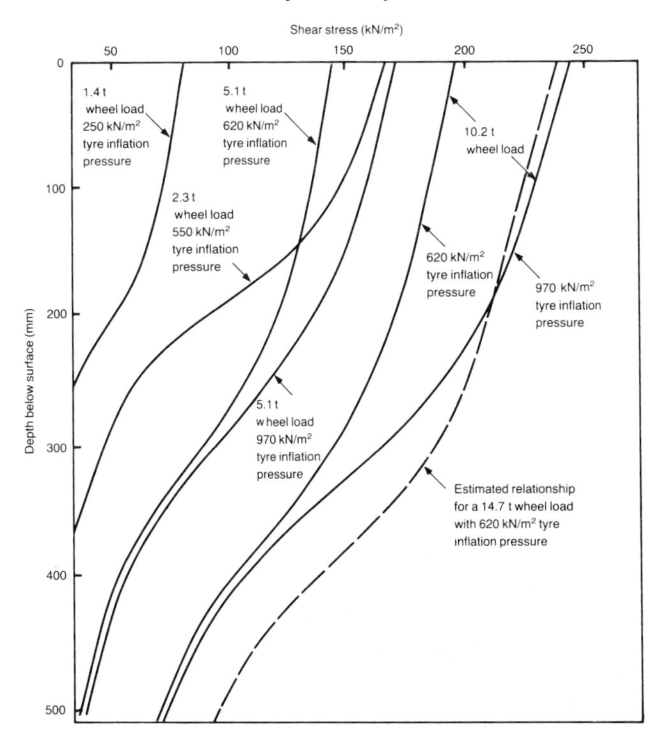

Figure 5.20 Relations between maximum shear stress and depth for various roller wheel loads and tyre inflation pressures

Figures 5.21 and 5.22 Relations between shear strength of the two clay soils and moisture content obtained for various values of relative compaction

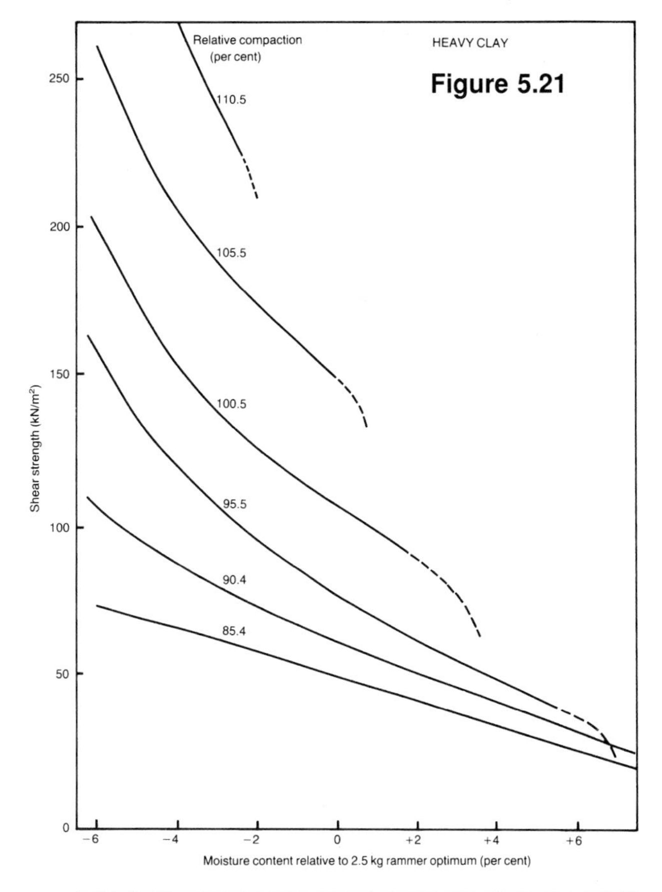

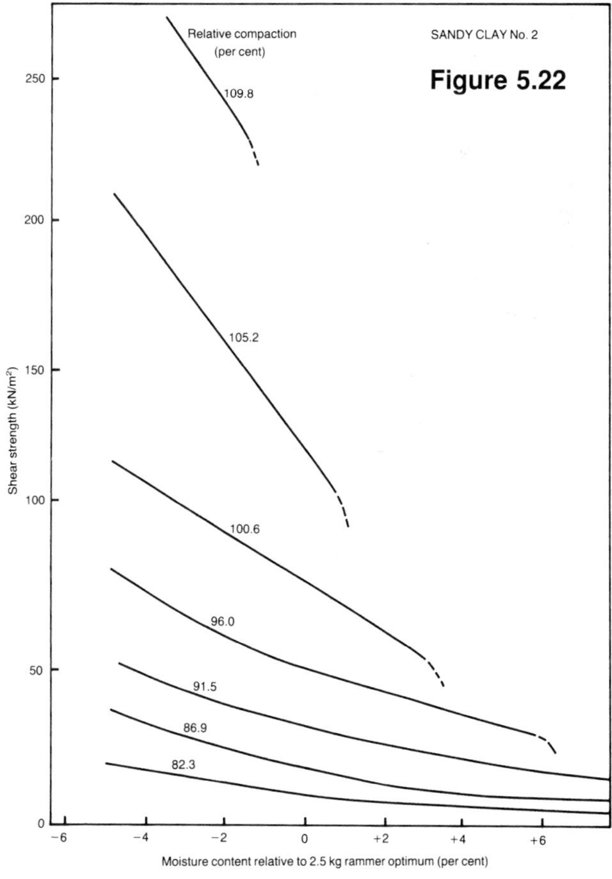

the curves were constructed by Lewis (1959). Using these curves the relative compaction value that gives a shear strength equal to the shear stress (or average shear stress) can be read off for any given moisture content value. For the calculation of the relation between relative compaction and moisture content, as the soil approaches the saturated state a very small amount of air will remain entrapped. After this condition has been attained, the assumption was made that the relative compaction-moisture content relation follows a constant air void line. This assumption was based on the results of the full-scale investigations with the cohesive soils which indicated that, on average, at moisture contents above the optimum for the machines, about 2 per cent of air remained in the soil. The peak portions of the curves were also rounded off to a slight extent to conform to the usual shape obtained in the compaction tests.

5.31 Relations between relative compaction value and moisture content, for the heavy clay

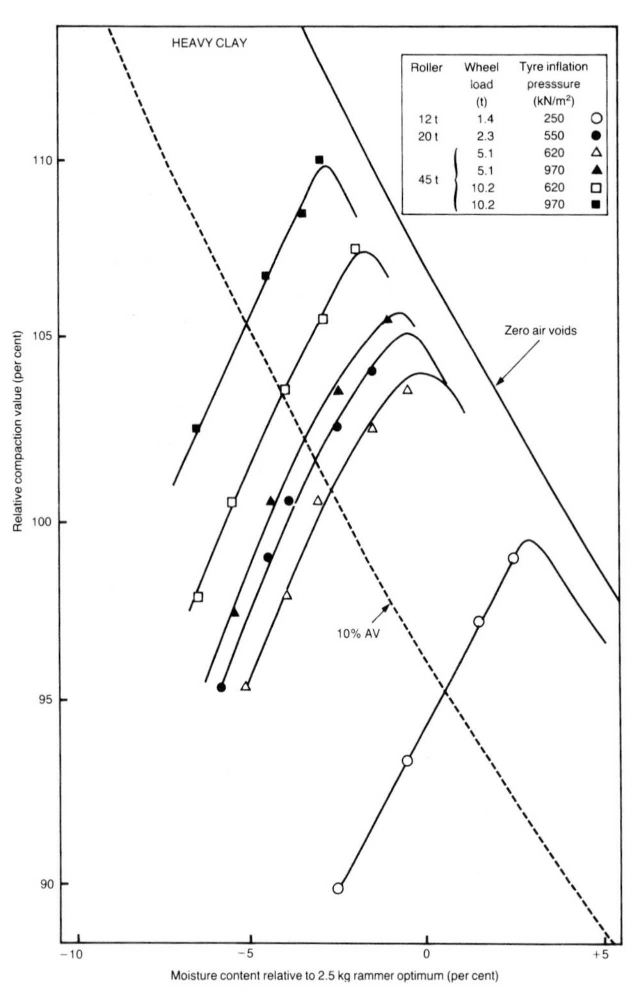

Figure 5.23 Calculated relations between relative compaction and moisture content for the top 150 mm of heavy clay for the various wheel loads and tyre inflation pressures of the pneumatic-tyred rollers

and sandy clay No 2, calculated by the procedure described above using the average shear stress for the top 150 mm of soil (Table 5.6), are given in Figures 5.23 and 5.24. On comparison of these relations with the experimental results obtained in Figures 5.1 and 5.3, fairly good agreement can be seen to have been achieved, despite the simplifying assumptions made in the calculations. Possible sources of error in the calculations include (i) the different rate of strain in the laboratory unconfined compression test compared with the rate under the wheels of the rollers, (ii) the assumption that the layer of compacted soil is a homogeneous, isotropic, semi-infinite elastic medium, and (iii) the use of an equivalent circular contact area to calculate the shear stress in the soil. Relations between relative compaction value and depth in the compacted layer for the same soils, calculated by the use of the shear stress distributions given in Figure 5.20, are given in Figures 5.25 and 5.26. Comparisons of these relations with the equivalent experimental results in Figures 5.15 and 5.16(a) show that broad agreement was obtained, although the calculated curves show somewhat smaller changes in relative compaction with depth than were actually measured; it was considered that the cause of this was the effect of density gradients on the stress distribution through the layer.

Granular soils

5.32 A somewhat different approach to that outlined for cohesive soils had to be employed with granular soils. With these soils two parameters, the cohesion and the angle of shearing resistance, are needed to define the strength properties, and the resistance to shear of the soils is a function of the stresses on the shear planes. Lewis (1959) decided that it was not possible, therefore, to employ the method suggested for cohesive soils and a simplified approach was developed in which the ultimate bearing capacity of the compacted soil was equated to the contact pressure applied by the rubber tyre. Thus it was assumed that the soil continues to be compacted until it develops a bearing capacity equal to the applied pressure.

5.33 The ultimate bearing capacity for a circular loaded area on the surface for local shear failure of soil (Terzaghi and Peck, 1948) can be expressed as:–

$$q = 0.87\, c\, N'_c + 0.6\, \gamma\, r\, N'_\gamma$$

Figure 5.24 Calculated relations between relative compaction and moisture content for the top 150 mm of sandy clay No 2 for the various wheel loads and tyre inflation pressures of the pneumatic-tyred rollers

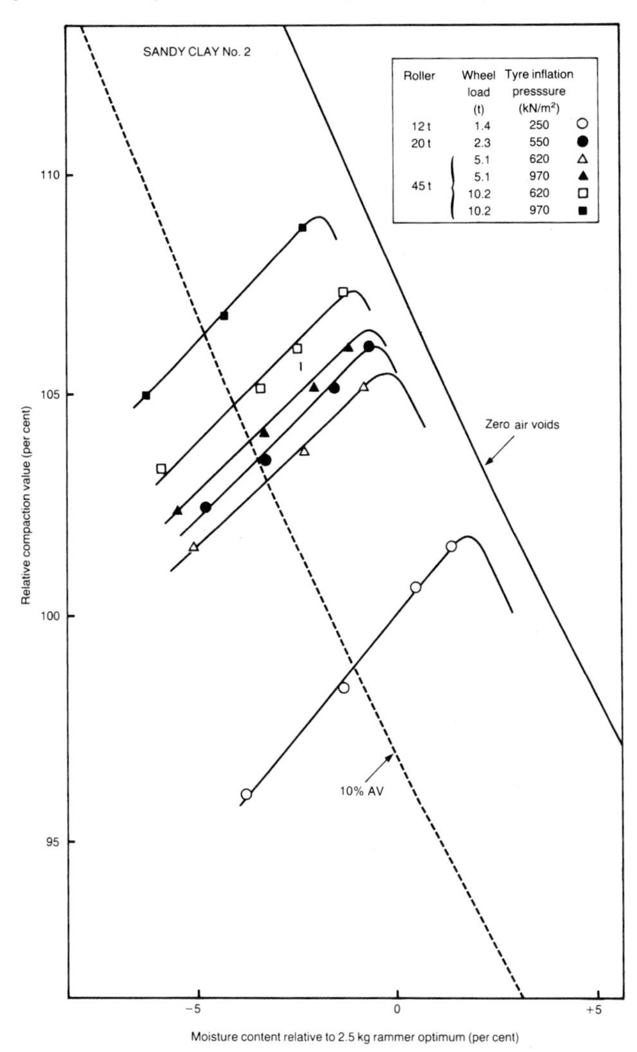

where q=ultimate bearing capacity (kN/m²)
c=cohesion of soil (kN/m²)
r=radius of loaded area (m)
N'_c, N'_γ= bearing capacity factors for local shear, which vary with the angle of shearing resistance
γ= bulk density of soil (kN/m³).

5.34 The radius of the equivalent circular area of contact of a loaded tyre is small and the second term of the expression can be neglected in comparison with the magnitude of the first term if the soil has a cohesion of more than about 15 kN/m². In these circumstances the bearing capacity can be expressed as:–

$$q = 0.87\ c\ N'_c$$

5.35 Tests were carried out on specially compacted specimens of the well-graded sand

Figure 5.25 Calculated relations between relative compaction and depth below the surface of the compacted layer of heavy clay for the various wheel loads and tyre inflation pressures of the pneumatic-tyred rollers

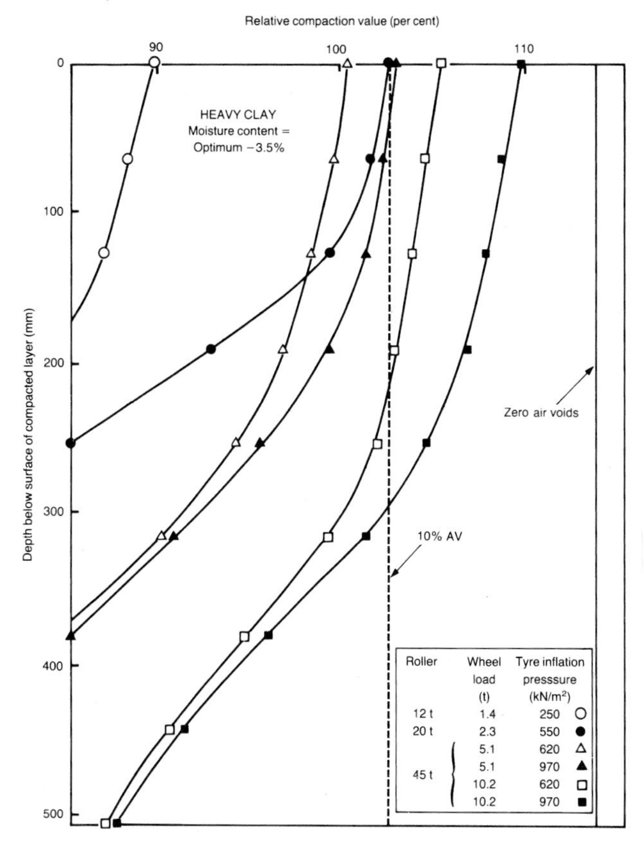

Figure 5.26 Calculated relations between relative compaction and depth below the surface of the compacted layer of sandy clay No 2 for the various wheel loads and tyre inflation pressures of the pneumatic-tyred rollers

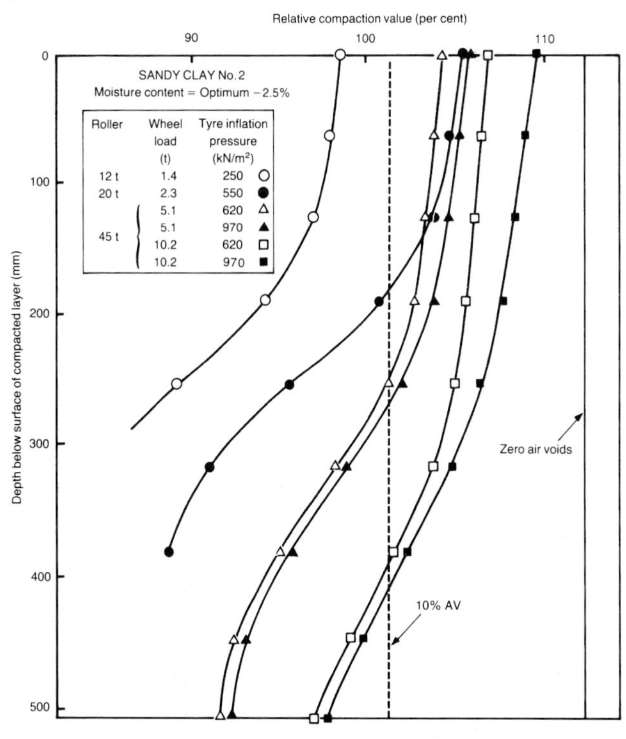

Figure 5.27 Relations between cohesion and angle of shearing resistance of well-graded sand and moisture content obtained for various values of relative compaction

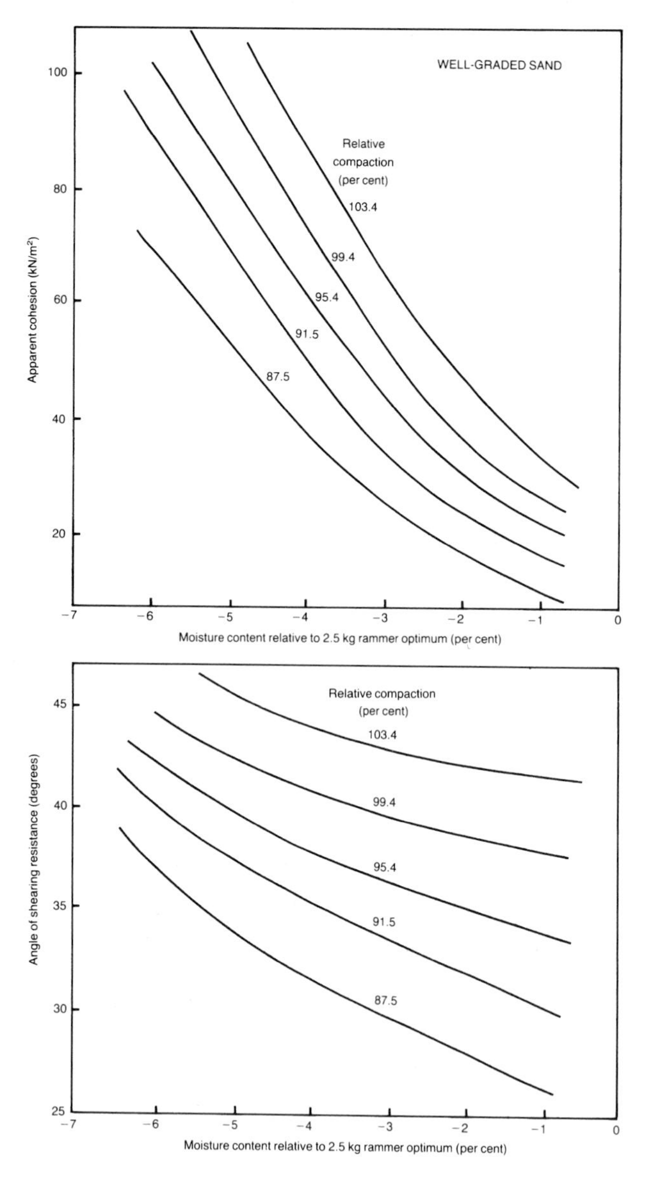

and gravel-sand-clay to determine relations between cohesion and moisture content and between angle of shearing resistance and moisture content for various relative compaction values. The results are given in Figures 5.27 and 5.28. The values of relative compaction used are direct conversions of the dry density values for which the curves were constructed by Lewis (1959). From values obtained from these Figures, relations between bearing capacity, moisture content and relative compaction were constructed and are given in Figures 5.29 and 5.30. The bearing capacity was then equated to the mean contact pressure of the tyre (Table 5.1) and the relative compaction value likely to be produced at any given moisture content was read directly from Figures 5.29 and 5.30. To complete the relations between relative compaction and moisture content for moisture contents above the optimum for the rollers, the assumption was made, as for cohesive soils, that the curve will follow a constant air void line. With the two granular soils the full-scale tests showed that about 4 per cent of air remained in the soil and the relations were completed on this basis. The calculated relations between relative compaction value and moisture content determined by this method for the two granular soils are given in Figures 5.31 and 5.32, and these may be compared with the equivalent experimental results in Figures 5.4 and 5.5. In general, the calculated values of relative

Figure 5.28 Relations between cohesion and angle of shearing resistance of gravel-sand-clay and moisture content obtained for various values of relative compaction

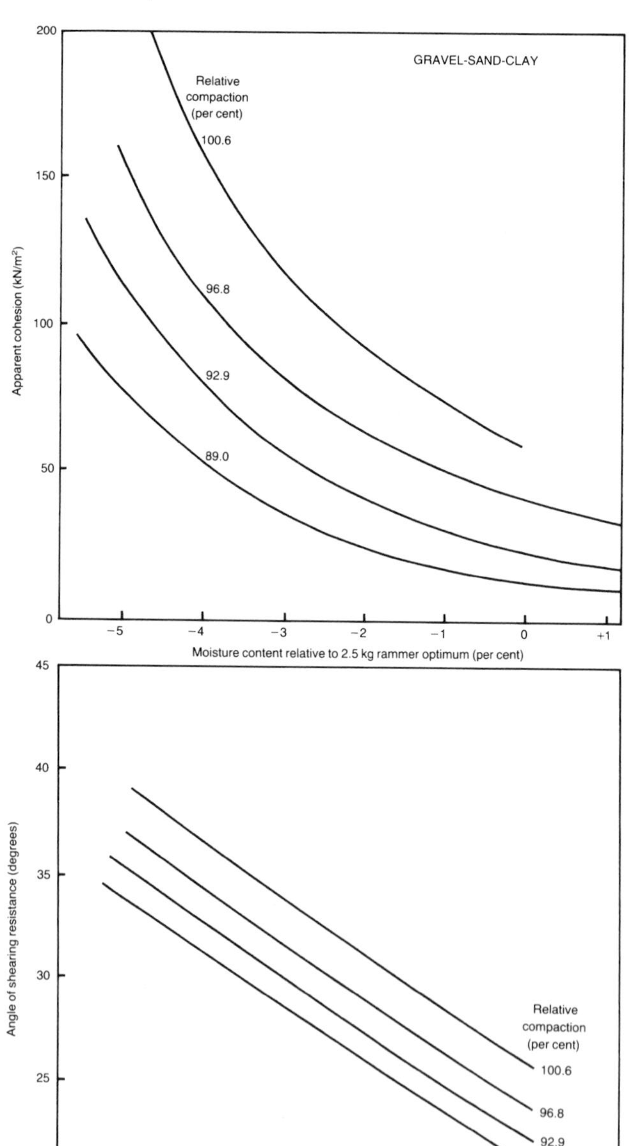

compaction are lower than the results obtained experimentally and it was concluded by Lewis (1959) that the calculated results obtained using the bearing capacity method should ideally relate to the average relative compaction over the depth likely to be stressed by the roller; this depth would be well in excess of the 150 mm for which the experimental results were obtained.

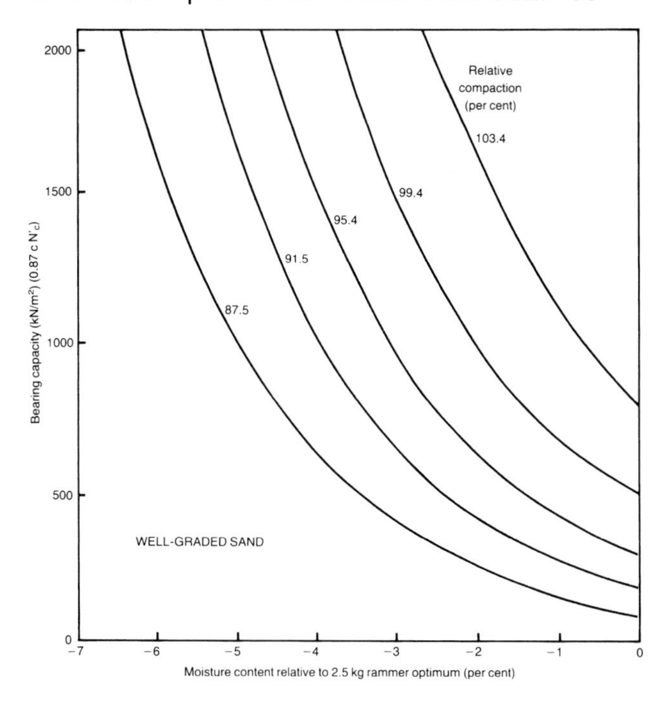

Figure 5.29 Relations between bearing capacity of the well-graded sand and moisture content for various values of relative compaction

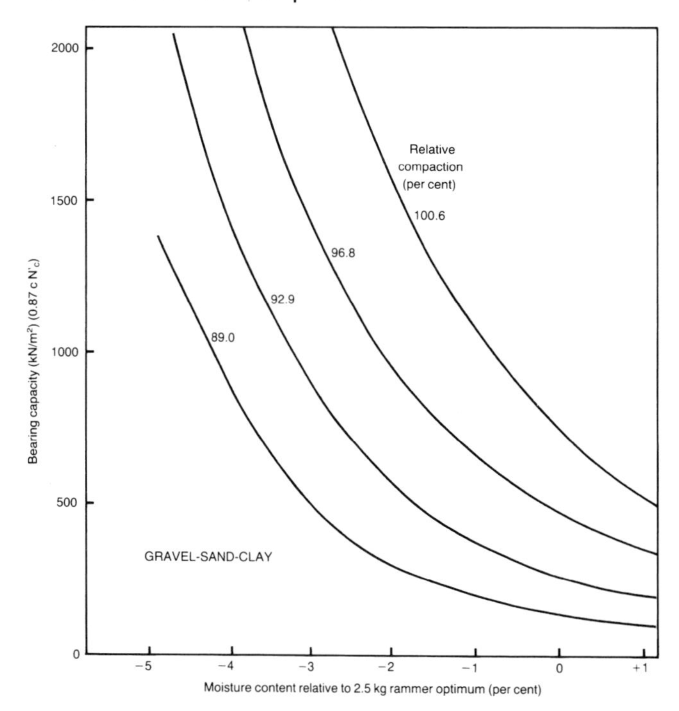

Figure 5.30 Relations between bearing capacity of the gravel-sand-clay and moisture content for various values of relative compaction

Figure 5.31 Calculated relations between relative compaction and moisture content for well-graded sand for the various wheel loads and tyre inflation pressures of the pneumatic-tyred rollers

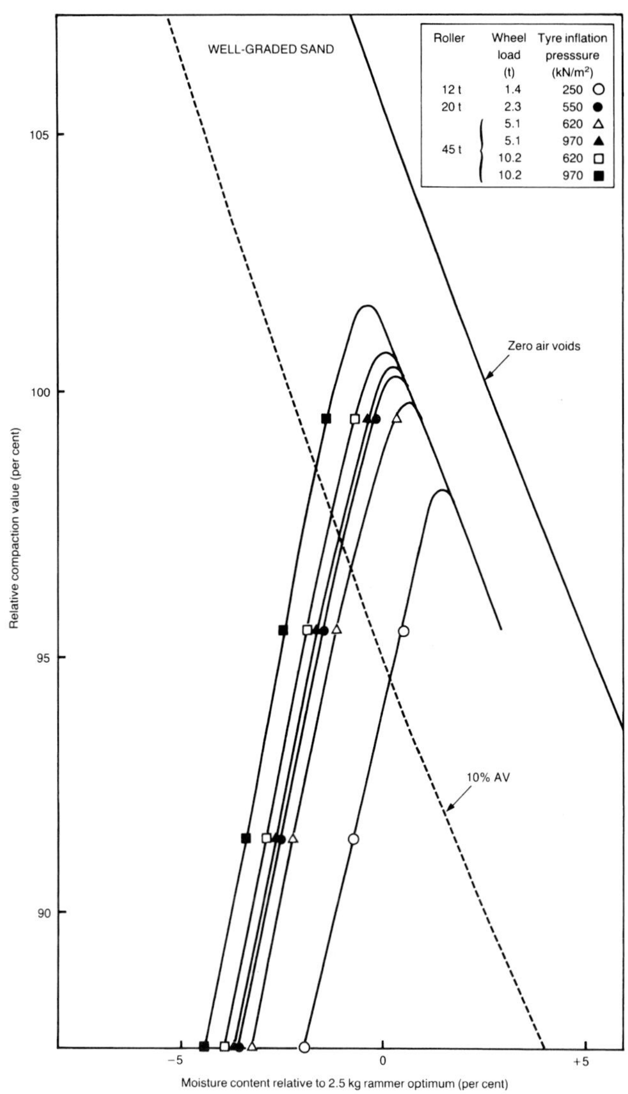

Prediction of performance

5.36 The theoretical approach made by Lewis (1959) provides a useful insight to the mechanism of the compaction process, but the limitations of the method lie in the simplifying assumptions and resulting sources of error (see Paragraph 5.31) and the application only to the 'compaction to refusal' condition. The prediction of number of required passes or of potential output to achieve a specified level of compaction under given moisture content conditions cannot be made by the methods used by Lewis and resort must be made to empirical relations, such as those established earlier in this Chapter, derived from regression analyses of experimental results.

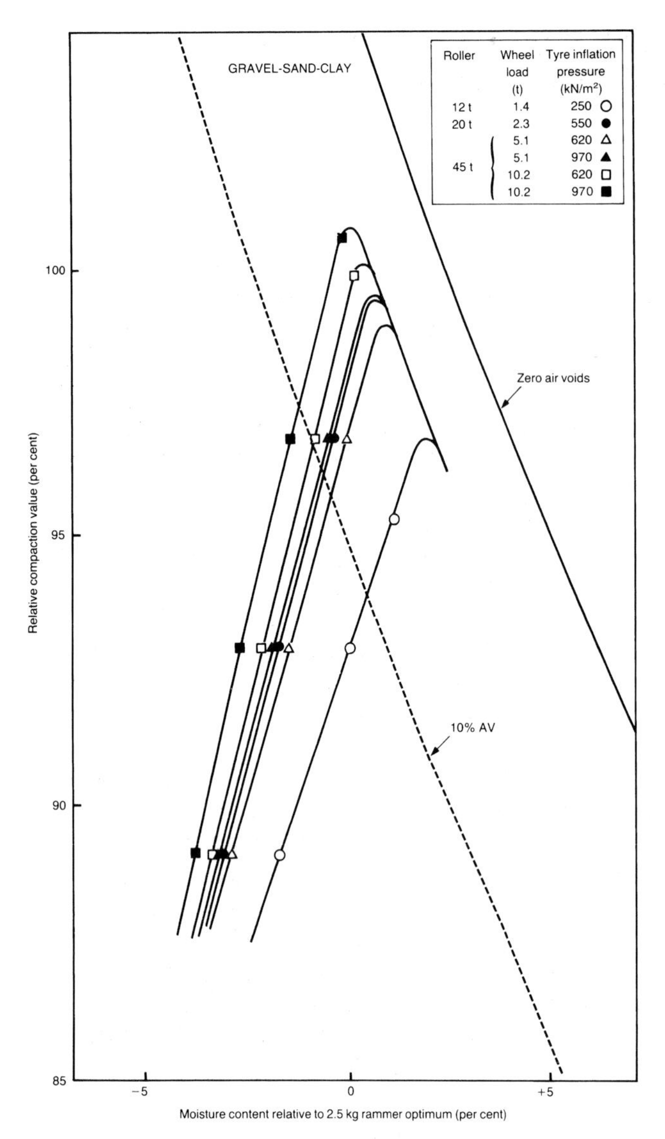

Figure 5.32 Calculated relations between relative compaction and moisture content for gravel-sand-clay for the various wheel loads and tyre inflation pressures of the pneumatic-tyred rollers

References

British Standards Institution (1990). British Standard methods of test for soils for civil engineering purposes: BS 1377: Part 4 Compaction related tests. BSI, London.

Jurgenson, L (1934). The application of theories of elasticity and plasticity to foundation problems. J. Boston Soc. Civ. Engrs, 21 (3), 206–41.

Lewis, W A (1954). Further studies in the compaction of soil and the performance of compaction plant. *Road Research Technical Paper* No 33. HM Stationery Office, London.

Lewis, W A (1959). Investigation of the performance of pneumatic-tyred rollers in the compaction of soil. *Road Research Technical Paper* No 45. HM Stationery Office, London.

Terzaghi, K and Peck, R B (1948). Soil mechanics in engineering practice. John Wiley and Sons Inc, New York.

Williams, F H P and Maclean, D J (1950). The compaction of soil: a study of the performance of plant. *Road Research Technical Paper* No 17. HM Stationery Office, London.

CHAPTER 6 COMPACTION OF SOIL BY DEAD-WEIGHT SHEEPSFOOT, TAMPING AND GRID ROLLERS

Description of machines

6.1 This Chapter describes the results of investigations into the performance of non-vibrating compactors described as sheepsfoot, tamping and grid rollers.

6.2 Sheepsfoot rollers can be towed or self-propelled, and the rolls consist of cylindrical shells with protruding 'feet' which provide areas of high contact pressure under the machine. The feet can have numerous shapes and terms such as 'taper foot' and 'club foot' are used. Because of the small contact area of the sheepsfoot roller it requires a significant number of passes to provide even one complete coverage of an area of soil. The stress distribution with depth under the small area of the foot is poor but is compensated by the high penetration of the feet during the initial passes, the feet finally 'walking out' at completion of compaction if the soil is sufficiently dry and strong.

6.3 Tamping rollers are modifications of the sheepsfoot roller principle, with feet protruding from a roll shell, the feet having a much larger area of contact than is the case with the sheepsfoot roller. One definition of a tamping roller is that the projected end area of each foot should exceed 0.01 m^2 and the sum of the areas of the feet should exceed 15 per cent of the area of the cylinder swept by the ends of the feet (Department of Transport, 1986). The contact pressure under a tamping roller is reduced in comparison with the sheepsfoot roller and as a result the tamping roller can be expected to be better suited to wetter soil conditions.

6.4 Grid rollers are usually designed to be towed by a tractor and they have a roll or rolls comprising heavy steel mesh of square pattern. They are particularly appropriate for the compaction of shallow layers of soft rock due to their ability to crush particles with the high peak stresses under the points of contact of the steel mesh (Hyster Company, 1970). A grid roller is capable of being ballasted to provide a considerable range of mass, eg 6 to 14 tonnes.

6.5 During the course of investigations of the performance of compaction plant at the Laboratory a range of machines in the above categories was tested, including two vibrating sheepsfoot rollers which were briefly tested with the vibrating mechanism inoperative, thus providing data on the performance of dead-weight machines. Details of the various machines are given in Table 6.1 and they are illustrated in Plates 6.1 to 6.8.

6.6 The 5 t club-foot and 4.5 t taper-foot sheepsfoot rollers (Plates 6.1 and 6.2) were towed machines with twin rolls mounted side-by-side in a frame with a tow bar for attachment to an appropriate tractor. In the investigations in 1946–47 the rollers were towed by the electrically powered lorry on the circular test

Plate 6.1 5 t club-foot sheepsfoot roller investigated in 1946–47

Plate 6.2 4.5 t taper-foot sheepsfoot roller investigated in 1946–47

Plate 6.3 4.3 t towed vibrating sheepsfoot roller which, when used without vibration, contributed data on the performance of sheepsfoot rollers; investigated in 1961

Plate 6.4 5.1 t towed vibrating sheepsfoot roller which, when used without vibration, contributed data on the performance of sheepsfoot rollers; investigated in 1961–62

track (see Chapter 3). The mass of the machines was raised to the level used in the investigations by filling the roll shells with water. The difference in shape of the feet of the two rollers is readily apparent in the photographs.

6.7 The 4.3 t and 5.1 t vibrating sheepsfoot rollers (Plates 6.3 and 6.4) were both towed machines, each with a power unit mounted on the chassis to drive the vibrating mechanism contained within the single roll. With the 4.3 t roller (Plate 6.3) each protruding foot was in the form of a truncated cone with a cylindrical extension at the end; the 5.1 t roller (Plate 6.4) had protruding feet of an asymmetrical design although, again, the contact area was circular. Both machines were equipped with scrapers, attached to the chassis and protruding between the roll feet, to remove any soil remaining on the roll.

6.8 The 11 t tamping roller (Plate 6.5) consisted of a box-girder framework within which three rolls were mounted in line. The centre roll was mounted in a separate sub-frame attached to the main frame at the front only and pivoted so that the roll could move up and down in relation to the two outer rolls. Each protruding foot was rectangular in plan view but asymmetrical in side elevation. The rolls were hollow and capable of being filled with sand ballast. Cleaning bars, attached to the rear of the frame, projected between the tamping feet. The frame was attached through a 'gooseneck' to a two-wheeled tractor unit, the steering of the machine being implemented by hydraulic jacks mounted on either side of the gooseneck pivot.

Table 6.1 Details of the various machines contributing data on the performance of non-vibrating sheepsfoot, tamping and grid rollers

Compactor	Overall roll width (m)	Diameter of roll shell (m)	Diameter to ends of feet (m)	Contact area per foot (m^2)	Coverage*	Mass on roll (kg)	Mass per unit overall width (kg/m)	Mass per unit effective width† (kg/m)
5 t club-foot sheepsfoot roller	2.44	1.06	1.42	7.74×10^{-3}	0.091	4994	2047	22470
4.5 t taper-foot sheepsfoot roller	2.44	1.22	1.61	3.12×10^{-3}	0.047	4572	1874	40210
4.3 t towed vibrating sheepsfoot roller	1.45	1.19	1.60	4.95×10^{-3}	0.067	4344	2996	45050
5.1 t towed vibrating sheepsfoot roller	1.83	1.07	1.47	8.11×10^{-3}	0.147	5080	2776	18910
11 t tamping roller	2.36	1.32	1.68	13.5×10^{-3}	0.388	11010	4666	12020
(Centre roll only)	0.79					3836	4874	12550
17 t tamping roller								
(Front rolls)	1.93	1.03	1.41	10.9×10^{-3}	0.153	7340	3800	24840
(Rear rolls)	1.93	1.03	1.41	10.9×10^{-3}	0.153	9940	5150	33660
Grid roller – 6.8 t	1.59	1.69	–	–	0.561#	6642	4177	7443
13.6 t	1.59	1.69	–	–	0.561#	13390	8419	15000

*Coverage is the sum of the end areas of the feet expressed as a proportion of the area of the cylinder swept by the ends of the feet
†Effective width = Overall width×Coverage
#Proportion of total area of roll cylinder occupied by grid mesh

Plate 6.5 11 t tamping roller investigated in 1965

Plate 6.6 17 t tamping roller investigated in 1972

Plate 6.7 Grid roller ballasted to 6.8 t; investigated in 1959

Plate 6.8 Grid roller ballasted to 13.6 t; investigated in 1959

6.9 The 17 t tamping roller (Plate 6.6) was a self-propelled machine based on a conventional four-wheeled bulldozer design, but having tamping rolls instead of pneumatic-tyred wheels. The main frame was articulated between the front and rear rolls to provide steering which was implemented by two double-acting hydraulic cylinders. Each roll carried a total of 60 tamping feet arranged in five rows of 12, the rows of feet on the front and rear rolls being exactly in line. The projected end area of each foot was rectangular, and in side profile the end of each foot was triangular. Each roll was equipped with a set of cleaner bars to remove material adhering between the rows of feet.

6.10 The grid roller investigated (Plates 6.7 and 6.8) consisted of a box-girder framework within which two grid rolls were mounted. The core of each roll was a conical container with its minimum diameter to the outside of the roller. The cores could be used to contain sand or water ballast and they were so shaped that soil which fell through the grid was ejected outside the path of the roller. This cleaning action was assisted by internal scrapers fitted to the frame and mounted close to the cores. External scrapers were also fixed beneath the frame at the front and rear of the rolls. Specially designed concrete ballast weights could be added to the chassis of the grid roller to add about 6.5 t to the total mass of the machine (compare Plates 6.7 and 6.8). The machine was normally towed in the tests by a heavy crawler tractor, but for some limited tests at high speeds of rolling in the Laboratory grounds (ie outside the pilot-scale facility) a large wheeled tractor was used as the prime mover (Plate 6.9).

6.11 Details of the machines described above are given in Table 6.1, and include values for the contact area of each foot (for sheepsfoot and tamping rollers) and the coverage. These values

Plate 6.9 Grid roller, mass 13.6 t, with large wheeled tractor for high speed tests over a prepared strip of soil in the Laboratory grounds

illustrate the wide range of characteristics of these machines, with contact areas varying by a factor of four and coverage, excluding the grid roller, by a factor of eight. Coverage of the grid roller has been estimated from the dimensions of the mesh of the grid and exceeds the coverage of the other machines by a considerable margin. Another characteristic given in Table 6.1 is the mass per unit effective width, defined in the Table; this provides a measure of the potential contact pressure under the feet. Values range from 7000 to 45000 kg/m. The definition of a tamping roller given in Paragraph 6.3 above is also seen to be complied with in Table 6.1. The 17 t tamping roller has a contact area per foot just exceeding 0.01 m^2 and a coverage just exceeding the 15 per cent required to comply with the definition of a tamping roller. By contrast, the 5.1 t vibrating sheepsfoot roller has a contact area per foot just below 0.01 m^2 and a coverage just below 15 per cent.

6.12 The results presented in this Chapter have been taken from Williams and Maclean (1950), Lewis (1954), Parsons (1962), Cross (1962), Parsons and Toombs (1965), Toombs (1973) and Parsons (1959).

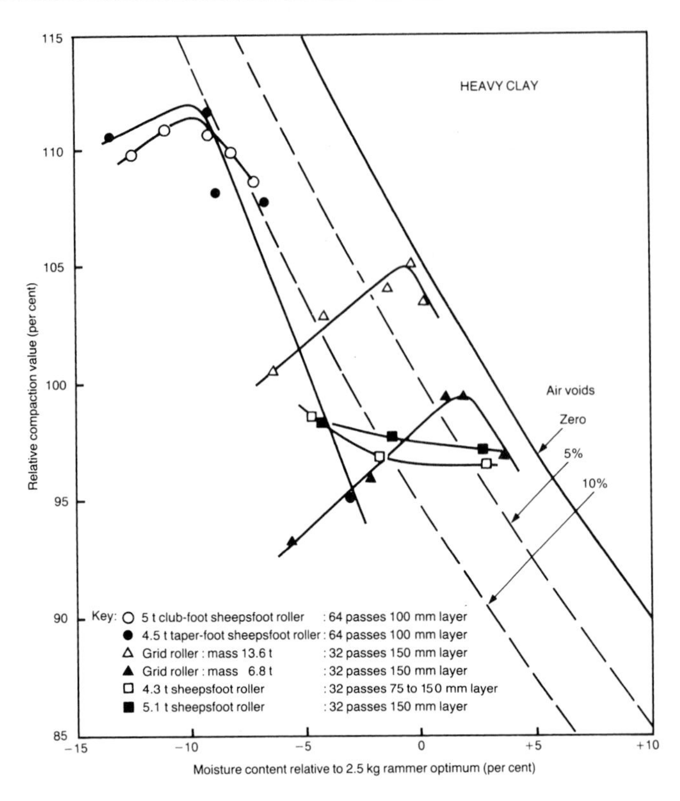

Figure 6.1 Relations between dry density and moisture content for sheepsfoot and grid rollers on heavy clay, expressed in terms of relative compaction with the 2.5 kg rammer test

Relations between state of compaction and moisture content

6.13 Relations between the states of compaction produced by sheepsfoot, tamping and grid rollers and the moisture content of the soil are given in Figures 6.1 to 6.7. As in earlier chapters the dry density has been expressed as a relative compaction value and the moisture content as a difference from optimum moisture content, both in relation to the results of the 2.5 kg rammer compaction test at the relevant period. Air void lines are based on average values for the results in each figure. It should be noted that the majority of the results from the early investigations, for the 5 t club-foot and the 4.5 t taper-foot sheepsfoot rollers, were obtained on 100 mm thick layers, rather than the 150 mm layers used in most of the remainder of the tests. The main reason for the use of the 100 mm layers for these particular machines was the large penetration of the feet of the rollers and the resulting depth of loosened soil (loose mulch) which had to be removed from the surface of the layer prior to the determination of the dry density. With the same two machines the relations were determined for 64 passes, in contrast to the 32 passes that were used in all other cases.

6.14 The results for heavy clay soil, given in Figure 6.1, were obtained with the four different sheepsfoot rollers and with the grid roller ballasted to two different test weights. The results for the grid roller follow the conventional pattern (see Figure 2.2), with clearly defined maximum relative compaction values and optimum moisture contents at low air void contents. The club-foot and taper-foot sheepsfoot rollers also produced well defined curves of conventional shape, but the air content at the optimum for the machines was in excess of 10 per cent. The two vibrating sheepsfoot rollers, used in this case without vibration, produced unconventional relations and taking into account the mass per unit effective width given in Table 6.1 (the 4.3 t roller is shown to have had the highest value of all the machines, and the 5.1 t roller had a value of about 85 per cent of that of the club-foot roller), it can be considered that the range of moisture contents tested with the 4.3 t and 5.1 t machines did not extend to sufficiently dry conditions to encompass the maximum dry densities.

6.15 Figures 6.2 and 6.3, for silty clay and sandy clay No 1, contain results for the club-foot and taper-foot sheepsfoot rollers only. Unlike the results for heavy clay, air contents as low as about 5 per cent were achieved near the optimum moisture contents for the machines.

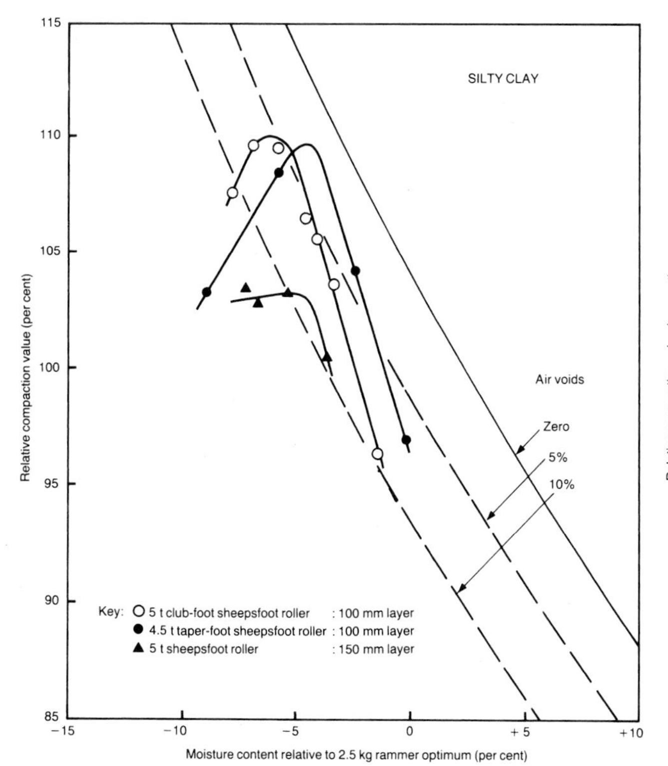

Figure 6.2 Relations between dry density and moisture content for sheepsfoot rollers on silty clay, expressed in terms of relative compaction with the 2.5 kg rammer test. Number of passes = 64

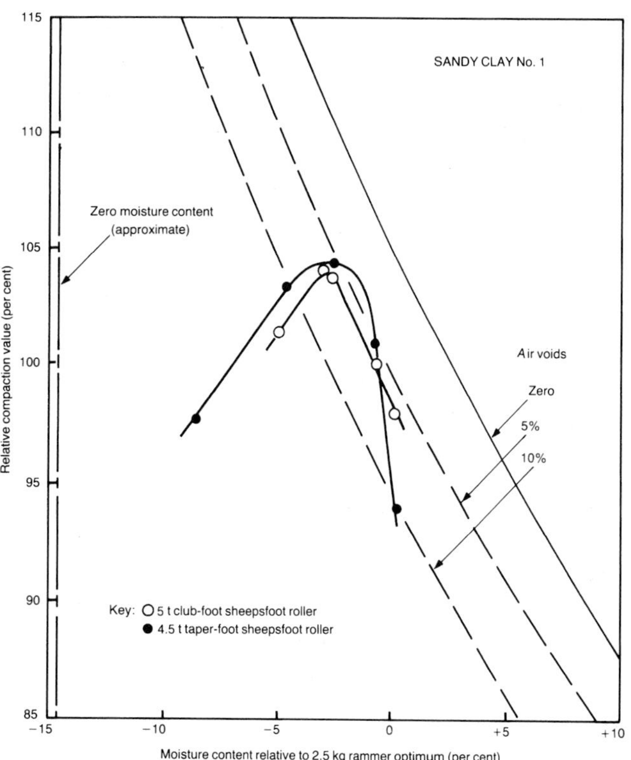

Figure 6.3 Relations between dry density and moisture content for sheepsfoot rollers on sandy clay No 1, expressed in terms of relative compaction with the 2.5 kg rammer test. Layer thickness = 100 mm, number of passes = 64

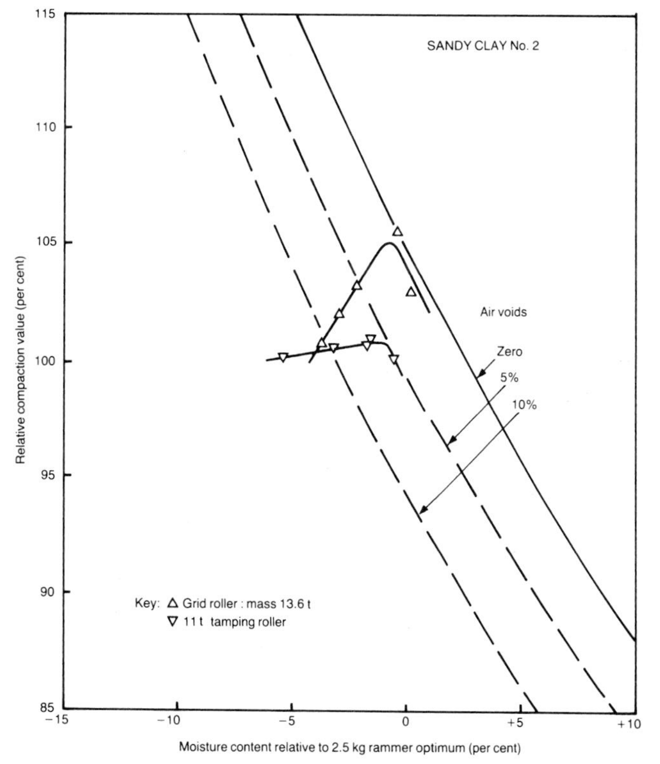

Figure 6.4 Relations between dry density and moisture content for the grid roller and a tamping roller on sandy clay No 2, expressed in terms of relative compaction with the 2.5 kg rammer test. Layer thickness = 150 mm, number of passes = 32

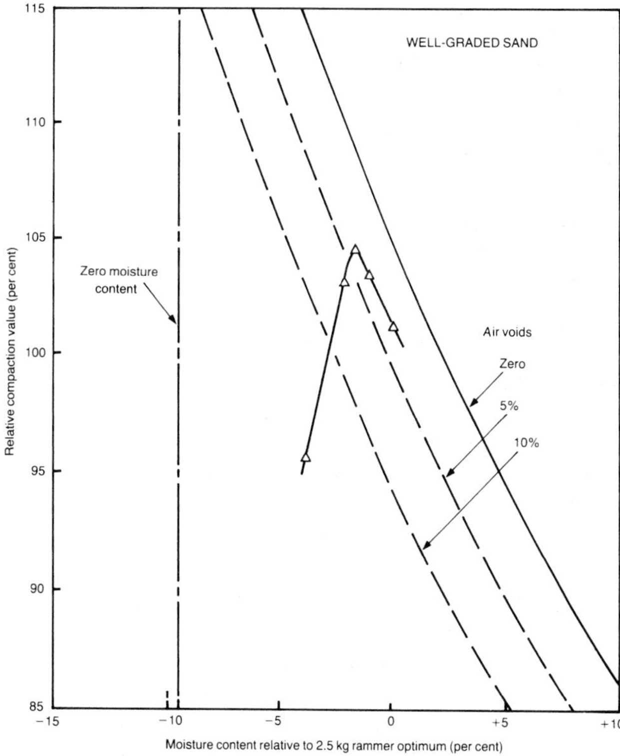

Figure 6.5 Relation between dry density and moisture content for the grid roller (mass = 13.6 t) on well-graded sand, expressed in terms of relative compaction with the 2.5 kg rammer test. Layer thickness = 150 mm, number of passes = 32

Figure 6.6 Relations between dry density and moisture content for sheepsfoot and grid rollers on gravel-sand-clay, expressed in terms of relative compaction with the 2.5 kg rammer test

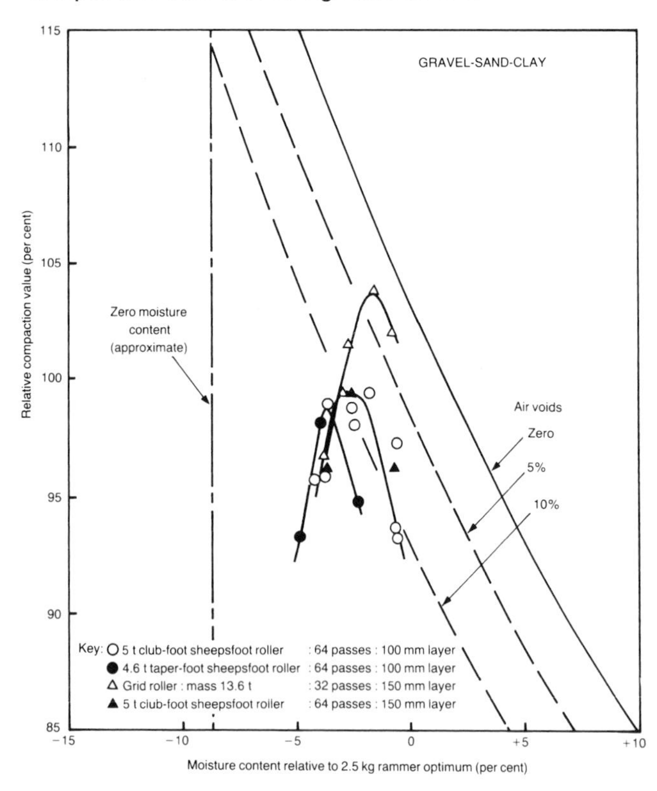

Figure 6.7 Relation between dry density and moisture content for the grid roller (mass = 6.8 t) on uniformly graded fine sand, expressed in terms of relative compaction with the 2.5 kg rammer test. Layer thickness = 150 mm, number of passes = 32

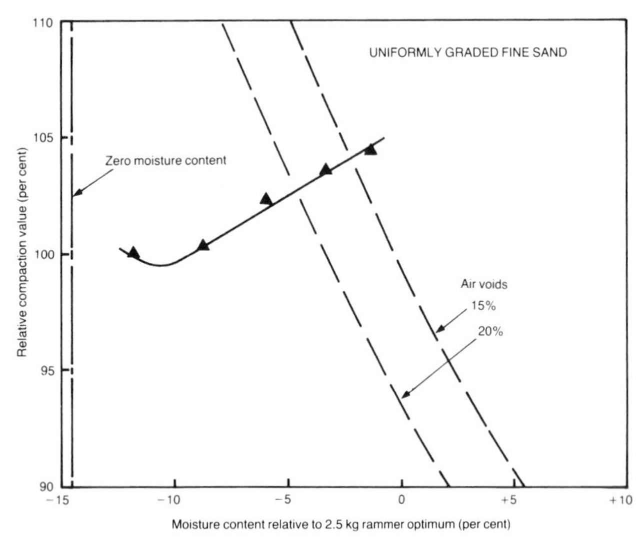

With the silty clay, results were also obtained with 150 mm thick compacted layers using the club-foot roller, and these reveal a considerable reduction in maximum relative compaction value (110 per cent to about 103 per cent) and an increase in air content at optimum. Thus the marked contrast between the results for the club-foot and taper-foot sheepsfoot rollers and the grid roller in Figure 6.1 might be considered to be exaggerated by the differences in thickness of layer used.

6.16 Results for the grid roller ballasted to 13.6 t and the 11 t tamping roller may be compared in Figure 6.4 for the sandy clay No 2. The results for the tamping roller apply to soil compacted by the centre roll only (see Table 6.1 for details applying to the centre roll) thus avoiding the effects of the pneumatic-tyred wheels of the prime mover (Plate 6.5).

6.17 Comparisons of the results obtained with the 13.6 t grid roller on heavy clay (Figure 6.1), sandy clay No 2 (Figure 6.4), well-graded sand (Figure 6.5) and gravel-sand-clay (Figure 6.6) show that very similar maximum relative compaction values (104 to 105 per cent) were achieved for the same compactive effort. Thus the relative efficiencies of the 2.5 kg rammer compaction method in the laboratory and the grid roller ballasted to 13.6 t in full-scale work remain constant over a wide spectrum of soil types.

6.18 Results for the club-foot and taper-foot sheepsfoot rollers on gravel-sand-clay are also included in Figure 6.6. In marked contrast to the results obtained on heavy clay (Figure 6.1) the sheepsfoot rollers produced significantly lower maximum relative compaction values than the 13.6 t grid roller. In addition, the results for 150 mm layers with the club-foot roller on gravel-sand-clay were similar to those on 100 mm layers. It is clear that with the more granular soil the material was being overstressed and, therefore, penetrated by the feet of the sheepsfoot roller, preventing the achievement of higher relative compaction values. It is noteworthy that at moisture contents below about optimum – 3 per cent, the grid and sheepsfoot rollers produced very similar states of compaction.

6.19 Results are given in Figure 6.7 for the compaction of uniformly graded fine sand by the grid roller. With this soil severe overstressing occurred when the roller was used at 13.6 t, but extremely satisfactory results were achieved with the roller at 6.8 t. As is usual with this type of soil, no clearly defined optimum moisture content was determined for the machine and the air content remaining in the soil was very high.

6.20 In all the results discussed above at least 32 passes of the rollers were used. The question of adequate coverage of the soil surface (see

Paragraph 6.2) would not arise with such large numbers of passes and the performance of the machines can be expected to relate solely to the mass per unit effective width, values of which are given in Table 6.1. Figure 6.8 gives relations between maximum relative compaction value and mass per unit effective width and between optimum moisture content (relative to the 2.5 kg rammer optimum) and the same parameter. No allowance has been made for the use of 100 mm compacted layers with the club-foot and taper-foot sheepsfoot rollers (mass per unit effective width 22500 and 40200 kg/m respectively). As a result of regression analyses, the soils were divided into three groups and relations determined. The equations and correlation coefficients are given in Table 6.2. Although trends in the results are fairly clear, some of the correlations are poor and the different thicknesses of layer involved in the tests are clearly a contributory cause of this. With the well-graded sand and gravel-sand-clay the reduction in maximum relative compaction value with increase in mass per unit effective width is a marked feature, but it should be remembered that a very limited amount of data was available. With such small quantities of data the trends should be taken as an indication only of the relative performances of the various machines.

Relations between state of compaction produced and number of passes

6.21 Results of investigations to determine the effect on compaction of variations in the number of passes of sheepsfoot, tamping and grid rollers are given in Figures 6.9 to 6.15. As in earlier chapters, the results for each soil are illustrated under a separate figure number, whilst different values of moisture content are in separate graphs within each Figure.

6.22 The taper-foot sheepsfoot roller was tested at a number of different moisture contents on silty clay (Figure 6.10 (a), (b), (c) and (d)) and sandy clay No 1 (Figure 6.11 (a), (c) and (d)) and at two moisture contents on heavy clay (Figure 6.9 (a) and (b)). The taper-foot sheepsfoot roller is noteworthy for its very small coverage and very high mass per unit effective width (Table 6.1). The pattern of imprints left by the feet after various numbers of passes are shown in Plate 6.10, and the number of passes necessary to ensure uniform compaction of a complete area of soil would clearly be fairly high. Another result of these characteristics is that the soil was severely overstressed at the higher

Table 6.2 Equations and correlation coefficients for the relations given in Figure 6.8

Soil	Regression equation*	Number of points	Correlation coefficient
Heavy clay & silty clay	$c = 57.4\ m^{.0632}$ $w = 580\ m^{-.370} - 20$	6 6	0.90 0.72
Sandy clay Nos 1 and 2	$c = 86.3\ m^{.0184}$ $w = 51.8\ m^{-.187} - 10$	4 4	0.54 0.75
Well-graded sand & gravel-sand-clay	$c = 174\ m^{-.0536}$ $w = 250\ m^{-.351} - 10$	4 4	0.89 1

*c = maximum dry density for the roller as percentage of the maximum dry density obtained in the 2.5 kg rammer compaction test

w = difference between optimum moisture content for the roller and optimum moisture content obtained in the 2.5 kg rammer compaction test (per cent)

m = mass per unit effective width as defined in Table 6.1

The above results apply in general to 32 to 64 passes of the rollers on 100 to 150 mm compacted layers

Figure 6.8 Maximum dry densities and optimum moisture contents related to the mass per unit effective width of sheepsfoot, tamping and grid rollers. Results from Figures 6.1 to 6.6

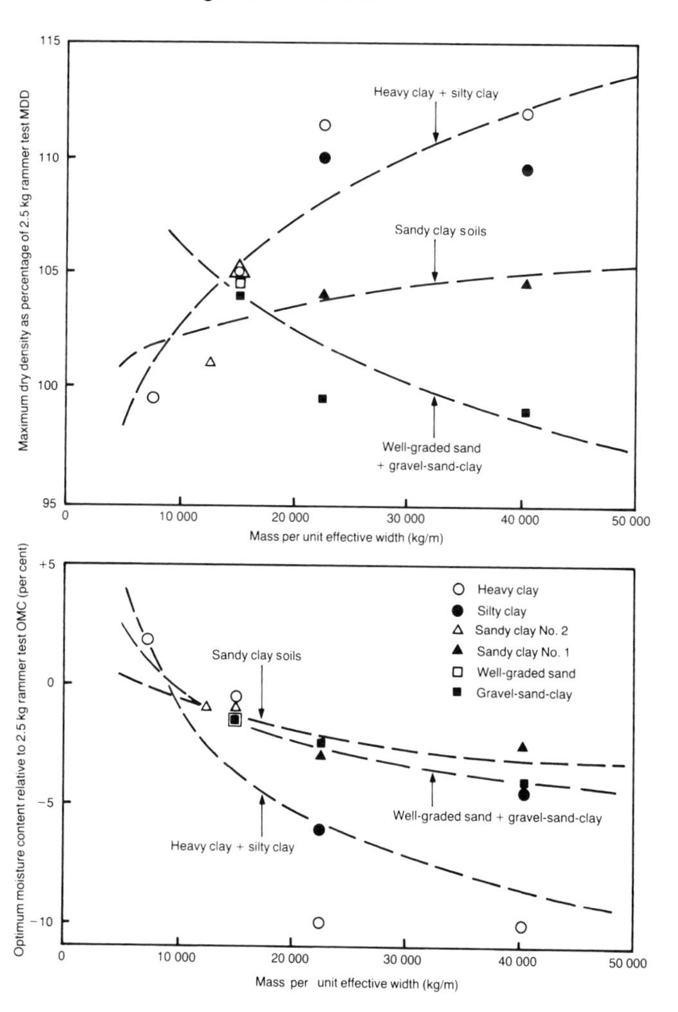

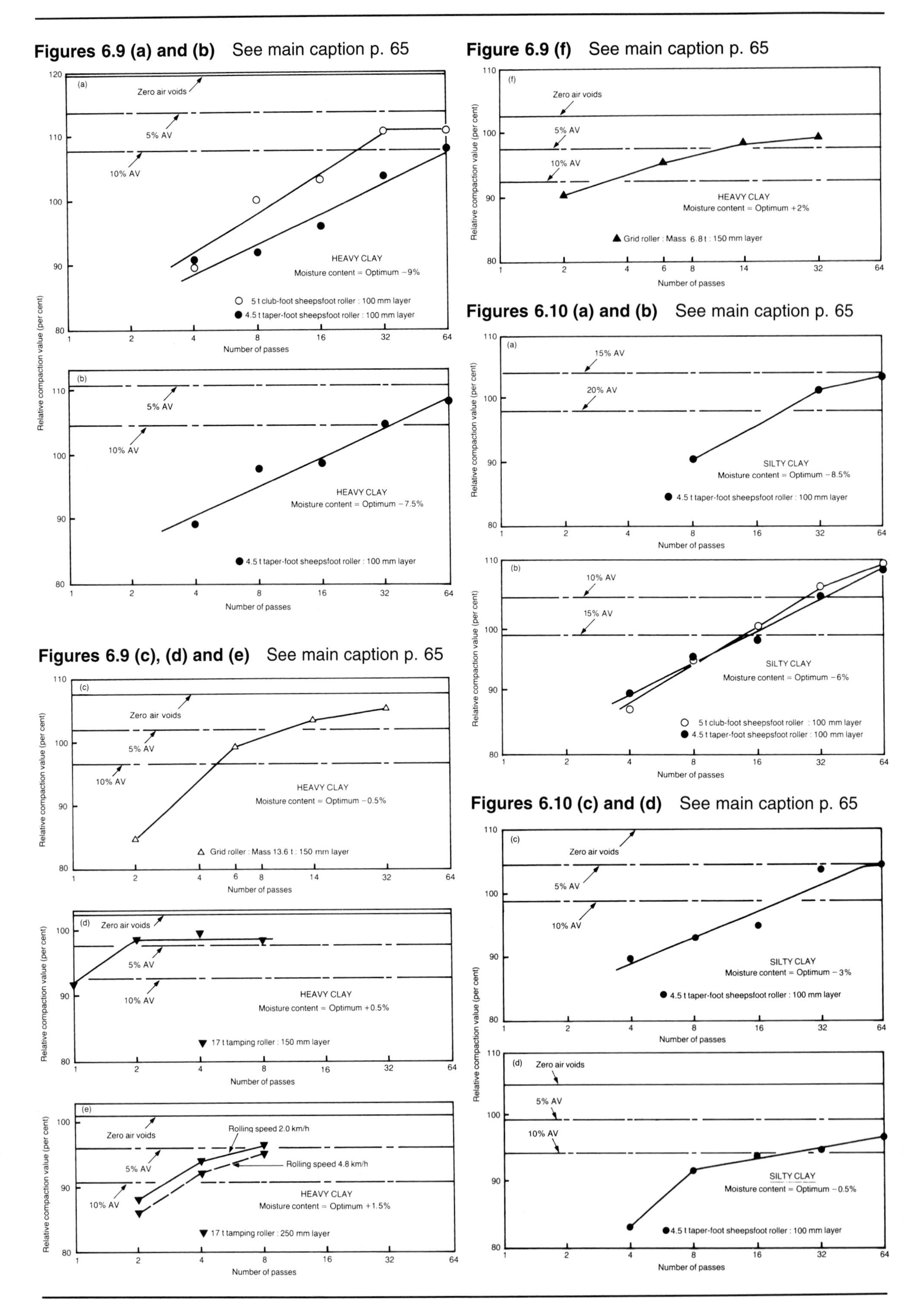

Figures 6.9 (a) and (b) See main caption p. 65

Figures 6.9 (c), (d) and (e) See main caption p. 65

Figure 6.9 (f) See main caption p. 65

Figures 6.10 (a) and (b) See main caption p. 65

Figures 6.10 (c) and (d) See main caption p. 65

Figures 6.11 (a) and (b) See main caption below

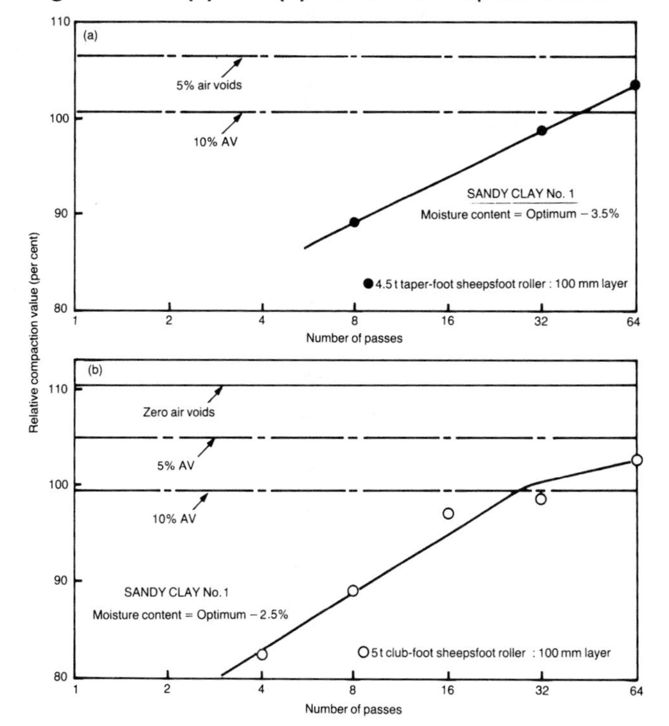

Figures 6.11 (c) and (d) See main caption below

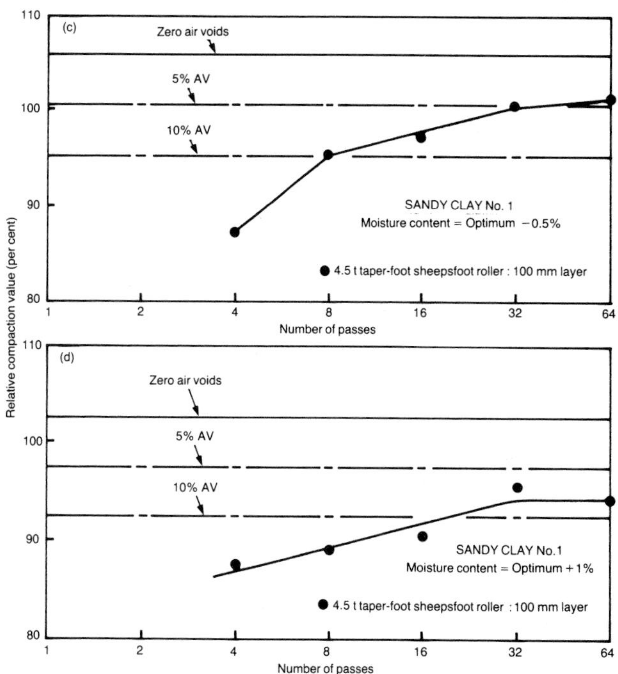

Figure 6.9 Relations between relative compaction and number of passes of sheepsfoot, tamping and grid rollers on heavy clay at six different values of moisture content. All values are related to the maximum dry density and optimum moisture content obtained in the 2.5 kg rammer compaction test (Figure 3.13)

Figure 6.10 Relations between relative compaction and number of passes of sheepsfoot rollers on silty clay at four different values of moisture content. All values are related to the maximum dry density and optimum moisture content obtained in the 2.5 kg rammer compaction test (Figure 3.13)

Figure 6.11 Relations between relative compaction and number of passes of sheepsfoot rollers on sandy clay No 1 at four different values of moisture content. All values are related to the maximum dry density and optimum moisture content obtained in the 2.5 kg rammer compaction test (Figure 3.13)

moisture contents, the feet penetrating the soil and producing a loosening effect. The evidence of this has already been discussed where high air contents remained in the soil at the optimum moisture content (see Figure 6.1). The number of passes of the taper-foot sheepsfoot roller necessary to achieve a relative compaction value of 95 per cent in the clay soils is shown plotted against relative moisture content in Figure 6.16. There is clearly a minimum in the

Figure 6.12 Relations between relative compaction and number of passes of tamping and grid rollers on sandy clay No 2 at three different values of moisture content. All values are related to the maximum dry density and optimum moisture content obtained in the 2.5 kg rammer compaction test (Figure 3.13)

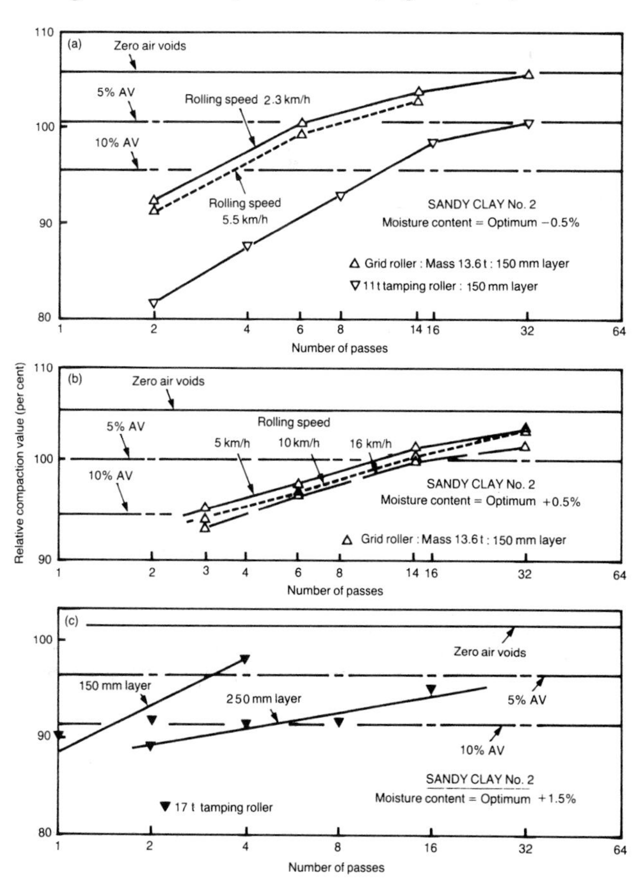

Figure 6.13 Relations between relative compaction and number of passes of tamping and grid rollers on well-graded sand. All values are related to the maximum dry density and optimum moisture content obtained in the 2.5 kg rammer compaction test (Figure 3.13)

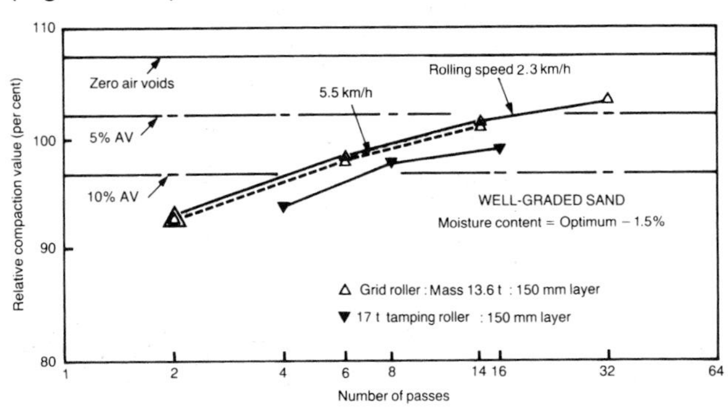

Figure 6.14 Relations between relative compaction and number of passes of sheepsfoot, tamping and grid rollers on gravel-sand-clay at two different values of moisture content. All values are related to the maximum dry density and optimum moisture content obtained in the 2.5 kg rammer compaction test (Figure 3.13)

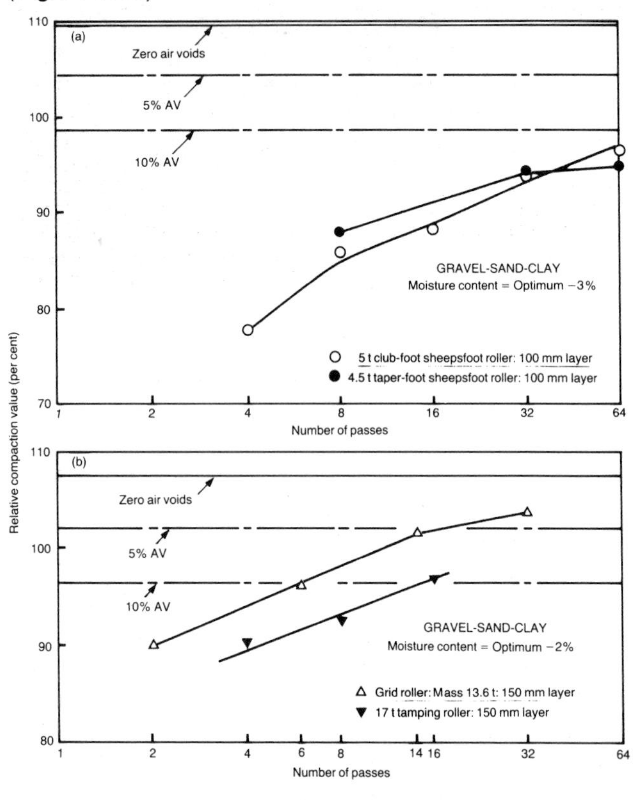

required number of passes on silty clay at a moisture content of about optimum − 5 per cent. At moisture contents higher than this the number of passes increased, contrary to normal expectations, and the efficiency of the roller was reduced. The results for heavy clay and sandy clay No 1 are inconclusive but it is possible that with the sandy clay the machine was operating effectively to higher moisture contents in relation to the optimum of the 2.5 kg rammer compaction test than with the silty clay soil.

Figure 6.15 Relations between relative compaction and number of passes of tamping and grid rollers on uniformly graded fine sand at two different values of moisture content. All values are related to the maximum dry density and optimum moisture content obtained in the 2.5 kg rammer compaction test (Figure 3.13)

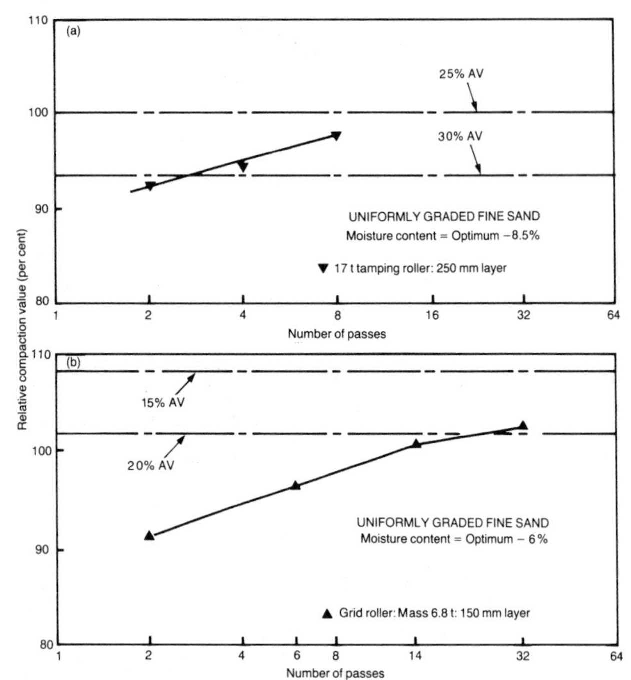

Figure 6.16 Relations between number of passes and moisture content to achieve a relative compaction value of 95 per cent with the 4.5 t taper-foot sheepsfoot roller. Thickness of compacted layer = 100 mm

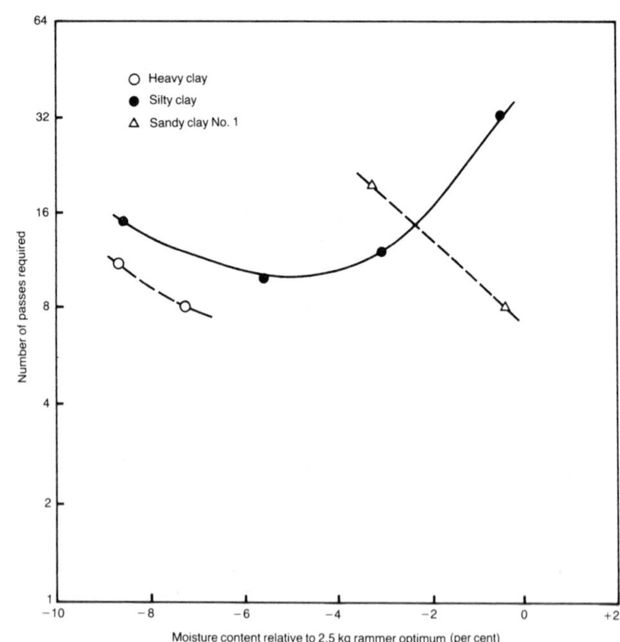

Plate 6.10 Coverage obtained with the taper-foot sheepsfoot roller after 4, 8, 16, 32 and 64 passes (25 mm grid)

Plate 6.11 Coverage obtained with the club-foot sheepsfoot roller after 4, 8, 16, 32 and 64 passes (25 mm grid)

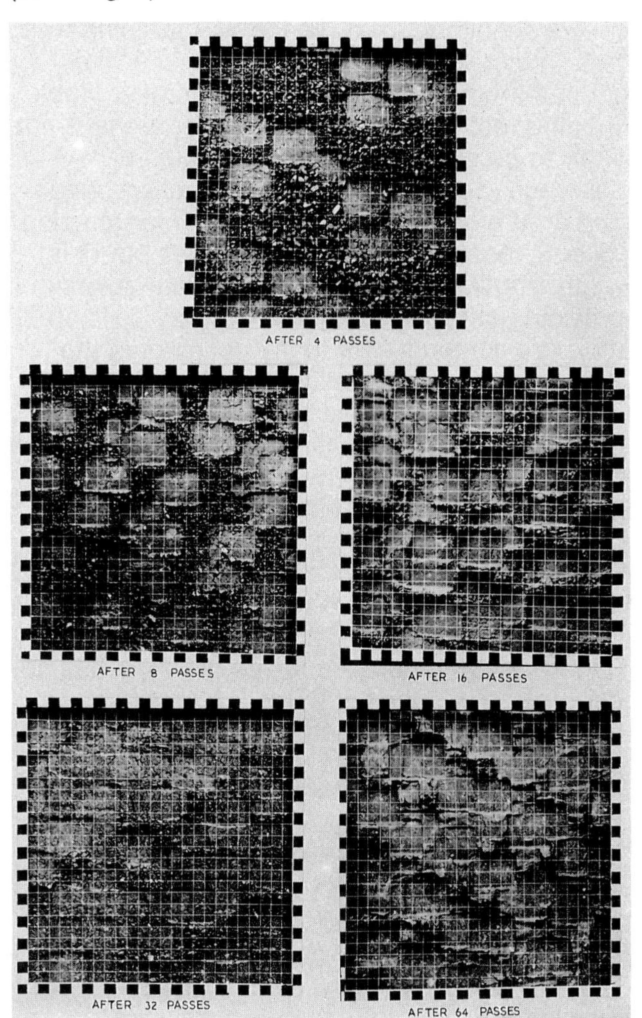

6.23 The club-foot sheepsfoot roller had a higher coverage and lower mass per unit effective width than the taper-foot machine; the increases in coverage and end area of foot are clearly shown in Plate 6.11, which may be directly compared with Plate 6.10. Interesting comparisons can be made where the two machines were tested at similar moisture contents. Figure 6.9(a) indicates that about half the number of passes were required with the club-foot sheepsfoot roller to achieve an equal level of compaction to the taper-foot roller, whereas in Figure 6.10 (b) the results obtained with the two rollers were almost identical. With the gravel-sand-clay, Figure 6.14(a), results were again fairly similar, but generally poor due to the overstressing of the granular soil (see Paragraph 6.18).

6.24 The results for the sheepsfoot rollers confirm that they are most effective on cohesive soils in dry conditions, with moisture contents below the value of optimum – 5 per cent in the case of the silty clay.

6.25 The tamping and grid rollers, with higher coverage and larger contact areas required fewer passes to achieve given levels of compaction and operated effectively at higher moisture contents, on granular soils as well as cohesive soils. An interesting comparison to illustrate this lies in Figures 6.9(d) and 6.10(d). In the former figure the 17 t tamping roller compacting heavy clay at a moisture content of optimum + 0.5 per cent produced a low air content, less than 5 per cent air voids, in a 150 mm layer after four passes of the machine (8 roll passes as this machine is effectively a tandem roller, see Plate 6.6). In Figure 6.10(d) results for the 4.5 t taper-foot sheepsfoot roller on silty clay at a moisture content of optimum – 0.5 per cent show that it just achieved 10 per cent air voids in a 100 mm layer after about 22 passes. Thus, with the clay soils at moisture contents close to the optimum of the 2.5 kg rammer compaction test, tamping rollers (as defined in Paragraph 6.3) rather than sheepsfoot rollers should be used. As the 17 t tamping roller has, in fact, a higher mass per unit effective width than the club-foot sheepsfoot roller (Table 6.1) the restriction of that parameter alone is not likely to be sufficient. The combination of coverage (above 0.15 for tamping rollers) and end area of foot (more than 0.01 m^2 for tamping rollers), together with adequate mass per unit width, are likely to be the determining factors in providing effective compaction of clay soils at moisture contents near to the optimum of the 2.5 kg rammer compaction test.

6.26 Some of the results included in Figures 6.9 to 6.15 are relations between relative compaction value and number of passes at different speeds of travel; these are for the 17 t tamping roller on heavy clay (Figure 6.9(e)) and for the grid roller at 13.6 t on sandy clay No 2 (Figure 6.12(a) and (b)) and well-graded sand (Figure 6.13). Where the speed of rolling has not been noted, a speed of 2.0 to 2.5 km/h was used. The effect of speed of rolling on the state of compaction produced is discussed later in this chapter.

6.27 With the 17 t tamping roller the effect of varying the number of passes was determined for two different thicknesses of layer, 150 mm and 250 mm, on heavy clay (Figure 6.9(d) and (e)) and sandy clay No 2 (Figure 6.12(c)). The results for heavy clay are at two different moisture contents, so comparison is difficult, but with sandy clay No 2 the relative outputs using the different thicknesses can be determined. Thus, at the moisture content of 1.5 per cent above the optimum of the 2.5 kg rammer test (optimum + 1.5 per cent), three passes (six roll passes) were required to achieve a relative compaction value of 95 per cent in the 150 mm layer, and about 20 passes (40 roll passes) on the 250 mm layer. Thus, under these circumstances, the output of the 17 t tamping roller on 150 mm layers would be four times its output on the thicker 250 mm layer. If a 10 per cent air void criterion was employed, the output with the thinner layer (four roll passes) would be 1.5 times that with the thicker, 250 mm, layer (10 roll passes).

6.28 Only two machines have been tested on the uniformly graded fine sand (Figure 6.15), the 17 t tamping roller and the grid roller with a mass of 6.8 t. The grid roller was used at its lower test weight as the soil was badly overstressed when the machine was used at 13.6 t; the opportunity to vary the weight of the machine over such a large range clearly allows it to be used effectively over a wide range of soil types and moisture conditions. With both machines adequate states of compaction, 95 to 100 per cent relative compaction, were achieved using from 8 to 24 roll passes on 250 mm layers with the tamping roller and from 4 to 12 roll passes on 150 mm layers with the grid roller.

6.29 The factors likely to have an influence on the number of passes necessary to achieve given levels of compaction using sheepsfoot, tamping and grid rollers are:–

(i) The coverage (as defined in Table 6.1). As shown in Plates 6.10 and 6.11, where coverage is low a significant number of passes is required to ensure uniform compaction over an area; in addition, penetration of the feet during the initial passes restricts the thickness of compacted layer so that only after a high number of passes, when the machine has 'walked out', is an adequate thickness of compacted material produced. Thus low coverage would be expected to have an adverse effect and increase the number of passes required.

(ii) The mass per unit effective width, which is partially dependent on the coverage. This parameter increases with mass of the machine and decreases with increases in width of roll and coverage. As the value of mass per unit effective width increases the stress immediately beneath the points of contact will increase and, as demonstrated in Figure 6.8, for all but the granular soils the maximum state of compaction achieved at high numbers of passes increases. In the case of granular soils overstressing can occur to the detriment of the state of compaction produced.

(iii) The moisture content of the soil, related to the optimum of the 2.5 kg rammer compaction test.

6.30 Regression analyses were made of the results for 100 mm and 150 mm layers in Figures 6.1 to 6.6 and 6.9 to 6.14 (ie excluding the results for uniformly graded fine sand) using the mass per unit effective width, coverage and relative moisture content as independent variables. The number of passes required to achieve a relative compaction value of 95 per cent and an air content of 10 per cent were related to these variables. Where reasonable correlations were obtained, the results, in the forms of equations and the correlation coefficients, are given in Table 6.3. Correlations for silty clay and for the 10 per cent air voids criterion on all soils other than the sandy clays, were generally poor and have been omitted. Contributing factors to the poor correlations would have been the loss of efficiency of the sheepsfoot rollers at increasing values of moisture content, as illustrated in Figure 6.16, and the variation in thickness of layer from 100 to 150 mm.

Table 6.3 Equations and correlation coefficients to determine the number of passes required to achieve specified levels of compaction with sheepsfoot, tamping and grid rollers. Layer thickness 100 to 150 mm

Soil	Level*	Regression equation†	No of data points	Correlation coefficient
Heavy clay	(a)	P=4.56–.918 m–.0403 c –.0569 w	8	0.86
Sandy clay Nos 1 & 2	(a)	P=4.71–1.01 m–.647 c –.0899 w	10	0.96
	(b)	P=5.63–1.24 m–.877 c –.171 w	12	0.93
Well-graded sand & gravel-sand-clay	(a)	P=.753–.112 m–.493 c –.243 w	9	0.94

*Level (a)=95 per cent of the maximum dry density obtained in the 2.5 kg rammer compaction test
Level (b)=10 per cent air voids
†P=Log_{10} (Number of roll passes required)
m=Log_{10} (Mass per unit effective width in kg/m)
c=Log_{10} (Coverage)
w=Moisture content of soil, per cent, minus the optimum moisture content, per cent, obtained in the 2.5 kg rammer compaction test

6.31 It must be concluded that it is not possible to provide general guidance on the potential performance of the complete range of machines investigated and the individual relations given in the various figures provide the most useful information.

Relations between state of compaction and depth within the compacted layer

6.32 Results of investigations of the variation of density with depth in layers of soil compacted by sheepsfoot, tamping and grid rollers are given in Figures 6.17 to 6.21. Only three machines were tested and in almost all cases only two increments of depth were used. Using the same criteria as in earlier chapters to assess the profiles of density with depth, ie 90 per cent relative compaction or 15 per cent air voids at the bottom of the layer, accurate determinations of the thickness of layer compacted by the grid roller from results in Figures 6.17(a), 6.18(b) and 6.19 are impossible. With gravel-sand-clay, however (Figure 6.20) at a moisture content 1.5 per cent below the optimum of the 2.5 kg rammer test, a depth of about 140 mm satisfies both criteria.

6.33 The 17 t tamping roller was investigated at various numbers of passes on both heavy clay and sandy clay No 2 (Figures 6.17(b) and 6.18(c)) and, on heavy clay, also at two speeds of rolling. Taking the results for a speed of rolling of 2 to 2.4 km/h, the thickness of layer capable

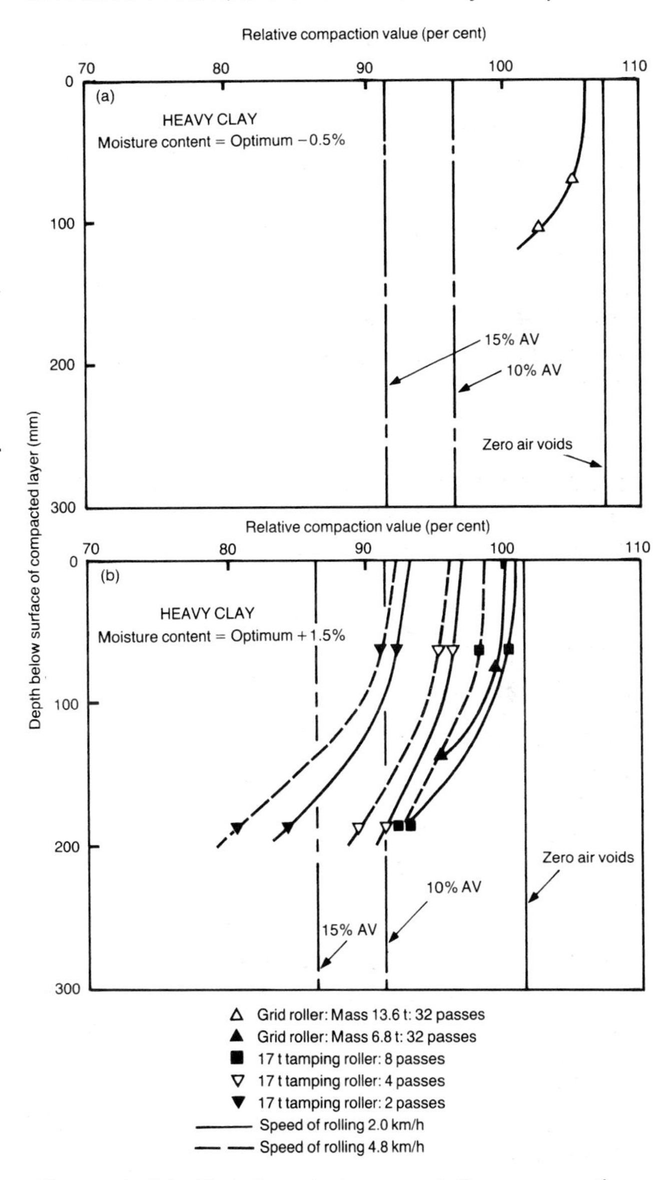

Figure 6.17 Relations between relative compaction and depth in the compacted layer of heavy clay obtained with tamping and grid rollers. All values are related to the maximum dry density and optimum moisture content obtained in the 2.5 kg rammer compaction test (Figure 3.13)

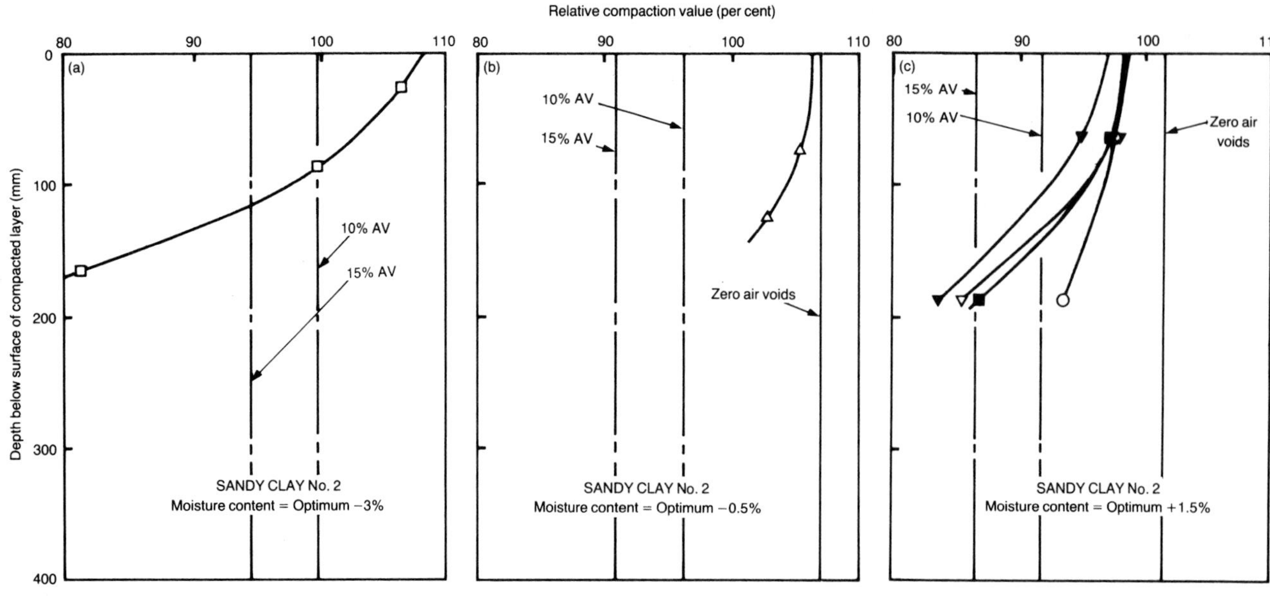

Figure 6.18 Relations between relative compaction and depth in the compacted layer of sandy clay No 2 obtained with sheepsfoot, tamping and grid rollers. All values are related to the maximum dry density and optimum moisture content obtained in the 2.5 kg rammer compaction test (Figure 3.13)

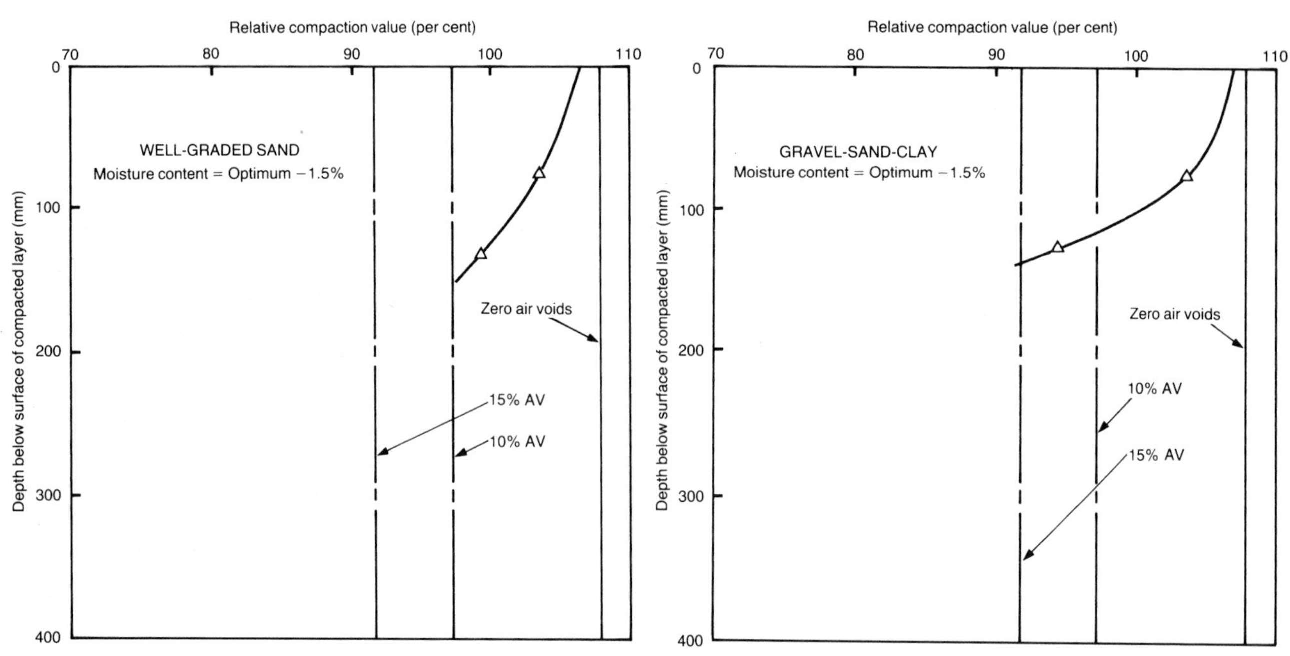

Figure 6.19 Relation between relative compaction and depth in the compacted layer of well-graded sand obtained with 32 passes of the 13.6 t grid roller. All values are related to the maximum dry density and optimum moisture content obtained in the 2.5 kg rammer compaction test (Figure 3.13)

Figure 6.20 Relation between relative compaction and depth in the compacted layer of gravel-sand-clay obtained with 32 passes of the 13.6 t grid roller. All values are related to the maximum dry density and optimum moisture content obtained in the 2.5 kg rammer compaction test (Figure 3.13)

of being compacted to the criteria given above has been related to the number of roll passes (for the 17 t tamping roller roll passes were double the machine passes) in Figure 6.22. The regression equations and correlation coefficients are given in Table 6.4. Thus, for the particular moisture contents used, 2.5 kg rammer optimum + 1.5 per cent, the 17 t tamping roller proved more effective on the heavy clay than on the sandy clay, with an increase in thickness of layer by a factor of at least 1.2 for 8 to 16 roll passes. The use of a layer thickness of 150 mm for tests on well-graded sand and gravel-sand-clay (Figures 6.13 and 6.14) are indicative that the

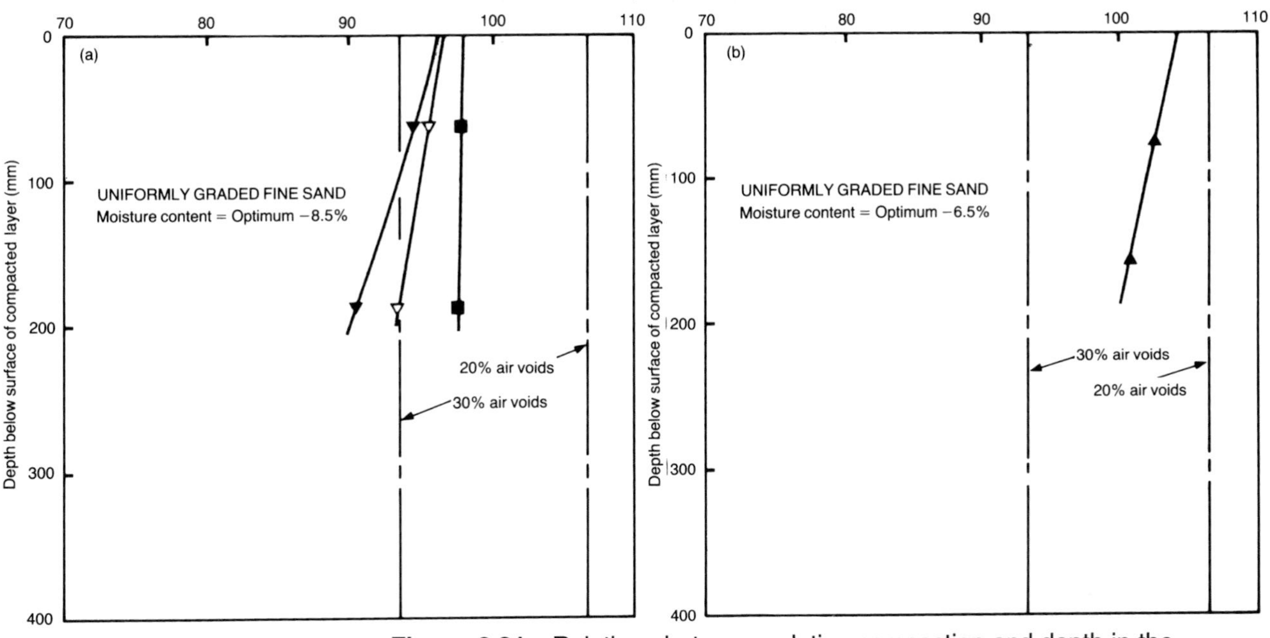

Figure 6.21 Relations between relative compaction and depth in the compacted layer of uniformly graded fine sand obtained with tamping and grid rollers. All values are related to the maximum dry density and optimum moisture content obtained in the 2.5 kg rammer compaction test (Figure 3.13)

trends demonstrated by the reducing depths of compaction from heavy clay to sandy clay continued with further reductions with granular soils. The exception to this trend is the performance of the 17 t tamping roller on uniformly graded fine sand (Figure 6.21(a)) where, for two machine passes (four roll passes) a depth of 200 mm was compacted to the criterion of 90 per cent relative compaction at the bottom of the layer. For the numbers of passes in excess of two the depths of layer capable of being compacted to the same criterion are indeterminate from the test data illustrated, but they are clearly in the region of 300 mm or more. It should be stressed, however, that the criterion of 90 per cent relative compaction at the bottom of the layer may not be adequate for uniformly graded soils if they are located in areas susceptible to inundation.

6.34 The results for the 4.3 t sheepsfoot roller on sandy clay No 2 (Figure 6.18(a)) confirm that machines of low coverage and small end areas of feet have a limited compaction depth, even though high penetration of the feet may initially occur. Thus, using the 90 per cent relative compaction and 15 per cent air voids criteria, the depth compacted with 32 passes was 130 and 110 mm respectively at a moisture content of optimum (2.5 kg rammer) minus 3 per cent.

Figure 6.22 Relations between depth of layer compacted to two different standards and number of roll passes of the 17 t tamping roller for clay soils at optimum + 1.5 per cent. From Figures 6.17 (b) and 6.18 (c)

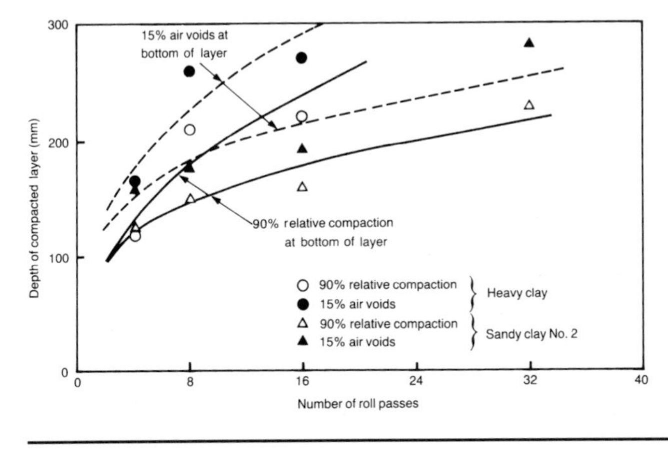

Table 6.4 Equations and correlation coefficients for the relations given in Figure 6.22 for the 17 t tamping roller

Soil	Level*	Regression equation†	No of data points	Correlation coefficient
Heavy clay	(a)	$D=71.3\ P^{.437}$	3	0.90
	(b)	$D=108\ P^{.356}$	3	0.90
Sandy clay No 2	(a)	$D=80.9\ P^{.283}$	4	0.96
	(b)	$D=107\ P^{.253}$	4	0.92

*Level (a)=Achievement at the bottom of the layer of 90 per cent of the maximum dry density obtained in the 2.5 kg rammer test

Level (b)=Achievement at the bottom of the layer of 15 per cent air voids

†D=Maximum depth of layer to achieve stated level of compaction (mm)

P=Number of roll passes of 17 t tamping roller

Effect of rolling speed on the performance of tamping and grid rollers

6.35 The effect of the speed of rolling was investigated with the 13.6 t grid roller (Figures 6.12(a) and (b) and 6.13) and the 17 t tamping roller (Figures 6.9(e) and 6.17(b)). The relation between relative compaction value and number of passes was determined with the grid roller on sandy clay No 2 at various speeds of rolling between 2.3 and 16 km/h, using a large wheeled tractor as the towing unit for the higher speed tests (Plate 6.9); with well-graded sand, speeds of rolling of 2.3 and 5.5 km/h were used (Figure 6.13). Although two different moisture contents were used with the sandy clay, it is clear that the number of passes required to achieve given specified levels of compaction tended to increase with increase in speed, as shown in Table 6.5. The results can be expected to be affected to some extent by moisture content and the soil type involved, but from the results given in Table 6.5 an average increase in output of about 120 per cent was achieved by an increase in rolling speed from 2.3 to 5.5 km/h (140 per cent increase in speed) and output increased by 110 per cent when speed was further increased from 5 to 16 km/h (220 per cent increase in speed).

Table 6.5 Results of tests at various rolling speeds with the 13.6 t grid roller compacting 150 mm layers

Soil	Moisture content	Speed of rolling (km/h)	No. of passes required to achieve:–	
			95% relative compaction	10% air voids
Sandy clay No 2	Optimum–0.5%	2.3	3	4
		5.5	4	4
	Optimum+0.5%	5	3	3
		10	4	4
		16	5	4
Well-graded sand	Optimum–1.5%	2.3	4	5
		5.5	4	5

6.36 The effect of speed of rolling of the tamping roller was determined on heavy clay with speeds of about 2 and 5 km/h being used to determine relations between relative compaction and number of passes (Figure 6.9(e)); this work yielded the relations between relative compaction and depth in the layer given in Figure 6.17(b). Combinations of numbers of passes and thicknesses of layer to produce specified levels of compaction can be derived and are shown in Table 6.6. Thus with this particular soil type and moisture content the output of the 17 t tamping roller was increased by amounts varying from 60 to 110 per cent for an increase in speed from about 2 to 5 km/h (a 140 per cent increase in speed).

Table 6.6 Results of tests at two rolling speeds with the 17 t tamping roller on heavy clay at a moisture content of optimum+1.5 per cent

(a) Constant depth of layer

Thickness of layer (mm)	No. of roll passes required to achieve:–			
	95 % relative compaction		10 % air voids	
	2.0 km/h	4.8 km/h	2.0 km/h	4.8 km/h
250	11	16	6	7

(b) Specific numbers of passes

No. of roll passes	Depth of layer (mm) to achieve at the bottom:–			
	90% relative compaction		15 % air voids	
	2.0 km/h	4.8 km/h	2.0 km/h	4.8 km/h
4	120	90	160	130
8	210	180	260	220

6.37 These results indicate that, compared with pneumatic-tyred rollers (Paragraphs 5.25 and 5.26) slightly larger adjustments in number of passes are required to allow for variations in speed of rolling of grid and tamping rollers. For the grid roller the results plotted in Figures 6.12(a) and (b) and 6.13 indicate that the number of passes should be increased by about 15 per cent for a doubling of the speed of travel to achieve a given level of compaction. With the 17 t tamping roller the results in Figure 6.9(e) indicate that an increase in number of passes of 30 per cent for a doubling of speed would be most appropriate. Except where the speed of rolling is noted on the figures in this Chapter it can be assumed that the speed of operation of the various rollers was in the range of 2 to 2.5 km/h.

Surface of layer following compaction by sheepsfoot, tamping and grid rollers

6.38 Because of the areas of high stress beneath the rolls, the small contact areas of some of the machines, and the large penetration of the feet into the soil on initial passes over loose soil and also on the wetter soils, a so-called 'loose mulch' is left on the surface of the layer. Examples of the compacted surfaces left after compaction by the 11 t tamping roller, the 17 t tamping roller, the grid roller ballasted to

Plate 6.12 Compacted surface of sandy clay No 2 at a moisture content of optimum – 0.5 % after 32 passes of the 11 t tamping roller

Plate 6.13 Compacted surface of sandy clay No 2 at a moisture content of optimum + 1% after 8 passes (16 roll passes) of the 17 t tamping roller

Plate 6.14 Compacted surface of heavy clay at a moisture content equal to the 2.5 kg rammer optimum after 32 passes of the 13.6 t grid roller

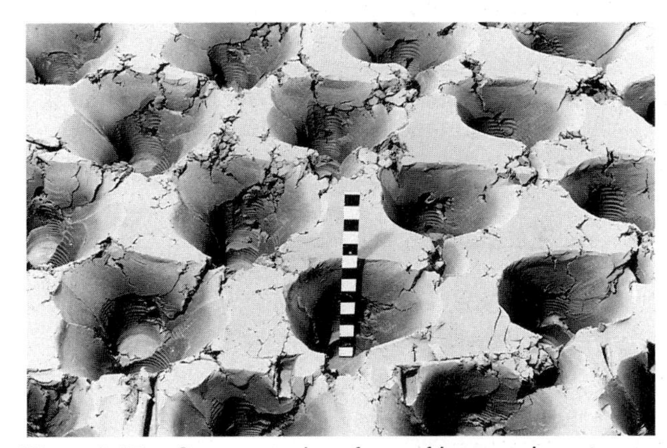

Plate 6.15 Compacted surface of heavy clay at a moisture content equal to the 2.5 kg rammer optimum after 32 passes of the 4.3 t sheepsfoot roller. This photograph was taken after compaction using the machine with vibration, but it illustrates the surface disturbance created by this type of machine at higher values of moisture content

13.6 t and the 4.3 t sheepsfoot roller are shown in Plates 6.12 to 6.15. These types of surfaces necessitate a laborious amount of preparation before any measurements can be made of the in-situ dry density, by whatever method. It is essential that all the 'loose mulch' is removed; the topmost level of any determination of in-situ state of compaction must be the level of the deepest imprint at the location of the test.

6.39 Depths of loose mulch were recorded during the course of some of the tests described in this Chapter and examples of the effects on the depth of loose mulch of moisture content and number of passes are shown in Figures 6.23 and 6.24 respectively.

6.40 The relations between the depth of loose mulch and moisture content (Figure 6.23) are

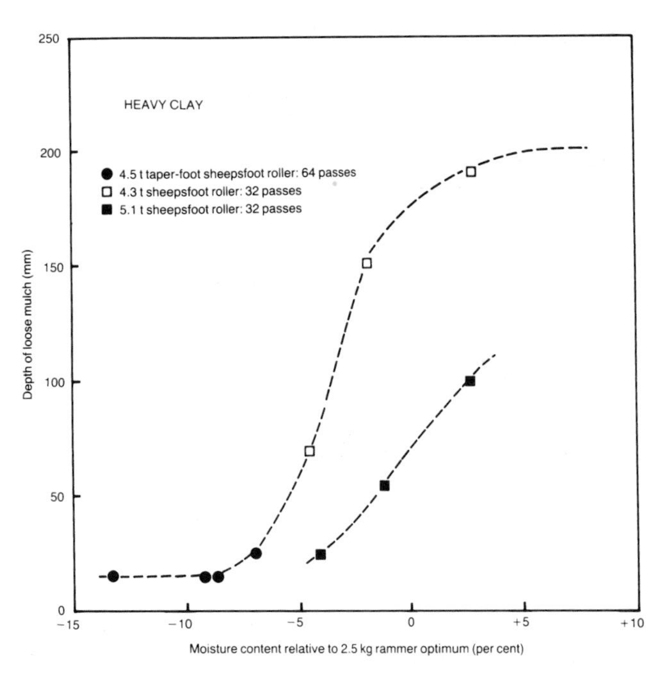

Figure 6.23 Relations between depth of disturbed heavy clay on surface of compacted layer and moisture content for compaction to refusal by sheepsfoot rollers

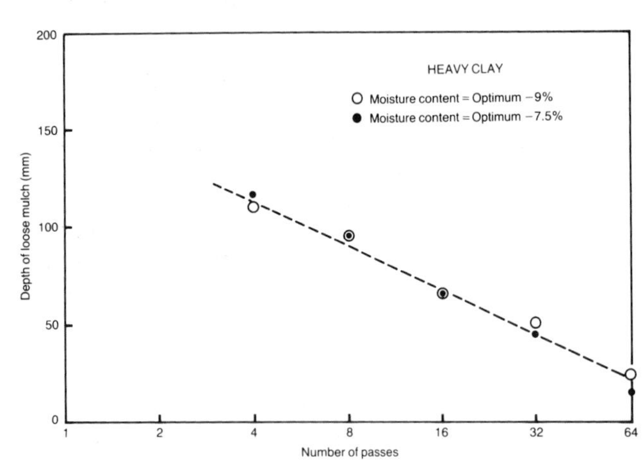

Figure 6.24 Relation between depth of disturbed heavy clay on surface of compacted layer and number of passes of the 4.5 t taper-foot sheepsfoot roller

given for three machines operating on heavy clay. The states of compaction of the material below the level of the disturbed surface are given in Figure 6.1 for the same machines (using the same symbols). A single relation has been drawn through the points produced by the 4.5 t taper-foot sheepsfoot roller and the 4.3 t sheepsfoot roller. It can be seen in Table 6.1 that the contact areas of the feet, the coverage and the mass per unit effective width were roughly the same for these two machines. The depth of loose mulch was 25 mm or less at moisture contents below the value of optimum minus 7 per cent. With increasing moisture content above this value the depth of disturbed material increased very rapidly until at moisture contents wet of the optimum of the 2.5 kg rammer test the feet were completely penetrating the soil and the roll shell was bearing on the surface of the layer of disturbed material. This condition is illustrated in Plate 6.15; the particular photograph applies to the 4.3 t sheepsfoot roller using vibration but the appearance is similar to that for the non-vibrating machine, although the total penetration of the feet occurred at lower moisture contents when vibration was used.

6.41 The increased contact area and coverage and reduced mass per unit effective width of the 5.1 t sheepsfoot roller (Table 6.1) reduced the depth of disturbed soil at any given moisture content (Figure 6.23). With this machine the depth of loose mulch exceeded 25 mm at moisture contents in excess of optimum – 4 per cent on the heavy clay. It could be considered that the moisture content at which the depth of loose mulch exceeds 25 mm is indicative of the soil condition at which the machine failed to walk out and compaction with sheepsfoot and tamping rollers would only be effective, therefore, at moisture contents below that value. It has already been demonstrated that the limiting moisture content (relative to the 2.5 kg rammer optimum) will vary with machine characteristics and it can also be expected to vary with soil type (see Paragraph 6.43).

6.42 The walking out of the taper-foot sheepsfoot roller with increasing numbers of passes on heavy clay soil at moisture contents within the limits of effective operation, ie dry of optimum – 7 per cent, is illustrated in Figure 6.24. After four passes the loose mulch had a depth of 110 to 115 mm, but this reduced to 15 to 25 mm after 64 passes. This ability to 'walk out' is essential to the production of a compacted layer of a practical thickness at an adequate state of compaction. The depths of loose mulch given in Figure 6.24 relate to the results for the taper-foot roller given in Figure 6.9 (a) and (b).

6.43 Insufficient reliable data are available to determine the effect of moisture content of different soil types on the ability of non-vibrating sheepsfoot rollers to 'walk out'. However, Figure 6.16, discussed in Paragraph 6.22, showed that with silty clay the number of passes necessary to achieve 95 per cent relative compaction were at a minimum at a moisture content of optimum – 5 per cent. This may well have been the moisture content of that soil above which the taper-foot roller failed to walk out, and compares with optimum – 7 per cent for the heavy clay.

General observations

6.44 The sheepsfoot rollers, with their high contact pressures but low coverage, were most effective at low values of moisture content, but even then required a large number of passes to achieve adequate compaction. They are clearly likely to be most suitable for use in arid climates where cohesive soils occur at moisture contents well dry of the optimum moisture content of the 2.5 kg rammer compaction test. As the contact pressures are reduced by increased contact area of the feet and increased coverage, so the moisture content of a soil at which the compactor is most effective increases and for this reason the tamping roller is more appropriate for the compaction of cohesive soils in climatic conditions where they exist at moisture contents close to the optimum of the 2.5 kg rammer test.

6.45 The grid roller provides even greater coverage than the tamping roller and is, therefore, also appropriate for compaction in the wetter climatic conditions. One feature of the grid roller that has been observed is that the grids tended to become blocked with soil when mixtures of wet and dry lumps of clay soil were being compacted; the benefits of the high peak stresses would be lost as a result of this.

References

Cross, J E (1962). An investigation of the performance of a 5-ton vibrating sheepsfoot roller in the compaction of a heavy clay soil. *D.S.I.R., RRL Laboratory Note* No LN/166/JEC (Unpublished).

Department of Transport (1986). Specification for highway works. 6th edition. HM Stationery Office, London.

Hyster Company (1970). Guide to Hyster compaction. Hyster Overseas, Brentford.

Lewis , W A (1954). Further studies in the compaction of soil and the performance of compaction plant. *Road Research Technical Paper* No 33. HM Stationery Office, London.

Parsons, A W (1959). An investigation of the performance of a 13½-ton grid roller in the compaction of soil. *D.S.I.R., RRL Research Note* No RN/3563/AWP (Unpublished).

Parsons, A W (1962). An investigation of the performance of a 4¼-ton vibrating sheepsfoot roller for compacting soil. *D.S.I.R., RRL Laboratory Note* No LN/211/AWP (Unpublished).

Parsons, A W and Toombs, A F (1965). An investigation of an 11-ton "tamping" roller for compacting soil. *RRL Laboratory Note* No LN/896/ AWP.AFT (Unpublished).

Toombs, A F (1973). The performance of a Caterpillar 815 17-Mg tamping roller in the compaction of soil. *DoE, TRRL Report* LR 529. Transport and Road Research Laboratory, Crowthorne.

Williams, F H P and Maclean, D J (1950). The compaction of soil: a study of the performance of plant. *Road Research Technical Paper* No 17. HM Stationery Office, London.

CHAPTER 7 COMPACTION OF SOIL BY VIBRATING SMOOTH-WHEELED ROLLERS

General description of vibrating smooth-wheeled rollers

7.1 This Chapter gives the results of investigations into the performance of vibrating smooth-wheeled rollers.

7.2 Vibrating rollers enhance the performance of static-weight machines by the addition of dynamic forces. These dynamic forces are usually achieved by the use of a rotary eccentrically weighted shaft through the centre of the roll. Variations on the method of generating vibration can occur, however, and the range of methods available include:–

(a) Eccentrically weighted shaft through centre of roll;

(b) Twin eccentric shafts in one roll, phased to produce resultant vertical forces only;

(c) Steel balls driven round a ball race inside the roll shell;

(d) Eccentrically weighted shaft or shafts outside the roll but installed in the chassis to which the roll is attached.

7.3 The vibrating system is isolated by a flexible suspension system from the main body of the machine, the power unit, operator's controls or compartment, and other ancillary components. A power unit is installed on the vibrating roller to drive the vibratory mechanism, and in the case of the self-propelled machines this power unit also drives the transmission system providing the necessary traction.

7.4 Vibrating smooth-wheeled rollers have a number of configurations, the principal of which are:–

(a) Self-propelled, pedestrian-operated. These can have a single vibrating roll or twin vibrating rolls arranged in tandem. The operator walks with the machine or, alternatively, some machines have 'ride-on' attachments (Plate 7.6);

(b) Self-propelled, single roll. These machines generally have a single vibrating roll with propulsion provided by pneumatic-tyred wheels on a second axle (Plate 7.14);

(c) Self-propelled, tandem. These have two rolls in tandem, only one of which is fitted with a vibrating mechanism;

(d) Self-propelled, double vibrating rollers. Similar to tandem vibrating rollers except that both rolls are equipped with vibrating mechanisms (Plate 7.12);

(e) Towed. These machines have a single vibrating roll. A separate towing unit is necessary to provide traction.

7.5 Various factors can be expected to influence the performance of vibrating rollers, including total mass, the masses of the vibrating and non-vibrating components, width of roll(s), frequency of vibration, eccentricity and mass of the out-of-balance system, and the characteristics of the flexible suspension between the vibrating and non-vibrating components. However, as discussed later in Paragraphs 7.61 to 7.66, the installed dynamic forces depend to a large extent on the mass of the machine. With the range of vibrating rollers available at the time of writing mass per unit width of vibrating roll is used, therefore, as an indicator of performance (Department of Transport, 1986).

Description of vibrating smooth-wheeled rollers included in investigations

7.6 During the investigations of the performance of compaction plant at the Laboratory a wide range of vibrating smooth-wheeled rollers was tested, from very early experimental or prototype machines to fairly recent models representative of the types in widespread use in earthwork construction in the 1980s. Details of the various machines are given in Table 7.1 and they are illustrated in Plates 7.1 to 7.16.

7.7 For some of the machines tested before 1962 the values for centrifugal force of the vibrating mechanisms are not available and are listed as NA in Table 7.1. It was customary for manufacturers not to release such details and values given for the early machines are taken from private communications.

Table 7.1 Details of the various machines contributing data on the performance of vibrating smooth-wheeled rollers

Compactor	Total mass	Front roll				Rear roll				Vibrating mechanism			Normal speed of rolling in tests
		Width	Diameter	Mass on roll	Mass per unit width	Width	Diameter	Mass on roll	Mass per unit width	Location	Normal frequency	Nominal centrifugal force	
	(kg)	(m)	(m)	(kg)	(kg/m)	(m)	(m)	(kg)	(kg/m)		(Hz)	(kN)	(km/h)
220 kg pedestrian-operated vibrating roller	220	0.61	0.53	220	360	–	–	–	–	Roll	75	10	NA Hand-propelled
350 kg pedestrian-operated vibrating roller	350	0.71	0.57	350	490	–	–	–	–	Roll	75	12	1.1–1.4
990 kg tandem vibrating roller	990	0.81	0.66	310	380	0.81	0.66	680	840	Rear roll	67	16	1.0–1.5
1.0 t pedestrian-operated double vibrating roller	1000	0.75	0.48	500	660	0.75	0.48	500	660	Both rolls	47	20 per roll	1.0
1.5 t pedestrian-operated double vibrating roller	1530	0.90	0.58	615	680	0.90	0.58	615	680	Both rolls	47	29 per roll	0.9
1.7 t pedestrian-operated double vibrating roller	1730	0.84	0.66	710	850	0.84	0.66	710	850	Both rolls	60	15 per roll	0.9
2.2 t tandem vibrating roller	2200	0.85	0.74	1030	1210	0.90	0.74	1170	1300	Rear roll	50	20	1.1–2.0
2.4 t tandem vibrating roller	2440	0.81	0.76	990	1220	0.81	0.76	1450	1790	Front roll	80	NA	0.3–0.7
3.9 t tandem vibrating roller	3880	1.00	0.89	2790	2790	0.70	0.76	1090	1560	Front roll	30–49	NA	0.6–1.7
3.9 t towed vibrating roller	3910	1.83	1.22	3880	2120	–	–	–	–	Roll	39	124	2.3
4.9 t towed vibrating roller	4900	1.91	1.19	4860	2550	–	–	–	–	Roll	27	98	2.4
6.9 t double vibrating roller	6900	1.25	1.10	3680	2940	1.25	1.10	3220	2580	Both rolls	42	59 per roll	2.2
7.3 t double vibrating roller	7280	1.90	0.80	3640	1910	1.90	0.80	3640	1910	Both rolls	43	156 per roll	1.1
7.7 t single-roll vibrating roller	7670	1.83	1.37	4140	2260	Pneumatic-tyred wheels		3530	–	Roll	40	142	2.4
8.6 t towed vibrating roller	8600	1.91	1.60	8580	4490	–	–	–	–	Roll	27	NA	1.2
12 t towed vibrating roller	11800	2.08	1.83	11800	5670	–	–	–	–	Roll	30	327	2.4

Plate 7.1 220 kg pedestrian-operated vibrating roller investigated in 1949

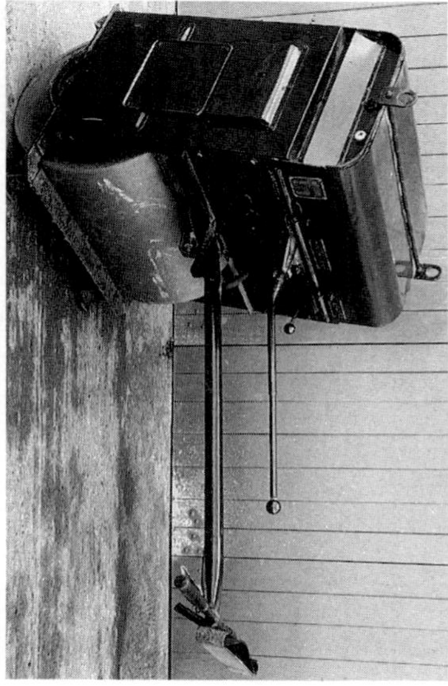

Plate 7.2 350 kg pedestrian-operated vibrating roller investigated in 1956

Plate 7.3 990 kg tandem vibrating roller investigated in 1958

7.8 The speeds of rolling given in Table 7.1 are those used in the majority of tests, excluding any where the effects of varying the speed of rolling were studied. With tandem self-propelled machines and pedestrian-operated single roll self-propelled machines the effect of the vibrating mechanism was to cause the machines to travel faster in one direction than the other, with either positive or negative slip of the driving roll over the soil surface. Such machines were operated in their lowest travelling gear and, apart from the 350 kg pedestrian-operated vibrating roller, in the forward and reverse directions on alternate passes. The ranges of speeds given in Table 7.1 represent the differences between forward and reverse travel for these machines.

Plate 7.4 1.0 t pedestrian-operated double vibrating roller investigated in 1971

Plate 7.5 1.5 t pedestrian-operated double vibrating roller investigated in 1971

Plate 7.6 1.7 t pedestrian-operated double vibrating roller investigated in 1968

7.9 The various machines are listed in Table 7.1 and are illustrated in Plates 7.1 to 7.16 in order of increasing total mass. However, descriptively they are best described in groups relating to their general type.

7.10 Pedestrian-operated, single-roll machines are represented by the 220 kg and 350 kg

Plate 7.7 2.2 t tandem vibrating roller investigated in 1965

Plate 7.8 2.4 t tandem vibrating roller investigated in 1949–50

Plate 7.9 3.9 t tandem vibrating roller investigated in 1960

Plate 7.10 3.9 t towed vibrating roller investigated in 1956–57

Plate 7.11 4.9 t towed vibrating roller investigated in 1965

vibrating rollers (Plates 7.1 and 7.2). The 220 kg machine was a very early model, tested in 1949. It was hand-propelled, its single-cylinder engine being used solely to drive an eccentric shaft mounted through the centre of the roll. The larger, 350 kg machine, was self-propelled; the engine, mounted above the roll, provided the drive to the eccentric shaft in the centre of the

Plate 7.12 6.9 t double vibrating roller investigated in 1973

Plate 7.13 7.3 t double vibrating roller investigated in 1971

Plate 7.14 7.7 t single-roll vibrating roller investigated in 1968–69

Plate 7.15 8.6 t towed vibrating roller investigated in 1960–61

Plate 7.16 12 t towed vibrating roller investigated in 1969

roll and the traction for travel in the forward and reverse directions at two operating speeds. Both machines were tested occasionally without vibration and are, therefore, also included and illustrated in Chapter 4.

7.11 Pedestrian-operated double vibrating rollers that have been investigated were the 1.0 t, 1.5 t and 1.7 t machines (Plates 7.4, 7.5 and 7.6 respectively). These machines were tested in the more recent phase of the investigations. The 1.0 t and 1.5 t machines differed in one important respect from the 1.7 t machine in that the eccentric shafts, one in each roll, of the two lighter machines rotated in the same direction but were 180° out-of-phase. To achieve this the drive to the vibrating mechanism was by toothed belt. With the 1.7 t machine the phasing of the two eccentric shafts was not controlled, the drive being by Vee-belt. The centrifugal forces installed in the 1.0 t and 1.5 t machines were higher than that in the 1.7 t machine (Table 7.1). Traction was provided by both rolls, and two speeds of travel were available in the forward and reverse directions. The lower speed was generally used for the tests with each machine. The 1.5 t and 1.7 t machines were equipped with steering units, each consisting of two pneumatic-tyred wheels attached to an arm pivoted below the frame extension (Plates 7.5 and 7.6). The loading of these wheels was adjustable by a hydraulic pump in the case of the 1.5 t roller and maintained constant by a hydro-pneumatic constant-pressure system in the 1.7 t machine. The load supported by the steering wheels gave rise to a discrepancy between the total mass and the sum of the masses on the two rolls with each of these machines (Table 7.1).

7.12 Tandem vibrating rollers are represented by the 990 kg, 2.2 t, 2.4 t, and 3.9 t machines (Plates 7.3, 7.7, 7.8 and 7.9 respectively). The 2.4 t machine (Plate 7.8) was a very early experimental prototype. The vibratory mechanism in all the machines consisted of an eccentric shaft through the centre of one of the rolls. With the 2.4 t machine vibration was applied to the front roll while traction was applied to the rear roll; with the other three machines vibration and traction were applied to the same roll, the rear one in the 990 kg and 2.2 t machines and the front roll in the 3.9 t roller. The 2.4 t machine was unique in this group in that the drive for the vibrating system and for traction was by means of a hydraulic system; this provided a continuously variable speed of travel up to a maximum of about 5 km/h. The 990 kg, 2.2 t and 3.9 t machines all had mechanical transmissions with two speeds of travel in both forward and reverse directions. A continuously variable frequency of vibration within the range given in Table 7.1 was available on the 3.9 t roller, achieved by means of a belt drive and expanding pulley system. The 2.4 t and 3.9 t machines were occasionally tested without vibration and are therefore also included and illustrated in Chapter 4.

7.13 Towed vibrating rollers investigated were the 3.9 t, 4.9 t, 8.6 t and 12 t machines (Plates 7.10, 7.11, 7.15 and 7.16 respectively). Each machine had a steel chassis within which the roll was installed (the 8.6 t machine had twin rolls on a common axle system) on flexible mountings. Vibration was produced by an eccentrically weighted shaft system in the roll, driven by the rear-mounted engine, except in the case of the 4.9 t machine, in which the vibrating system consisted of steel balls which were driven round races, contained within the roll, by a drive shaft through the centre of the roll. The 8.6 t roller was modified during the course of the test programme to comply with later production standards, the modified machine having larger dynamic forces than the machine in its original form. All of these machines were towed during the tests by a crawler tractor of appropriate size. The 3.9 t and 8.6 t machines were occasionally tested without vibration and are, therefore, also included and illustrated in Chapter 4.

7.14 The two large double vibrating rollers tested, the 6.9 t and 7.3 t machines (Plates 7.12 and 7.13) had distinctly different characteristics. The 6.9 t machine (Plate 7.12) had two vibrating rolls in tandem, an articulated main frame, hydrostatic transmission, and a vibration system designed to produce vertically directed forces only. The vibrating system in each roll comprised twin eccentric shafts rotating in opposite directions and appropriately phased to produce the vertically directed forces. A continuously variable speed of travel up to a maximum of 10 km/h was available. The 7.3 t machine (Plate 7.13) had four rolls, mounted in a chassis in tandem pairs. The vibrational forces were generated by a rotating eccentric shaft in each of the rolls. The shafts were all coupled by a toothed-belt drive system, all rotating in the same direction with the front pair 180° out-of-phase with the rear pair. The rolls on each side of the machine were driven independently to facilitate steering. Two speeds of travel in the forward and reverse directions were available.

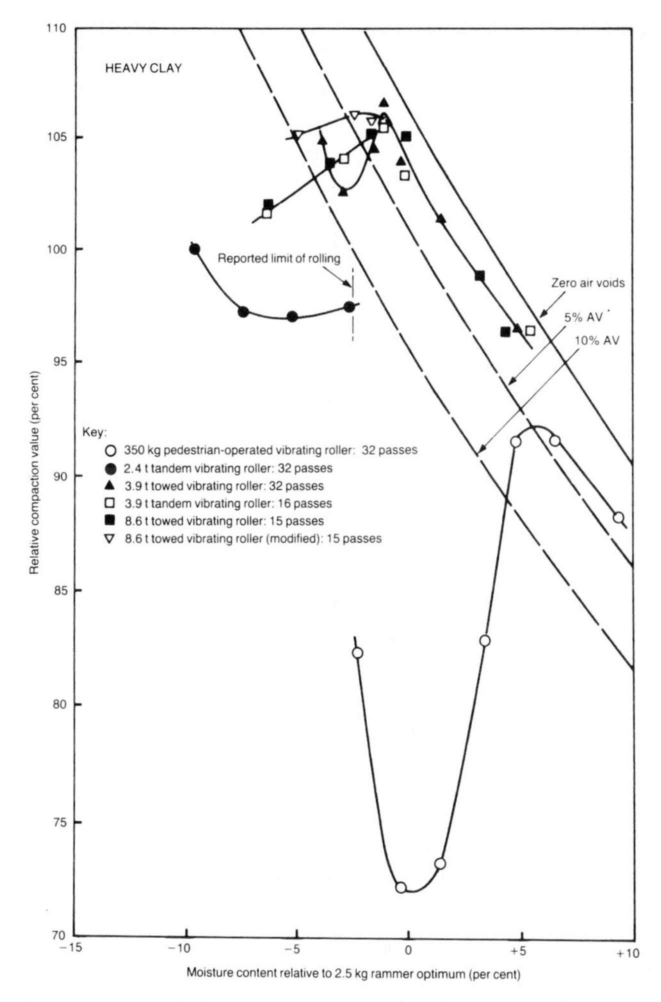

Figure 7.1 Relations between dry density and moisture content for vibrating smooth-wheeled rollers on heavy clay, expressed in terms of relative compaction with the 2.5 kg rammer test. Layer thickness = 150 mm

7.15 The sole example of a self-propelled single-roll vibrating roller, the 7.7 t machine (Plate 7.14) was driven by two steerable pneumatic-tyred wheels at the rear. The freely rotating vibrating roll was mounted at the front of the machine, with an eccentric shaft to produce the dynamic forces mounted through its centre. Drive to the rear wheels included a hydrostatic transmission unit so travel speed was continuously variable up to a maximum of 19 km/h, the high maximum speed allowing rapid transit between different areas of work. This machine was occasionally tested without vibration and was, therefore, also included and illustrated in Chapter 4.

7.16 The machine descriptions given above and the results presented in this Chapter have been taken from the following:– Lewis (1954), Lewis (1961), Parsons (1966), Parsons et al (1962) and Toombs (1966), (1968), (1969), (1970), (1972) and (1973).

Relations between state of compaction and moisture content

7.17 Relations between the states of compaction produced by vibrating smooth-wheeled rollers and the moisture content of the soils are given in Figures 7.1 to 7.7. As in earlier chapters the dry density has been expressed as a relative compaction value and the moisture content as a difference from optimum moisture content, both in relation to the results of the 2.5 kg rammer compaction test at the relevant period. Air void lines are based on average values for the results in each figure.

7.18 In Figures 7.1 to 7.6 the thickness of layer used was 150 mm in all the tests, but in Figure 7.7, for the uniformly graded fine sand, layer thickness was increased to 300 mm with the 3.9 t and 8.6 t towed vibrating rollers. All the results in Figures 7.1 to 7.7 are for an intended 'compaction to refusal' condition, ie a large number of passes was used to provide an indication of the maximum dry density likely to be achieved. Thus, 32 passes were used with the majority of the machines, although 15

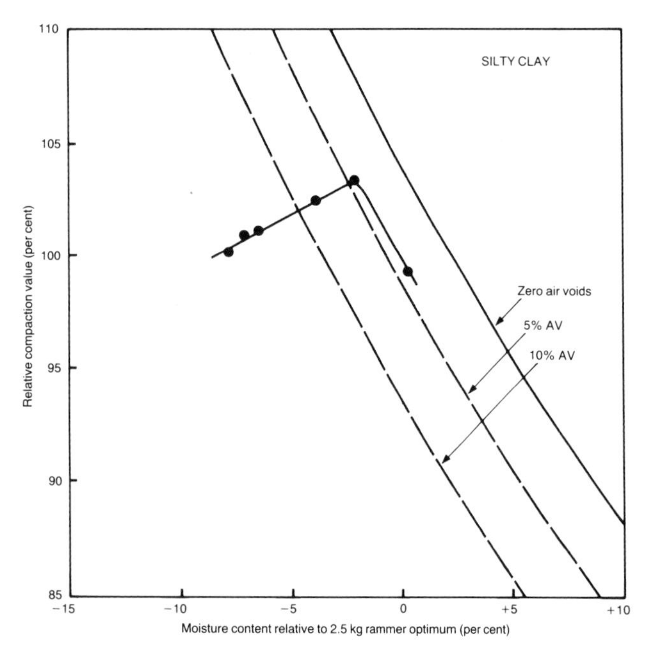

Figure 7.2 Relation between dry density and moisture content for the 2.4 t tandem vibrating roller on silty clay, expressed in terms of relative compaction with the 2.5 kg rammer test. Number of passes = 32; layer thickness = 150 mm

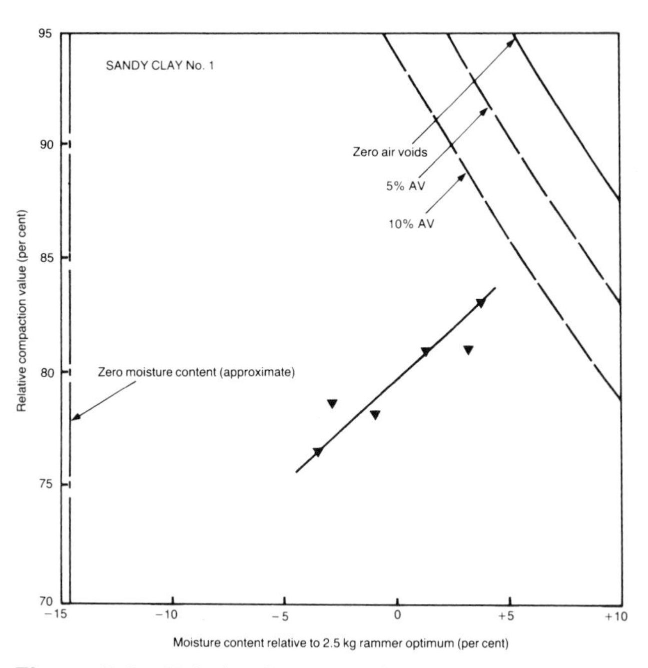

Figure 7.3 Relation between dry density and moisture content for the 220 kg pedestrian-operated vibrating roller on sandy clay No 1, expressed in terms of relative compaction with the 2.5 kg rammer test. Number of passes = 32; layer thickness = 150 mm

passes were used with the 8.6 t towed vibrating roller (16 passes with uniformly graded fine sand) and 16 passes with the 3.9 t tandem vibrating roller. Results are included for the 8.6 t towed vibrating roller in both its original and modified forms, the latter introducing an increase in the centrifugal force of the vibratory system (see Paragraph 7.13).

7.19 For the heavy clay (Figure 7.1) only a limited number of machines was tested, the lighter vibrating rollers generally being considered unsuitable for the compaction of this type of soil. However, the 350 kg pedestrian-operated vibrating roller was tested and a low air content was achieved, although at a very high moisture content of 4 per cent or more above the optimum of the 2.5 kg rammer compaction test. The tests with the 2.4 t tandem vibrating roller, an early prototype machine (Plate 7.8) were curtailed at a moisture content about 2.5 per cent below the 2.5 kg rammer optimum, reported to be the limit of rolling of the machine (Lewis, 1954). In contrast, the 3.9 t tandem vibrating roller operated to moisture contents of 5 per cent above the 2.5 kg rammer optimum (Figure 7.1) so the reported limit of rolling for the 2.4 t machine must be treated with some reserve. The results achieved with the 2.4 t machine on both the heavy clay and the silty clay (Figure 7.2) are surprisingly good for such an early design of vibrating roller and, weight for weight, good

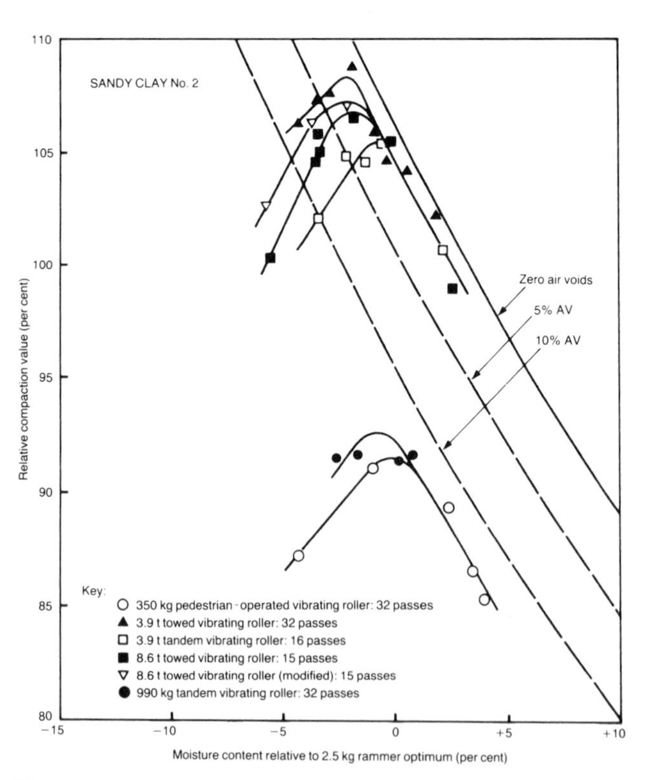

Figure 7.4 Relations between dry density and moisture content for vibrating smooth-wheeled rollers on sandy clay No 2, expressed in terms of relative compaction with the 2.5 kg rammer test. Layer thickness = 150 mm

Figure 7.5 Relations between dry density and moisture content for vibrating smooth-wheeled rollers on well-graded sand, expressed in terms of relative compaction with the 2.5 kg rammer test. Layer thickness = 150 mm

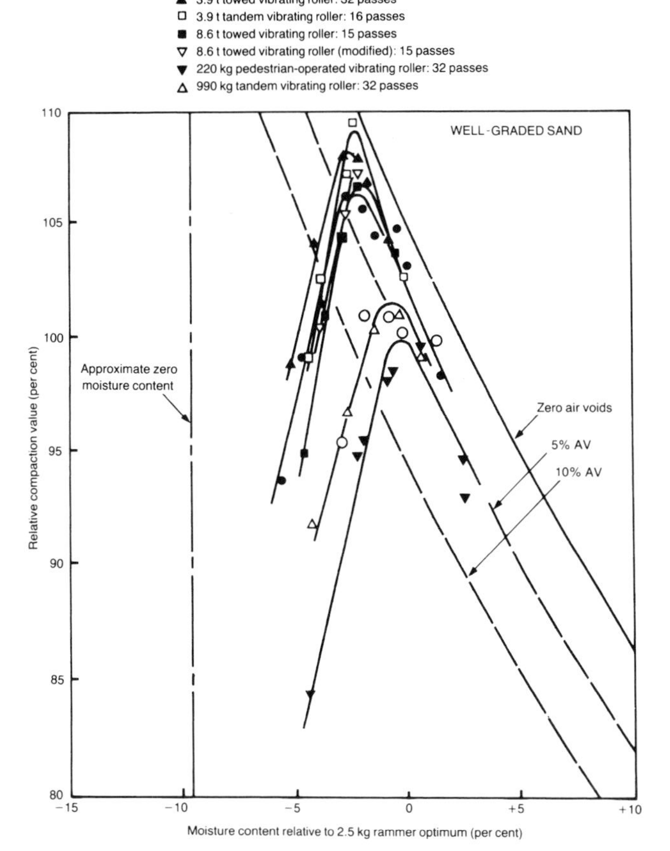

Figure 7.6 Relations between dry density and moisture content for vibrating smooth-wheeled rollers on gravel-sand-clay, expressed in terms of relative compaction with the 2.5 kg rammer test. Layer thickness = 150 mm

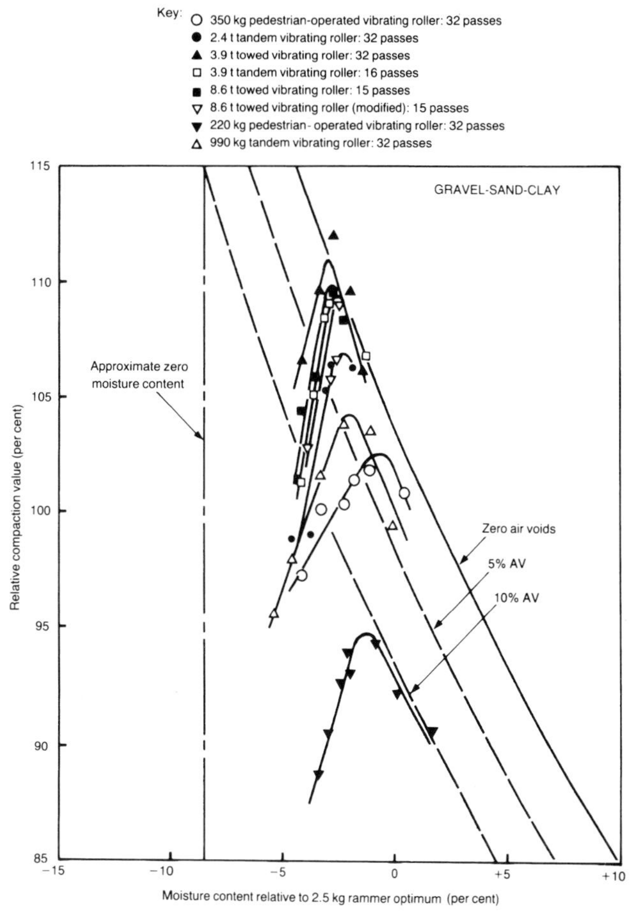

comparisons with the more modern machines were also obtained on well-graded sand and gravel-sand-clay (Figures 7.5 and 7.6).

7.20 The results shown for the sandy clay soils (Figures 7.3 and 7.4) indicate that the achievement of low air contents with the small vibrating rollers, 220 kg and 350 kg pedestrian-operated and the 990 kg tandem machines, was impossible at moisture contents less than 5 per cent above the 2.5 kg rammer optimum. These machines become effective, however, on the more granular soils (Figures 7.5 to 7.7).

7.21 The more effective vibrating rollers were the 3.9 t towed, 3.9 t tandem, and 8.6 t towed machines, ie the heaviest machines tested. Maximum values of relative compaction achieved ranged from 106 per cent on heavy clay (Figure 7.1) and 108 per cent on sandy clay No 2 (Figure 7.4) to around 110 per cent on well-graded sand and gravel-sand-clay (Figures 7.5 and 7.6). Maximum values of relative compaction with the heavier vibrating rollers on uniformly graded fine sand (Figure 7.7) were restricted by the overstressing of the surface of the layer (see Paragraph 2.16). The lower portions of the 300 mm layers compacted by the 3.9 t and 8.6 t towed vibrating rollers were at a significantly higher state of compaction than the overall average plotted in the Figure (the relations between relative compaction and depth in the layer given in Figure 7.23 and discussed later illustrate this); the absence of any significant overstressing at the surface of the 150 mm layer compacted by the 350 kg pedestrian-operated roller resulted in a maximum relative compaction value of 104 per cent (Figure 7.7).

7.22 The moisture content for the achievement of the maximum relative compaction value obtained with the most effective machines varied from 1 to 2 per cent below the 2.5 kg rammer optimum for cohesive soils to 2 to 3 per cent below that optimum for the well-graded sand and gravel-sand-clay. With the uniformly graded fine sand the maximum relative compaction value was achieved at about 1 per cent below the 2.5 kg rammer optimum moisture content, but the variation of relative compaction values over a considerable range of moisture contents was fairly small.

Figure 7.7 Relations between dry density and moisture content for vibrating smooth-wheeled rollers on uniformly graded fine sand, expressed in terms of relative compaction with the 2.5 kg rammer test

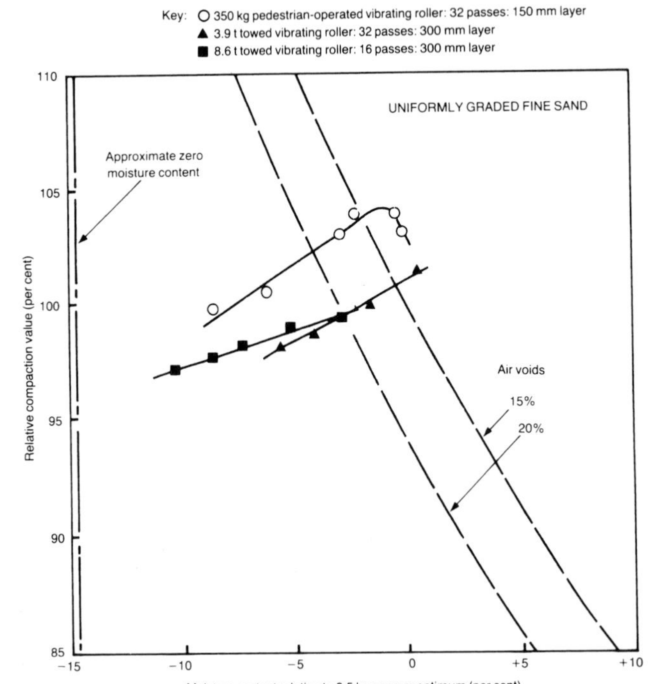

Table 7.2 Equations and correlation coefficients for the relations given in Figure 7.8

Soil	Regression equation*	Number of points	Correlation coefficient
Heavy clay	$w=80.9\ m^{-.274}-10$	6	0.96
Sandy clay No 2	$w=17.6\ m^{-.092}-10$	6	0.79
Clay soils	$c=61.0\ m^{.068}$	13	0.89
Well-graded sand	$w=17.7\ m^{-.101}-10$	9	0.85
Gravel-sand-clay	$w=14.7\ m^{-.087}-10$	8	0.82
Well-graded granular soils	$c=79.8\ m^{.038}$	17	0.85

*c =maximum dry density for the roller as percentage of the maximum dry density obtained in the 2.5 kg rammer compaction test

w=difference between optimum moisture content for the roller and optimum moisture content obtained in the 2.5 kg rammer compaction test (per cent)

m=mass per unit width of vibrating roll (kg/m)

The above results apply in general to 15 or more passes of the rollers on 150 mm layers

7.23 The dynamic forces are not available for all of the machines for which results are given in Figures 7.1 to 7.7. Thus the only parameter usable as an overall indicator of performance is the mass per unit width of vibrating roll. Figure 7.8 gives relations between maximum relative compaction value and mass per unit width of vibrating roll and between optimum moisture content (relative to the 2.5 kg rammer optimum) and the same parameter. Uniformly graded fine sand was not included in the regression analyses which were made to produce the relations. For the maximum relative compaction value soils were divided into two groups, clay soils and well-graded granular soils. To determine optimum moisture content for the machines, however, it was found necessary to treat each type of soil separately (Figure 7.8). The equations and correlation coefficients are given in Table 7.2.

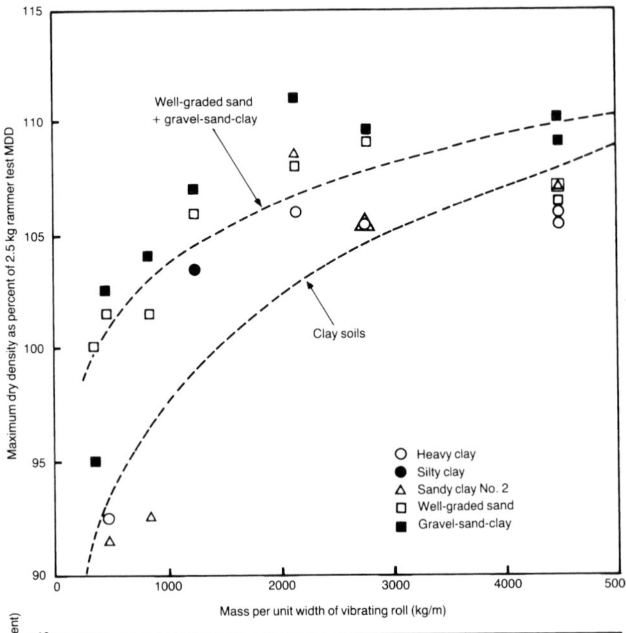

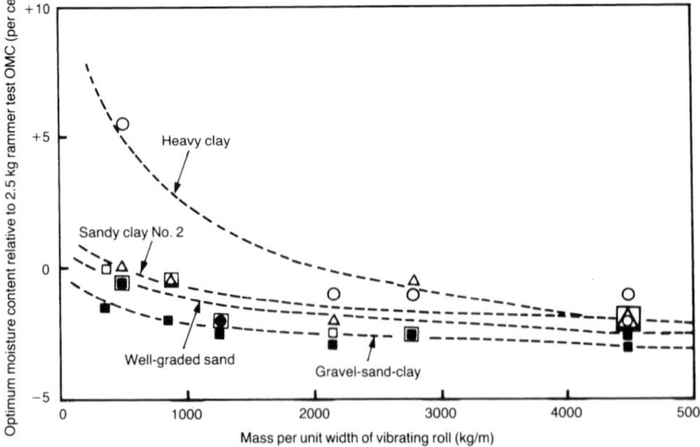

Figure 7.8 Maximum dry densities and optimum moisture contents related to the mass per unit width of vibrating roll for the vibrating smooth-wheeled rollers. Results from Figures 7.1 to 7.6

7.24 The trends of the relations in Figure 7.8 show that for the lighter machines the maximum states of compaction achievable with 15 or more passes on 150 mm layers were considerably higher with granular soils than with cohesive soils. The difference between the results for the two types of soil reduce as the machines become heavier. It can be concluded that machines with a mass per unit width of vibrating roll of less than 1000 kg/m are unsuitable for the compaction of clay soils.

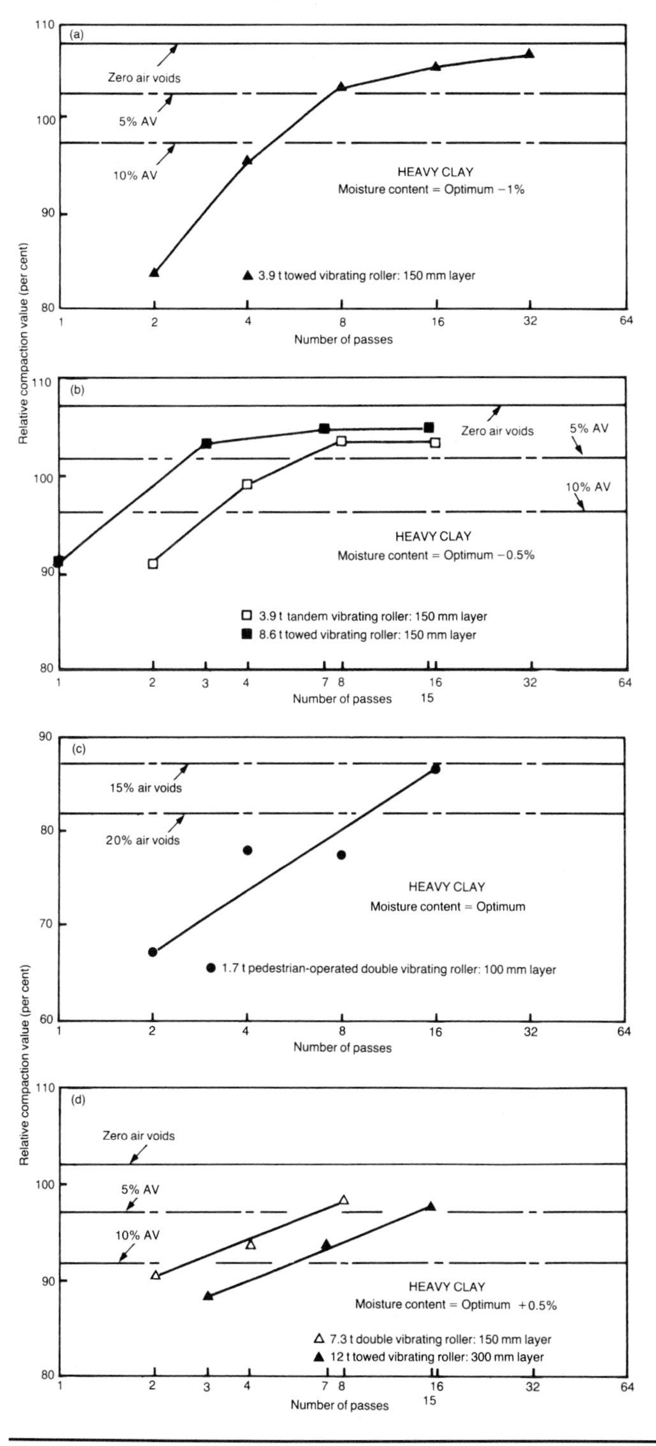

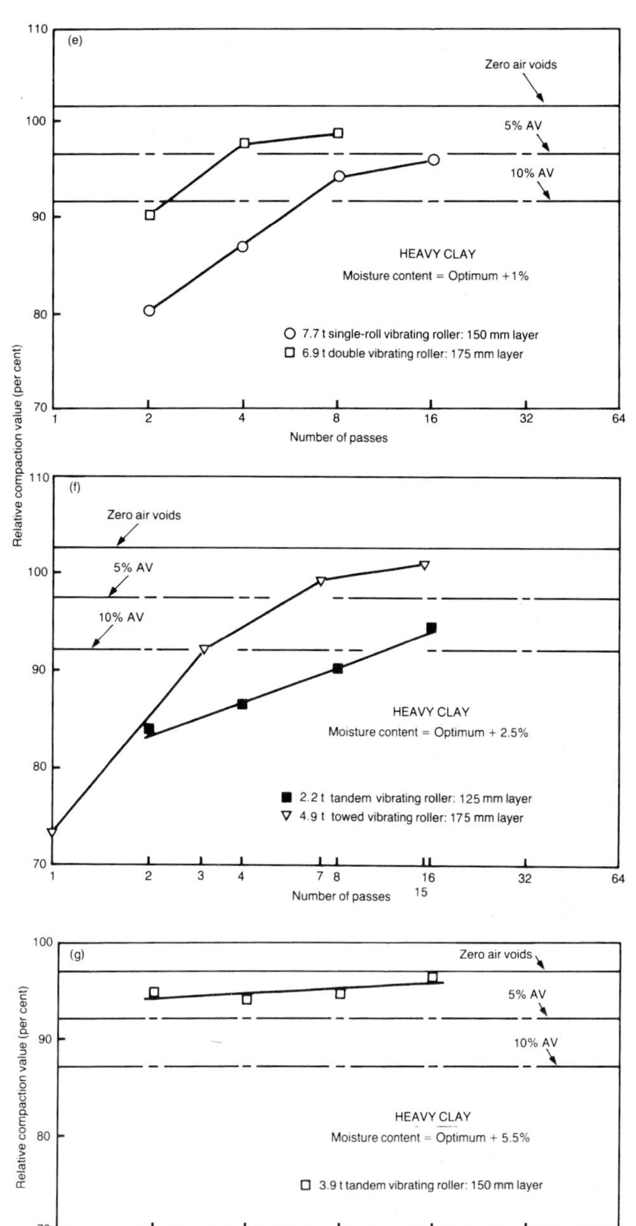

Figure 7.9 Relations between relative compaction and number of passes of vibrating smooth-wheeled rollers on heavy clay at seven different values of moisture content. All values are related to the maximum dry density and optimum moisture content obtained in the 2.5 kg rammer compaction test (Figure 3.13)

Relations between state of compaction produced and number of passes

7.25 Results of investigations to determine the effect on compaction of variations in the number of passes of vibrating smooth-wheeled rollers are given in Figures 7.9 to 7.13. As before, each soil type is allocated an individual figure number, whilst different values of moisture content are in separate graphs in each figure.

7.26 With the heavy clay soil the lightest machine tested was the 1.7 t pedestrian-operated double vibrating roller (Figure 7.9 (c)); poor levels of relative compaction, less than 90 per cent, and of air content, more than 15 per cent, were achieved at the test moisture content equal to the 2.5 kg rammer optimum, even after 16 passes with a layer thickness of 100 mm. With the same soil the 2.2 t tandem vibrating roller was tested (Figure 7.9(f)) and 95 per cent relative compaction would have been achieved in a 125 mm compacted layer after a little over 16 passes; however, the moisture content was at 2.5 per cent above the 2.5 kg rammer optimum.

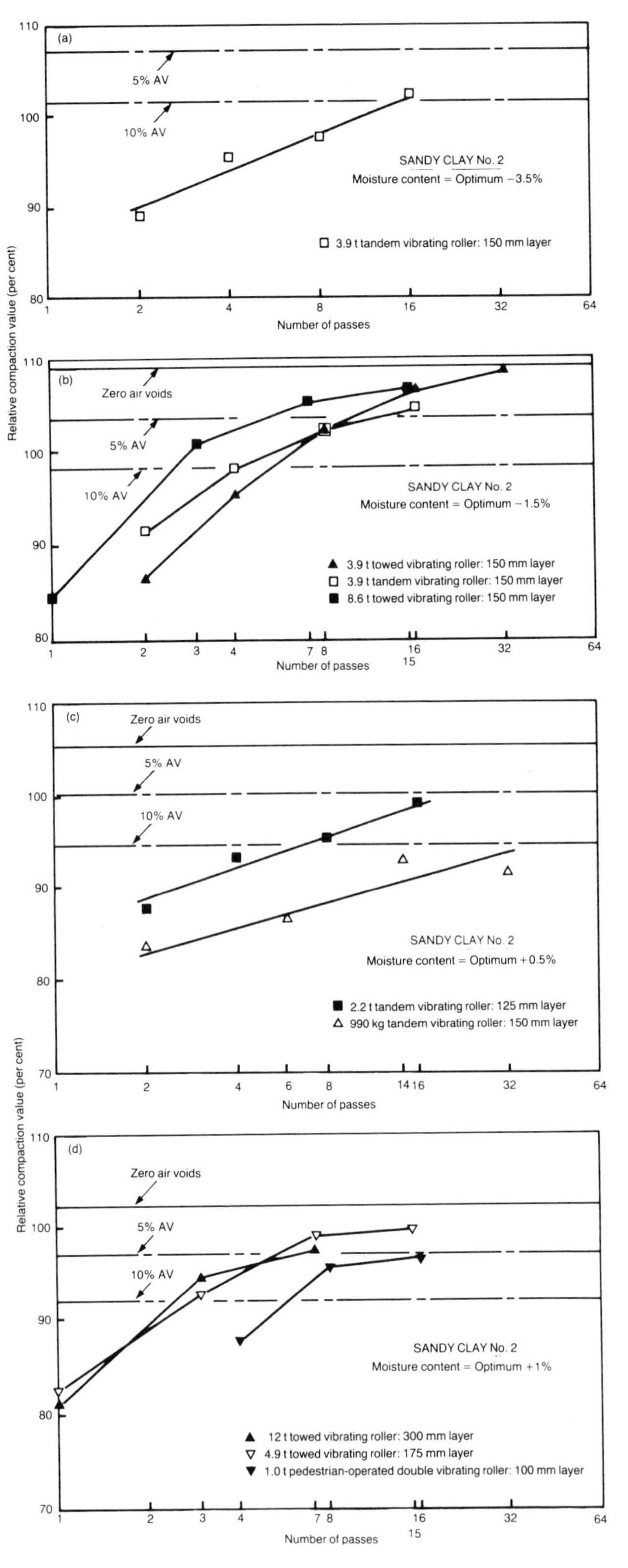

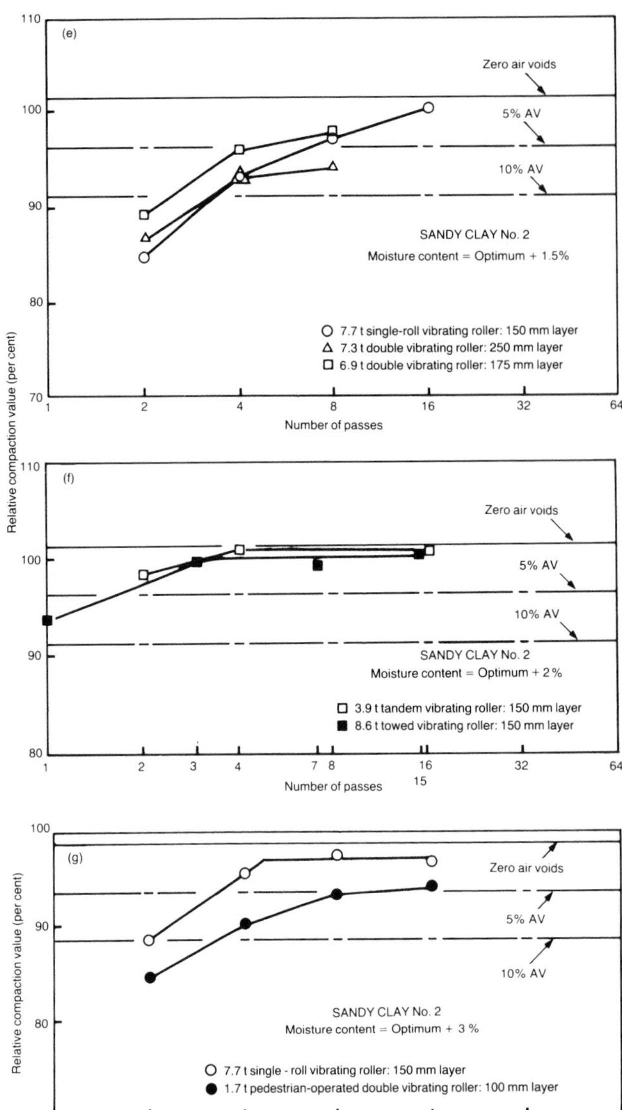

Figure 7.10 Relations between relative compaction and number of passes of vibrating smooth-wheeled rollers on sandy clay No 2 at seven different values of moisture content. All values are related to the maximum dry density and optimum moisture content obtained in the 2.5 kg rammer compaction test (Figure 3.13)

Other heavier machines all achieved good levels of compaction with the heavy clay soil (Figure 7.9).

7.27 With sandy clay No 2 the 2.2 t tandem vibrating roller achieved a satisfactory state of

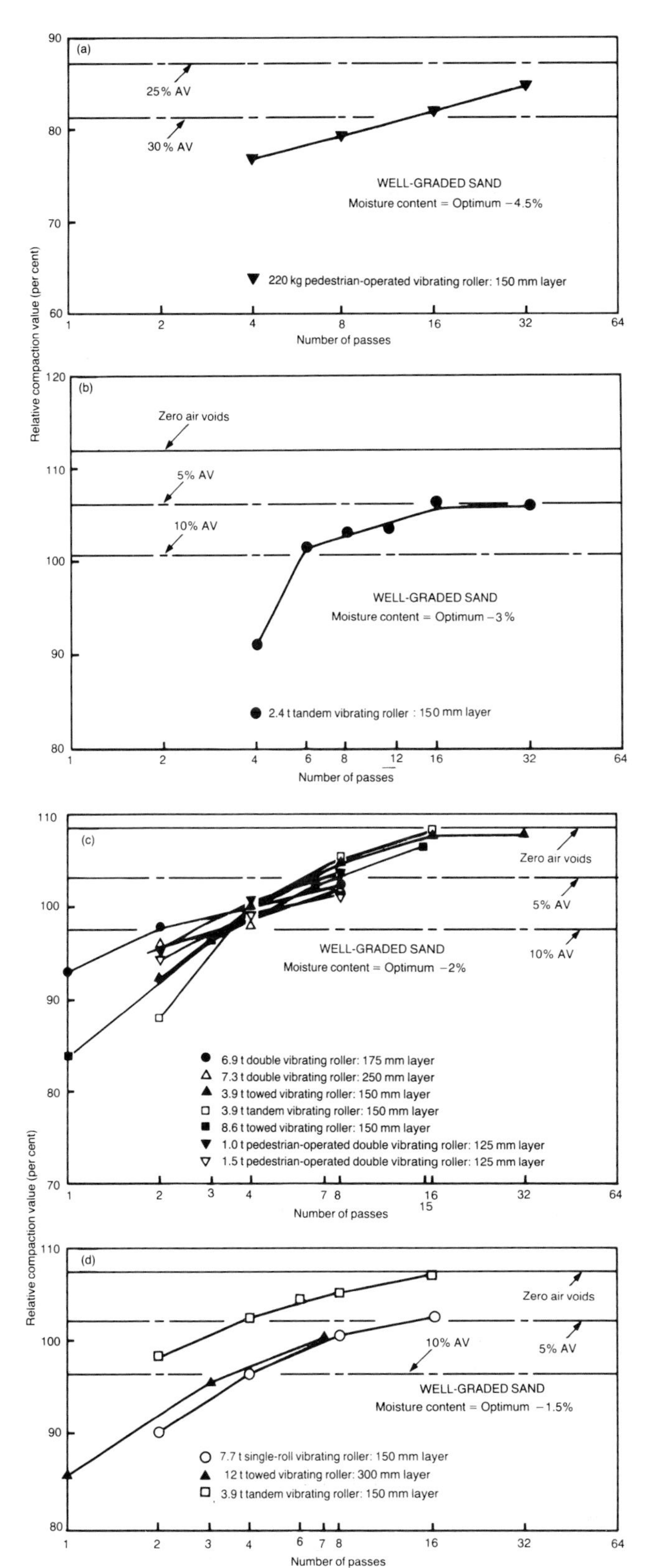

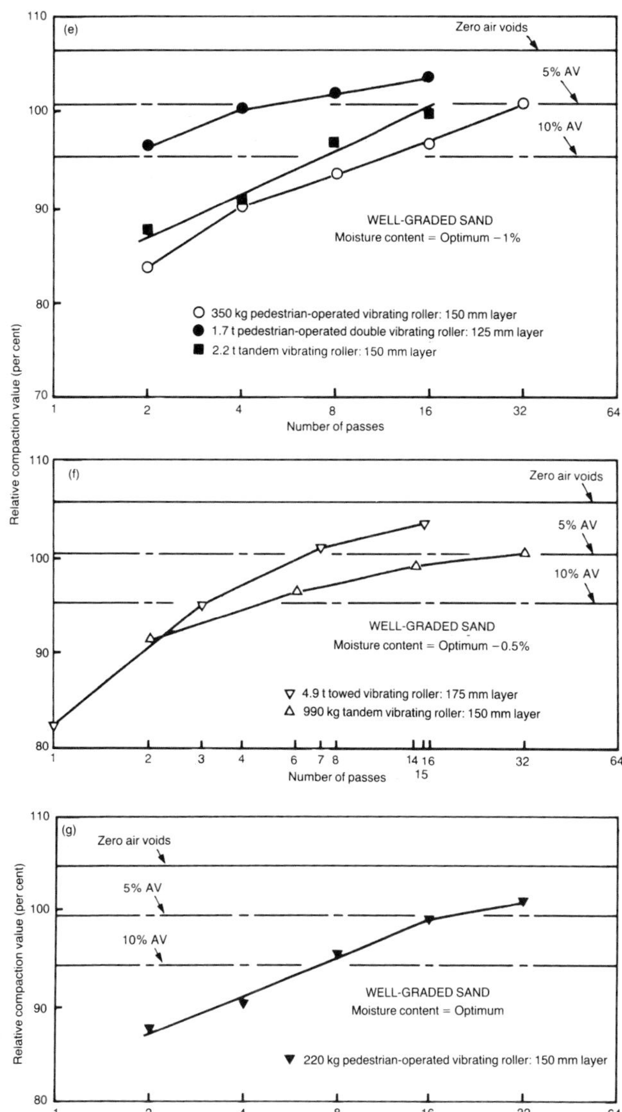

Figure 7.11 Relations between relative compaction and number of passes of vibrating smooth-wheeled rollers on well-graded sand at seven different values of moisture content. All values are related to the maximum dry density and optimum moisture content obtained in the 2.5 kg rammer compaction test (Figure 3.13)

compaction at a moisture content of 0.5 per cent above the 2.5 kg rammer optimum (Figure 7.10(c)), but lighter machines had difficulty in achieving a relative compaction value of 95 per cent (Figures 7.10(c), (d) and (g)).

7.28 With the well-graded sand and gravel-sand-clay (Figures 7.11 and 7.12) satisfactory levels of compaction were achieved by all machines. In the case of the 220 kg pedestrian-operated vibrating roller, however, tests were conducted on well-graded sand at two different

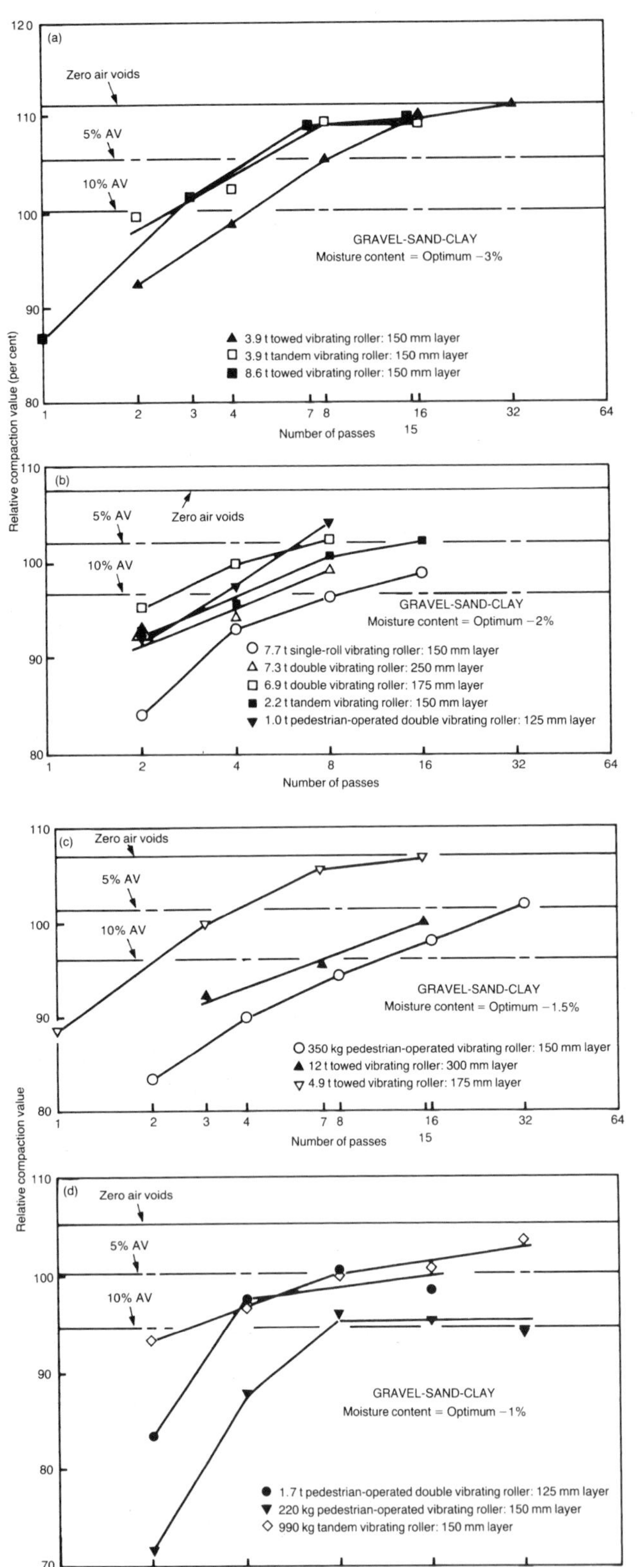

Figure 7.12 Relations between relative compaction and number of passes of vibrating smooth-wheeled rollers on gravel-sand-clay at four different values of moisture content. All values are related to the maximum dry density and optimum moisture content obtained in the 2.5 kg rammer compaction test (Figure 3.13)

moisture contents, at 2.5 kg rammer optimum minus 4.5 per cent (Figure 7.11(a)) and at 2.5 kg rammer optimum (Figure 7.11(g)); only in the latter instance was a relative compaction value of 95 per cent and an air content less than 10 per cent obtained.

7.29 Results for uniformly graded fine sand are given in Figure 7.13. This soil was much more susceptible to vibratory compaction than the other soils and in the majority of tests layer thicknesses in excess of 150 mm were used.

7.30 Using mass per unit width of vibrating roll as the readily available indicator of performance for all the machines, regression analyses were made of the results illustrated in Figures 7.1 to 7.6 and 7.9 to 7.13. For the cohesive and well-graded granular soils the analyses were restricted to the results pertaining to 150 mm compacted layers, but with the uniformly graded fine sand (Figure 7.13) the depth of layer was introduced as an additional factor. The results of the analyses are given in Figures 7.14 to 7.17 in the form of relations between the number of vibrating-roll passes required to achieve either 95 per cent relative compaction or 10 per cent air voids and mass per unit width of vibrating roll. It should be noted that with double vibrating rollers, two vibrating-roll passes are achieved with each pass of the machine. The equations and correlation coefficients obtained in the analyses are given in Table 7.3.

Table 7.3 Equations and correlation coefficients for the relations given in Figures 7.14 to 7.17

Soil	Level*	Regression equation†	No of data points	Correlation coefficient
Clay soils	(a)	P=6.35–1.68 m–.0560 w	18	0.92
	(b)	P=7.30–1.94 m–.163 w	22	0.92
Well-graded sand	(a)	P=3.44–.993 m–.232 w	19	0.91
	(b)	P=3.74–1.09 m–.350 w	17	0.90
Gravel-sand-clay	(a)	P=2.14–.613 m–.207 w	14	0.90
	(b)	P=3.27–1.12 m–.423 w	14	0.92
Uniformly graded fine sand	(a)	P=–2.05–.571 m–.00289 w +1.87 d	10	0.95

*Level (a)=95 per cent of the maximum dry density obtained in the 2.5 kg rammer compaction test
Level (b)=10 per cent air voids
†P=Log_{10} (Number of vibrating roll passes required)
m=Log_{10} (Mass per unit width of vibrating roll in kg/m)
w=Moisture content of soil, per cent, minus the optimum moisture content, per cent, obtained in the 2.5 kg rammer compaction test
d=Log_{10} (Thickness of compacted layer in mm)

7.31 The analyses showed that the results for all the clay soils could be taken together to produce the single set of relations given in Figure 7.14, whereas the granular soils each had to be taken separately. The good correlation coefficients with each of the regression equations given in Table 7.3 were obtained after exclusion of one or at most two of the more erratic results from each set of data. These high levels of correlation confirm that Figures 7.14 to 7.17 can be used with confidence to predict the likely number of vibrating-roll passes required to achieve the levels of compaction chosen as criteria in the analyses; they also support the view that mass per unit width of vibrating roll is a good indicator of the performance of vibrating smooth-wheeled rollers.

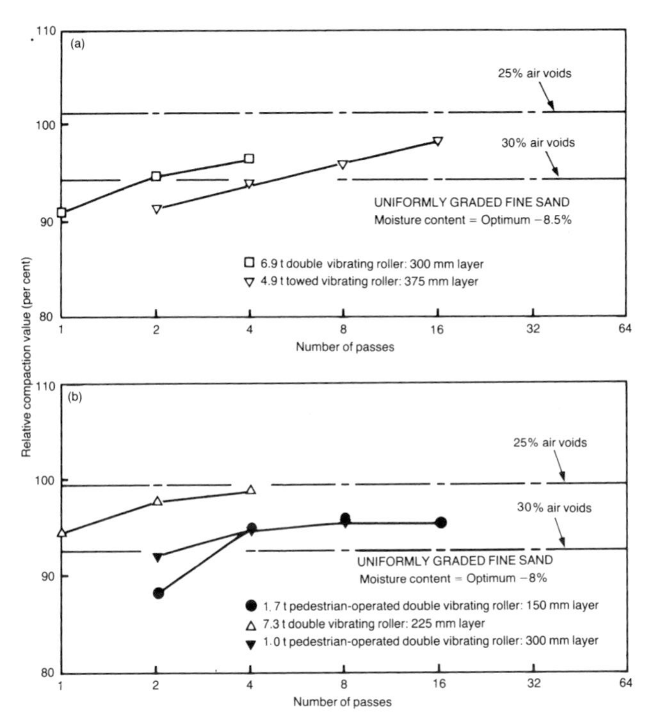

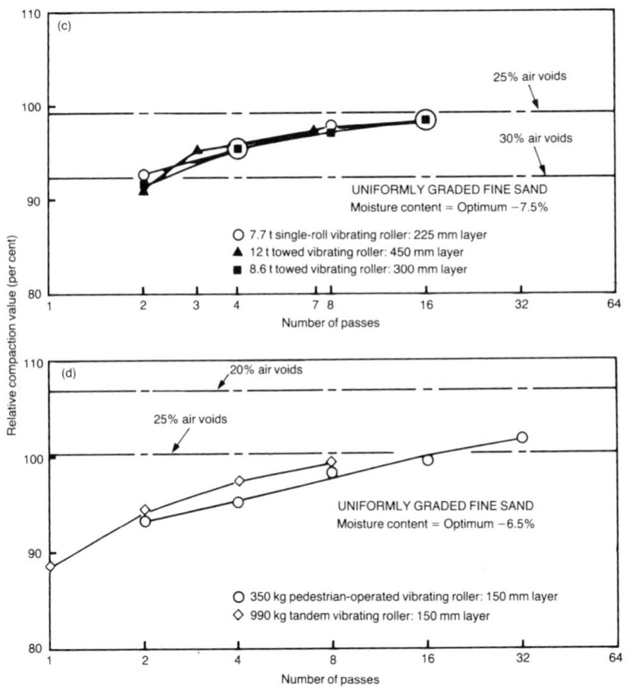

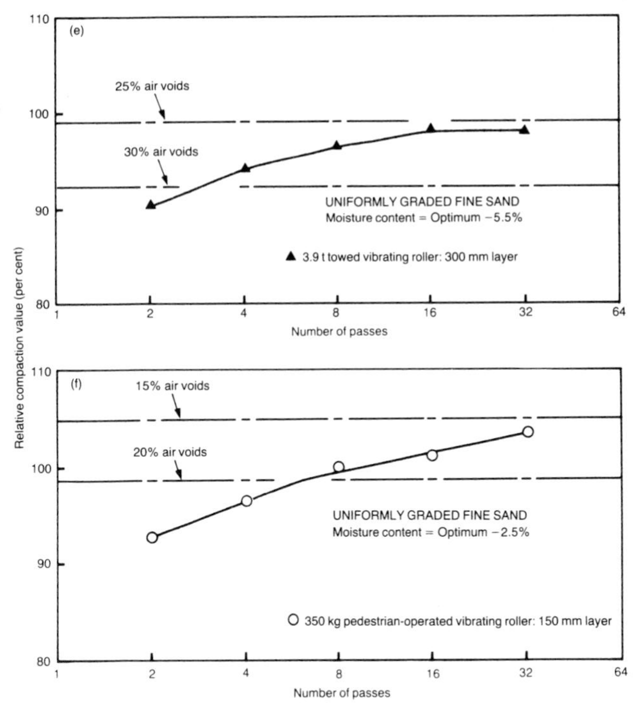

Figure 7.13 Relations between relative compaction and number of passes of vibrating smooth-wheeled rollers on uniformly graded fine sand at six different values of moisture content. All values are related to the maximum dry density and optimum moisture content obtained in the 2.5 kg rammer compaction test (Figure 3.13)

Relations between state of compaction and depth within the compacted layer

7.32 Results of tests to determine the variation of dry density throughout the depth of layers compacted by vibrating smooth-wheeled rollers are given in Figures 7.18 to 7.23. The results have been arranged in a similar manner regarding soil type and moisture content to that used in illustrating the relations between state of compaction and number of passes; dry density has been expressed as a relative compaction value, ie as a percentage of the maximum dry density achieved in the 2.5 kg rammer compaction test.

7.33 There are basically two sets of results contained within Figures 7.18 to 7.23. One set concerns tests carried out before 1962, in which a relatively large number of passes of each

Figure 7.14 Vibrating smooth-wheeled rollers on clay soils. Relations between number of vibrating-roll passes required and mass per unit width of vibrating roll to achieve (a) 95 per cent of the maximum dry density obtained in the 2.5 kg rammer compaction test and (b) 10 per cent air voids. These graphs were derived by regression analyses of the results for 150 mm compacted layers given in Figures 7.1, 7.3, 7.4, 7.9 and 7.10

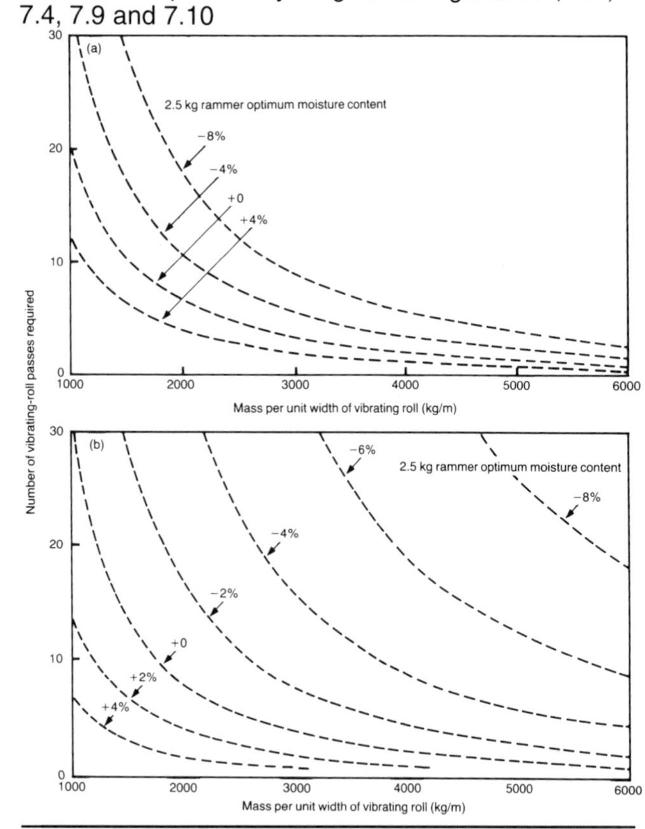

Figure 7.15 (a) and (b) See main caption

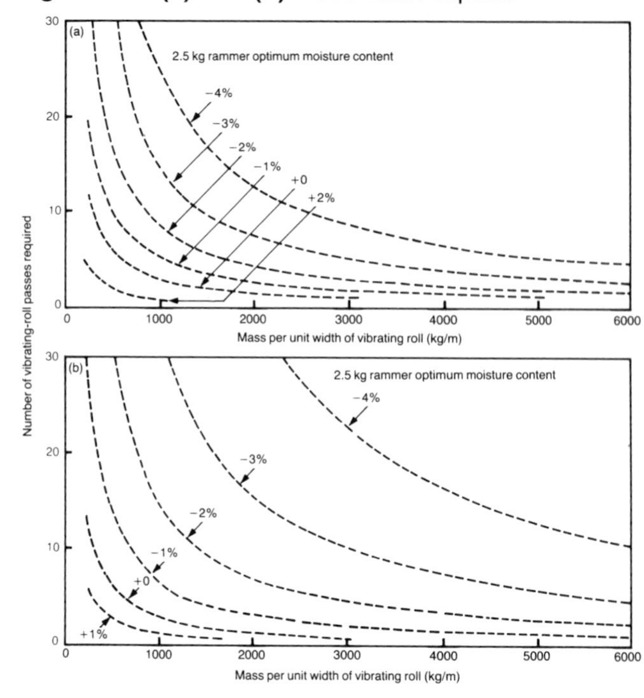

Figure 7.16 (a) and (b) See main caption

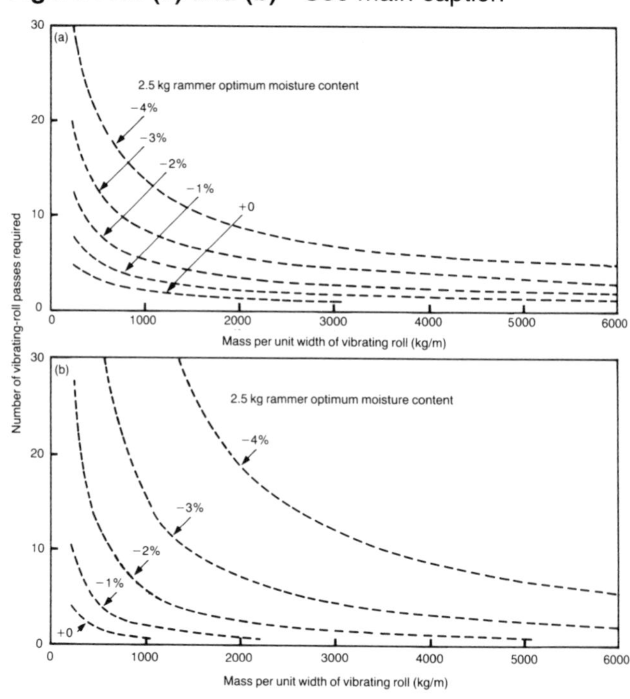

Figure 7.17 See main caption

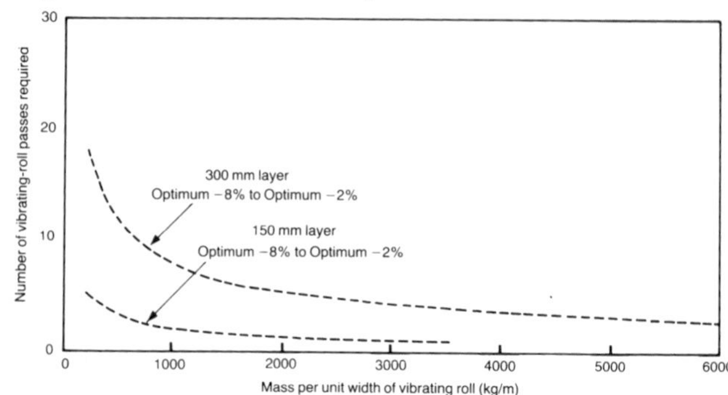

Figure 7.15 Vibrating smooth-wheeled rollers on well-graded sand. Relations between number of vibrating-roll passes required and mass per unit width of vibrating roll to achieve (a) 95 per cent of the maximum dry density obtained in the 2.5 kg rammer compaction test and (b) 10 per cent air voids. These graphs were derived by regression analyses of the results for 150 mm compacted layers given in Figures 7.5 and 7.11

Figure 7.16 Vibrating smooth-wheeled rollers on gravel-sand-clay. Relations between number of vibrating-roll passes required and mass per unit width of vibrating roll to achieve (a) 95 per cent of the maximum dry density obtained in the 2.5 kg rammer compaction test and (b) 10 per cent air voids. These graphs were derived by regression analyses of the results for 150 mm compacted layers given in Figures 7.6 and 7.12

Figure 7.17 Vibrating smooth-wheeled rollers on uniformly graded fine sand. Relations between number of vibrating-roll passes required and mass per unit width of vibrating roll to achieve 95 per cent of the maximum dry density obtained in the 2.5 kg rammer compaction test. These graphs were derived by regression analyses of the results given in Figure 7.13

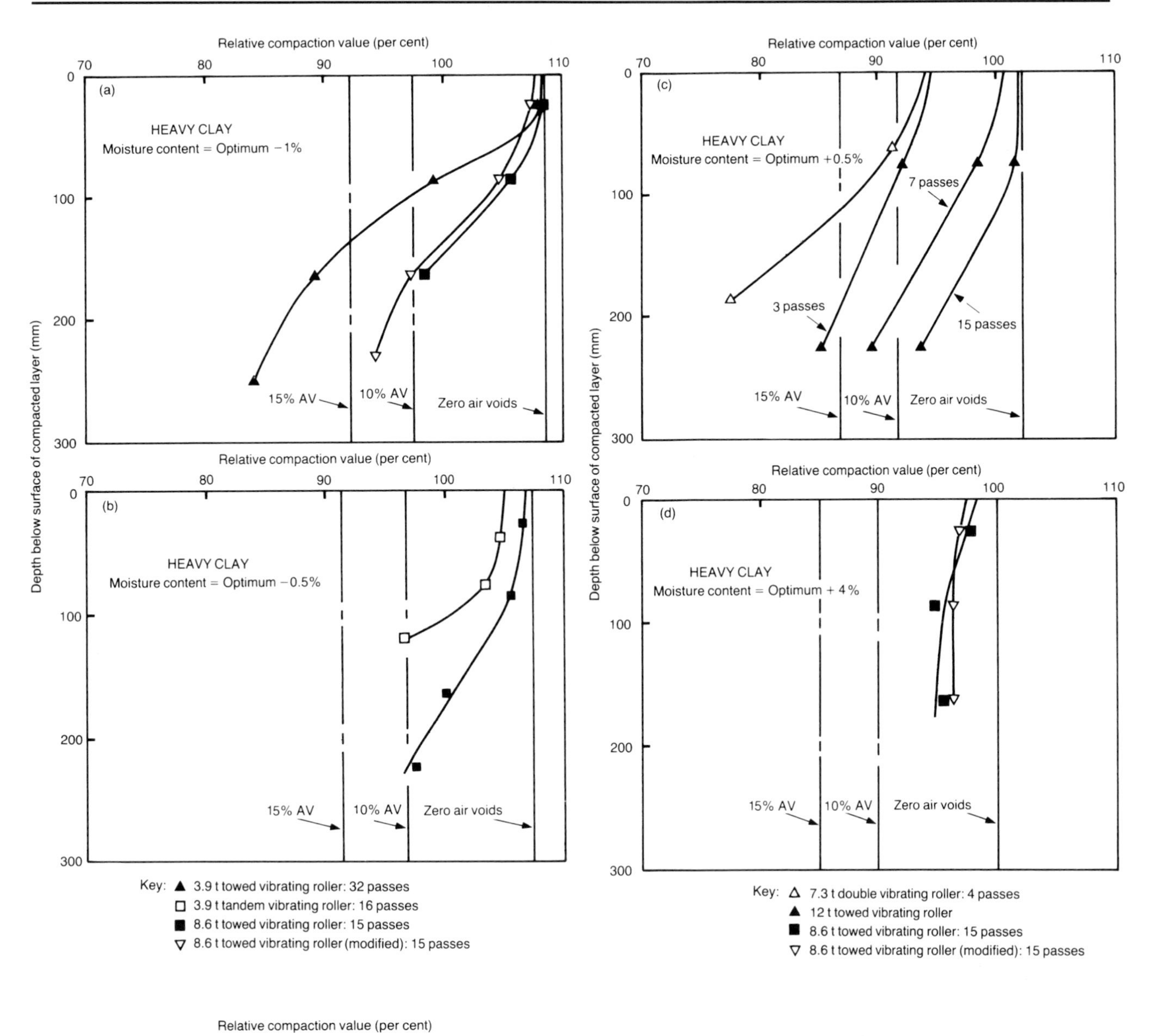

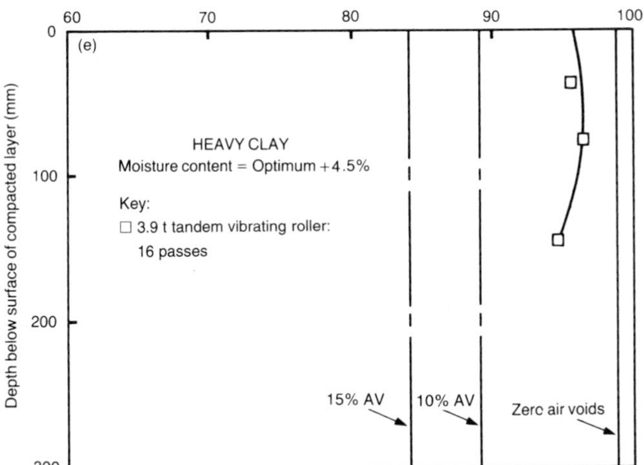

Figure 7.18 Relations between relative compaction and depth in the compacted layer of heavy clay obtained with vibrating smooth-wheeled rollers. All values are related to the maximum dry density and optimum moisture content obtained in the 2.5 kg rammer compaction test (Figure 3.13)

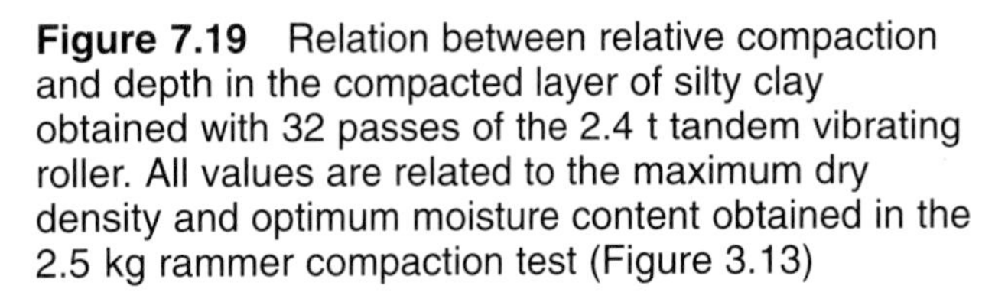

Figure 7.19 Relation between relative compaction and depth in the compacted layer of silty clay obtained with 32 passes of the 2.4 t tandem vibrating roller. All values are related to the maximum dry density and optimum moisture content obtained in the 2.5 kg rammer compaction test (Figure 3.13)

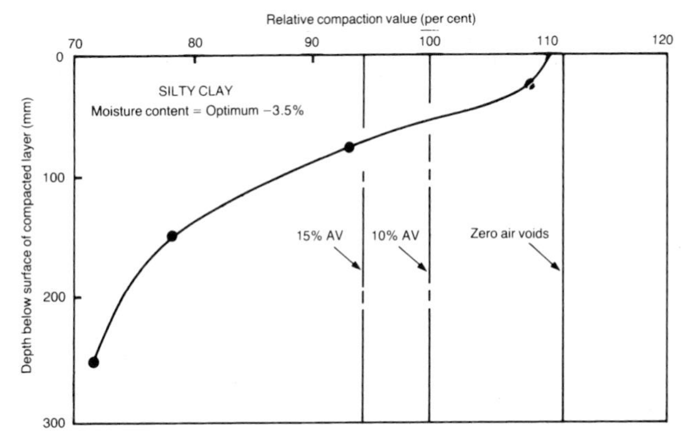

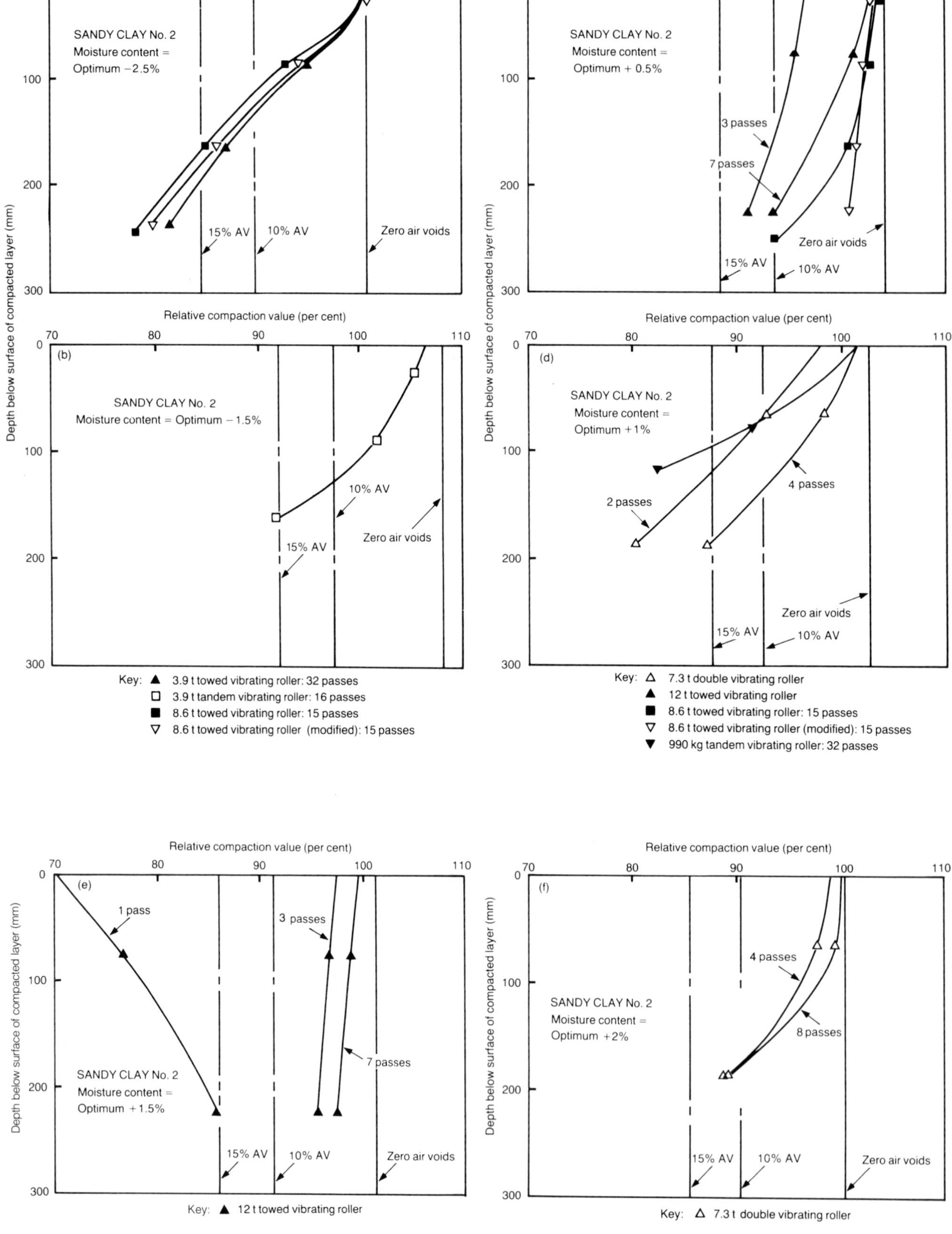

Figure 7.20 Relations between relative compaction and depth in the compacted layer of sandy clay No 2 obtained with vibrating smooth-wheeled rollers. All values are related to the maximum dry density and optimum moisture content obtained in the 2.5 kg rammer compaction test (Figure 3.13)

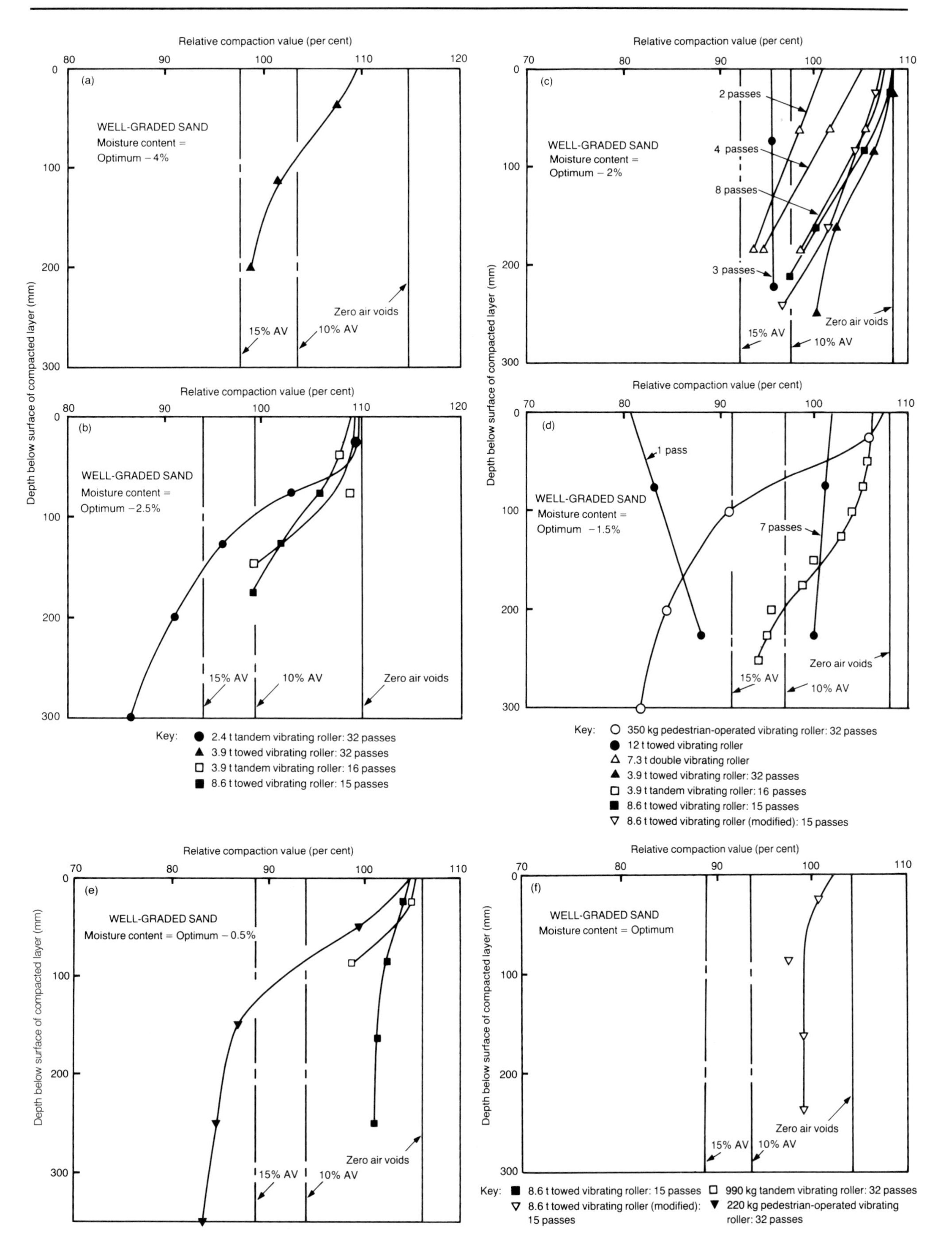

Figure 7.21 Relations between relative compaction and depth in the compacted layer of well-graded sand obtained with vibrating smooth-wheeled rollers. All values are related to the maximum dry density and optimum moisture content obtained in the 2.5 kg rammer compaction test (Figure 3.13)

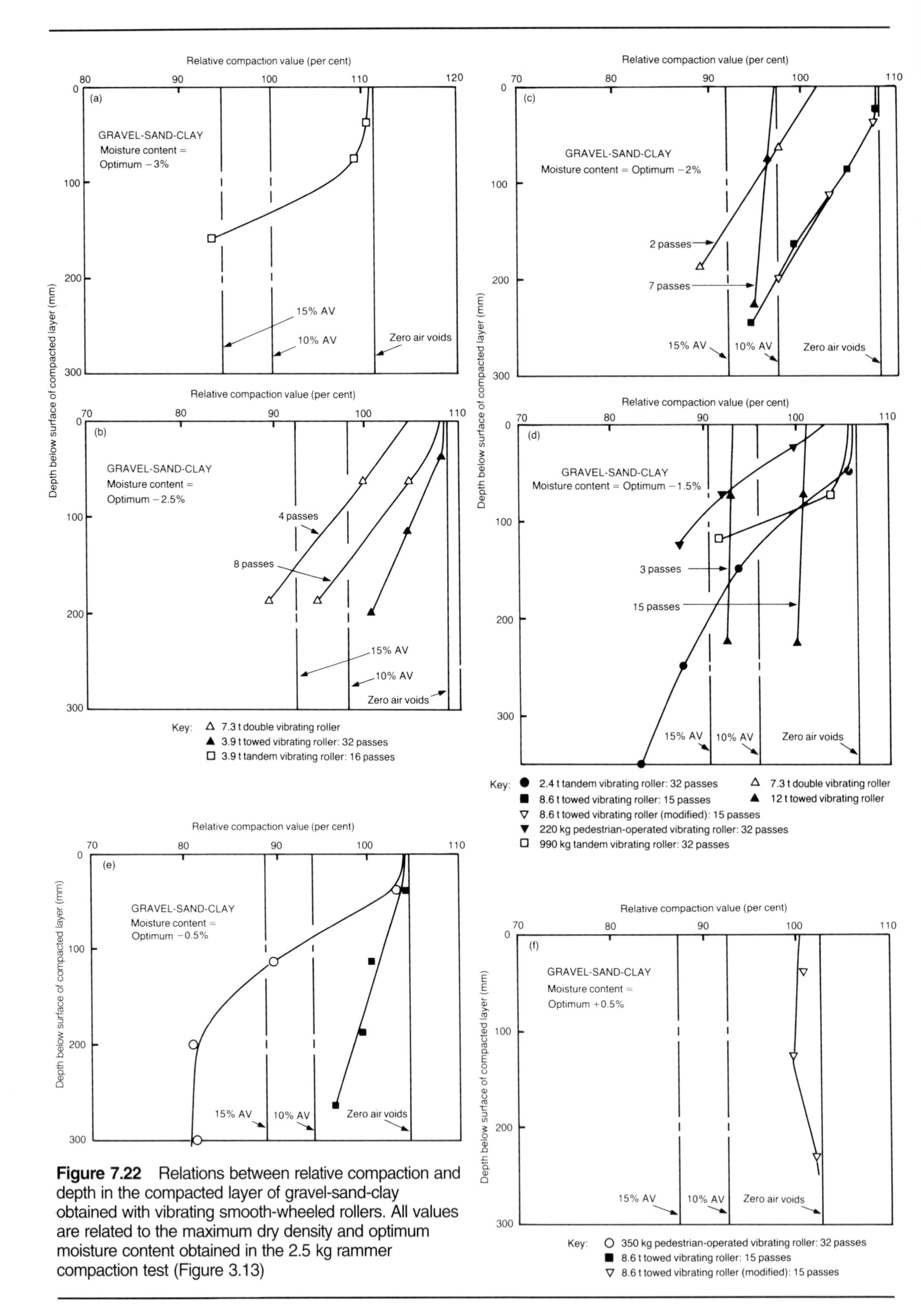

Figure 7.22 Relations between relative compaction and depth in the compacted layer of gravel-sand-clay obtained with vibrating smooth-wheeled rollers. All values are related to the maximum dry density and optimum moisture content obtained in the 2.5 kg rammer compaction test (Figure 3.13)

Figures 7.23 (a) and (b) See caption below Figure 7.23 (e) on page 96

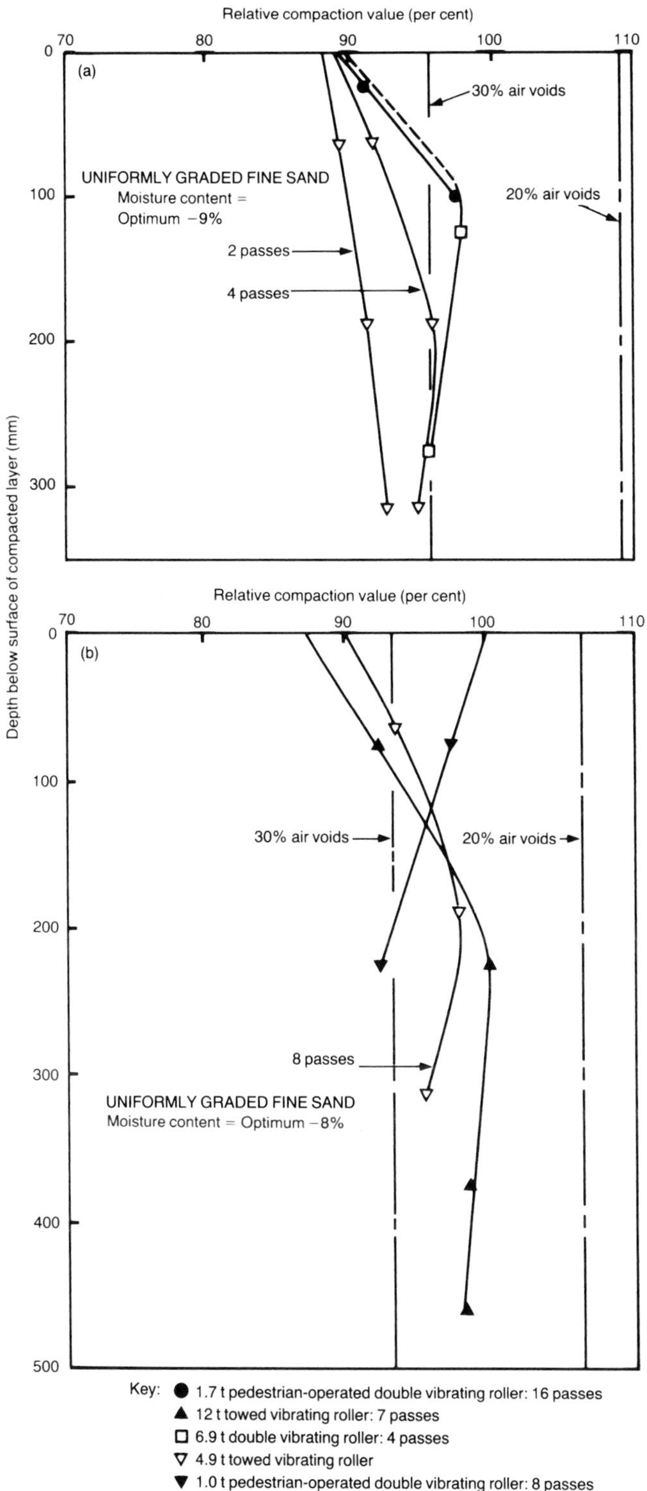

Figures 7.23 (c) and (d) See caption below Figure 7.23 (e) on page 96

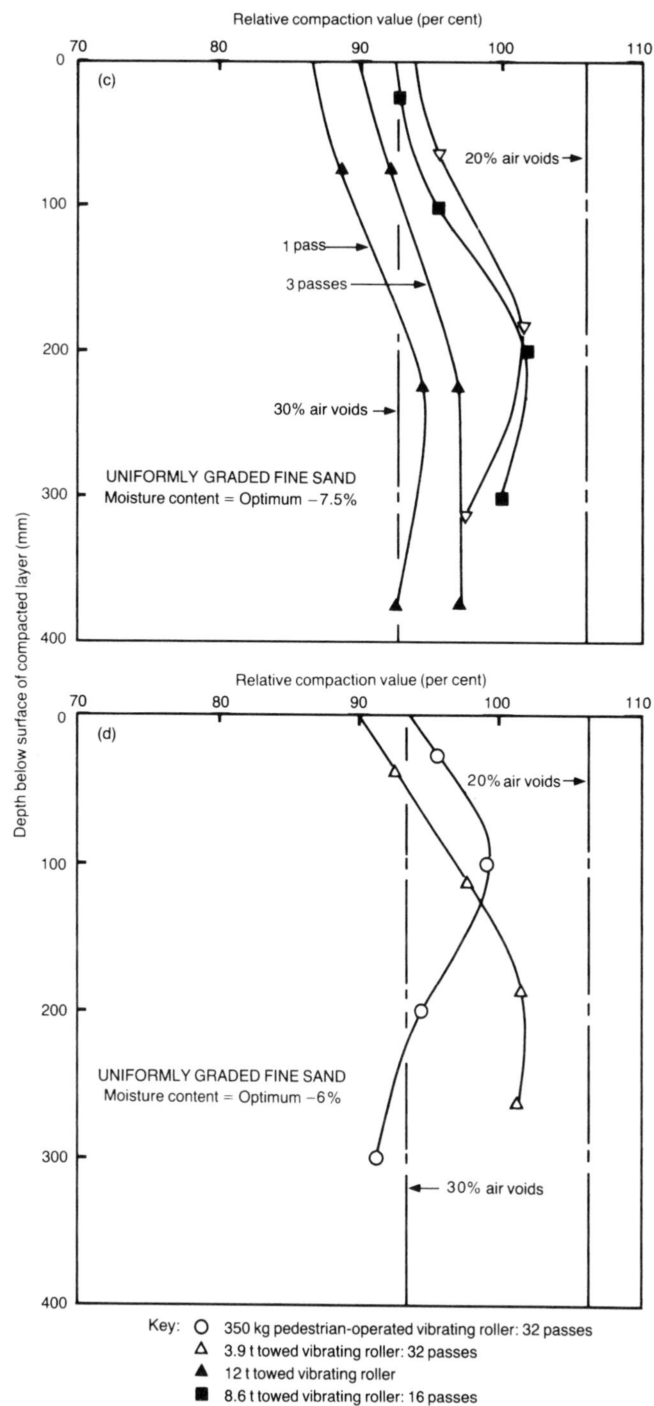

Figure continued overleaf

machine (at least 15) was used, and determinations of the state of compaction were made over relatively small increments of depth, 50 to 100 mm. The second set, concerning investigations made since 1962, are derived from tests to determine relations between dry density and number of passes, in which layers thicker than 200 mm were used; this applied generally to the more effective machines. The determination of the in-situ dry density, using the sand-replacement method (British Standards Institution, 1990), was made in these deeper layers, with all soils except the uniformly graded fine sand, in two depth increments, thus

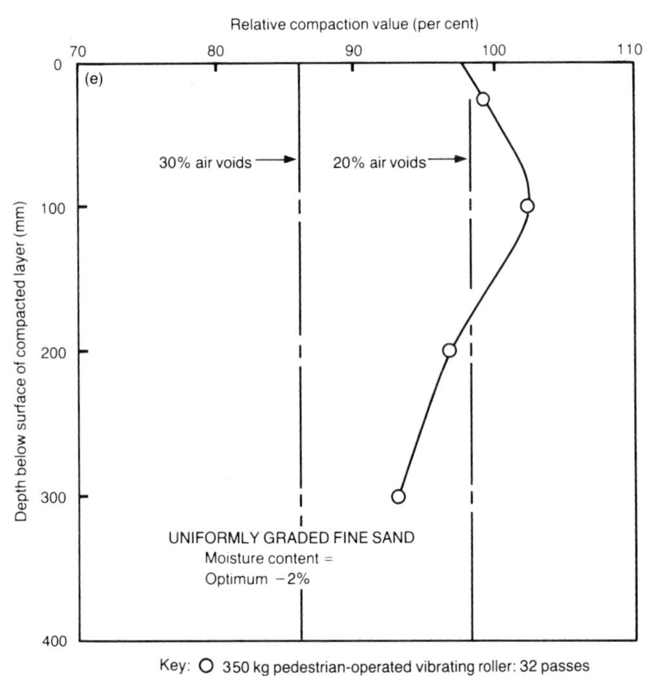

Figure 7.23 Relations between relative compaction and depth in the compacted layer of uniformly graded fine sand obtained with vibrating smooth-wheeled rollers. All values are related to the maximum dry density and optimum moisture content obtained in the 2.5 kg rammer compaction test (Figure 3.13) (Note Figure 7.23 (a), (b), (c) and (d) appear on page 95.)

providing a comparison between the state of compaction in the top and bottom halves of the layer for different numbers of passes. The machines to which this method applied in Figures 7.18 to 7.22 were the 7.3 t double vibrating roller and the 12 t towed vibrating roller. With the uniformly graded fine sand (Figure 7.23) the more recent tests included two or three depth increments at various numbers of passes.

7.34 Using the same criteria as in earlier chapters to determine the thicknesses of layer capable of being compacted, ie the achievement of 90 per cent relative compaction or 15 per cent air voids at the bottom of the layer, the thickness has been related to the mass per unit width of vibrating roll and moisture content. Only results for 15 or more vibrating-roll passes were included in the regression analyses so that the relations so determined can be considered to yield the maximum value for thickness of layer likely to be achieved. The results are illustrated in Figures 7.24 to 7.26, and the regression equations and correlation coefficients are given in Table 7.4.

7.35 It was found that the soils could be formed into three groups, clay soils, well-graded

Table 7.4 Equations and correlation coefficients for the relations given in Figures 7.24 to 7.26

Soil	Level*	Regression equation†	No of data points	Correlation coefficient
Clay soils	(a)	D=-.061+.679 m+.0156 w	14	0.96
	(b)	D=.009+.656 m+.0490 w	15	0.95
Well-graded granular soils	(a)	D=.807+.464 m-.0033 w	23	0.92
	(b)	D=.802+.473 m+.0354 w	24	0.93
Uniformly graded fine sand	(a)	D=2.09+.184 m+.0208 w	6	0.91

*Level (a)=Achievement at the bottom of the layer of 90 per cent of the maximum dry density obtained in the 2.5 kg rammer test

(b)=Achievement at the bottom of the layer of 15 per cent air voids

†D=Log_{10} (Maximum depth of layer to achieve stated level of compaction in mm)

m=Log_{10} (Mass per unit width of vibrating roll in kg/m)

w=Moisture content of soil, per cent, minus the optimum moisture content, per cent, obtained in the 2.5 kg rammer compaction test

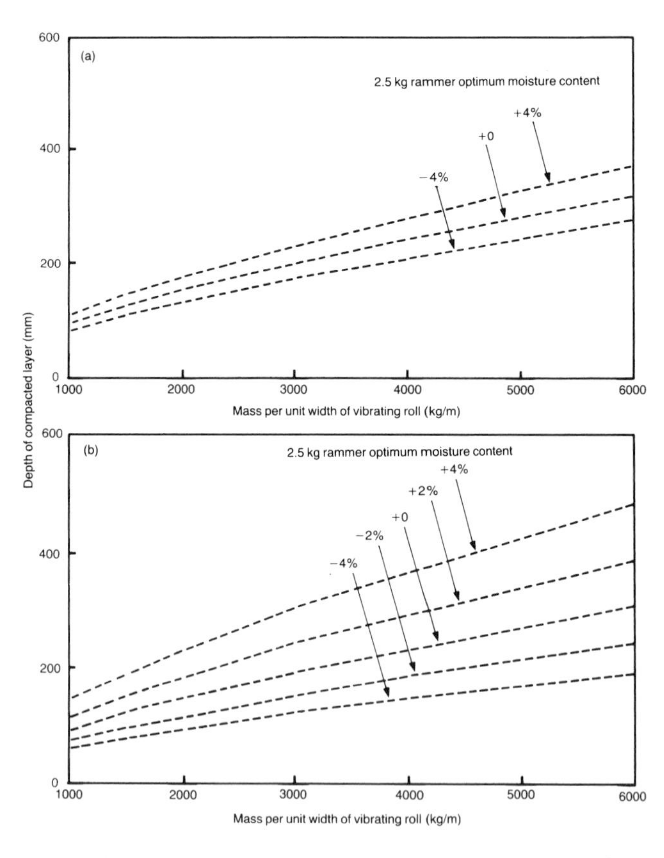

Figure 7.24 Relations between depth of layer and mass per unit width of vibrating roll for clay soils compacted to a standard of (a) 90 per cent relative compaction and (b) 15 per cent air voids at the bottom of the layer. These graphs were derived by regression analyses of the results in Figures 7.18 to 7.20 for 15 or more vibrating-roll passes

granular soils comprising well-graded sand and gravel-sand-clay, and uniformly graded fine sand. In all cases the correlations were very good, with correlation coefficients ranging from 0.91 to 0.96 (Table 7.4). This is further evidence to support the view that mass per unit width of vibrating roll is a good guide to the likely performance of this type of machine.

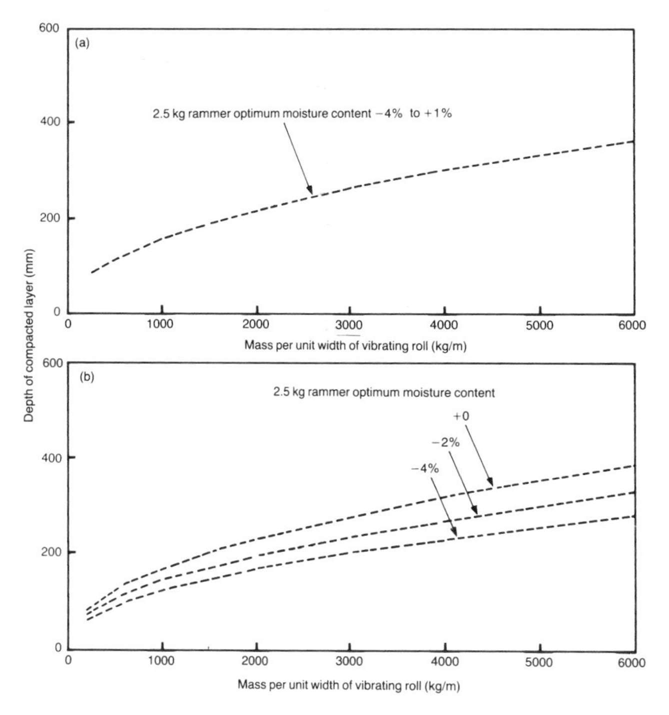

Figure 7.25 Relations between depth of layer and mass per unit width of vibrating roll for well-graded sand and gravel-sand-clay compacted to a standard of (a) 90 per cent relative compaction and (b) 15 per cent air voids at the bottom of the layer. These graphs were derived by regression analyses of the results in Figures 7.21 and 7.22 for 15 or more vibrating-roll passes

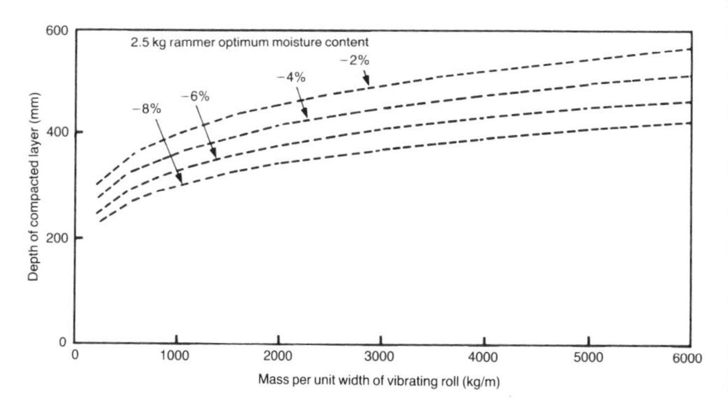

Figure 7.26 Relations between depth of layer and mass per unit width of vibrating roll for uniformly graded fine sand compacted to achieve 90 per cent relative compaction at the bottom of the layer. These relations were derived by regression analyses of the results in Figure 7.23 for 15 or more vibrating-roll passes

7.36 For clay soils (Figure 7.24) it can be assumed that with masses per unit width of vibrating roll less than 1000 kg/m, the depth of compacted layer will reduce very rapidly and this roll loading (1000 kg/m) might be considered the practical minimum for this type of soil. The apparent anomaly of a small negative factor from the effect of moisture content for well-graded granular soils, level (a), in Table 7.4 is indicative of the negligible effect of moisture content on the results using the criterion of 90 per cent relative compaction at the bottom of the layer. A single relation is shown in Figure 7.25(a) to represent the results for moisture contents from 4 per cent below to 1 per cent above the 2.5 kg rammer optimum. The potential for compacting thicker layers of uniformly graded fine sand is clearly illustrated by comparing the relations in Figure 7.26 with those in Figures 7.24 and 7.25; however, the capability to traffic the thicker layers may be severely restricted in the case of self-propelled vibrating rollers (see Paragraph 7.72).

7.37 The results portrayed in Figures 7.24 to 7.26 can be regarded as providing useful guidance on the potential maximum thickness of layer which can be compacted by vibrating smooth-wheeled rollers when at least 15 vibrating-roll passes are applied.

Relations between the average state of compaction in the layer and the total depth of the layer

7.38 A few tests were carried out with vibrating smooth-wheeled rollers using a relatively small number of passes, either 2 or 4, to determine relations between the average state of compaction in a compacted layer and the thickness of the layer. The results of these tests are given in Figures 7.27 to 7.30. All items of plant tested in this way were double vibrating rollers, so the number of vibrating-roll passes would be double the number of machine passes quoted on the figures. Nonetheless, the total number of passes are well below the minimum of 15 vibrating roll passes to which Figures 7.24 to 7.26 apply. As the results are in terms of the average relative compaction values through the layers, criteria for assessing the depth of layer capable of being adequately compacted must be the same as those used in determining the number of vibrating roll passes required (see Paragraph 7.30), ie either a relative compaction value of 95 per cent or an air content of 10 per cent, although the latter criterion cannot be applied to uniformly graded fine sand.

7.39 In view of the small number of results no statistical analyses could be made of the data. The thicknesses of layer capable of being compacted by the machines to the criteria given above, determined from Figures 7.27 to 7.30, are summarised in Table 7.5. Only the result for the lower of the two speeds of travel of the 6.9 t double vibrating roller (Figure 7.29) has been included in Table 7.5, but the difference between the relations for that machine at the two speeds of 2.2 km/h and 1.5 km/h is a good illustration of the effect of the speed of rolling on the performance of this type of machine (see Paragraph 7.47).

Figure 7.27 Relation between the average relative compaction of the compacted layer and its total thickness for 4 passes of the 7.3 t double vibrating roller on heavy clay

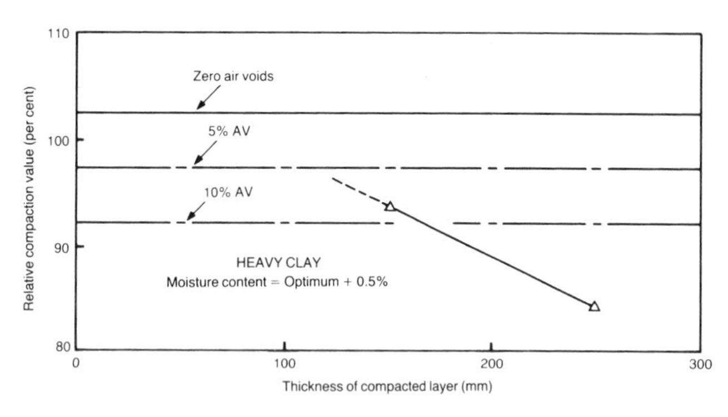

Figure 7.28 Relation between the average relative compaction of the compacted layer and its total thickness for 2 passes of the 7.3 t double vibrating roller on sandy clay No 2

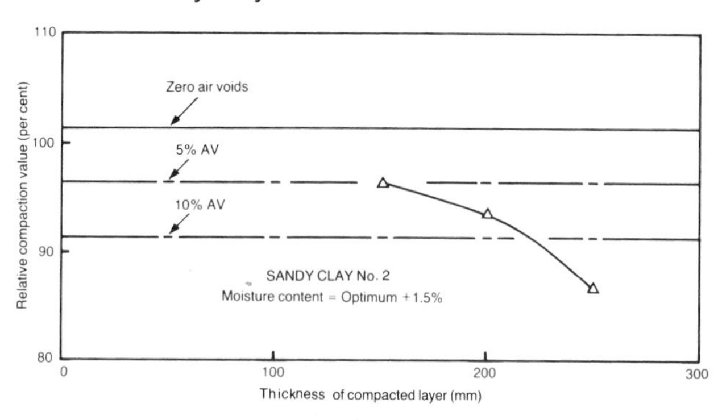

Figure 7.29 Relations between the average relative compaction of the compacted layer and its total thickness for vibrating smooth-wheeled rollers on well-graded sand

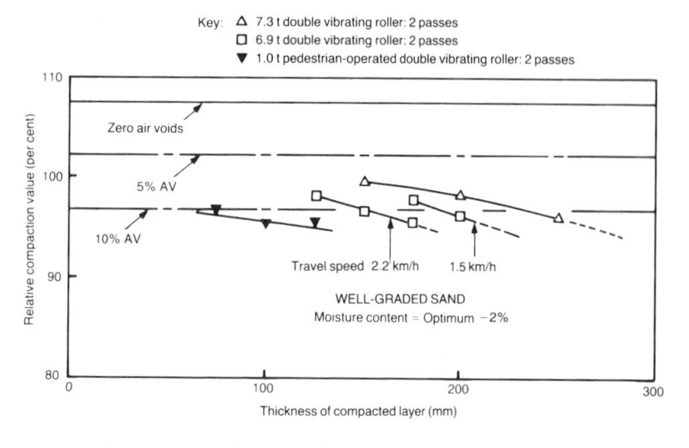

Table 7.5 Thicknesses of layer capable of being compacted by three types of vibrating smooth-wheeled roller using the two criteria of 95 per cent relative compaction and 10 per cent air voids. Taken from Figures 7.27 to 7.30

Soil	Moisture content relative to 2.5 kg rammer optimum (per cent)	Compactor	Number of vibrating-roll passes	Depth of layer (mm) for average of:– 95% relative compaction	10% air voids
Heavy clay	+0.5	7.3 t double vibrating roller	8	140	170
Sandy clay No 2	+1.5		4	180	220
Well-graded sand	−2.0		4	270	235
		6.9 t double vibrating roller (1.5 km/h)	4	220	190
		1.0 t pedestrian-operated double vibrating roller	4	120	60
Uniformly graded fine sand	−7.5		4	260	–

7.40 Because of the small number of vibrating-roll passes applied, the thicknesses of layer given in Table 7.5 would be expected to be smaller than those derived from Figures 7.24 to 7.26. However, with the 7.3 t double vibrating

Figure 7.30 Relation between the average relative compaction of the compacted layer and its total thickness for 2 passes of the 1.0 t pedestrian-operated double vibrating roller on uniformly graded fine sand

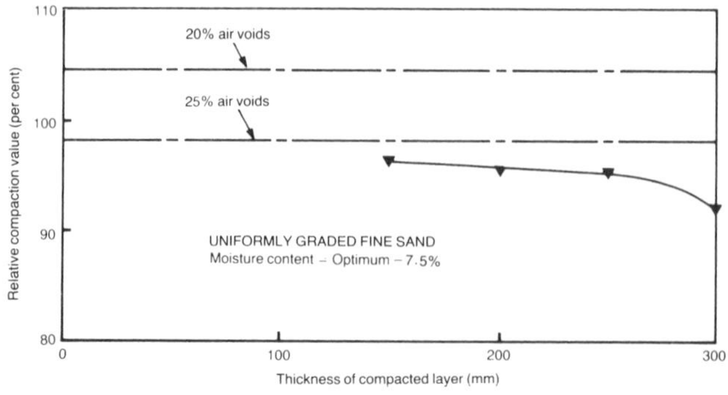

roller this is not always the case; the main reason for this is likely to be that the two different sets of criteria used, 90 per cent relative compaction or 15 per cent air voids at the bottom of the layer as opposed to an average of 95 per cent relative compaction or 10 per cent air voids in the total layer, may not be entirely compatible. It is also possible, however, that the 7.3 t double vibrating roller was more effective, for its particular mass per unit width of vibrating roll, than the other machines included in the analyses.

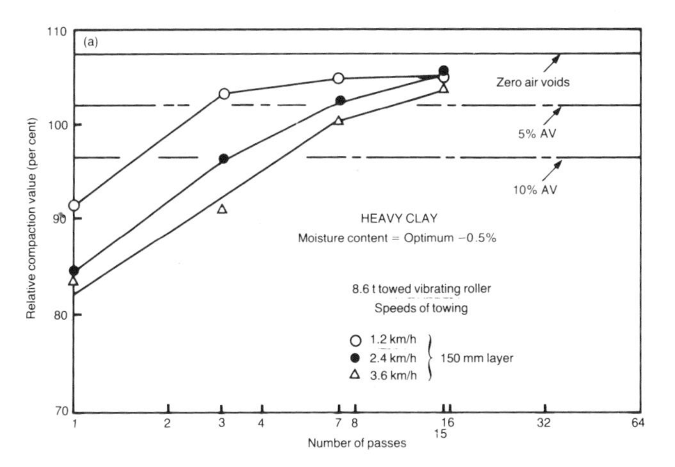

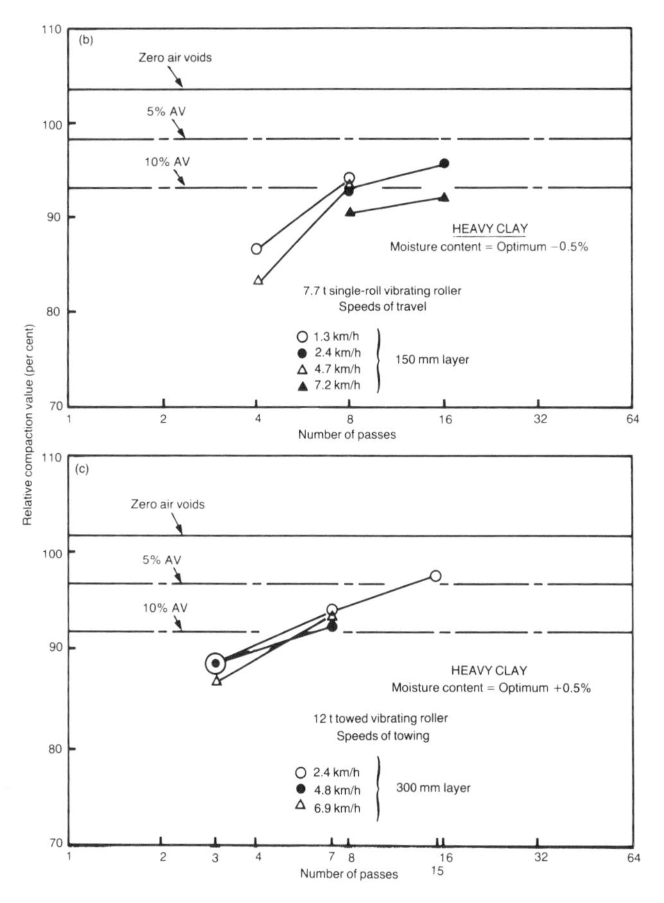

Figure 7.31 Relations between relative compaction and number of passes of vibrating smooth-wheeled rollers using various speeds of rolling on heavy clay

Effect of speed of rolling on the performance of vibrating smooth-wheeled rollers

7.41 The effect of speed of rolling was investigated with a number of the machines, usually in the course of determining the effect of number of passes on the state of compaction produced. The relations between relative compaction value obtained and the number of passes of the vibrating rollers given in Figures 7.9 to 7.13 included only the results for the normal speed of rolling used in the tests with each individual machine. These 'normal' speeds or the normal ranges of speeds are given in Table 7.1. Relations between relative compaction and number of passes for the tests where a number of different rolling speeds were involved are given in Figures 7.31 to 7.33; the relations at 'normal' speeds, already given in Figures 7.9 to 7.13, are also included.

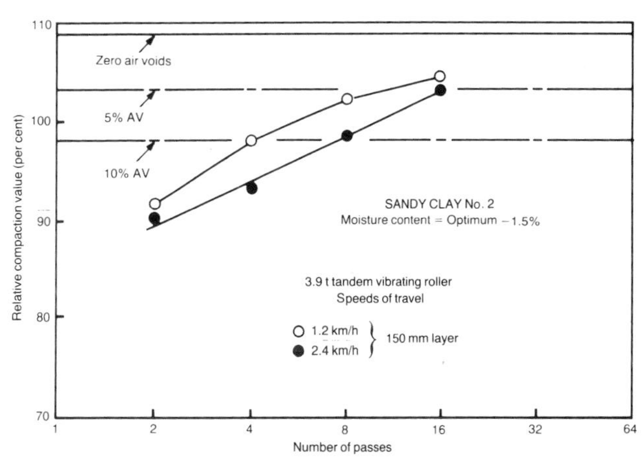

Figure 7.32 Relations between relative compaction and number of passes of the 3.9 t tandem vibrating roller using two speeds of travel on sandy clay No 2

7.42 For the towed rollers (3.9 t, 8.6 t and 12 t) and for self-propelled machines with hydrostatic transmissions (6.9 t and 7.7 t) at least three speeds of rolling were tested. For self-propelled machines with mechanical transmissions (1.7 t, 3.9 t tandem and 7.3 t) the number of different speeds of rolling was restricted by the number of gears available and only two speeds of rolling were tested in each case.

7.43 It can be readily seen that as the speed of rolling was increased the number of passes required to achieve a given state of compaction generally increased (Figures 7.31 to 7.33).

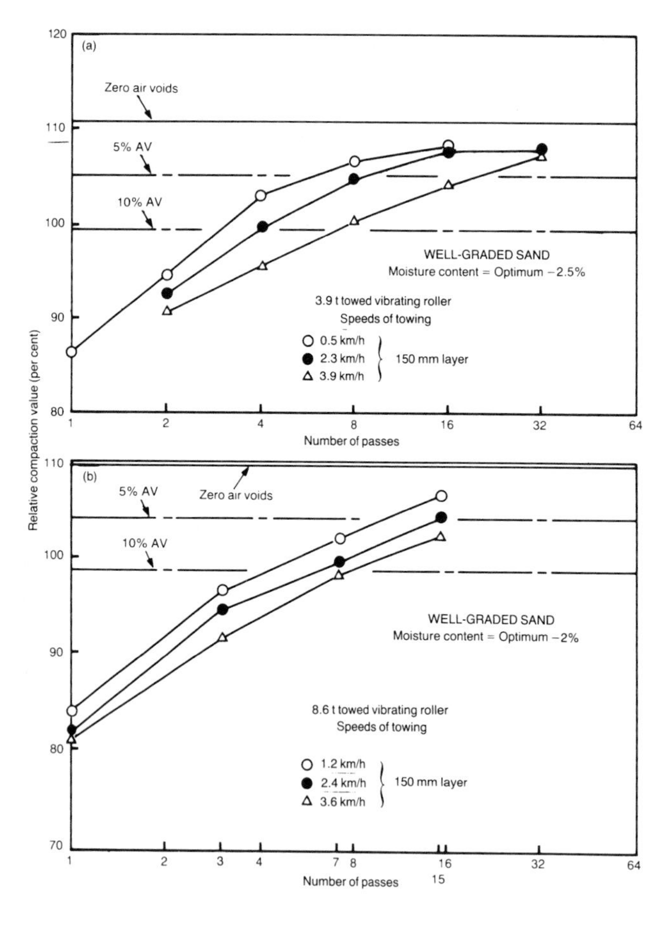

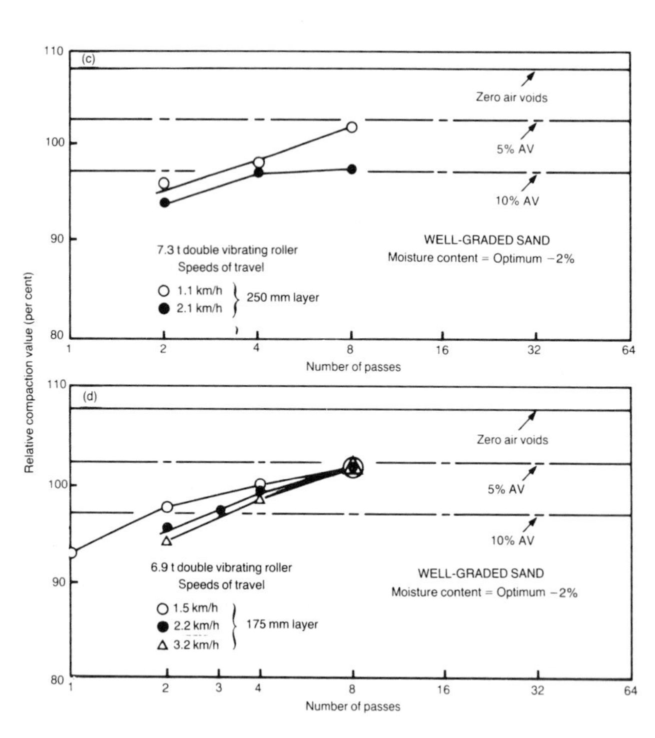

Figure 7.33 Relations between relative compaction and number of passes of vibrating smooth-wheeled rollers using various speeds of rolling on well-graded sand

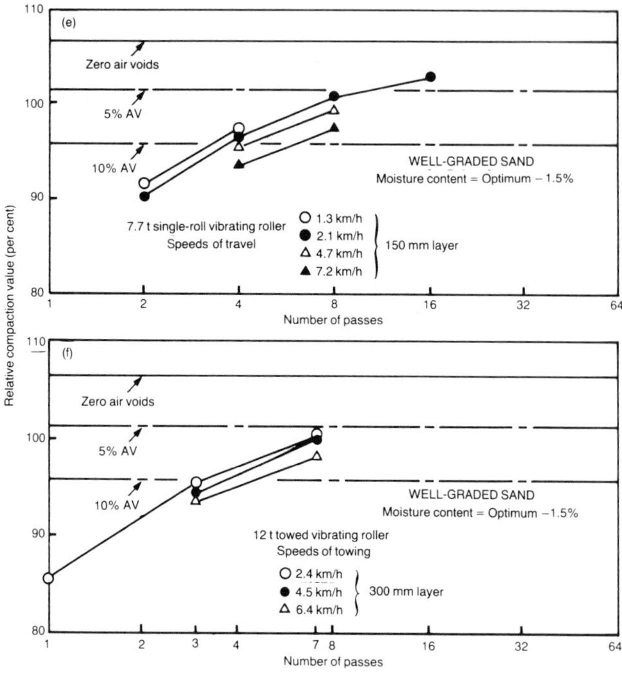

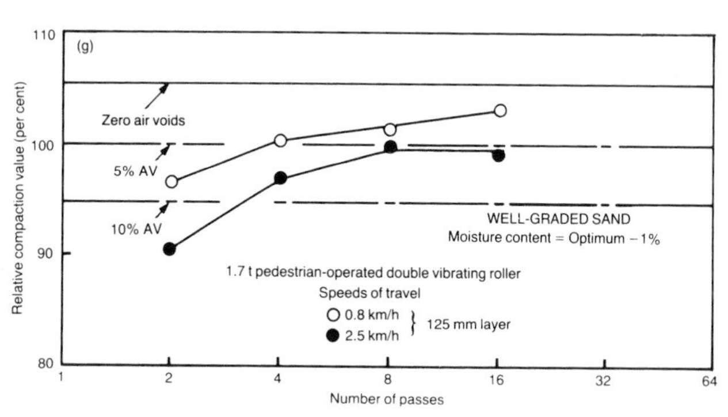

Differences in moisture content and in thickness of layer could well affect the rate at which the number of passes to achieve given levels of compaction could vary with variations in speed of rolling. To indicate the effect of the variation in speed of rolling relations between the number of passes to achieve 95 per cent relative compaction and 10 per cent air voids and speed of rolling are given for heavy clay in Figure 7.34 and for well-graded sand in Figure 7.35. The numbers of passes have been taken from the results given in Figures 7.31 and 7.33 respectively.

7.44 The normal speed of rolling with vibrating smooth-wheeled rollers is generally about 1 km/h for self-propelled machines with mechanical transmissions and about 2 to 2.5 km/h for most other machines (Table 7.1). The results obtained with heavy clay (Figure 7.34) for both the achievement of 95 per cent relative compaction and of 10 per cent air voids indicate that about

Figure 7.34 Relations between number of passes required to achieve (a) 95 per cent relative compaction and (b) 10 per cent air voids and speed of rolling of vibrating smooth-wheeled rollers on heavy clay

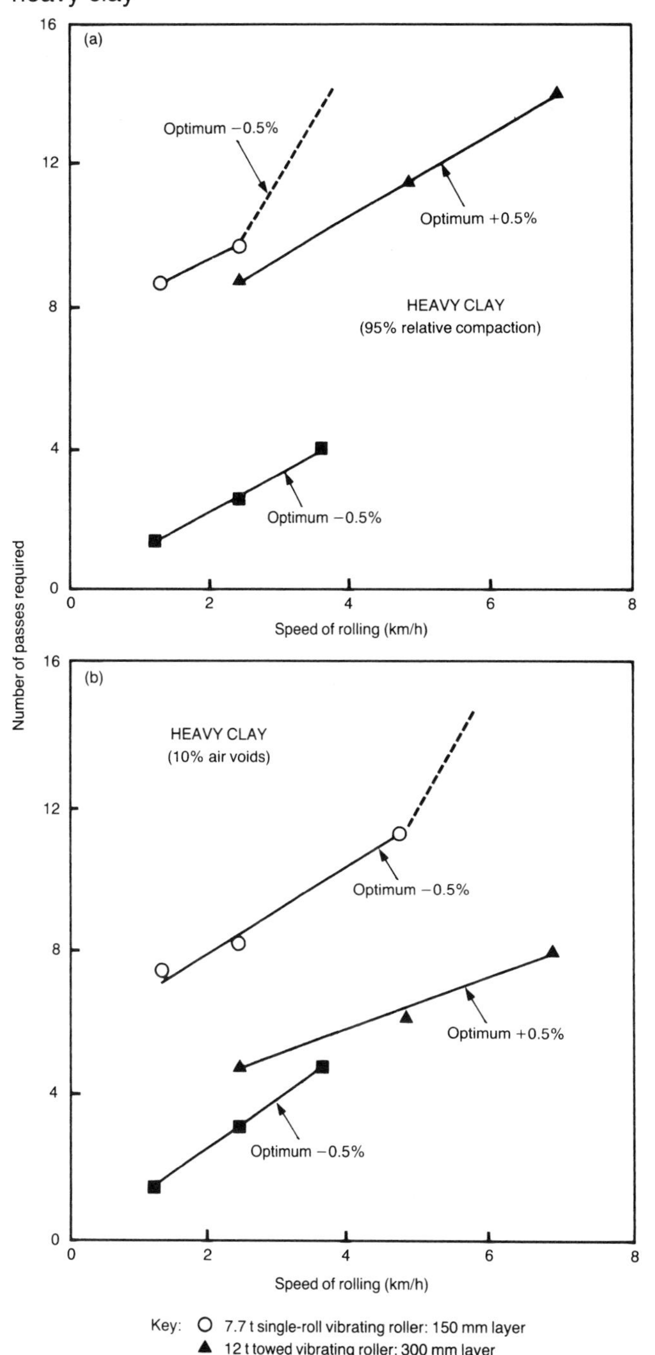

one additional pass was required for each extra 1 km/h in excess of the normal speed of rolling as given in Table 7.1. With well-graded sand (Figure 7.35) the effect of speed was quite variable, with one additional pass required for each extra 0.5 to 4 km/h in the speed of rolling above the normal given in Table 7.1.

7.45 The effect on the output of compacted soil to the required standard will also be variable,

Figure 7.35 Relations between number of passes required to achieve (a) 95 per cent relative compaction and (b) 10 per cent air voids and speed of rolling of vibrating smooth-wheeled rollers on well-graded sand

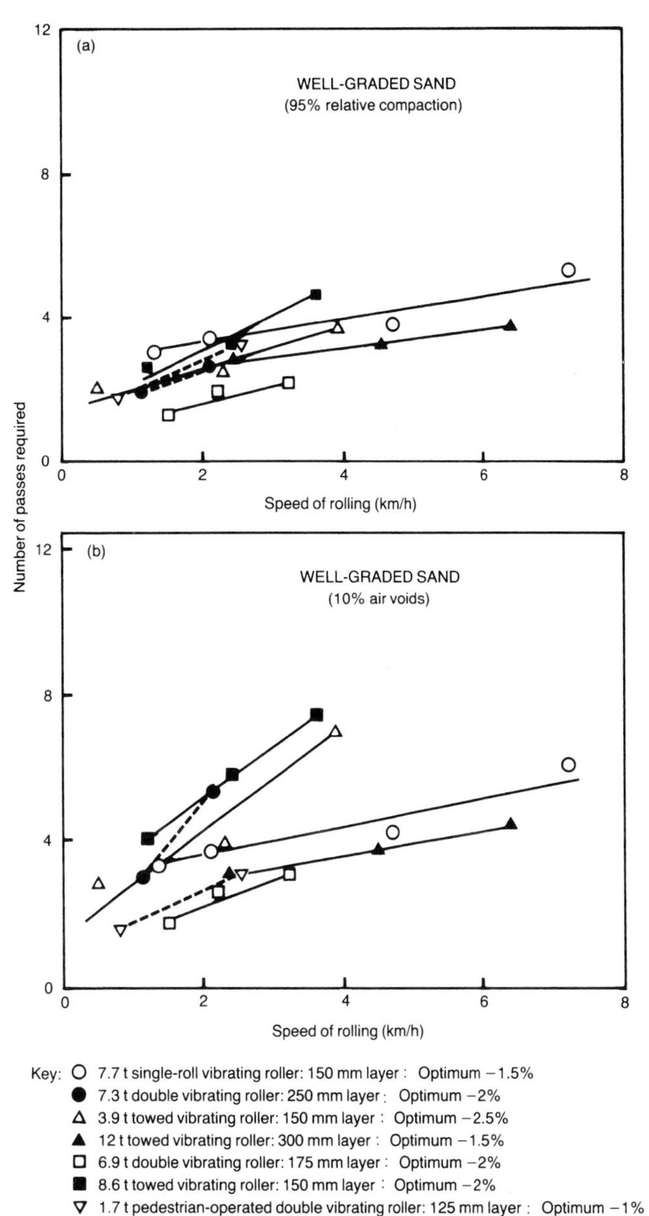

therefore, with some machines having constant output (number of passes proportional to speed) while other machines apparently had increases in output similar to those yielded by increases in speed of rolling of static-weight compactors (see Paragraph 5.25 for the effect with pneumatic-tyred rollers).

7.46 It could be considered that the compaction effect of vibrating smooth-wheeled rollers might be solely attributable to the number of vibration impulses as the roll traverses the soil. If this is so then, at a constant frequency of vibration, a doubling of the speed of rolling would result in only half the number of impulses being applied to a given area of soil on each pass of the machine. Thus the number of passes would have to be increased in proportion to the increase in speed of rolling to achieve the same level of compaction. Machines which tended to produce results which support this hypothesis are the 8.6 t towed vibrating roller (Figures 7.31(a) and 7.33(b)), the 3.9 t tandem vibrating roller (Figure 7.32), the 3.9 t towed vibrating roller (Figure 7.33(a)) and the 6.9 t double vibrating roller (Figure 7.33(d)). In Figures 7.34 and 7.35 the relations would approximate to straight lines through the origin to satisfy this concept. Other machines, however, show the need for a relatively small increase in number of passes with substantial increases in speed of rolling and, in such situations, the output for the machine would increase significantly with higher speeds of rolling; the maximum effect was obtained with the 12 t towed vibrating roller, which doubled its output by increasing its speed of rolling by a factor of about 2.4. Thus no general conclusion can be reached on the effect of speed of rolling on the performance of vibrating smooth-wheeled rollers. A conservative approach would be to assume that any increase in speed of rolling beyond that for which the required number of passes is known should necessitate increasing the number of passes in proportion to the speed, ie

$$n_2 = (V_2/V_1).n_1$$

where n_1 and n_2 are the number of passes required at speeds of rolling of V_1 and V_2 respectively.

7.47 The effect of speed of rolling on the thickness of layer capable of being compacted at a constant number of passes is illustrated in Figure 7.29. Relations between the average relative compaction of the layer of well-graded sand and its total thickness are given for speeds of rolling of 1.5 and 2.2 km/h with 2 passes of the 6.9 t double vibrating roller. The increase in speed of about 50 per cent resulted in the thickness of layer capable of being compacted to an average of 95 per cent relative compaction being reduced from 220 mm to 180 mm, ie a reduction of about 20 per cent on the original thickness. The net effect of this on output, for the particular moisture content used, 2 per cent below the 2.5 kg rammer optimum, is a 20 per cent increase for the 50 per cent increase in speed of rolling.

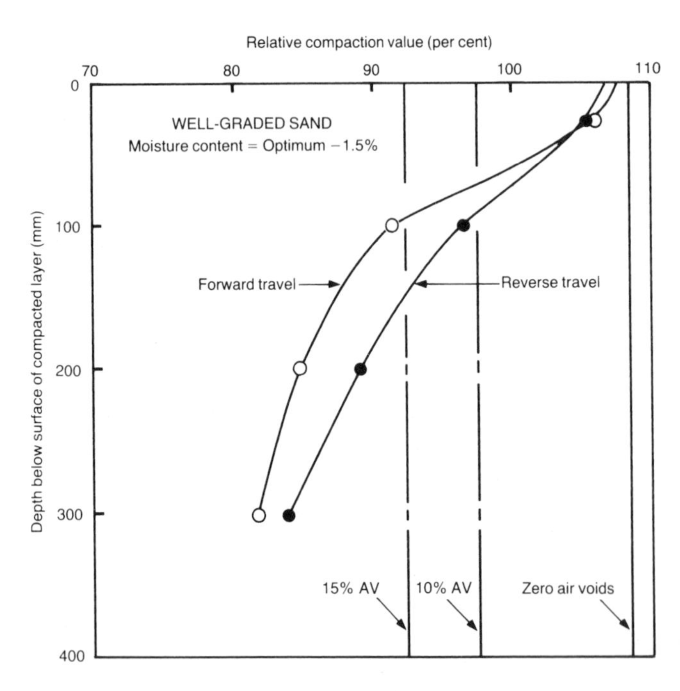

Figure 7.36 Relations between relative compaction and depth in the compacted layer of well-graded sand obtained with 32 passes of the 350 kg pedestrian-operated vibrating roller for the two different directions of travel

Effect of direction of travel on the performance of vibrating smooth-wheeled rollers

7.48 The 350 kg pedestrian-operated vibrating roller was normally operated in the forward direction only during the tests, but comparative tests were made with the machine operating in reverse to produce relations between relative compaction value and depth in the compacted layer (Figure 7.36). The relations given indicate that there is a significant difference between the performance in the forward and reverse directions; for instance, the thickness of layer capable of compaction to achieve 90 per cent relative compaction at the bottom of the layer was about 110 mm using the forward direction of travel and almost 190 mm in the reverse direction (32 passes of the machine were used in these tests).

7.49 It was considered that the difference in performance was achieved by virtue of the influence of the vibration system on the speed of travel. With most vibrating rollers the vibrating mechanisms consist of eccentrically weighted shafts rotating in the same direction regardless of the direction of travel. The horizontal component of the vibration force will have most influence on the speed of travel when the vertical component has minimised the contact pressure of the roll, so that travel is assisted in one direction and opposed in the other. The practical effect of this is that some self-propelled machines in which traction is applied through the vibrating roll tend to travel faster when the eccentric shaft is rotating in the same direction as the roll. Most of the machines listed in Table 7.1 for which ranges of speeds are given are those affected by this influence; they comprise the pedestrian-operated single roll and tandem vibrating rollers.

7.50 The 350 kg machine had a speed of rolling of about 1.4 km/h in the forward direction and about 0.7 km/h in the reverse direction when compacting well-graded sand (as the 350 kg vibrating roller was used in the forward direction only in the majority of tests the range of speeds in Table 7.1 for this machine is for that direction only). Thus the effect shown in Figure 7.36 can be explained by the variations in speed of travel, the effect being similar to that described in Paragraphs 7.41 to 7.47, especially in Paragraph 7.47 with regard to the tests with the 6.9 t double vibrating roller. For the results given in Figure 7.36, with 32 passes of the machine, the higher speed and reduced thickness of layer for the forward direction represent a net increase in output of about 25 per cent (soil compacted to achieve 90 per cent relative compaction at the bottom of the layer) for a 100 per cent increase in speed of rolling.

7.51 It must be remembered that output is not always increased with increases in speed of rolling (see Paragraph 7.46) and no general conclusion should be drawn from the results obtained with the 350 kg pedestrian-operated machine.

Effect of variations in the frequency of vibration on the state of compaction produced

7.52 A number of investigations were carried out where the frequency of vibration was varied. These involved the 3.9 t and 8.6 t towed vibrating rollers, where the frequency could be varied solely by adjustment of the speed of operation of the engine, mounted on each roller, which drove the eccentrically weighted shaft; the 3.9 t tandem vibrating roller was also included as it had a means of continuously varying the frequency (within the limits given in Table 7.1) at constant engine speed by means of a belt drive and expanding pulley system. The results are shown in Figures 7.37 to 7.40 in the form of

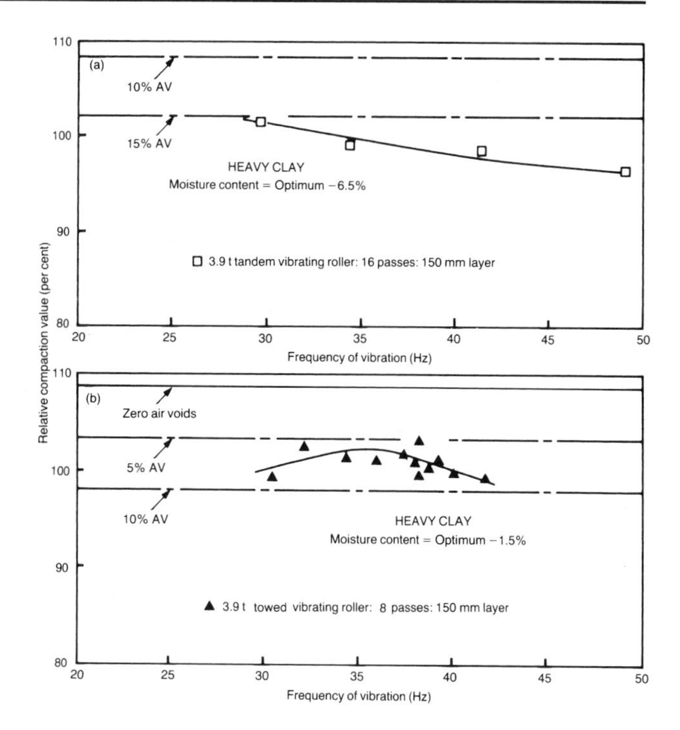

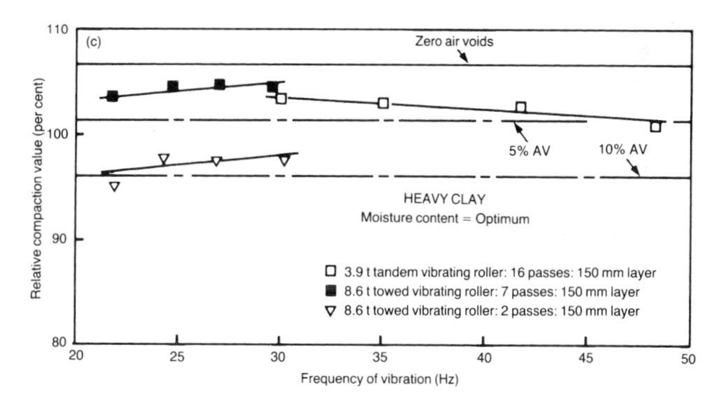

Figure 7.37 Relations between relative compaction and frequency of vibration of vibrating smooth-wheeled rollers on heavy clay

relations between relative compaction value and frequency of vibration.

7.53 With heavy clay (Figure 7.37) all three machines were tested and in all cases, for the numbers of passes and thickness of layer used, as given in the Figure, the effect of variation in frequency was very small. With sandy clay No 2 (Figure 7.38) only the 3.9 t tandem vibrating roller was tested, but at five different values of moisture content; as with heavy clay the effect was again fairly small. With well-graded sand and gravel-sand-clay (Figures 7.39 and 7.40) the effect was more pronounced, especially with the 3.9 t towed vibrating roller, where an improvement in relative compaction value of 4 to 6 per cent was achieved by varying the frequency from about 30 Hz to a maximum of about 41 Hz. It should be noted that the normal

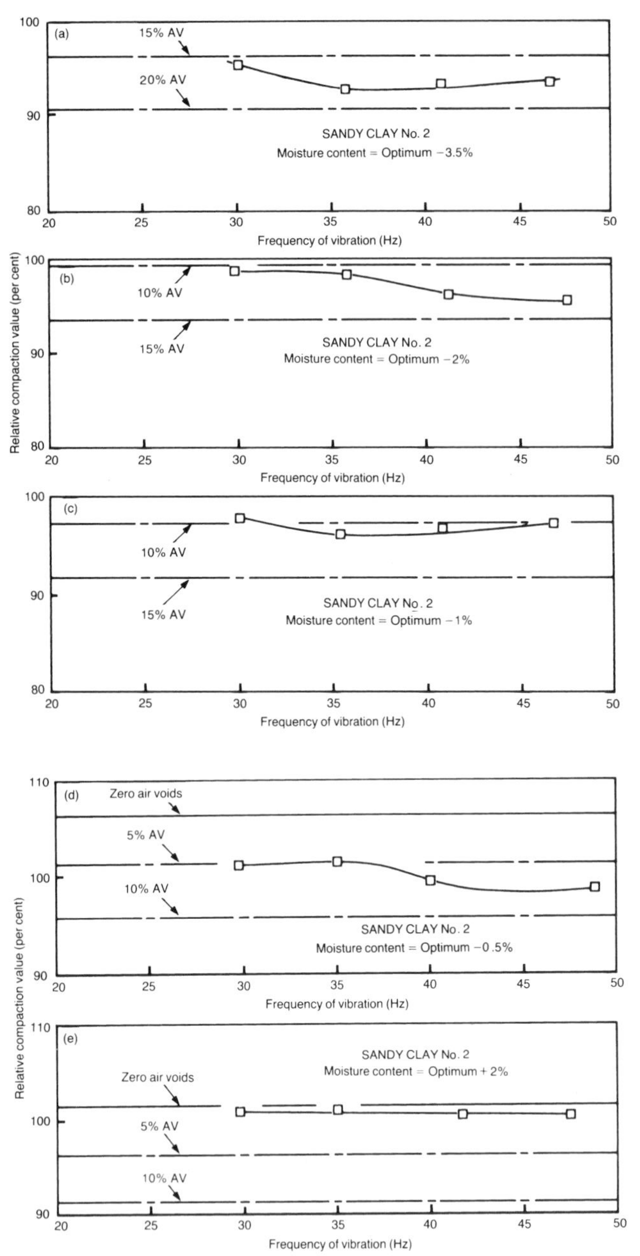

Figure 7.38 Relations between relative compaction and frequency of vibration of the 3.9 t tandem vibrating roller on sandy clay No 2. Number of passes = 4; depth of layer = 150 mm

operating frequency of vibration with this machine was 39 Hz (Table 7.1).

7.54 For the 3.9 t towed vibrating roller complete relations between relative compaction value of gravel-sand-clay and number of passes were determined for frequencies near the upper and lower limits used in Figure 7.40(b). These relations are given in Figure 7.41; they confirm that the difference in performance at 8 passes was repeated over the full range of applied passes. The net effect was that the number of

Figure 7.39 Relations between relative compaction and frequency of vibration of vibrating smooth-wheeled rollers on well-graded sand

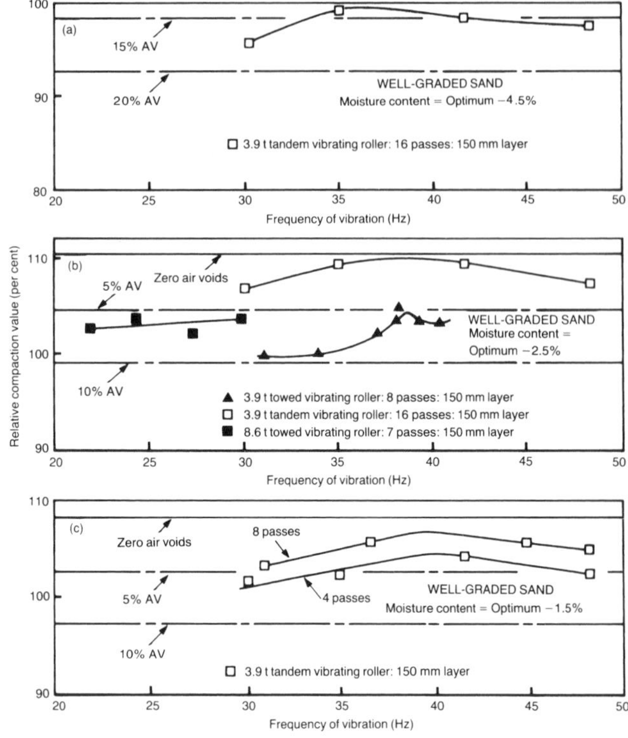

Figure 7.40 Relations between relative compaction and frequency of vibration of vibrating smooth-wheeled rollers on gravel-sand-clay

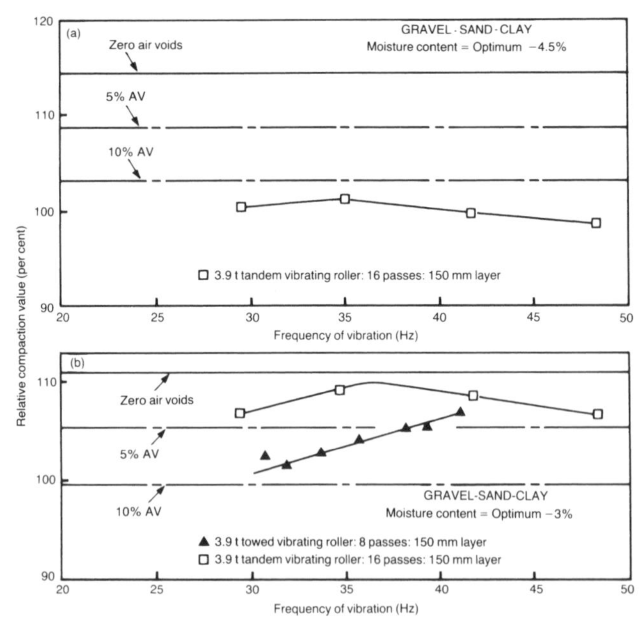

passes required to achieve an air content of 5 per cent was doubled when the lower frequency was used (Figure 7.41).

7.55 The effects of variation in frequency exhibited in Figures 7.37 to 7.40 are not clear and conflict in some cases with the increase in dynamic forces which must result from an increase in frequency of vibration. Thus, as frequency of vibration is increased:–

(a) The centrifugal force increases in proportion to the square of the frequency (the frequency increase for the 3.9 t tandem vibrating roller from the minimum of 30 Hz to the maximum of 49 Hz represents an increase in centrifugal force by a factor of 2.5);

(b) The number of vibration pulses applied to the soil during each pass of the machine at constant speed of rolling will increase in proportion to the frequency.

7.56 The net effect of these influences is that the relative compaction would be expected to increase with frequency. The results for the 8.6 t towed vibrating roller in Figures 7.37(c) and 7.39(b) all show this to some extent and those for the 3.9 t towed vibrating roller on gravel-sand-clay (Figure 7.40(b) and 7.41) show the trend to a significant extent. Where results are produced with low air voids, as in many of the tests with the 3.9 t tandem vibrating roller (Figures 7.37(c), 7.38(e), 7.39(b) and (c) and 7.40(b)), possible increases in relative compaction value with increase in frequency of vibration would be expected to be inhibited as the soil would be already in a near-fully compacted condition. Thus it is the results which indicate decreases in state of compaction as the frequency increased, or which produced apparent maximum states of compaction at an optimum frequency, associated with air contents in excess of about 5 per cent, which are significant and indicate that factors other than those given in (a) and (b) above were involved. The results for sandy clay No 2 at moisture contents from 3.5 per cent to 1 per cent below the 2.5 kg rammer optimum (Figures 7.38 (a), (b) and (c)), for well-graded sand, Figure 7.39(a) for the tandem roller and Figure 7.39(b) for the 3.9 t towed roller, and for gravel-sand-clay (Figure 7.40(a)) all indicate that an optimum frequency, possibly associated with resonance of the soil-roller system, does exist. The differences for clay soils do not appear important but for the granular soils an optimum frequency in the range of 36 to 40 Hz, possibly depending to some extent on size of vibrating roller and moisture content, is indicated. The use of a frequency other than that recommended by the manufacturers is not advisable, however, because of the potential risk of mechanical damage if higher values are used and the possible reduction in the state of compaction produced at the lower frequencies. Where a range of frequencies is available, the results of the investigations described above indicate that, with granular soils in particular, some form of trial may be justified to determine the most suitable frequency of vibration for the particular site conditions.

Figure 7.41 Relations between relative compaction and number of passes of the 3.9 t towed vibrating roller on gravel-sand-clay, using two different frequencies of vibration. Layer thickness = 150 mm

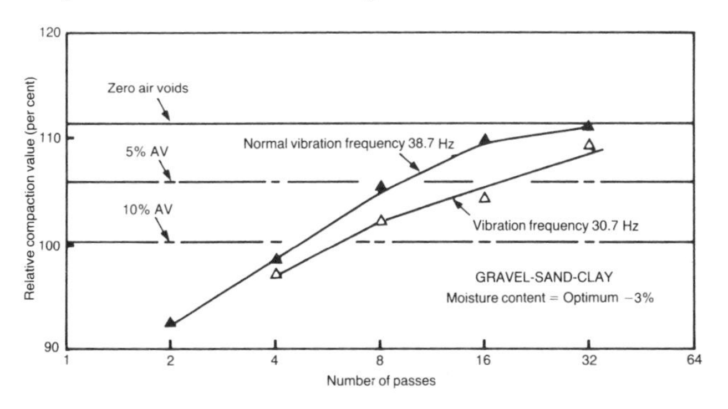

Effect on performance of double vibrating rollers of operating with one roll only vibrating

7.57 Relations between relative compaction value and number of passes were determined for the 7.3 t double vibrating roller and the 1.0 t pedestrian-operated double vibrating roller using both the normal operating mode and with the vibrating mechanism in the rear roll(s) disengaged so that each machine effectively became, by definition (see Paragraph 7.4), a tandem vibrating roller. The results of this study are given in Figure 7.42.

7.58 The study was carried out mainly to determine if the 'pitching' action induced by the 180° out-of-phase vibration of the front and rear

Figure 7.42 Relations between relative compaction and number of passes on well-graded sand showing the effect of operating double vibrating rollers with the vibrating mechanism inoperative in one roll

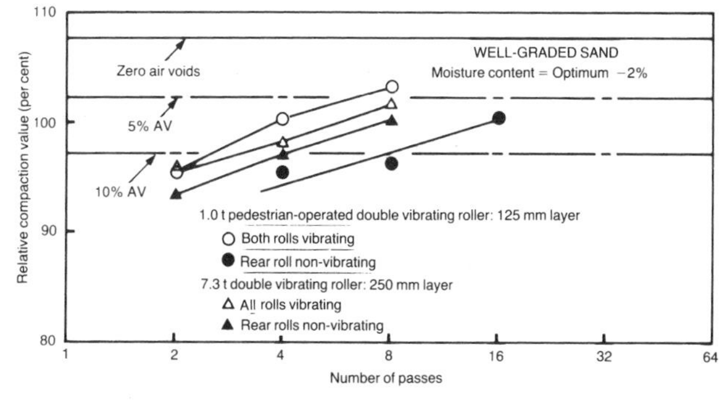

rolls was contributing to the compaction effect of the machines. Thus, if the number of passes required to produce a given state of compaction when compacting with one roll only vibrating was more than double that when both rolls were vibrating, positive evidence that the pitching action was a contributing factor would have been obtained. Normally it could be expected that the second, non-vibrating, roll would have made some minor further contribution to the compaction process, thus reducing the number of passes required to less than double the number required with both rolls vibrating.

7.59 With the 1.0 t pedestrian-operated vibrating roller (Figure 7.42) the number of passes required to achieve 95 per cent relative compaction in a 125 mm layer of well-graded sand was two when operating normally and about five when one roll only was vibrating. Thus, with this machine, there is clear evidence that the phasing of the eccentric shafts added to the performance. With the 7.3 t double vibrating roller, however, 2 passes were required to achieve the same state of compaction in a 250 mm layer of the soil when operating normally and 3 passes when the front rolls only were vibrating. With this second machine, therefore, no clear indication is given as to the effect of the phasing of the eccentric shafts.

7.60 For unphased eccentric shafts the double vibrating roller can be assumed to have the output of two single vibrating rollers of the same roll loading; where, however, the eccentric shafts are 180° out-of-phase, the results described above indicate that the machine may have an output more than that of two single vibrating rollers, although this may only apply to the lighter, pedestrian-operated machines.

Consideration of centrifugal force as an influence on performance

7.61 It has already been stated that the mass per unit width of vibrating roll is not the sole factor influencing the performance of a vibrating roller (Paragraph 7.5), and the dynamic forces involved might also be expected to have a significant effect. To determine the relative effects of mass and centrifugal force, regression analyses were made of the results produced by all the machines for which values of centrifugal force are available (Table 7.1). The results were taken from Figures 7.1 to 7.6 and 7.9 to 7.13. The regression equations used were of the form shown in Table 7.3, ie, using the number of vibrating-roll passes required to achieve an average state of compaction of either 95 per cent relative compaction or 10 per cent air voids as the dependent variable. In addition to the independent variables used in Table 7.3 (mass per unit width of vibrating roll and moisture content), centrifugal force per unit width of roll and thickness of compacted layer were used. (The thickness of layer was included as an independent variable to allow an increased number of results to be analysed.) All variables except moisture content were used in logarithmic format as in Table 7.3. The regression analyses were combined with analyses of variance to determine the relative significance of the independent variables and the most significant combinations of variables are given in Table 7.6.

Table 7.6 Effects of various factors on the determination of the number of passes required to achieve given levels of compaction using vibrating smooth-wheeled rollers. Results of regression and variance analyses

Soil	Level*	Variable‡				No of data points
		m	F	w	d	
Heavy clay	(a)	†			†	8
	(b)	†				8
Sandy clay No 2	(a)	†		†	†	11
	(b)	†		†		11
Well-graded sand	(a)		†	†	†	17
	(b)	†	†	†	†	17
Gravel-sand-clay	(a)	†		†	†	14
	(b)	†		†	†	16
Uniformly graded fine sand	(a)	†			†	10

*Level (a)=95 per cent of the maximum dry density obtained in the 2.5 kg rammer compaction test
(b)=10 per cent air voids
‡Variable: m=$\log_{10}$ (Mass per unit width of vibrating roll in kg/m)
F= $\log_{10}$ (Centrifugal force per unit width of vibrating roll in kN/m)
w=Moisture content of soil, per cent, minus the optimum moisture content, per cent, obtained in the 2.5 kg rammer compaction test
d= $\log_{10}$ (Depth of compacted layer in mm)
†Indicates the most significant combination of variables to determine the required number of passes

7.62 It can normally be expected that moisture content and depth of compacted layer would have a significant effect on the number of passes required to achieve specified levels of compaction. Where these factors are not shown to be significant in Table 7.6, the causes are either the small range of moisture content values in the data, or the high level of correlation between thickness of layer and mass per unit width of vibrating roll (the thicknesses of layer in later tests were increased with increase in size of machine to comply, in general, with the Department of Transport's method specification for compaction of earthwork materials (Department of Transport, 1986)). The absence of centrifugal force per unit width of roll as a significant variable in all but two sets of results in Table 7.6 (well-graded sand, levels (a) and (b)), is caused by the high level of correlation between mass per unit width and centrifugal force per unit width. The Table indicates that in virtually all cases the mass per unit width provides the better indication of performance. The relation between the two factors, for all machines where both values are given in Table 7.1, is shown in Figure 7.43. The correlation coefficient achieved was 0.92.

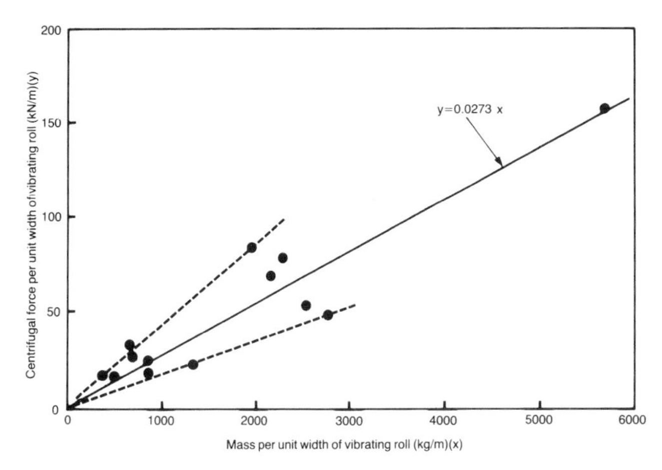

Figure 7.43 Relation between centrifugal force and mass for the vibrating smooth-wheeled rollers in Table 7.1 for which data are available

7.63 An attempt has also been made to assess the relative performances of machines tested in different periods. Regression analyses were made to determine relations similar to those given in Table 7.3 for vibrating smooth-wheeled rollers tested on 150 mm layers of well-graded sand in successive 10-year periods from 1946. No significant difference could be determined between the relations obtained for the three periods concerned and there was certainly no trend towards a reduction in the required number of passes with the more modern machines.

7.64 The effect of increasing the frequency of vibration, thus increasing centrifugal force by a factor of 2.5 with the 3.9 t tandem vibrating roller, as discussed in Paragraph 7.55, indicates that performance cannot be automatically enhanced by increasing dynamic forces solely by increasing the frequency of vibration. The improvement in performance of the 8.6 t towed vibrating roller in its modified form, with increased centrifugal force (see Paragraph 7.13) over that of the original machine (Figures 7.1, 7.4 to 7.6, 7.18(a) and (d), 7.20(a) and (c), 7.21(c) and 7.22(c)) was fairly small, although the extent to which the centrifugal force had been increased is unknown. All these results show that the effect of dynamic force is relatively small within the range of dynamic forces normally used in vibrating rollers.

7.65 By the use of an experimental machine, such as the 'Vibrex', tested by the French Laboratoires des Ponts et Chaussees (Quibel and Froumentin, 1980) independent variations of the various parameters can be explored. Following the work in the early 1970s with the 'Vibrex' the French specifications for earthwork compaction categorises vibrating rollers in terms of mass per unit width (Ministere de l'Equipement, 1976), the strong correlation of centrifugal force with mass in manufactured machines (Figure 7.43) making the method reliable for all practical purposes.

7.66 It can be concluded with a reasonable degree of confidence that the mass per unit width of vibrating roll is a practical measure of the performance of a vibrating roller manufactured in the period 1950 to the time of writing. Relations such as those given in Figures 7.8, 7.14 to 7.17 and 7.24 to 7.26, represent the best summary information that can be obtained from the results of the numerous investigations carried out at the Laboratory. In using these relations to predict the performance of a given vibrating smooth-wheeled roller a useful criterion should be that the centrifugal force per unit width-mass per unit width coordinates lie within the indicated envelope for the relation given in Figure 7.43. As mentioned in Paragraph 7.63, there appears to be no significant difference between the performances of the machines, tested in the various periods, which contributed to the determination of the summary relations.

Pressures produced in compacted soil by vibrating smooth-wheeled rollers

7.67 During early investigations of the 2.4 t tandem vibrating roller, in 1949 to 50, piezoelectric pressure cells were used to determine the vertical pressures produced at two different depths in silty clay at a moisture content about 3.5 per cent below the 2.5 kg rammer optimum. The depths at which the cells were installed were about 100 mm and 200 mm below the surface of the compacted layer. Typical results obtained with the pressure cells placed at a depth of 200 mm, during the passage of the vibrating and non-vibrating rolls, are given in Figure 7.44. The pressure curve for the vibrating roll is a double curve, the two lines representing the upper and lower limits between which the pressure oscillated sinusoidally. The lower line is in the region of zero pressure while the upper limit rises to a maximum which is about twice the value which would have been obtained without vibration. During repetitive tests a considerable scatter of results was obtained and the result given in Figure 7.44 should be regarded as a general indication only of the behaviour of the vibrating roller. The scatter of results was mainly attributed to the unevenness of the compacted surface, and the pressure cells set at a shallow depth of 100 mm were far more severely affected than those at the depth of 200 mm. The gradient of vertical pressure through the compacted layer could not, therefore, be determined.

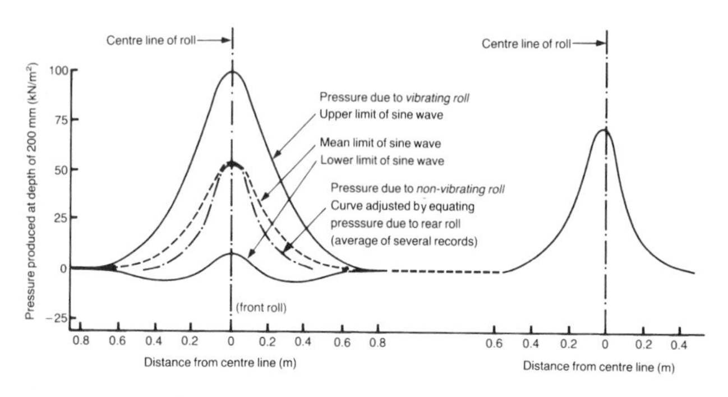

Figure 7.44 Pressures produced in silty clay by the 2.4 t tandem vibrating roller

7.68 In more recent tests in 1971 a similar type of exercise was carried out with the 1.0 t pedestrian-operated double vibrating roller and the 7.3 t double vibrating roller. This was associated with the study of the effects of operating the machines with only one roll vibrating, as described in Paragraphs 7.57 to 7.60. Typical results are shown in Figures 7.45 and 7.46 for the 1.0 t and 7.3 t machines respectively. The pressure cells used were the type operating on the inductive principle with the results registered by a high-speed ultra-violet galvanometer recorder. The close proximity of the pressure traces for the front and rear rolls indicate the close coupling of the roll axles as shown in Plates 7.4 and 7.13. Even with the vibrating mechanism in the rear rolls disconnected some vibration from the other roll was experienced by transfer through the connecting chassis, as shown by the continuation of the double trace, representing the maximum and minimum pressures, in the pressure records for the non-vibrating rolls in the lower halves of the figures. With both machines (Figures 7.45 and 7.46) the vibrating roll produced peak pressures about three times those under the non-vibrating roll. Apart from the differences in the factor by which vibration increased the vertical pressure, the traces obtained in the 1971 tests were very similar to those obtained in 1949–50 (Figure 7.44). A further interesting feature of the more recent pressure measurements was that there was no overlap of pressures from the front and rear rolls, the pressure dropped to zero between the rolls with no evidence of an additive effect, despite the close proximity of the rolls.

7.69 The inclusion of pressure cells in a soil can be expected to cause an aberration in the

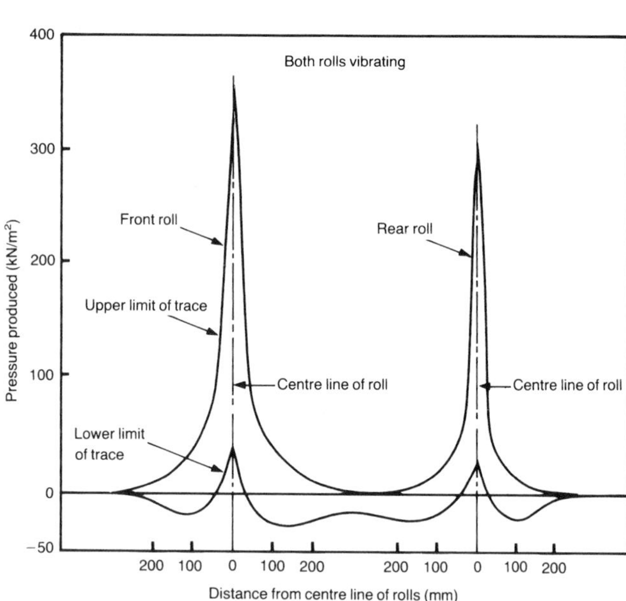

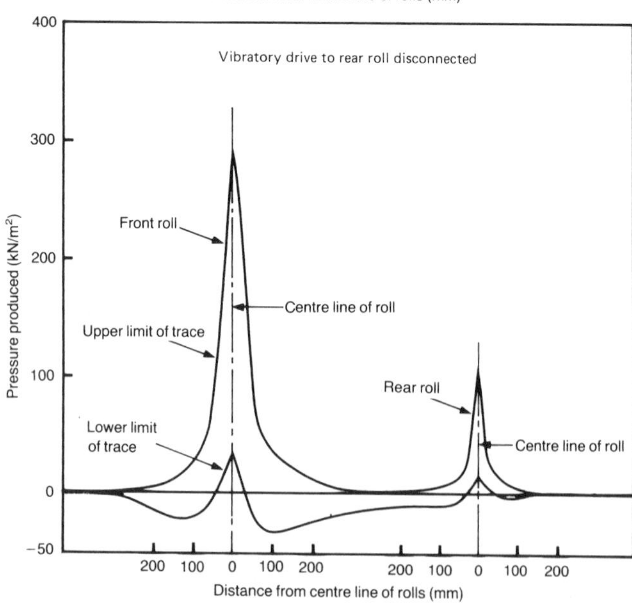

Figure 7.45 Typical pressures recorded in the well-graded sand at a depth of 75 mm with the 1.0 t pedestrian-operated double vibrating roller

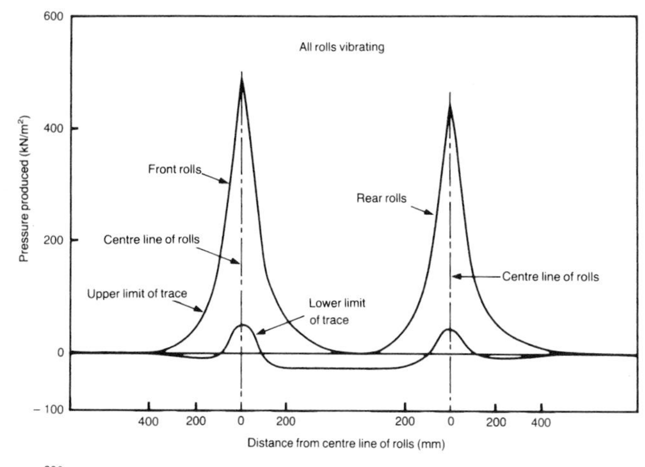

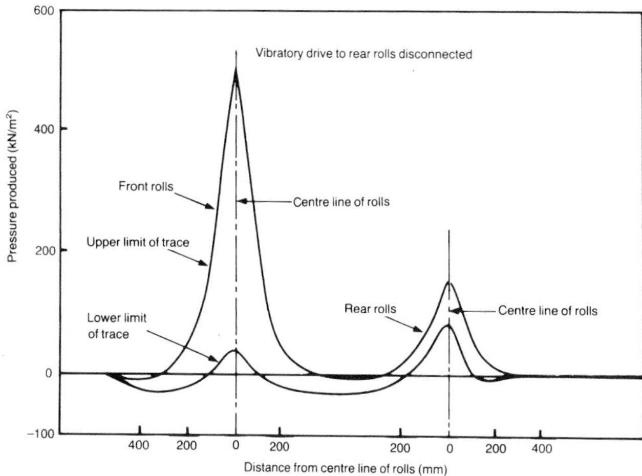

Figure 7.46 Typical pressures recorded in the well-graded sand at a depth of 130 mm with the 7.3 t double vibrating roller

normal pressure distribution under the vibrating rollers. Thus the values of pressure registered may not be strictly accurate and the results in Figures 7.44 to 7.46 should be regarded only as general illustrations of the trends in vertical pressures exerted by the machines.

Surface characteristics of soil compacted by vibrating smooth-wheeled rollers

7.70 The differences in speed of travel in the forward and reverse directions of self-propelled tandem vibrating rollers and pedestrian-operated single-roll machines was attributed to the action of the vibratory system, which gave rise to a positive or negative slip of the driving roll (Paragraphs 7.48 to 7.51); such slippage of the roll also occurred with towed vibrating rollers, the roll rotating at a higher or lower speed than that compatible with the speed of travel. This action gives rise to narrow transverse cracks in the surface of non-cohesive materials; the potential for low strength at the surface of, for instance, cement stabilised granular soils, cannot be ignored in these circumstances.

7.71 With clay soils at moisture contents generally dry of the plastic limit, polishing of the surface of the compacted layer occurred. During initial passes of towed vibrating smooth-wheeled rollers on clay soils severe transverse cracking arose as shown in Plate 7.17. With further passes, however, the cracks became closed up and polishing resulting from roll slip occurred to produce the type of surface shown in Plate 7.18. In situations where the keying together of successive layers is essential, eg in embankment dams, it is possible that serious problems could be created by the production of such potential planes of weakness and zones of seepage. The use of vibrating smooth-wheeled rollers should be considered carefully, therefore, for such applications.

Plate 7.17 The surface of a strip of sandy clay soil after compaction by one pass of the 12 t towed vibrating roller

Plate 7.18 The polished surface of sandy clay soil after several passes of the 3.9 t towed vibrating roller

Plate 7.20 The 7.7 t single-roll vibrating roller after 'bogging down' on a loose layer of uniformly graded fine sand

Mobility of vibrating smooth-wheeled rollers on uniformly graded fine sand

7.72 With uniformly graded fine sand difficulties were often experienced in operating self-propelled vibrating rollers over the initial loose layer. Plates 7.19 and 7.20 illustrate the 2.2 t tandem vibrating roller and the 7.7 t single-roll vibrating roller after they completely lost traction on the loose layers of material. Whenever a machine failed to operate over the loose soil during the investigations, the procedure was to pre-compact the uniformly graded fine sand by systematically traversing the area with the tracks of one of the crawler tractors fitted with rotary cultivators (shown in Plates 3.8 and 3.11). This action was effective in providing a sufficiently firm surface for compaction to proceed with machines such as the 7.7 t machine, with its pneumatic-tyred traction wheels (Plate 7.20) and double vibrating rollers in which both rolls were driven. Tandem vibrating rollers, such as illustrated in Plate 7.19, were still not capable of operating even after the pre-compaction process.

7.73 With towed vibrating smooth-wheeled rollers a large wave of soil was created ahead of the roll during the first pass over loose uniformly graded fine sand (Plate 7.21) and it was found beneficial to carry out the first pass with the vibrating mechanism inoperative.

7.74 With all types of vibrating roller severe shearing of the uniformly graded fine sand occurred in the near-surface material, even after several passes, and low states of compaction were registered in the sheared zone (see relations between relative compaction and depth in the compacted layer in Figure 7.23). The zone of sheared soil deepened with increasing size of vibrating roller. Lightweight, pedestrian-operated double and single roll vibrating rollers operated very effectively on the uniformly graded fine sand, but with self-propelled machines with

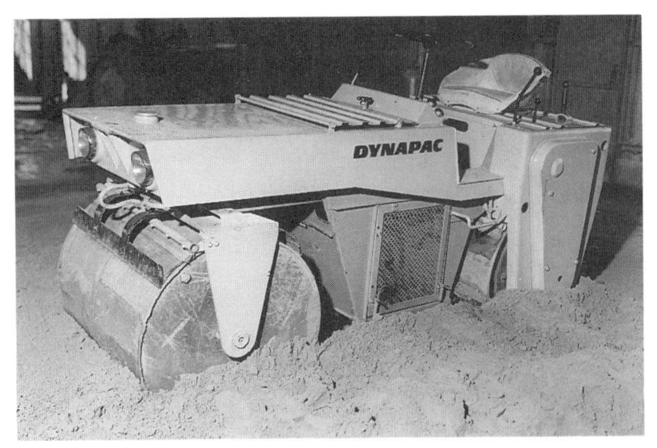

Plate 7.19 The 2.2 t tandem vibrating roller with trafficability problems on a loose layer of uniformly graded fine sand

Plate 7.21 The 12 t towed vibrating roller during the first pass over a loose layer of uniformly graded fine sand

masses in excess of about 2 t, operation on the soil was always difficult. In practice, with soils of such characteristics, machines of 2 t or more should be of a type towed by crawler tractors to ensure that efficient compaction operations are sustained.

References

British Standards Institution (1990). British Standard methods of test for soils for civil engineering purposes: BS 1377: Part 9 In situ tests. BSI, London.

Department of Transport (1986). Specification for highway works. 6th edition. HM Stationery Office, London.

Lewis, W A (1954). Further studies in the compaction of soil and the performance of compaction plant. *Road Research Technical Paper* No 33. HM Stationery Office, London.

Lewis, W A (1961). Recent research into the compaction of soil by vibratory compaction equipment. *Proc.5th Internat. Conf. Soil Mech.* Vol II. Dunod, Paris.

Ministere de l'Equipement (1970). Recommendations for the construction of road earthworks (in French). Setra, Laboratoire Central des Ponts et Chaussees, Paris.

Parsons, A W (1966). An investigation into the performance of a Winget Dynapac CG 11 2¼-ton tandem vibrating roller in the compaction of soil. *Ministry of Transport, RRL Laboratory Report* No 15. Road Research Laboratory, Harmondsworth.

Parsons, A W, Krawczyk, J and Cross, J E (1962). An investigation of the performance of an 8½-ton vibrating roller for the compaction of soil. *D.S.I.R., Road Research Laboratory Note* No LN/64/ AWP.JK.JEC (Unpublished).

Quibel, A and Froumentin, M (1980). The influence of the parameters of a vibrating roller on its efficiency (in French). *International Conference on Compaction,* Editions Anciens ENPC, Paris, Vol II, 677–82.

Toombs, A F (1966). An investigation into the performance of a Winget Dynapac CH 44 4¾-ton towed vibrating roller in the compaction of soil. *Ministry of Transport, RRL Laboratory Report* No.13. Road Research Laboratory, Harmondsworth.

Toombs, A F (1968). The performance of a Tramac Vibras 85/10 CP 1700 kg (1¾-ton) double vibrating roller in the compaction of soil. *Ministry of Transport, RRL Laboratory Report* LR 183. Road Research Laboratory, Crowthorne.

Toombs, A F (1969). The performance of an Aveling Barford VP 7.7 Mg self-propelled vibrating roller in the compaction of soil. *Ministry of Transport, RRL Laboratory Report* LR 257. Road Research Laboratory, Crowthorne.

Toombs, A F (1970). The performance of a Stothert and Pitt Vibroll T 208 12 Mg towed vibrating roller in the compaction of soil. *Ministry of Transport, RRL Laboratory Report* LR 341. Road Research Laboratory, Crowthorne.

Toombs, A F (1972). The performance of Bomag BW 75S and BW 200 double vibrating rollers in the compaction of soil. *Department of the Environment, TRRL Laboratory Report* LR 480. Transport and Road Research Laboratory, Crowthorne.

Toombs, A F (1973). The performance of a Clark Scheid DV 60 6.9 Mg double vibrating roller in the compaction of soil. *Department of the Environment, TRRL Laboratory Report* LR 590. Transport and Road Research Laboratory, Crowthorne.

CHAPTER 8 COMPACTION OF SOIL BY VIBRATING SHEEPSFOOT ROLLERS

Description of machines

8.1 This Chapter describes the results of investigations into the performance of vibrating rollers where the rolls have the configuration of sheepsfoot or tamping rolls as described in Chapter 6.

8.2 In general, machines of this type fit the descriptions given in Paragraphs 6.2 and 6.3 for the dead-weight sheepsfoot and tamping rollers, but in addition they have vibrating mechanisms installed similar to those used in vibrating smooth-wheeled rollers and described in Paragraph 7.2.

8.3 Vibrating sheepsfoot and tamping rollers are generally of two main configurations:–

(a) Towed. The machines have a single roll, with a separate tractor to provide traction, and

(b) Self-propelled, single roll. These machines have a single roll with propulsion provided by pneumatic-tyred wheels on a second axle (on some machines propulsion is provided by drive to the vibrating roll as well as to the wheels).

Plate 8.1 700 kg pedestrian-operated double vibrating sheepsfoot roller, fitted with 600 mm rolls, investigated in 1987

Plate 8.2 4.3 t towed vibrating sheepsfoot roller investigated in 1961

Plate 8.3 5.1 t towed vibrating sheepsfoot roller investigated in 1961–62

A third category is the pedestrian-operated machine as illustrated in Plate 8.1, used for small-scale compaction tasks such as trench reinstatements.

8.4 The three machines investigated at the Laboratory which belong to this class of compactor are illustrated in Plates 8.1 to 8.3, and details are given in Table 8.1.

8.5 The 700 kg pedestrian-operated double vibrating sheepsfoot roller is shown in Plate 8.1. It was a self-propelled machine intended basically for the compaction of backfill in trenches. It had four sheepsfoot (tamping) rolls mounted in tandem pairs, and rolls were available in four different widths to match the particular dimensions of the trench. These rolls provided overall compacted widths, between the outer edges of the rolls, of 400, 450, 500 and 600 mm, but only the 400 and 600 mm rolls were used in the tests. The machine relied upon a hydraulic transmission system, traction being provided by a hydraulic motor within the hub of each roll. Directional control of the machine was achieved by a 'skid steering' process, either by applying drive to the rolls on one side only of the machine, or by driving the rolls on each side in opposite directions. Vibration was applied to all rolls using a single, hydraulically driven, eccentric shaft mounted transversely in the chassis between the front and rear rolls. A unique feature was that the vibrator shaft could be operated in either direction; it was recommended by the manufacturer that it should always revolve in the same direction as the rolls. This method of operation ensured that the machine maintained maximum traction capability under difficult soil conditions. The engine compartment and operator's controls were isolated from the vibrating chassis and rolls by a series of flexible mountings. It should be noted that the coverage for this machine (defined in Table 8.1) was 0.185; in this respect, therefore, it complies with the definition of a tamping roller as detailed in Paragraph 6.3, but for simplicity it will be referred to as a vibrating sheepsfoot roller in this Chapter.

Table 8.1 Details of the machines contributing data on the performance of vibrating sheepsfoot rollers

	700 kg pedestrian-operated double vibrating sheepsfoot roller		4.3 t towed vibrating sheepsfoot roller	5.1 t towed vibrating sheepsfoot roller
	400 mm rolls	600 mm rolls		
Overall roll width (m)	0.37	0.57	1.45	1.83
Diameter of roll shell (m)	0.32	0.32	1.19	1.07
Diameter to ends of feet (m)	0.35	0.35	1.60	1.47
Contact area per foot (m^2)	0.80×10^{-3}	0.80×10^{-3}	4.95×10^{-3}	8.11×10^{-3}
Coverage*	0.185	0.185	0.067	0.147
Mass per roll (kg)	350	370	4340	5080
Mass per unit overall width of roll (kg/m)	950	650	3000	2780
Mass per unit effective width of roll (kg/m)†	5110	3510	45050	18910
Vibrator system				
Location	Chassis between rolls	Chassis between rolls	Roll	Roll
Frequency of vibration (Hz)	33	33	25	39
Nominal centrifugal force (kN)	36	36	98	170
Centrifugal force per unit effective width of roll (kN/m)	263	171	1009	632
Normal speed of rolling during tests (km/h)	1.1	1.1	2.4	2.4

*Coverage is the sum of the end areas of the feet expressed as a proportion of the area of the cylinder swept by the ends of the feet
†Effective width=Overall width×coverage

8.6 The 700 kg vibrating sheepsfoot roller was usually tested on layers of soil in a trench excavated in the test soil. The width of the trench was adjusted to suit the particular width of roll in use. The machine was tested, therefore, in similar conditions to those in which it would normally be used in practice. Plate 8.4 shows the operation of the machine in a trench of uniformly graded fine sand.

Plate 8.4 The 700 kg pedestrian-operated double vibrating sheepsfoot roller, fitted with 600 mm rolls, being operated in a test trench in uniformly graded fine sand

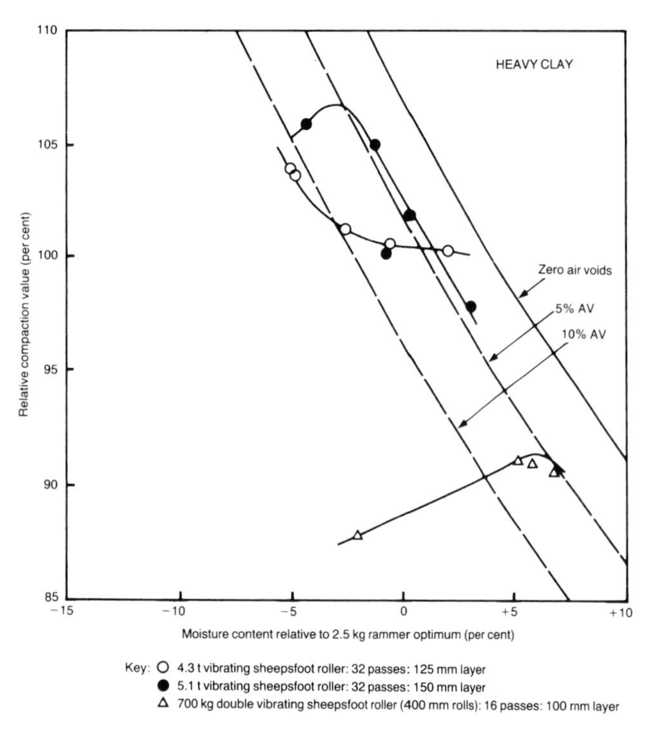

Figure 8.1 Relations between dry density and moisture content for vibrating sheepsfoot rollers on heavy clay, expressed in terms of relative compaction with the 2.5 kg rammer test

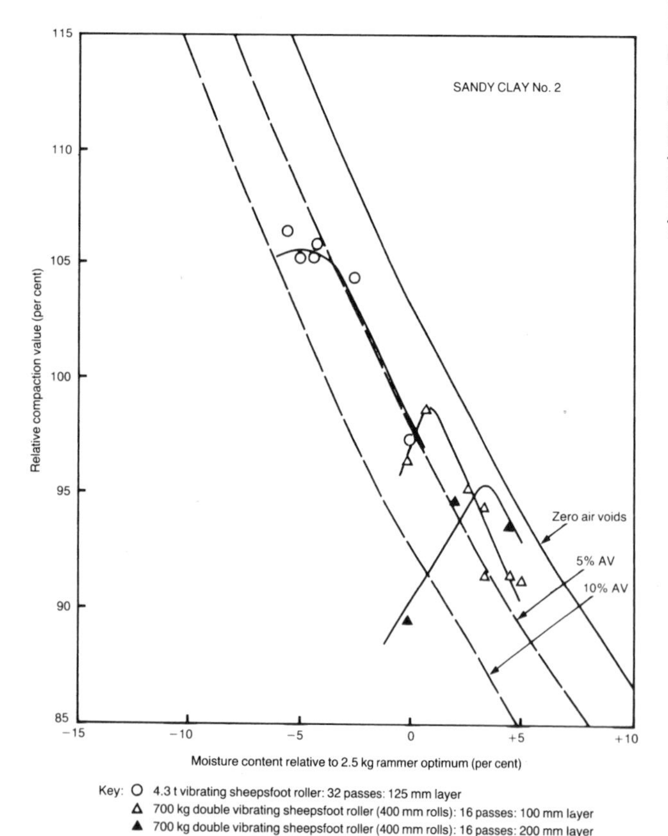

Figure 8.2 Relations between dry density and moisture content for vibrating sheepsfoot rollers on sandy clay No 2, expressed in terms of relative compaction with the 2.5 kg rammer test

8.7 The 4.3 t and 5.1 t towed vibrating sheepsfoot rollers are illustrated in Plates 8.2 and 8.3. Both these machines were occasionally tested in the non-vibrating mode and have also been included in Chapter 6. The rolls and protruding sheepsfeet have already been described in Paragraph 6.7 and it only remains to describe the vibratory system. With the 4.3 t roller the vibrating forces were mainly produced by steel balls which were driven round races contained within the roll, with the additional effect of out-of-balance weights attached to the drive shaft through the centre of the roll. The vibrating mechanism was driven via a centrifugal clutch and Vee-belts by the engine mounted on the rear of the chassis. With the 5.1 t machine the vibrating mechanism consisted of an eccentrically weighted shaft through the centre of the roll, driven via a centrifugal clutch and a flat belt by the engine mounted on the rear of the chassis.

8.8 The results presented in this Chapter have been taken from Cross (1962), Parsons (1962), and unpublished records.

Relations between state of compaction and moisture content

8.9 Relations between the states of compaction produced by the vibrating sheepsfoot rollers and the moisture content of the soil are given in Figures 8.1 to 8.5. Dry density is expressed in terms of the relative compaction value and the

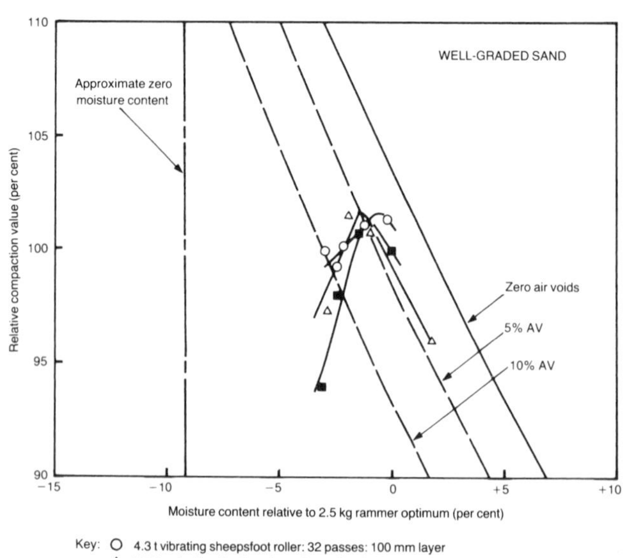

Figure 8.3 Relations between dry density and moisture content for vibrating sheepsfoot rollers on well-graded sand, expressed in terms of relative compaction with the 2.5 kg rammer test

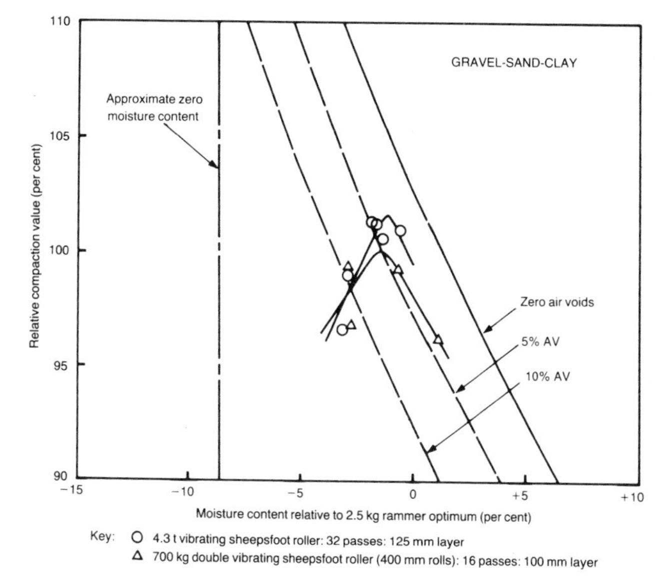

Figure 8.4 Relations between dry density and moisture content for vibrating sheepsfoot rollers on gravel-sand-clay, expressed in terms of relative compaction with the 2.5 kg rammer test

moisture content as an arithmetical difference from the optimum moisture content, both in relation to the results of the 2.5 kg rammer compaction test at the relevant period. The 5.1 t machine was tested on the heavy clay only (Figure 8.1) and the 700 kg machine fitted with 600 mm rolls was tested on the well-graded sand and uniformly graded fine sand only

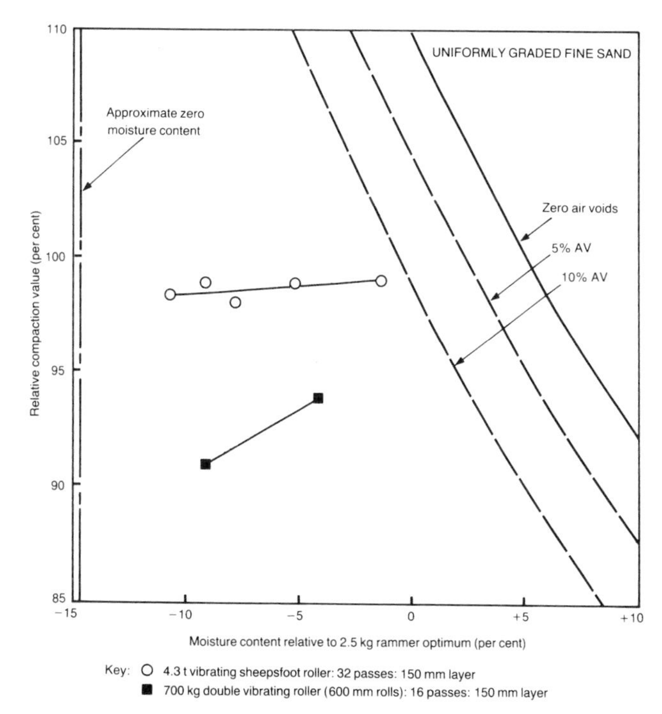

Figure 8.5 Relations between dry density and moisture content for vibrating sheepsfoot rollers on uniformly graded fine sand, expressed in terms of relative compaction with the 2.5 kg rammer test

(Figures 8.3 and 8.5). The 4.3 t machine was tested on all the soils and results for this machine appear on each of the figures.

8.10 The results obtained with 32 passes of the 4.3 t vibrating sheepsfoot roller on heavy clay (Figure 8.1) do not follow the normal pattern in that no maximum relative compaction value or optimum moisture content was obtained. The results for 32 passes of the 5.1 t machine on the same soil exhibit a maximum relative compaction value of about 107 per cent. With the sandy clay (Figure 8.2) a maximum of 106 per cent was achieved with the 4.3 t machine. On the well-graded sand and gravel-sand-clay (Figures 8.3 and 8.4) the 4.3 t machine achieved maximum relative compaction values of 102 per cent, which is considerably lower than the maximum values achieved on the clay soils.

8.11 The 700 kg double vibrating sheepsfoot roller was tested using 16 passes, which was equivalent to 32 vibrating-roll passes. Maximum relative compaction values for the machine fitted with the 400 mm rolls, compacting 100 mm layers, ranged from 91 per cent on heavy clay (Figure 8.1) to 102 per cent on well-graded sand (Figure 8.3). Thus with this machine the best performance was achieved on the more granular soils, as might be expected with a vibratory compactor. Results for the machine fitted with the 600 mm rolls showed only a marginal deterioration in maximum relative compaction value on the well-graded sand from that achieved with the narrower rolls (Figure 8.3). Results are included in Figure 8.2 for compaction of a 200 mm layer of sandy clay No 2 using the machine fitted with 400 mm rolls; as expected, a lower maximum relative compaction value was produced than that achieved using a 100 mm layer.

8.12 Because of the effect of penetration of the protruding feet and the resultant creation of a 'loose mulch' of variable depth, the thicknesses of layer used with the 4.3 t vibrating sheepsfoot roller were relatively small and, except with the uniformly graded fine sand (Figure 8.5), did not attain the desired thickness of 150 mm.

8.13 The results have been summarised, for layer thicknesses of 100 to 150 mm, in Figure 8.6. Relations are given between maximum relative compaction value and mass per unit effective width and also between optimum moisture content and mass per unit effective width; all the results relate to 32 vibrating-roll passes, ie a condition likely to represent compaction to refusal. Because

Figure 8.6 Maximum dry densities and optimum moisture contents related to the mass per unit effective width of vibrating sheepsfoot rollers. Results from Figures 8.1 to 8.4 for 100 to 150 mm layers

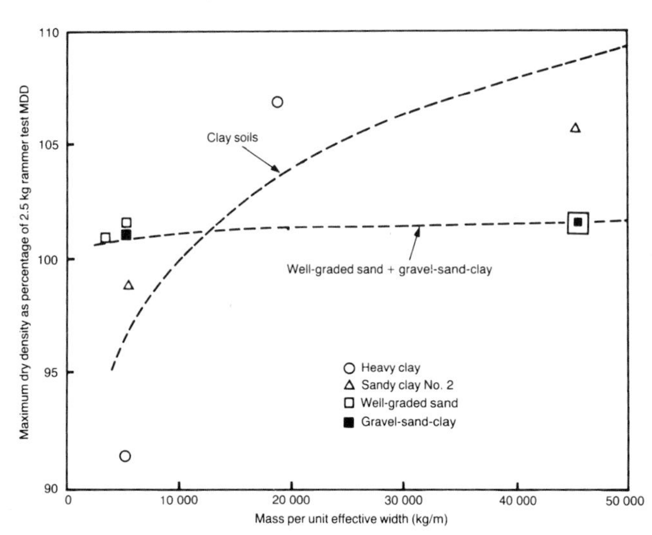

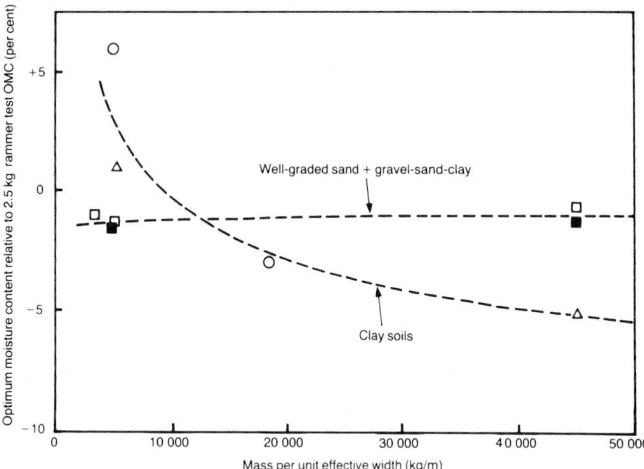

Table 8.2 Equations and correlation coefficients for the relations given in Figure 8.6

Soil	Regression equation*	Number of points	Correlation coefficient
Heavy clay & sandy clay No 2	$c = 60.0\ m^{.0555}$ $w = 607\ m^{-.450} - 10$	4 4	0.82 0.95
Well-graded sand & gravel-sand-clay	$c = 98.2\ m^{.00324}$ $w = 7.69\ m^{.0147} - 10$	5 5	0.58 0.52

*c = maximum dry density for the vibrating sheepsfoot roller as percentage of the maximum dry density obtained in the 2.5 kg rammer compaction test

w = difference between optimum moisture content for the vibrating sheepsfoot roller and optimum moisture content obtained in the 2.5 kg rammer compaction test (per cent)

m = mass per unit effective width as defined in Table 1

The above results apply in general to 32 vibrating-roll passes on 100 to 150 mm compacted layers

of the small amount of data it has been necessary to group together the two clay soils, and the same was the case with the two well-graded granular soils. The clay soils exhibit substantial increases in maximum relative compaction value with increasing mass per unit effective width, but there is little variation in the maximum values achieved on the well-graded granular soils. The equations and correlation coefficients are given in Table 8.2. No attempt has been made to introduce the dynamic force as an additional factor as, for the limited number of machines tested, there was a high level of correlation between mass per unit effective width and centrifugal force per unit effective width (Table 1).

8.14 It is of interest to compare Figure 8.6 with Figure 6.8, which gives similar relations for the non-vibrating sheepsfoot and tamping rollers. The relation for clay soils in Figure 8.6 lies very close to the mean of the relations for the two groups of clay soils shown in Figure 6.8, and so from the limited data it may be concluded that the application of vibration to sheepsfoot rollers did not improve their performance to any significant extent on clay soils. With well-graded sand and gravel-sand-clay, however, performance was improved slightly by vibration at the higher values of mass per unit effective width, but appear to be reduced at the lower values. (A glance at Figure 7.8 indicates very clearly that for these granular soils the vibrating smooth-wheeled roller is considerably more effective.)

8.15 The overstressing and disruption of granular soils by sheepsfoot rollers have already been discussed in Paragraph 6.18, and this effect would also apply to vibrating sheepsfoot rollers, thus creating the lack of improvement in performance with increasing mass per unit effective width illustrated for the well-graded sand and gravel-sand-clay in Figure 8.6.

Relations between state of compaction produced and number of passes

8.16 Results of investigations to determine the effect on the state of compaction achieved of variations in the number of passes of vibrating sheepsfoot rollers are given in Figures 8.7 to 8.11. As in earlier chapters, the results for each soil are illustrated under a separate figure number, whilst different values of moisture content are in separate graphs within each figure.

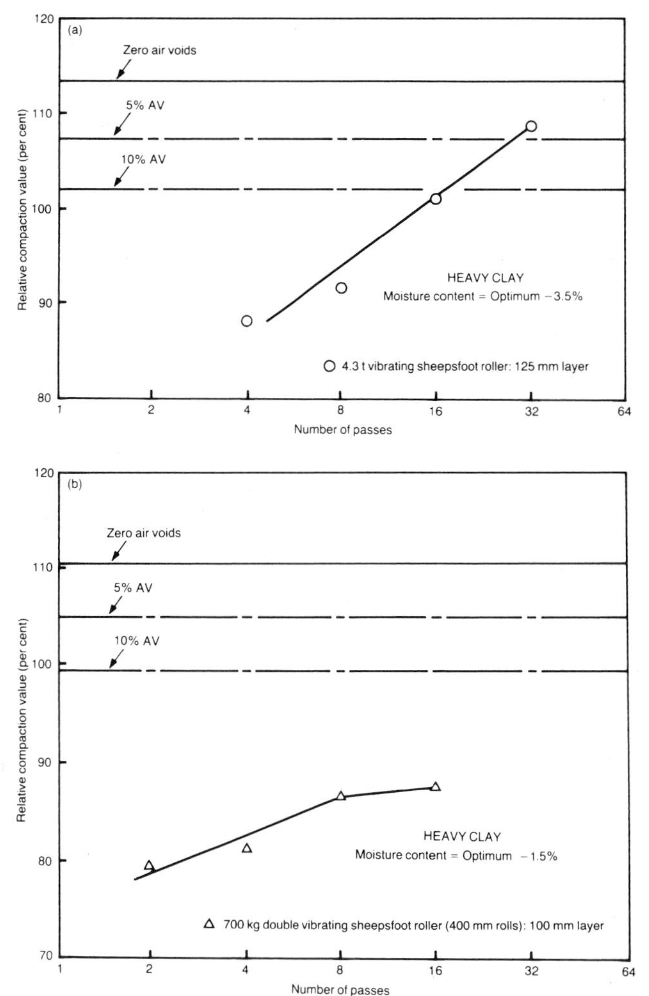

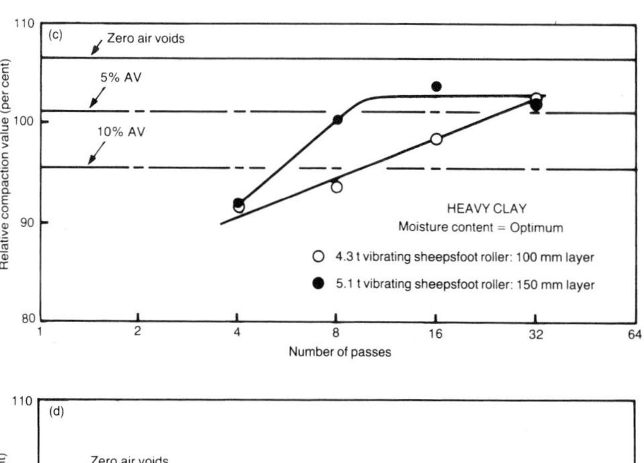

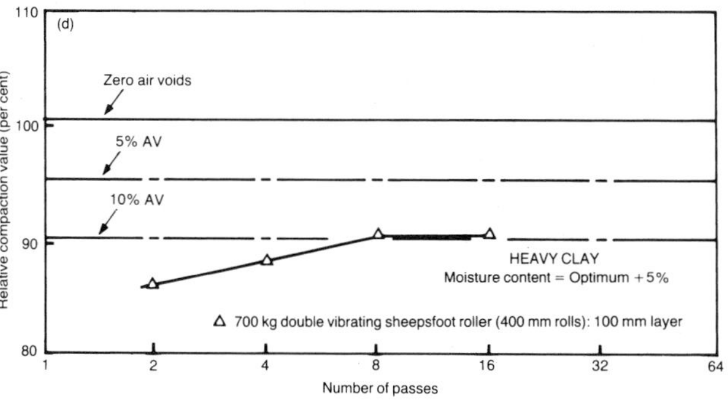

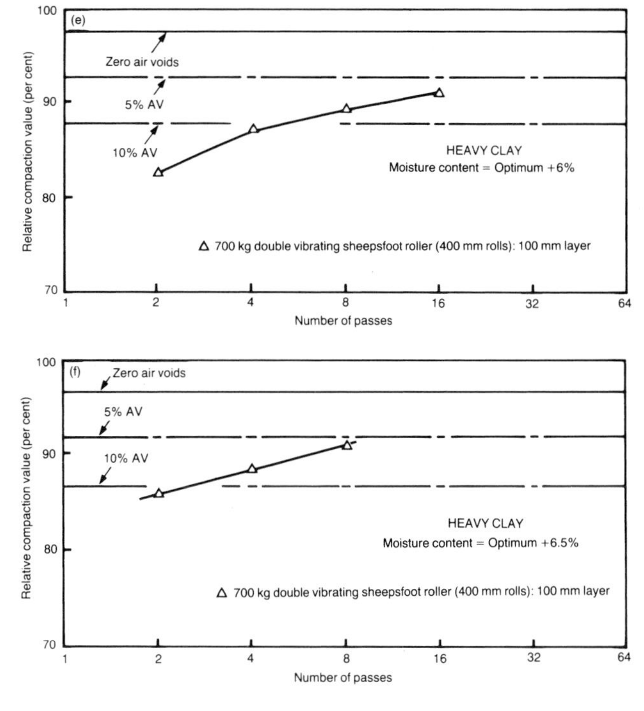

Figure 8.7 Relations between relative compaction and number of passes of vibrating sheepsfoot rollers on heavy clay at six different values of moisture content. All values are related to the maximum dry density and optimum moisture content obtained in the 2.5 kg rammer compaction test (Figure 3.13)

8.17 The 4.3 t vibrating sheepsfoot roller was tested on all the soils and results for the machine appear in Figures 8.7(a) and (c), 8.8(a), 8.9(d), 8.10(b) and 8.11(b). This machine had the highest loadings, static and dynamic, per unit effective width of the machines tested (Table 8.1) and the smallest coverage, with a value of 0.067 (Table 8.1). Thus a relatively high number of passes might be expected to be necessary to achieve complete coverage of an area of soil, although eventually high states of compaction would be produced, given sufficient passes of the roller. The effect of coverage is best illustrated in Figure 8.7(c) where results for the 4.3 t and 5.1 t vibrating sheepsfoot rollers are given for the same moisture content on the heavy clay. The 5.1 t machine, with a coverage of 0.147 (Table 8.1) achieved the maximum attainable state of compaction, between 3 and 4 per cent air voids, after about 10 passes, whereas the 4.3 t machine, with much higher loading per unit effective width but a coverage of only 0.069, required 32 passes to achieve the same low air content. If 95 per cent relative compaction is taken as an adequate level of compaction, the 4.3 t machine achieved such a level on the clay soils in about 9 to 10 passes at

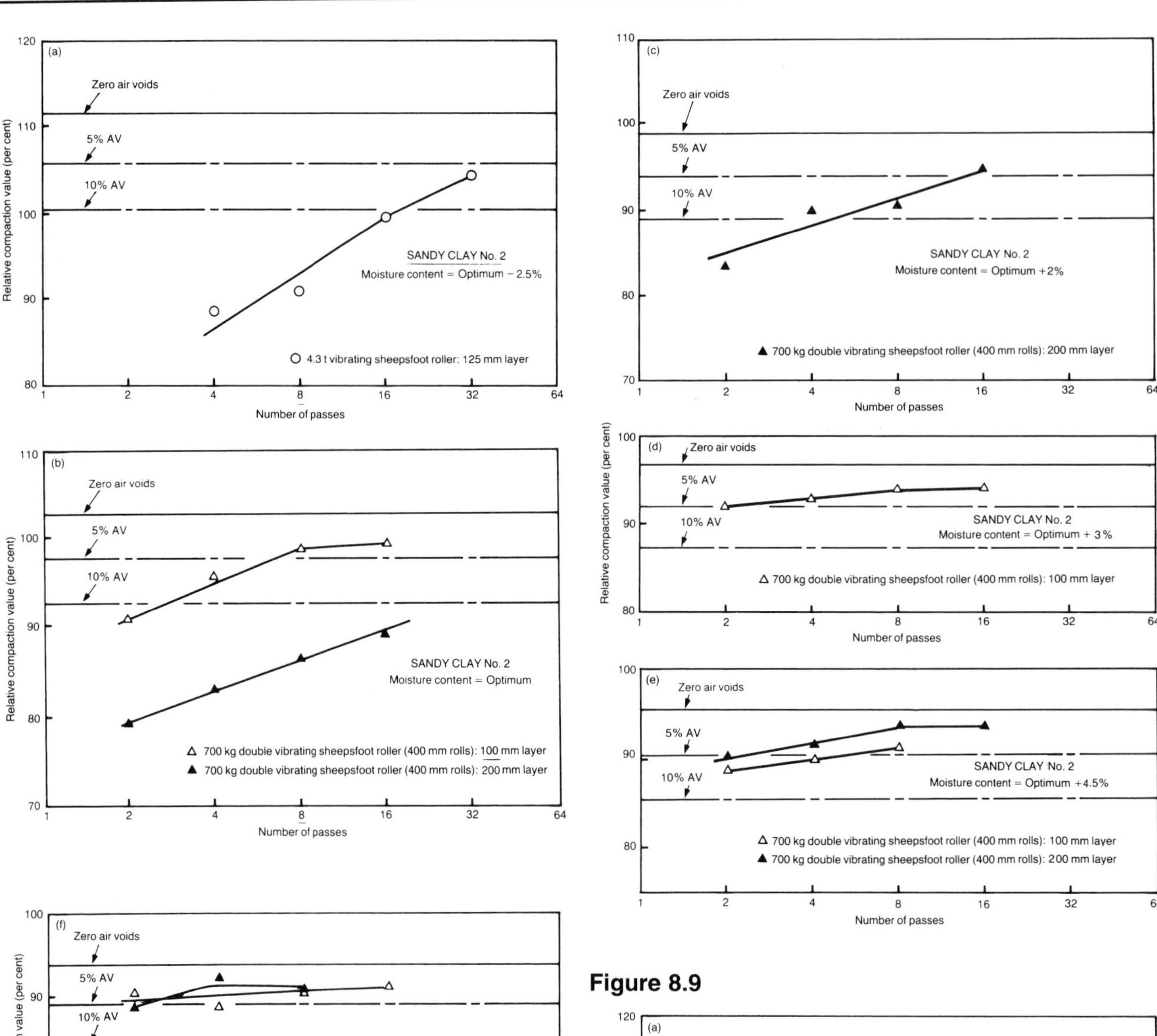

Figure 8.8 Relations between relative compaction and number of passes of vibrating sheepsfoot rollers on sandy clay No 2 at six different values of moisture content. All values are related to the maximum dry density and optimum moisture content obtained in the 2.5 kg rammer compaction test (Figure 3.13)

the moisture contents at which the tests were carried out; with the granular soils about 6 passes were required to reach the same level of compaction.

8.18 The 700 kg double vibrating sheepsfoot roller was tested on all soils and at a number of different moisture contents with each. Even with the relatively thin 100 mm layers an adequate

Figure 8.9

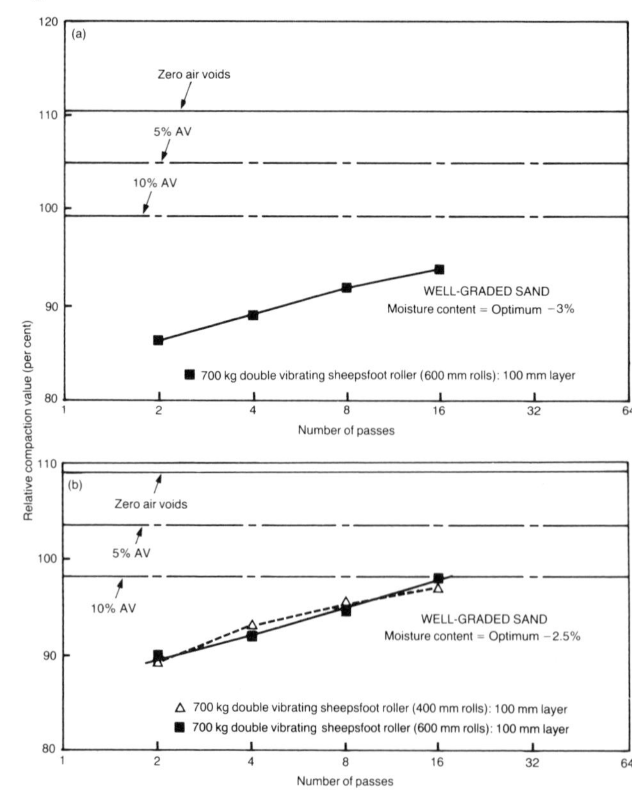

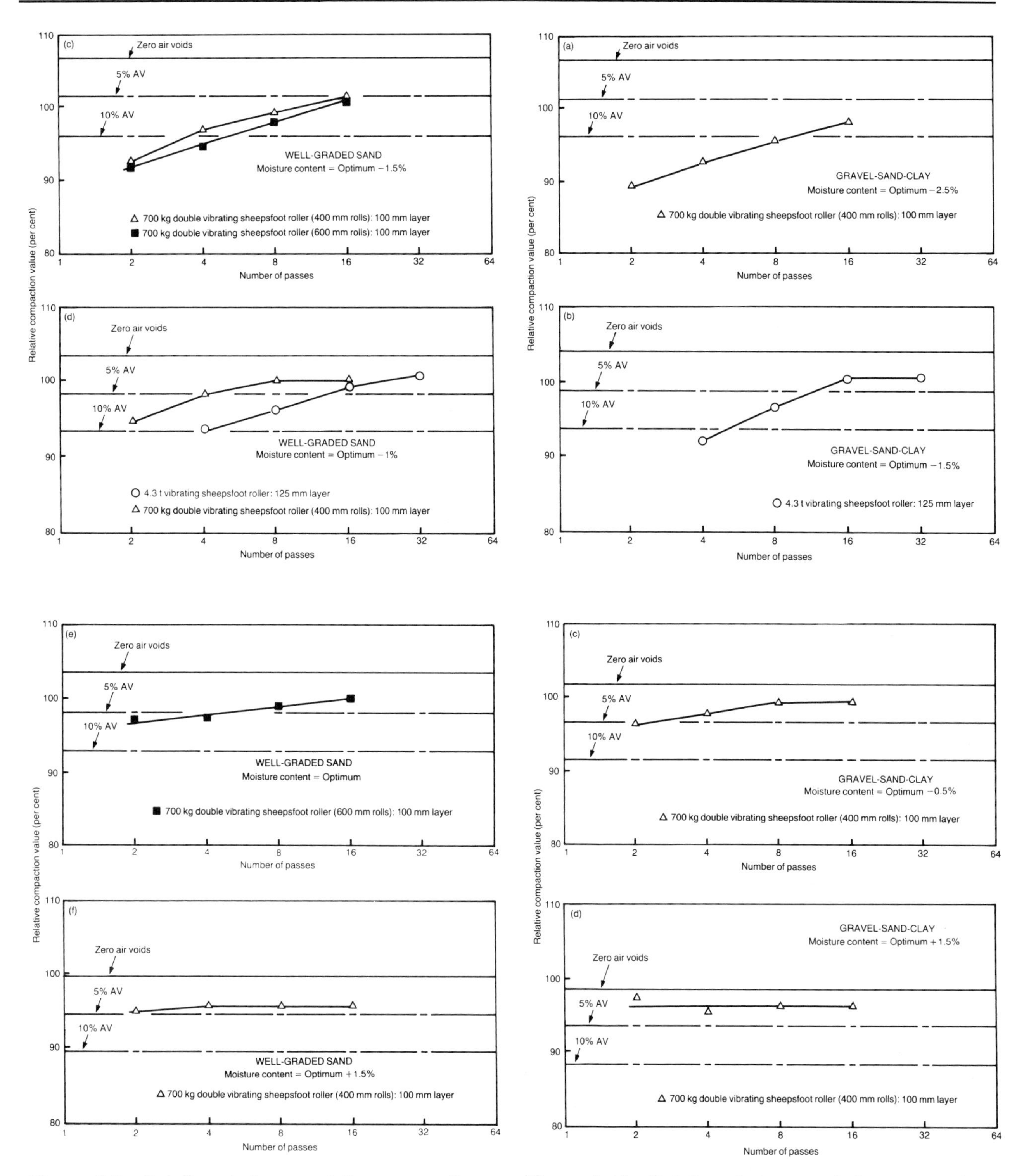

Figure 8.9 Relations between relative compaction and number of passes of vibrating sheepsfoot rollers on well-graded sand at six different values of moisture content. All values are related to the maximum dry density and optimum moisture content obtained in the 2.5 kg rammer compaction test (Figure 3.13)

Figure 8.10 Relations between relative compaction and number of passes of vibrating sheepsfoot rollers on gravel-sand-clay at four different values of moisture content. All values are related to the maximum dry density and optimum moisture content obtained in the 2.5 kg rammer compaction test (Figure 3.13)

state of compaction (95 per cent relative compaction) was not achieved on heavy clay at any moisture content, although 10 per cent air voids or less was attained at moisture contents more than 5 per cent above the optimum of the 2.5 kg rammer compaction test. Note that as this machine was a double vibrating roller each machine pass plotted in the figures represents

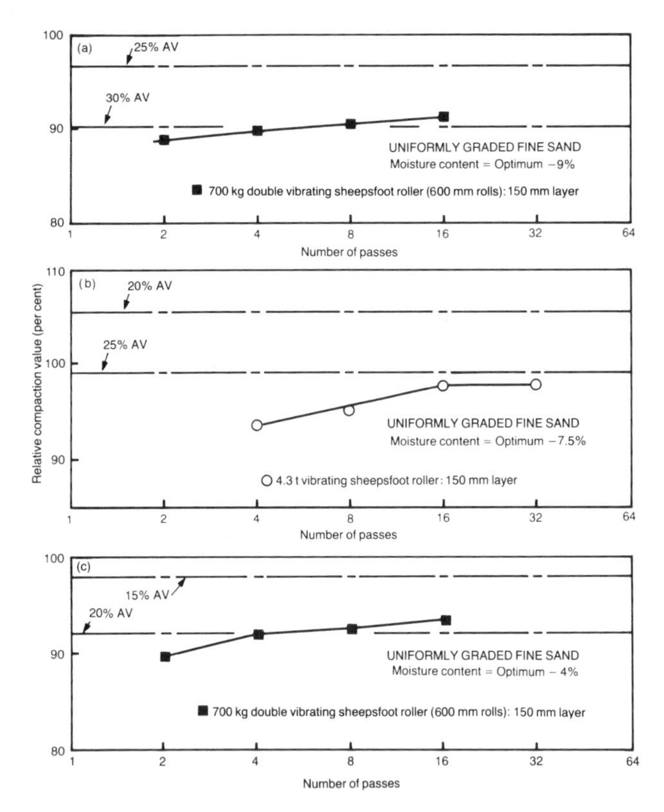

Figure 8.11 Relations between relative compaction and number of passes of vibrating sheepsfoot rollers on uniformly graded fine sand at three different values of moisture content. All values are related to the maximum dry density and optimum moisture content obtained in the 2.5 kg rammer compaction test (Figure 3.13)

two vibrating-roll passes. With the sandy clay difficulty was also experienced with this machine in reaching 95 per cent relative compaction, but over a limited range of moisture contents and with a high number of passes it could be achieved in 100 mm layers when fitted with the 400 mm rolls (Figures 8.2 and 8.8(b)).

8.19 In contrast, with well-graded sand and gravel-sand-clay the 700 kg machine performed well on 100 mm layers (Figures 8.9 and 8.10) over a wide range of moisture contents. As also demonstrated in Figure 8.3, there was little difference in the performance of the machine with either the 400 mm or 600 mm rolls on well-graded sand (see Figures 8.9(b) and (c) for direct comparisons).

8.20 With uniformly graded fine sand the 700 kg machine was used with the 600 mm rolls only, possibly with a view to restrict the degree of overstressing normally obtained on the surface of the compacted layer with this type of soil. However, as shown in Figure 8.11(a) and (c), a relative compaction value of 95 per cent was not achieved after 16 passes (32 vibrating-roll passes) on a 150 mm layer.

8.21 With the limited number of machines investigated it has not been possible to determine the effect of factors such as coverage (used for static-weight sheepsfoot, tamping and grid rollers in Table 6.3) and centrifugal force (as explored in Paragraphs 7.61 to 7.66) because of their high levels of correlation with mass per unit effective width. A summary of the results can only be presented, therefore, in terms of moisture content and mass per unit effective width. Regression analyses have been made of the results for 100 mm to 150 mm layers in Figures 8.1 to 8.4 and 8.7 to 8.10 (ie excluding the results for uniformly graded fine sand) using mass per unit effective width and moisture content as independent variables. The numbers of passes required to achieve a relative compaction value of 95 per cent and an air content of 10 per cent were related to these variables. The results, where reasonable correlations were obtained, are given in Table 8.3. The limited data for the clay soils resulted in only one correlation being sufficiently good to be included in the Table, that for the achievement of 95 per cent relative compaction on sandy clay No 2. However, only two machines were involved and so even this, despite its seemingly perfect correlation, cannot be relied upon. More data were available for the granular soils, however, allowing the inclusion of results for the achievement of both 95 per cent relative compaction and 10 per cent air voids. Although

Table 8.3 Equations and correlation coefficients to determine the number of passes required to achieve specified levels of compaction with vibrating sheepsfoot rollers; layer thickness 100 to 150 mm

Soil	Level*	Regression equation†	No of data points	Correlation coefficient
Sandy clay No 2	(a)	$P=6.11-1.34m-.421w$	3	0.999
Well-graded sand & gravel-sand-clay	(a)	$P=1.35-.181m-.176w$	15	0.89
	(b)	$P=.584-.126m-.563w$	14	0.93

*Level (a)=95 per cent of the maximum dry density obtained in the 2.5 kg rammer compaction test
Level (b)=10 per cent air voids
†P=Log$_{10}$ (Number of vibrating-roll passes required)
m=Log$_{10}$ (Mass per unit effective width of roll in kg/m)
w=Moisture content of soil, per cent, minus the optimum moisture content, per cent, obtained in the 2.5 kg rammer compaction test

these latter equations in Table 8.3 give a fair representation of the results obtained in the investigations, the fact that they apply to tests with only four machines where the possible effects of important parameters such as coverage and centrifugal force could not be separated, indicate that caution must be

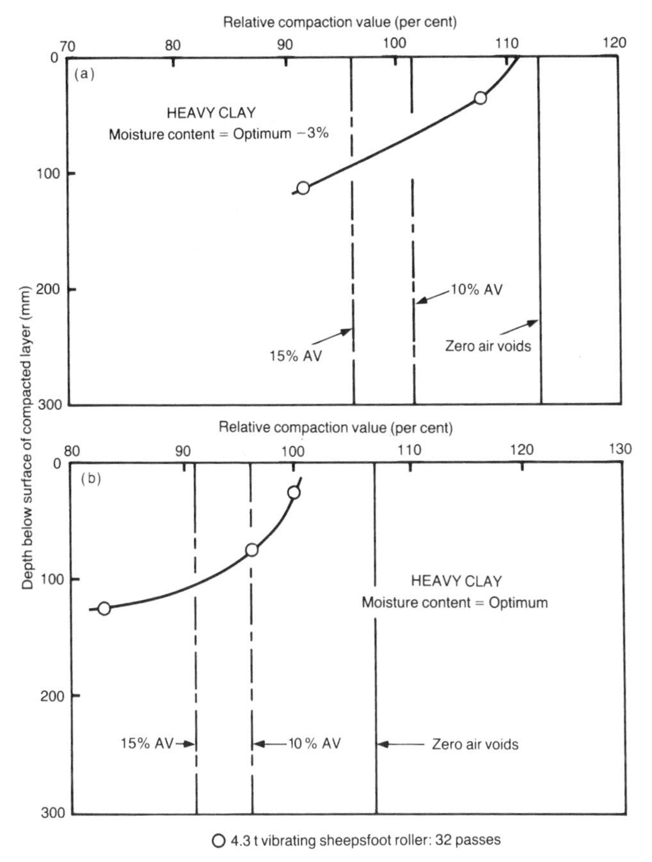

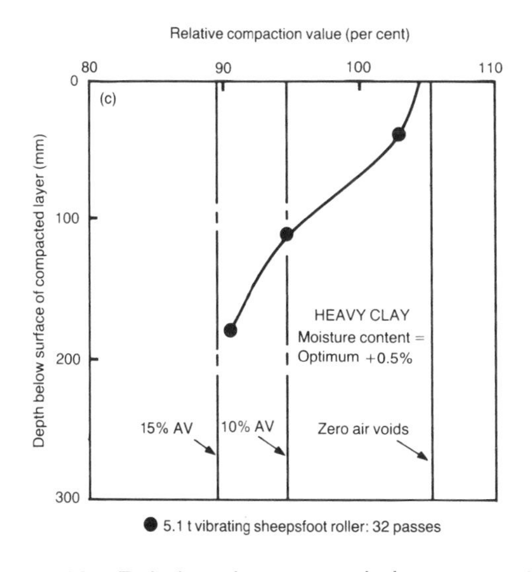

Figure 8.12 Relations between relative compaction and depth in the compacted layer of heavy clay obtained with vibrating sheepsfoot rollers. All values are related to the maximum dry density and optimum moisture content obtained in the 2.5 kg rammer compaction test (Figure 3.13)

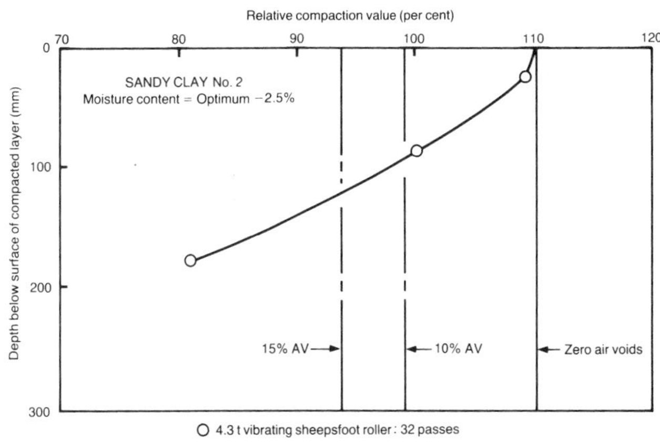

Figure 8.13 Relation between relative compaction and depth in the compacted layer of sandy clay No 2 obtained with the 4.3 t vibrating sheepsfoot roller. All values are related to the maximum dry density and optimum moisture content obtained in the 2.5 kg rammer compaction test (Figure 3.13)

exercised in any attempt to utilise the information to predict the performance of other types of vibrating sheepsfoot roller.

Relations between state of compaction and depth within the compacted layer

8.22 Results of investigations of the variation of density with depth in layers of soil compacted by vibrating sheepsfoot rollers are given in Figures 8.12 to 8.16. The results were obtained using the 4.3 t towed vibrating sheepsfoot roller in all cases except in Figure 8.12(c), where the results were obtained on heavy clay using the 5.1 t towed vibrating sheepsfoot roller. Because of the absence of variety in the type of machine used in these particular tests, therefore, no analyses can be made of the data to determine the influence of relevant machine parameters. The thicknesses of layer capable of being compacted by the 4.3 t machine at the test moisture contents, using the criteria of 90 per cent relative compaction and 15 per cent air voids to be achieved at the bottom of the layer, were about 100 to 140 mm with the clay soils and about 140 to 180 mm in the well-graded granular soils; these thicknesses were achieved using 32 passes of the roller and did not include the 'loose mulch' on the surface of the layer. It should be noted that the results obtained with uniformly graded fine sand (Figure 8.16) were inconclusive but it appears that in excess of 200 mm could have been compacted to achieve 90 per cent relative compaction at the bottom of the layer. With the 5.1 t vibrating sheepsfoot roller, the tests on heavy clay at a moisture content 0.5 per cent above the optimum of the 2.5 kg rammer

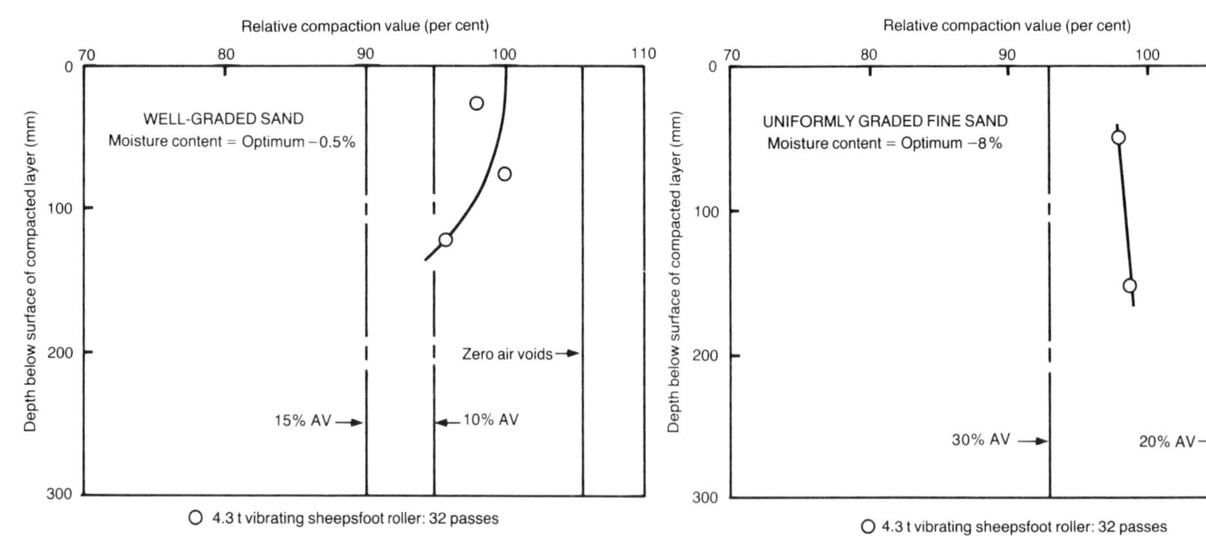

Figure 8.14 Relation between relative compaction and depth in the compacted layer of well-graded sand obtained with the 4.3 t vibrating sheepsfoot roller. All values are related to the maximum dry density and optimum moisture content obtained in the 2.5 kg rammer compaction test (Figure 3.13)

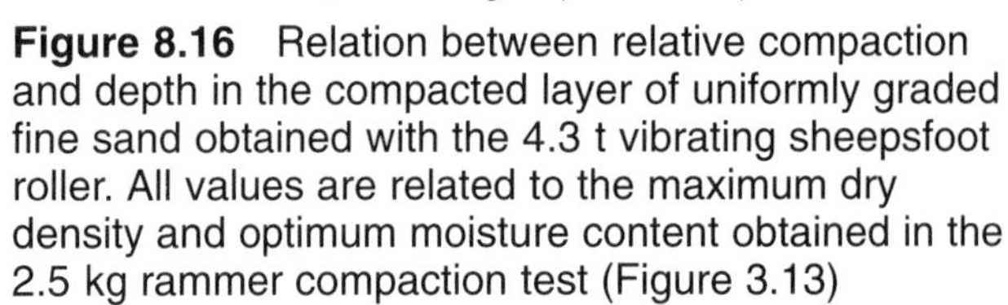

Figure 8.16 Relation between relative compaction and depth in the compacted layer of uniformly graded fine sand obtained with the 4.3 t vibrating sheepsfoot roller. All values are related to the maximum dry density and optimum moisture content obtained in the 2.5 kg rammer compaction test (Figure 3.13)

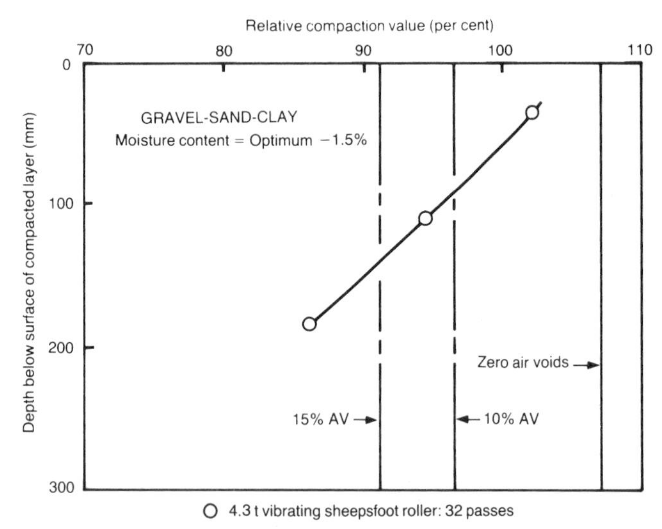

Figure 8.15 Relation between relative compaction and depth in the compacted layer of gravel-sand-clay obtained with the 4.3 t vibrating sheepsfoot roller. All values are related to the maximum dry density and optimum moisture content obtained in the 2.5 kg rammer compaction test (Figure 3.13)

compaction test (Figure 8.12(c)) indicate that a layer thickness of about 180 to 190 mm would be compacted to the same criteria.

Effect of speed of rolling on the performance of vibrating sheepsfoot rollers

8.23 The effect of varying the speed of rolling on the state of compaction produced was investigated with the 4.3 t towed vibrating sheepsfoot roller and the results are given, in the form of relations between relative compaction value and number of passes, in Figures 8.17 and 8.18. Three speeds of rolling were used on the heavy clay (Figure 8.17) and two speeds of rolling on sandy clay No 2 (Figure 8.18). The results for the normal speed of rolling, 2.4 km/h, have also been included in Figures 8.7(c) and 8.8(a). Although it is normally expected that an increase in speed of rolling would result in an increase in the number of passes required to achieve a given level of compaction, there was little difference in the results obtained at 2.4 and 3.6 km/h on both the heavy clay and sandy clay. In fact, with the heavy clay, the state of compaction for passes in excess of 6 was increased slightly by increasing the speed from 2.4 to 3.6 km/h. However, with further increases

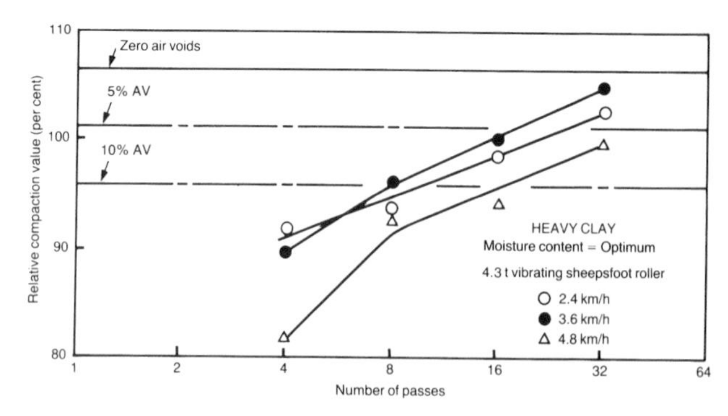

Figure 8.17 Relations between relative compaction and number of passes of the 4.3 t vibrating sheepsfoot roller on heavy clay at three different speeds of rolling. Values are related to the maximum dry density and optimum moisture content obtained in the 2.5 kg rammer compaction test (Figure 3.13)

in speed of rolling to 4.8 km/h the more usual reduction in state of compaction was obtained (Figure 8.17). It was concluded at the time that the improvement in the state of compaction on increasing the speed from 2.4 to 3.6 km/h was probably attributable to an improvement in the coverage of the soil at the higher speed, brought about by a reduction in the tendency for the feet to 'slot in' to previously made imprints (see Paragraph 8.35).

8.24 The results indicate, therefore, that with the 4.3 t towed vibrating sheepsfoot roller a speed of rolling of 3.6 km/h was more efficient on clay soils than the lower speed of 2.4 km/h normally used in the tests. Further increase in speed of rolling above 3.6 km/h resulted in substantial increases in the number of passes required to achieve given levels of compaction in heavy clay (Figure 8.17). In general, an increase in speed by a factor of 1.3 to 4.8 km/h resulted in a doubling of the required number of passes to achieve 95 per cent relative compaction or 10 per cent air voids.

Comparison of the performances of the 700 kg double vibrating sheepsfoot roller working in a trench and on a surface layer of soil

8.25 As mentioned in Paragraph 8.6, the 700 kg machine was normally operated within the confines of a trench excavated to a width to match the size of rolls being used (Plate 8.4). The purpose of this was to operate the machine in its normal environment where any effects on the layer being compacted of side-wall reaction and friction could be incorporated in the test results. A limited amount of testing was also

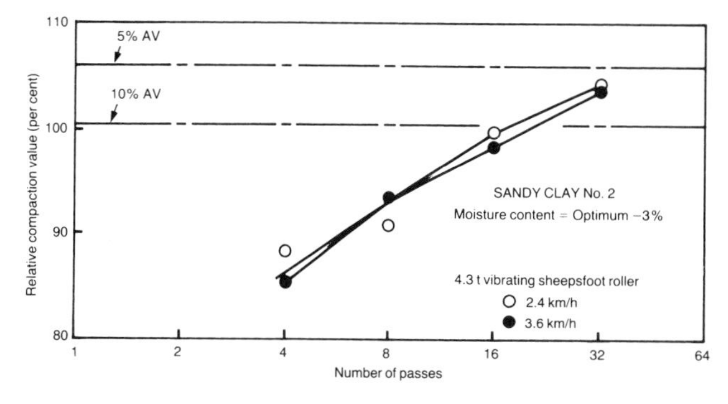

Figure 8.18 Relations between relative compaction and number of passes of the 4.3 t vibrating sheepsfoot roller on sandy clay No 2 at two different speeds of rolling. Values are related to the maximum dry density and optimum moisture content obtained in the 2.5 kg rammer compaction test (Figure 3.13)

carried out by the procedure normally adopted for determining the performance of compaction plant in the test facilities, compacting a prepared layer of loose soil at the surface level of the test soil in its 1 m deep pit (see Chapter 3). The results are given in Figure 8.19 in the form of relations between relative compaction value and moisture content of sandy clay No 2, for 2 and 16 passes of the machine on 100 mm layers in each of the two test locations. It is unfortunate that, as shown in the Figure, most of the results lie on the wet side of the optimum moisture content for the compaction efforts used, so that any differences between the performances in the two test locations would largely be masked. However, it is clear that for the results presented a common relation can be drawn for the two locations for both 2 and 16 passes. Thus it can be concluded that, for sandy clay soil at least, no detectable difference in the performance of this machine occurred in varying the location of the test from the trench to the surface.

8.26 This conclusion is of considerable importance, as many machines which are commonly used in the compaction of backfill in trenches have been tested over the years solely on surface layers, and the applicability of the results to the use of the machines in trenches can be questioned. A more detailed comparison of within-trench and surface compaction locations was made with a vibro-tamper and is described later in Chapter 10.

Effect on the performance of the 700 kg double vibrating sheepsfoot roller of operating the vibrator shaft in the opposite direction to the direction of travel

8.27 As described in Paragraph 8.5, the eccentric shaft, mounted in the chassis of the 700 kg machine, was hydraulically powered and could be operated rotating in either direction. The manufacturer recommended that it should normally rotate in sympathy with the direction of travel, ie rotating in the same direction as the rolls, and this was the normal mode of operation of the machine during the tests. Thus, when the direction of travel was reversed to return backwards along the area being compacted, the direction of rotation of the vibrator shaft was also reversed.

8.28 To determine the effect of operating the machine with the vibrator shaft rotating in the opposite direction to the rolls a limited amount of

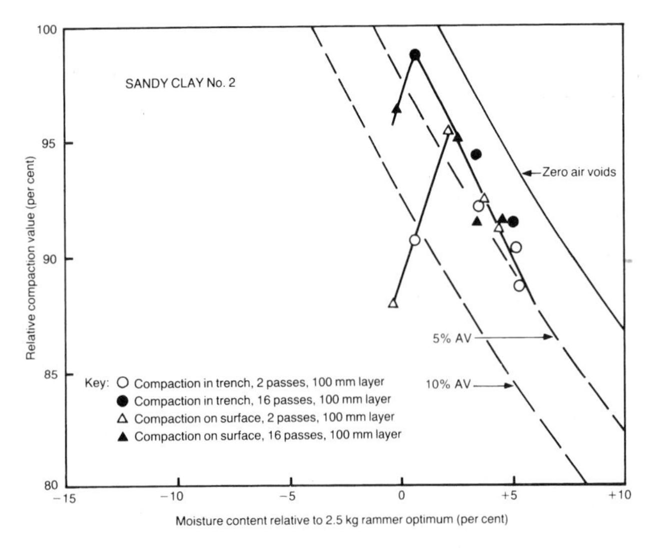

Figure 8.19 Relations between dry density and moisture content for the 700 kg double vibrating sheepsfoot roller (400 mm rolls) used on sandy clay No 2 in a trench and on the surface, expressed in terms of relative compaction with the 2.5 kg rammer test

testing was carried out on the well-graded sand. The results are given in Figure 8.20 in the form of relations between relative compaction value and moisture content for 4 and 16 passes of the machine when compacting 100 mm layers of soil. Forward rotation of the eccentric shaft is indicative of the normal mode of operation, and reverse rotation is that opposed to the direction of rotation of the rolls.

8.29 The results in Figure 8.20 show that an increase in the state of compaction resulted from the reversal of the eccentric shaft for both 4 and 16 passes. At moisture contents dry of the machine's optimum the increase amounted to about 1.5 to 2 per cent in the relative compaction value.

8.30 During these tests the speed of rolling was measured and the results are given in Figure 8.21 in relation to the moisture content of the soil. Considerable differences in speed of rolling were recorded for the two directions of rotation of the eccentric shaft. The effect of the vibrating mechanism on speed of travel of vibrating rollers has been explained in Paragraph 7.49, and this explanation also applies to the differences exhibited in Figure 8.21. The speed of rolling was also affected by changes in moisture content above a threshold value about 1 to 2 per cent below the 2.5 kg rammer optimum moisture content. With increases in moisture content above this threshold the speed of rolling decreased with forward rotation of the eccentric shaft and increased with the shaft rotating in the opposite direction, with differences in speeds of rolling becoming minimal at moisture contents about 1 per cent or more above the 2.5 kg rammer optimum (Figure 8.21). Reasons for this reduction in the differences in speed of rolling could be associated with:–

(a) The smaller amplitude of the rolls as the soil became weaker, thus reducing the potential for roll slip, or

(b) Sufficient penetration of the roll feet at the higher moisture contents to provide a keying-in of the roll with the soil to prevent roll slip.

8.31 The difference in performance of the 700 kg machine for the two modes of operation illustrated in Figure 8.20 can be explained,

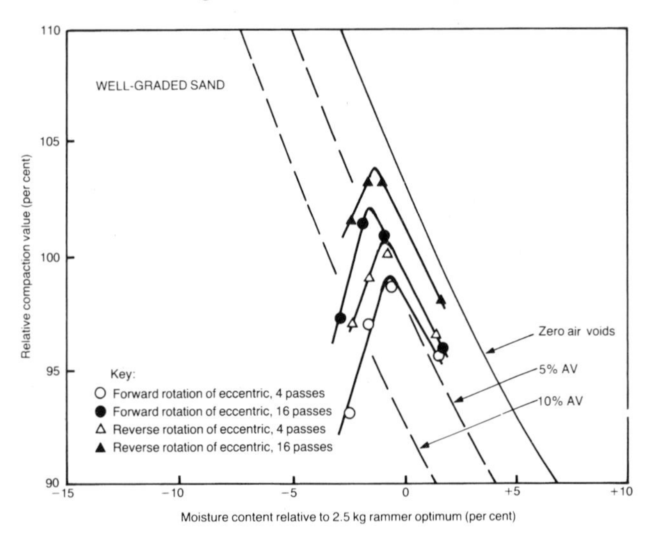

Figure 8.20 Relations between dry density and moisture content for the 700 kg double vibrating sheepsfoot roller (400 mm rolls) on a 100 mm layer of well-graded sand, expressed in terms of relative compaction with the 2.5 kg rammer test

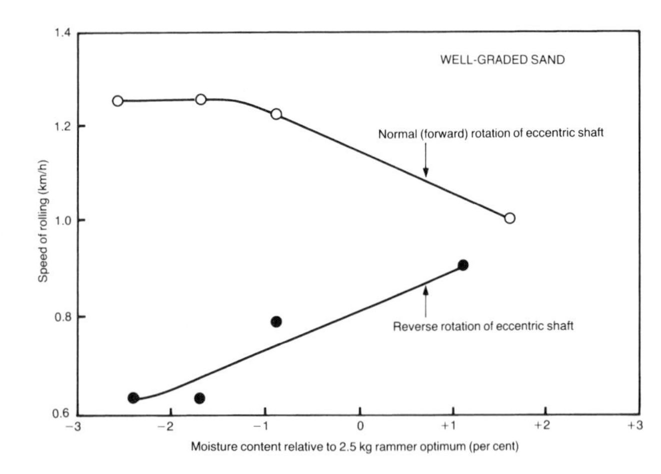

Figure 8.21 Relation between speed of rolling of the 700 kg double vibrating sheepsfoot roller and the moisture content of well-graded sand for different directions of rotation of the vibrator shaft

therefore, by the differences in speed of rolling in exactly the same way as the differences in performance of the 350 kg pedestrian-operated vibrating smooth-wheeled roller, using different directions of travel, were explained in Paragraphs 7.48 to 7.51. (It is of interest to note that the ratio of the speeds of travel of the 350 kg smooth-wheeled machine in the two directions, at 1.4 : 0.7 km/h, was almost identical to the ratio of speeds for the two directions of rotation of the eccentric shaft of the 700 kg sheepsfoot roller at the lower moisture contents at 1.25 : 0.63 km/h.)

8.32 The improved performance of the 700 kg vibrating sheepsfoot roller with the vibrator shaft rotating in the opposite direction to the rotation of the rolls also represents an improvement in output of compacted soil in the range of moisture contents where the speed differential was greatest. Thus, on the well-graded sand at moisture contents drier than 2.5 kg rammer optimum minus 1.5 per cent, the output of soil compacted to 95 per cent relative compaction or 10 per cent air voids is estimated to be about 1.3 times greater with the reverse operation of the vibrator shaft. However, it should be remembered that the forward operation of the shaft is intended to assist traction, as clearly evidenced by the increased speed of rolling. For instance, it was found during the investigations with the machine that it could only successfully climb the ramp out of the test trench when the vibrator shaft was rotating in the forward direction.

Surface of layer following compaction by vibrating sheepsfoot rollers

8.33 The disturbed surface created by the feet of sheepsfoot rollers has already been described and illustrated in Paragraphs 6.38 to 6.43. Plate 6.15 shows the surface of heavy clay at a relatively high moisture content after compaction by the 4.3 t vibrating sheepsfoot roller. In this instance the feet had completely penetrated the soil so that the roll shell bore on the surface of the so-called 'loose mulch'. At reduced values of moisture content such deep penetrations would only occur during the initial passes of the machine and, at sufficiently low moisture contents, 'walk-out' would occur, with reducing thickness of loose mulch with successive passes, as demonstrated for the static-weight sheepsfoot roller in Figure 6.24. Unfortunately, no relations between dry density and number of passes were determined at sufficiently low moisture contents to demonstrate such reductions in depth of loose mulch with vibrating sheepsfoot rollers.

8.34 The two large vibrating sheepsfoot rollers were, however, used to compact heavy clay soil over a range of moisture contents, using 32 passes to represent compaction to refusal. Relations between depth of loose mulch and moisture content resulting from these tests are shown in Figure 8.22; the equivalent relative compaction values in the underlying compacted layer are given in Figure 8.1. The depth of loose mulch left by the 4.3 t machine was considerably greater than that left by the 5.1 t machine over a wide range of moisture contents (Figure 8.22). This was clearly associated with the smaller end area of the roll foot of the 4.3 t machine (Table 8.1) and the resulting higher static and dynamic stresses exerted under the feet. It is interesting to compare the results in Figure 8.22 with those for the same machines used without vibration in Figure 6.23. The application of vibration generally caused the relation to move upwards, ie the depth of loose mulch increased at a given moisture content, by about 20 to 30 mm.

8.35 When using the 4.3 t machine on cohesive soil at moisture contents above the threshold value for effective walk-out it was found that the feet had a tendency to 'slot-in' to the imprints made during the previous pass (see Plate 6.15 for an illustration of the imprints), thus preventing uniformity of coverage when using a large number of passes. To overcome this tendency it was recommended (Parsons, 1962) that the direction of towing over an area of soil should be

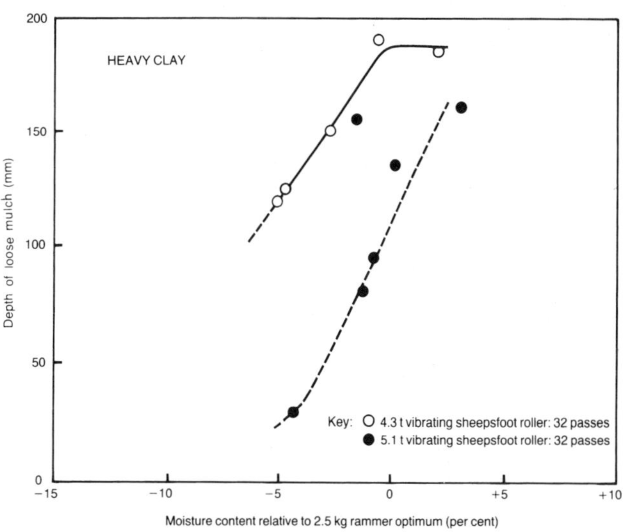

Figure 8.22 Relations between depth of disturbed heavy clay on surface of compacted layer and moisture content for compaction to refusal by vibrating sheepsfoot rollers

altered by a few degrees on each successive pass. This, in fact, was the technique applied in the test programme with this particular machine.

General observations on the application of heavy vibrating sheepsfoot rollers

8.36 If the walking-out of the sheepsfeet is judged to be complete when the depth of loose mulch has been reduced to about 25 mm (see Paragraph 6.41), it can be seen in Figure 8.22 that the 5.1 t vibrating sheepsfoot roller 'walked out' at moisture contents of 5 per cent or more below the 2.5 kg rammer optimum moisture content and, by extrapolation of the relation, the 4.3 t machine 'walked out' at moisture contents of 12 per cent or more below the 2.5 kg rammer optimum moisture content. Unfortunately, the states of compaction likely to be produced by the 4.3 t machine on heavy clay at such low moisture contents cannot be accurately determined from the results given in Figure 8.1, but it is clear that the 5.1 t machine was capable of producing relative compaction values well in excess of 95 per cent when working in the range of 5 to 10 per cent below the 2.5 kg rammer optimum moisture content.

8.37 Records of depths of loose mulch for 32 passes of the 4.3 t vibrating sheepsfoot roller on sandy clay No 2 show that a very similar relation to that for heavy clay was produced, with indications that moisture contents no wetter than 12 per cent below the 2.5 kg rammer optimum would be necessary for complete walk-out of the feet. It can be concluded, therefore, that the maximum moisture content for efficient operation of the 4.3 t machine on cohesive soil was 12 per cent below the 2.5 kg rammer optimum and for the 5.1 t machine was 5 per cent below the same optimum. With granular soils excessive depths of loose mulch were produced by the 4.3 t machine at all moisture contents tested and, coupled with the fairly poor results shown on the well-graded sand and gravel-sand-clay (Figure 8.3 and 8.4) it must be concluded that such soils could be more efficiently compacted using vibrating smooth-wheeled rollers of equivalent size.

References

Cross, J E (1962). An investigation of the performance of a 5-ton vibrating sheepsfoot roller in the compaction of a heavy clay soil. *D.S.I.R., RRL Laboratory Note* No LN/166/JEC (Unpublished).

Parsons, A W (1962). An investigation of the performance of a 4¼-ton vibrating sheepsfoot roller for compacting soil. *D.S.I.R., RRL Laboratory Note* No LN/211/AWP (Unpublished).

CHAPTER 9 COMPACTION OF SOIL BY VIBRATING-PLATE COMPACTORS

Description of machines

9.1 Vibrating-plate compactors have a flat plate in contact with the soil on which either one or two eccentrically weighted shafts are mounted. If two out-of-balance shafts are used they are counter-rotating and phased to provide a resultant force either just forward or just to the rear of vertical to give forward or reverse travel; where one eccentric shaft is used forward motion only is achieved by positioning the shaft ahead of the centre of gravity of the machine. The power unit and control handles for the pedestrian operator are attached to a chassis supported above the base-plate on flexible mountings, usually in the form of springs or rubber cushions. All machines are equipped with some form of wheeled undercarriage which can be used to assist transit between working areas.

9.2 The machines of this type which have been included in investigations at the Laboratory are illustrated in Plates 9.1 to 9.14, and details are given in Table 9.1. They appear in the Plates and are listed in the Table in ascending order of total mass. For a number of the machines tested in early investigations the dynamic forces were not published and no records can be found of centrifugal force where 'NA' appears in Table 9.1.

9.3 A total of eight of the vibrating-plate compactors listed in Table 9.1 were fitted with vibrator units comprising single eccentric shafts; these machines were therefore restricted to forward motion only. They are illustrated in Plates 9.1, 9.2, 9.4 to 9.7, 9.9 and 9.10. In all cases the vibration was actuated by increasing the engine speed to the full extent of the throttle lever, the vibrator being driven via a Vee-belt system by the engagement of a centrifugal clutch. As can be seen from the details given in Table 9.1, these machines differed considerably with respect to their mass, centrifugal force and contact area of base-plate. The 450 kg vibrating-plate compactor could, in addition, be fitted with extension plates to increase the area of the base-plate, as shown in Plate 9.15.

Plate 9.1 80 kg (410 kg/m²) vibrating-plate compactor investigated in 1985

Plate 9.2 80 kg (450 kg/m²) vibrating-plate compactor investigated in 1985

Plate 9.3 140 kg vibrating-plate compactor investigated in 1986

Plate 9.4 150 kg vibrating-plate compactor investigated in 1985

Plate 9.5 180 kg (470 kg/m^2) vibrating-plate compactor investigated in 1987

Plate 9.6 180 kg (710 kg/m^2) vibrating-plate compactor investigated in I986

9.4 The machines equipped with twin eccentric shafts differed also in their mass, centrifugal force and area of base-plate. In addition, the use made of the phase angle of the eccentric shafts or rotation of the vibrator unit to control speed and direction of travel varied between the various machines.

9.5 With the 140 kg vibrating-plate compactor (Plate 9.3) the phase angle of the shafts was varied hydraulically to produce a continuously variable speed of travel in forward and reverse directions. The tests with this machine were carried out using the maximum forward and reverse speeds.

9.6 The 240 kg machine (Plate 9.8) had twin eccentric shafts, one of which could be rotated about the pivot of the second shaft and could be locked in any one of five positions to give two forward speeds, two reverse speeds, or a vertical vibration, as required. The machine was normally used in the tests with the eccentrics locked in the 'full-forward' position.

Table 9.1 Details of the various machines contributing data on the performance of vibrating-plate compactors

Vibrating-plate compactor	Total mass (kg)	Width of base plate (m)	Contact area of base plate (m^2)	Mass per unit area (kg/m^2)	Vibrating mechanism: No of eccentric shafts	Frequency (Hz)	Nominal centrifugal force (kN)	Centrifugal force per unit area (kN/m^2)	Normal speed of travel (km/h)
80 kg (410 kg/m^2)	83	0.50	0.200	410	1	83	9.7	48	1.0
80 kg (450 kg/m^2)	80	0.40	0.176	450	1	100	13.5	77	1.0
140 kg	144	0.40	0.180	800	2	90	24.0	133	0.9
150 kg	151	0.49	0.194	780	1	72	13.7	71	0.9
180 kg (470 kg/m^2)	178	0.75	0.383	470	1	78	17.0	44	0.8
180 kg (710 kg/m^2)	181	0.59	0.254	710	1	67	16.5	65	1.1
180 kg (820 kg/m^2)	181	0.74	0.222	820	1	77	17.7	80	0.6
240 kg	241	0.38	0.184	1310	2	37	NA	–	0.6
330 kg	331	0.68	0.423	780	1	56	NA	–	0.5
450 kg	448	0.56	0.412	1090	1	53	35.5	86	1.1
670 kg	673	0.62	0.429	1570	2	20	41.2	96	1.0
710 kg	711	0.61	0.372	1910	2	25	NA	–	0.8
1.5 t	1497	0.91	0.572	2620	2	25	44.5	77	0.6
2.0 t	2000 approx.	0.86	1.091	1860	2	18	NA	–	0.5

Plate 9.7 180 kg (820 kg/m²) vibrating-plate compactor investigated in 1966

Plate 9.8 240 kg vibrating-plate compactor investigated in 1953

Plate 9.9 330 kg vibrating-plate compactor investigated in 1957

Plate 9.10 450 kg vibrating-plate compactor investigated in 1966

Plate 9.11 670 kg vibrating-plate compactor investigated in 1958

Plate 9.12 710 kg vibrating-plate compactor investigated in 1955

Plate 9.13 1.5 t vibrating-plate compactor investigated in 1951

Plate 9.14 2.0 t vibrating-plate compactor investigated in 1949

9.7 The 670 kg vibrating-plate compactor (Plate 9.11) had a means of controlling the phasing of the two eccentrically weighted shafts to provide two forward and two reverse speeds of travel. Frequency of vibration was controlled by the engine throttle lever. Three different dynamic forces could be obtained by moving auxilliary eccentric weights around the main eccentric shafts. Auxilliary weights bolted to the vibrating base-plate and weighing a total of 50 kg could be removed to lighten the machine, if required. Two or more machines could be linked together

Plate 9.15 The 450 kg vibrating-plate compactor fitted with extension plates (compare Plate 9.10)

Plate 9.16 Two 670 kg vibrating-plate compactors coupled together to work as a single unit

(Plate 9.16) by tie bars across the front and rear of the engine mountings. The normal mode of operation of this machine in the tests involved the use of the highest forward and reverse speeds of travel and the maximum available centrifugal force; the machine was always used with the auxilliary weights attached to the base-plate.

9.8 The 710 kg machine (Plate 9.12) had twin eccentric shafts mounted on the base-plate but no means of varying the resultant force. The machine was restricted, therefore, to forward travel only.

9.9 The 1.5 t machine (Plate 9.13) had twin eccentric shafts; the vibrator unit could be tilted on its mounting by rotation of the control wheel to provide variable forward and reverse speeds of travel. The compactor was operated in the tests at its maximum forward and reverse speeds.

9.10 The 2.0 t vibrating-plate compactor (Plate 9.14) was a very early design and was the first machine of this type to be investigated. The eccentric shaft system was contained in a drum at the front of the machine, and the records include the comment that this caused the front of the base-plate to act as a tamper as well as a vibrator. By rotating the drum containing the eccentric shafts the machine could be operated in a stationary position or over a range of forward speeds. The compactor was steered with the aid of wheels, one on each side of the base-plate; either wheel could be lowered by means of wire ropes controlled by the steering wheel at the rear of the machine and the compactor then pivoted about the lowered wheel. A capstan was also provided on the front of the compactor, and this could be used for pulling the machine over wet soil.

9.11 The above descriptions of the machines and the results presented in this Chapter have been largely taken from Lewis (1954) and (1961) and Toombs (1966a), (1966b) and (1989).

Relations between state of compaction produced and moisture content

9.12 Relations between the states of compaction produced by the vibrating-plate compactors and the moisture content of the soil are given in Figures 9.1 to 9.6. With the well-graded sand and gravel-sand-clay soils such a large number of relations were determined that a number of separate graphs have been produced, one for each combination of thickness of layer and number of passes (Figures 9.4 and 9.5). Dry density is expressed in terms of relative

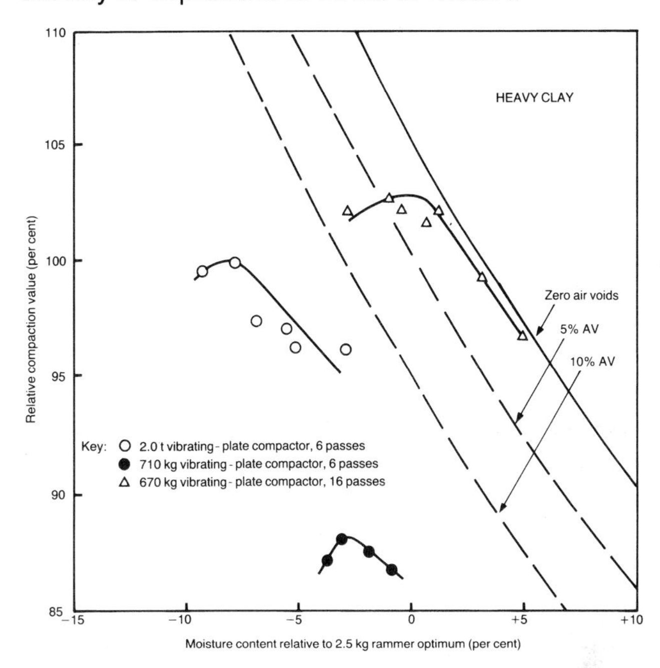

Figure 9.1 Relations between dry density and moisture content for vibrating-plate compactors on heavy clay, expressed in terms of relative compaction with the 2.5 kg rammer test. Layer thickness = 150 mm

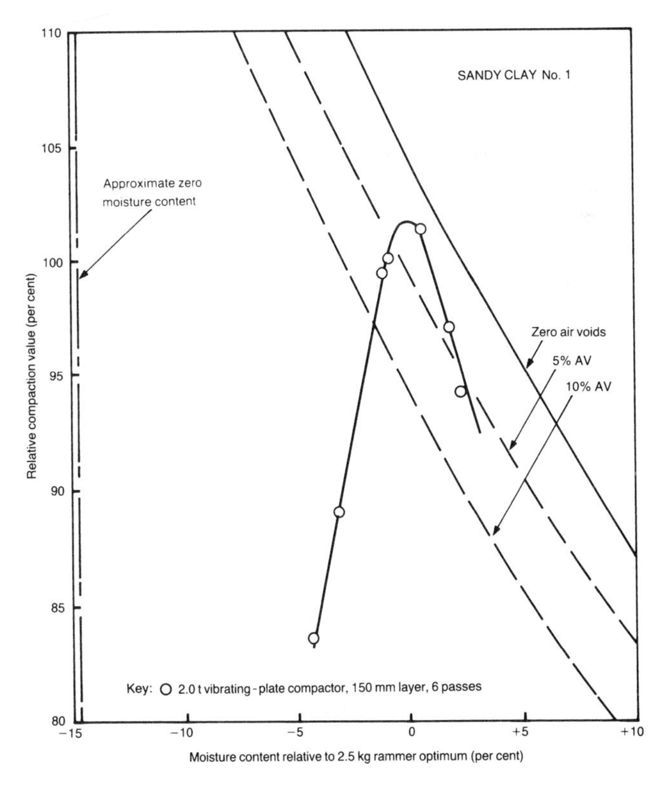

Figure 9.2 Relation between dry density and moisture content for a vibrating-plate compactor on sandy clay No 1, expressed in terms of relative compaction with the 2.5 kg rammer test

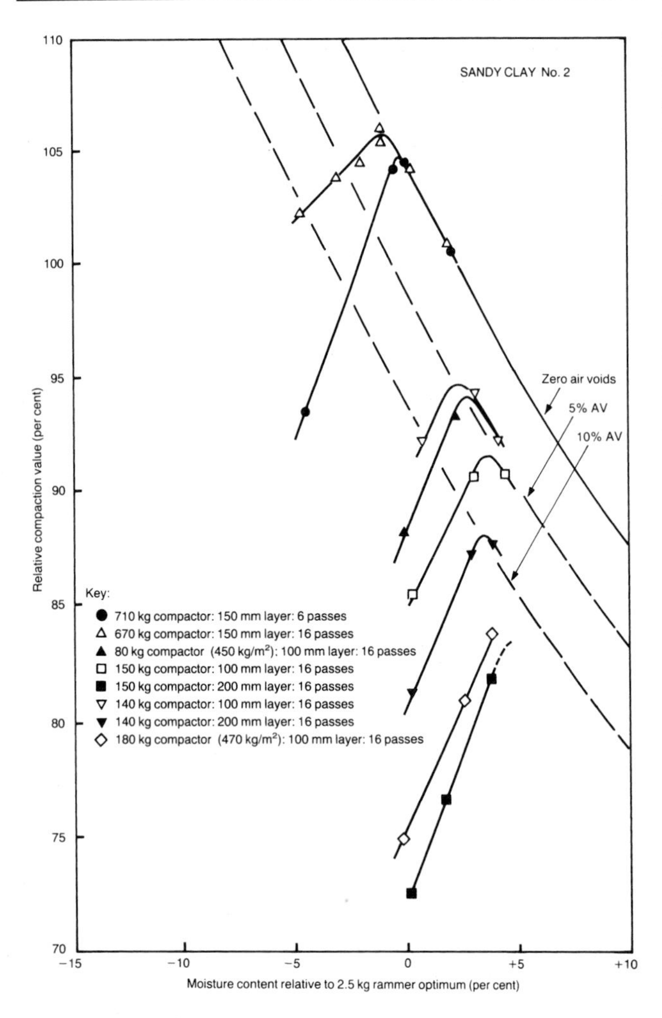

Figure 9.3 Relations between dry density and moisture content for vibrating-plate compactors on sandy clay No 2, expressed in terms of relative compaction with the 2.5 kg rammer test

compaction value (the percentage of the maximum dry density obtained in the 2.5 kg rammer compaction test) and the moisture content as the arithmetical difference from the optimum moisture content of the 2.5 kg rammer test; the results obtained in the 2.5 kg rammer compaction tests at the relevant period are shown in Figure 3.13.

Figure 9.4 (b) See caption below Figure 9.4 (e)

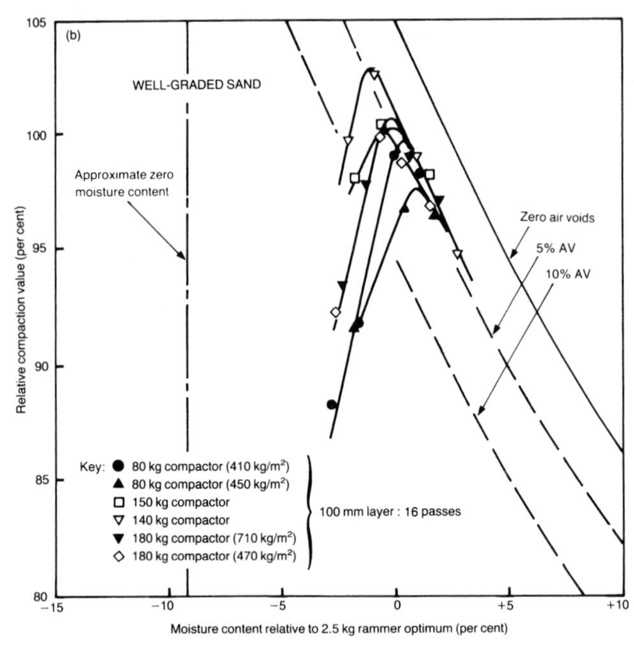

Figure 9.4 (a) See caption below Figure 9.4 (e)

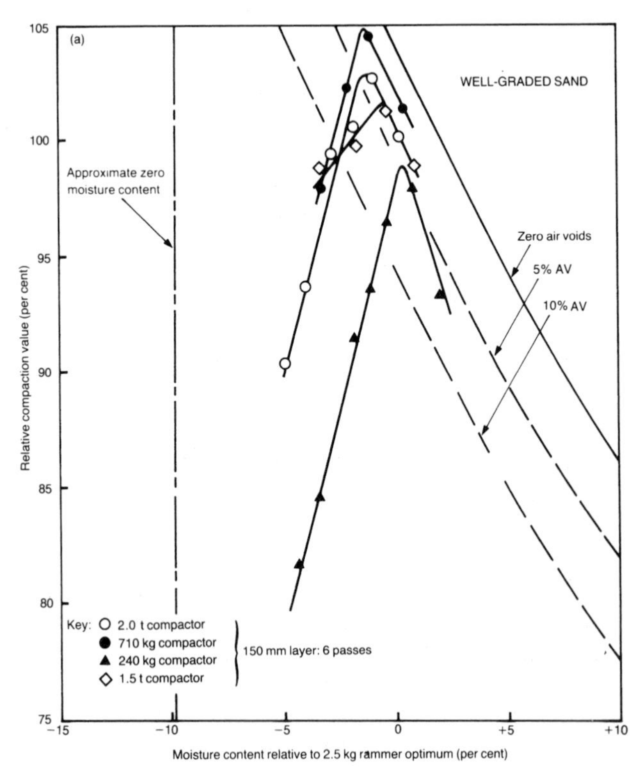

Figure 9.4 (c) See caption below Figure 9.4 (e)

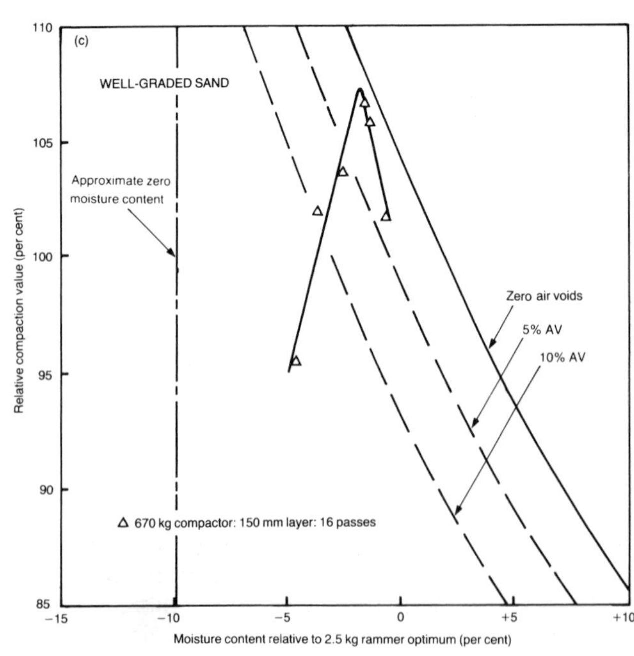

9.13 With the four machines tested before 1956, ie the 2.0 t, 1.5 t, 710 kg and 240 kg vibrating-plate compactors, relations were produced on 150 mm layers using 6 passes. In later investigations 16 passes were used on various thicknesses of layer which depended to a large extent on the capabilities of each individual machine. In the early tests the machines generally had lower speeds of travel than those tested more recently (Table 9.1) and it was considered at the time that 6 passes represented a compaction-to-refusal condition. To all intents and purposes, therefore, the relations shown in Figures 9.1 to 9.6 represent the maximum states of compaction that might be achieved with the various machines for the thicknesses of layer used.

Figure 9.4 (d) See caption below Figure 9.4 (e)

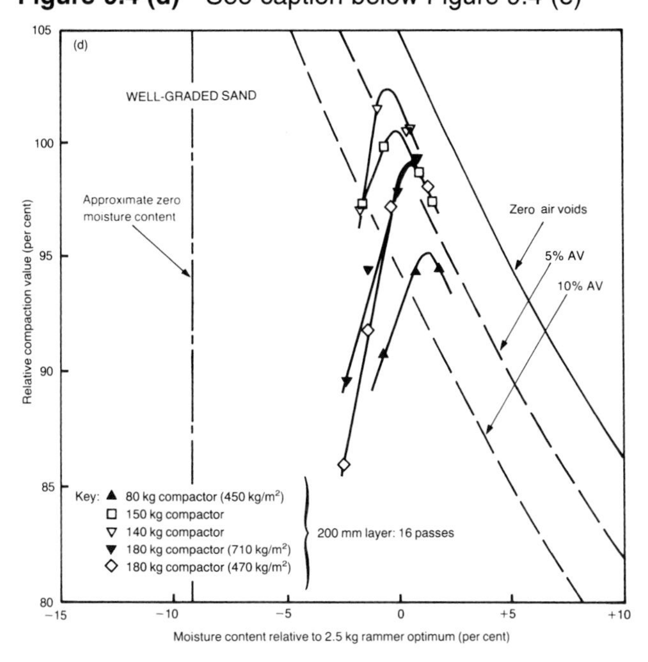

Figure 9.4 (e) See caption below

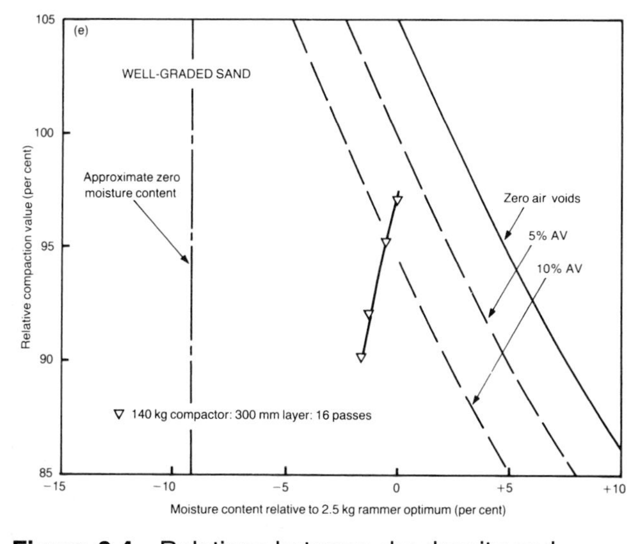

Figure 9.4 Relations between dry density and moisture content for vibrating-plate compactors on well-graded sand, expressed in terms of relative compaction with the 2.5 kg rammer test

Figure 9.5 (a) See caption below Figure 9.5 (e)

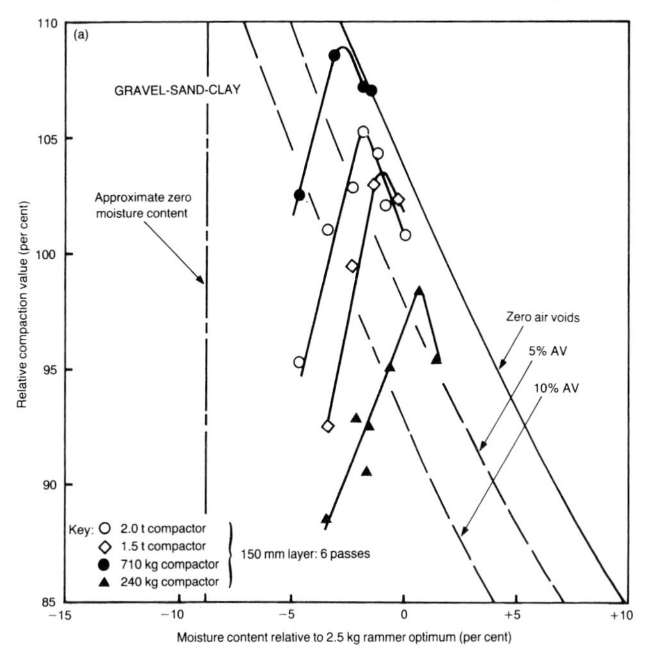

Figure 9.5 (b) See caption below Figure 9.5 (e)

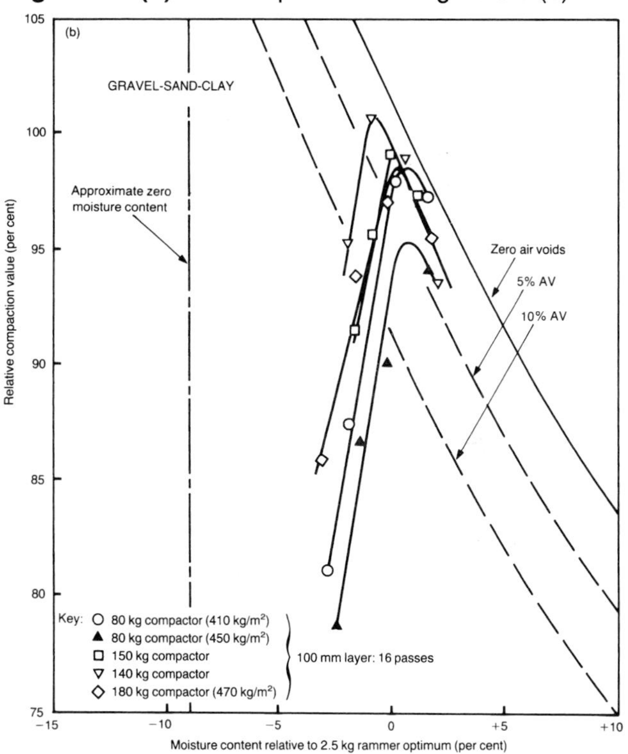

9.14 Only three machines were tested on heavy clay (Figure 9.1) and of these only the 670 kg vibrating-plate compactor succeeded in obtaining a low air content, with an optimum moisture content for the machine approximately equal to the optimum of the 2.5 kg rammer compaction

Figure 9.5 (c) See caption below Figure 9.5 (e)

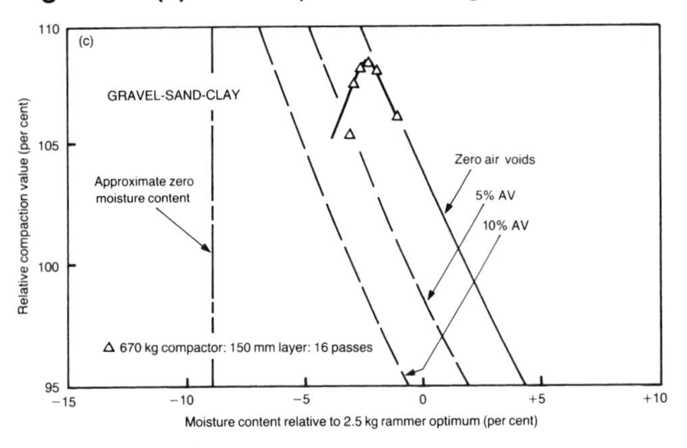

Figure 9.5 (d) See caption below Figure 9.5 (e)

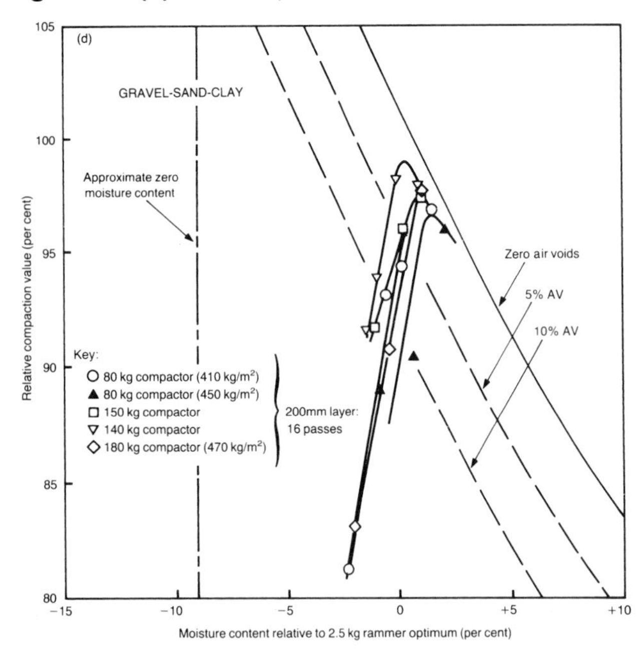

Figure 9.5 (e) See caption below

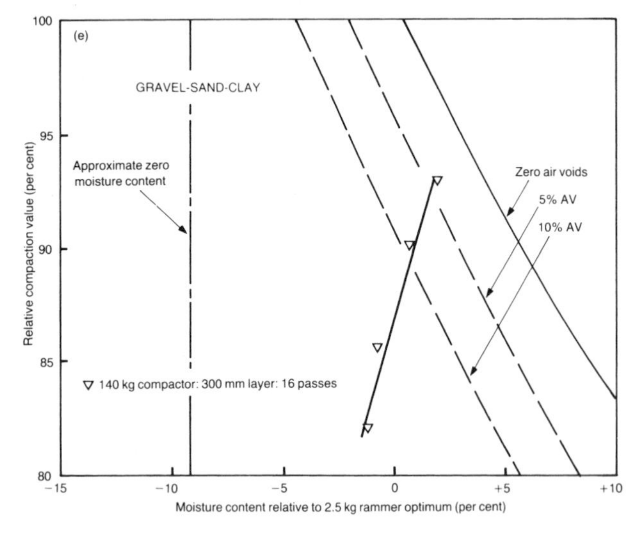

Figure 9.5 Relations between dry density and moisture content for vibrating-plate compactors on gravel-sand-clay, expressed in terms of relative compaction with the 2.5 kg rammer test

test. The 2.0 t and 710 kg machines appear to have been incapable of achieving low air contents in 150 mm layers of the soil, although the ranges of moisture contents for the tests were lower than those used with the successful machine.

9.15 In tests with the sandy clay soils (Figures 9.2 and 9.3) all the machines appear to have been capable of achieving low air contents for the particular thickness of layer used. With the well-graded sand and gravel-sand-clay (Figures 9.4 and 9.5) the machines also performed well, as might be expected, with remarkably small reductions in the states of compaction produced by the lighter machines in the majority of cases where the thickness of layer was increased from 100 mm to 200 mm.

9.16 With the uniformly graded fine sand the effect of variation in moisture content was only determined using the 670 kg vibrating-plate compactor (Figure 9.6). This machine was equipped with a means of adjusting the centrifugal force (see Paragraph 9.7) and two levels of dynamic force were used, the normal one of 20.6 kN and a reduced level of 9.8 kN. The reduced level was used to alleviate the overstressing of the surface of the soil and, as shown in Figure 9.6, the lower dynamic force resulted in an improvement in the state of compaction produced in a 150 mm layer of the soil. This effect is unique to this particular type of soil, and with well-graded soils the maximum dynamic force available can be expected to have produced the highest state of compaction achievable by the machine.

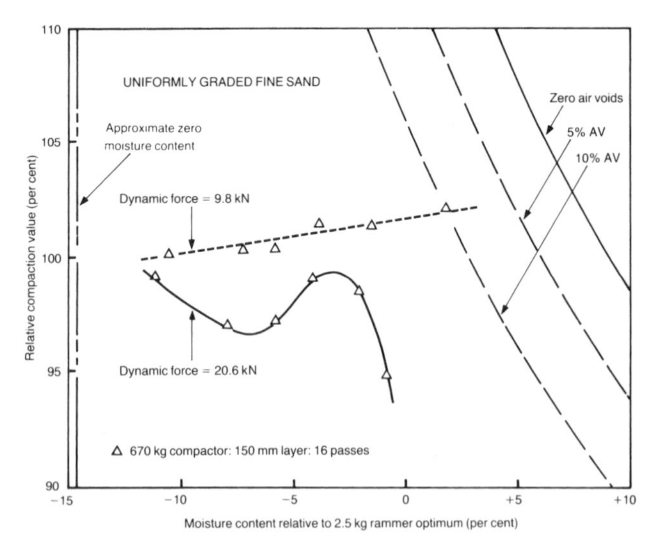

Figure 9.6 Relations between dry density and moisture content for a vibrating-plate compactor on uniformly graded fine sand, expressed in terms of relative compaction with the 2.5 kg rammer test

Figure 9.7 Maximum dry densities and optimum moisture contents related to the mass per unit area of base plate for vibrating-plate compactors. Results for 150 mm compacted layers from Figures 9.4 (a) and 9.5 (a)

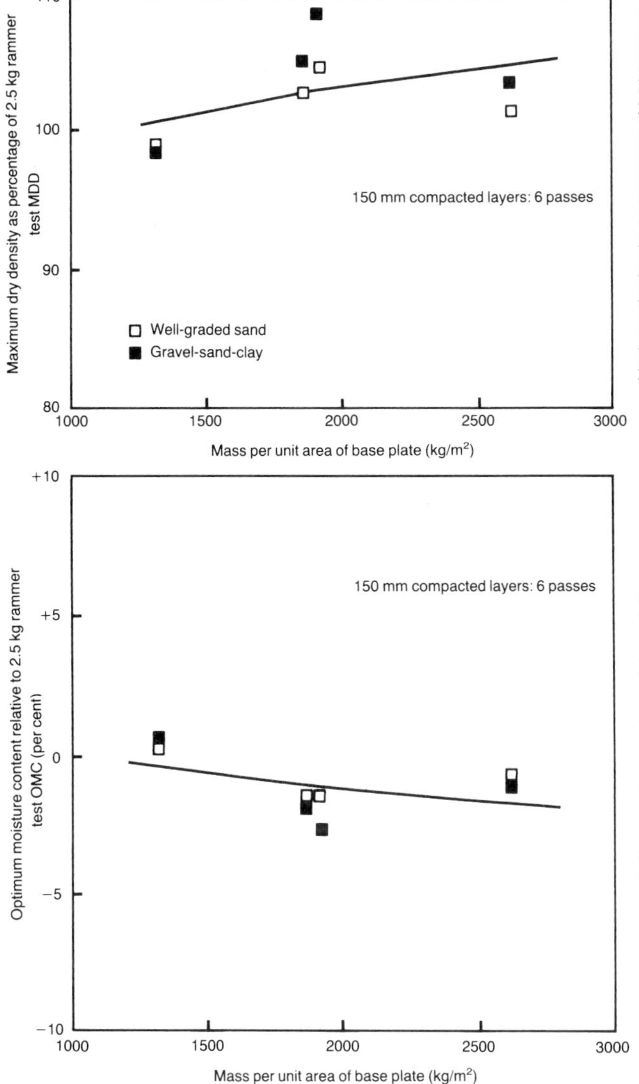

Figure 9.8 Maximum dry densities and optimum moisture contents related to the mass per unit area of base plate for vibrating-plate compactors. Results for 100 mm compacted layers from Figures 9.3, 9.4 (b) and 9.5 (b)

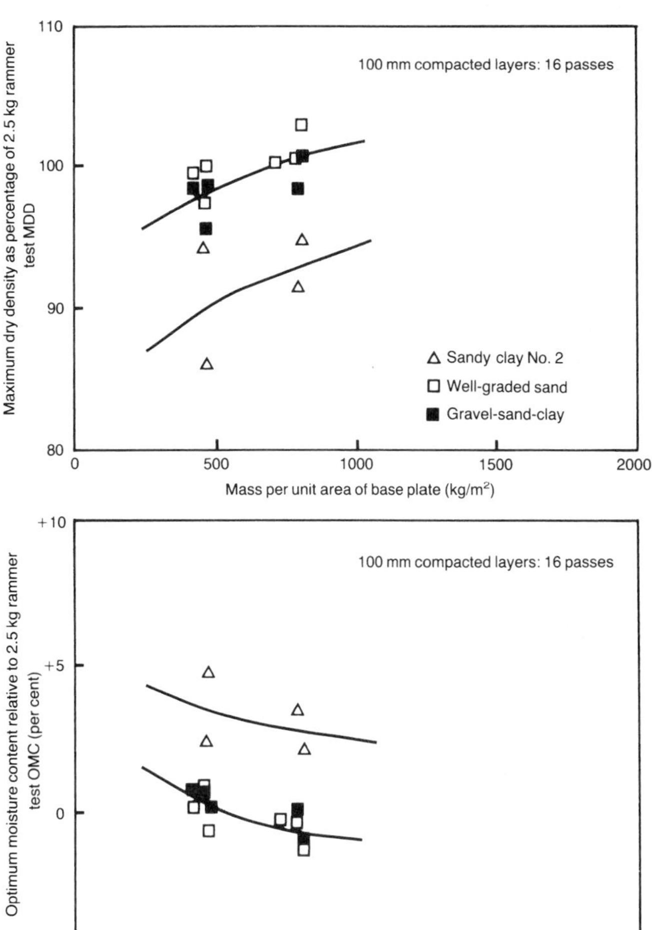

9.17 The results for sandy clay No 2, well-graded sand and gravel-sand-clay are summarised in Figures 9.7 to 9.9, where relations are given between maximum dry density and mass per unit area of base-plate and between optimum moisture content and mass per unit area of base-plate. Figure 9.7 shows the results for 6 passes on 150 mm layers of well-graded sand and gravel-sand-clay; a wide range of mass per unit area is included, extending from about 1300 kg/m^2 to about 2600 kg/m^2. Maximum relative compaction values range from 100 to 105 per cent and optimum moisture contents extend from the optimum of the 2.5 kg rammer compaction test to about 2 per cent below that optimum. Figures 9.8 and 9.9 contain the results for the lightweight vibrating-plate compactors with masses per unit area ranging from 400 to 800 kg/m^2, using 16 passes on 100 and 200 mm layers. Here the maximum relative compaction values range from 95 to 103 per cent for the granular soils with little change over the two thicknesses of layer used, and from 86 to 95 per cent for 100 mm layers of the sandy clay. Optimum moisture contents range from about 1 per cent above to about 1 per cent below the 2.5 kg rammer optimum for the granular soils and from 2 to 5 per cent above the 2.5 kg rammer optimum with the sandy clay. In all cases the correlation coefficients for the relations given in Figures 9.7 to 9.9 are poor and only the general trends of the results are given by the lines drawn through the points. Thus, in general, for a given thickness of layer, the maximum relative compaction values increased, and optimum moisture contents decreased, as the mass per unit area of base-plate increased. The possible effects of dynamic forces, where

Figure 9.9 Maximum dry densities and optimum moisture contents related to the mass per unit area of base plate for vibrating-plate compactors. Results for 200 mm compacted layers from Figures 9.4 (d) and 9.5 (d)

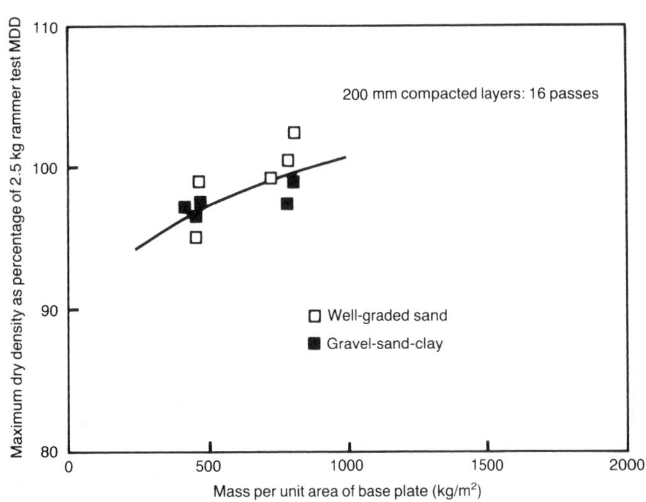

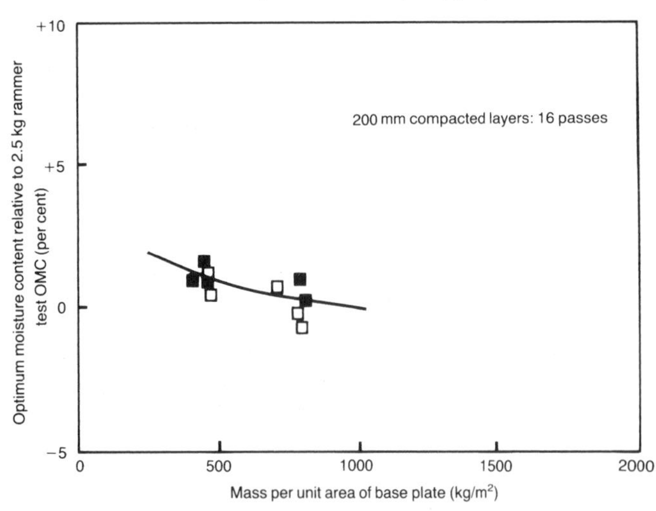

Figure 9.11 Relation between relative compaction and number of passes of the 2.0 t vibrating-plate compactor on sandy clay No 1. Values are related to the maximum dry density and optimum moisture content obtained in the 2.5 kg rammer compaction test (Figure 3.13)

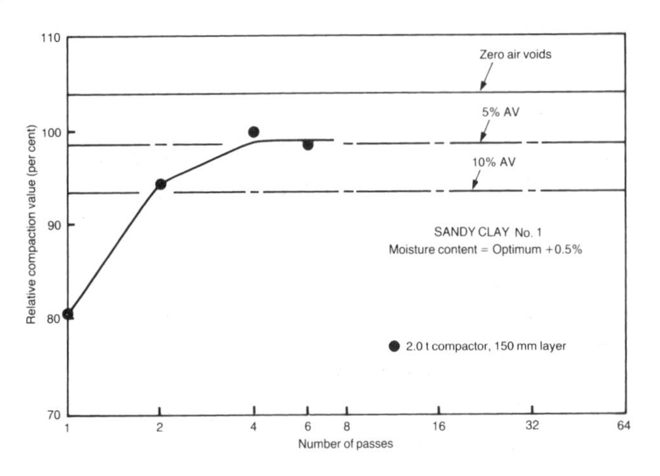

Figure 9.10 Relation between relative compaction and number of passes of the 670 kg vibrating-plate compactor on heavy clay. Values are related to the maximum dry density and optimum moisture content obtained in the 2.5 kg rammer compaction test (Figure 3.13)

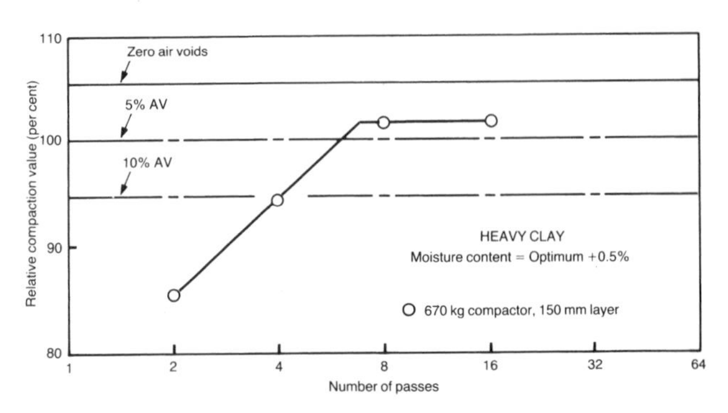

they are known, are discussed in Paragraphs 9.25 to 9.31.

Relations between state of compaction produced and number of passes

9.18 Results of investigations to determine the effect on the state of compaction achieved of variations in the number of passes of vibrating-plate compactors are given in Figures 9.10 to

Figure 9.12

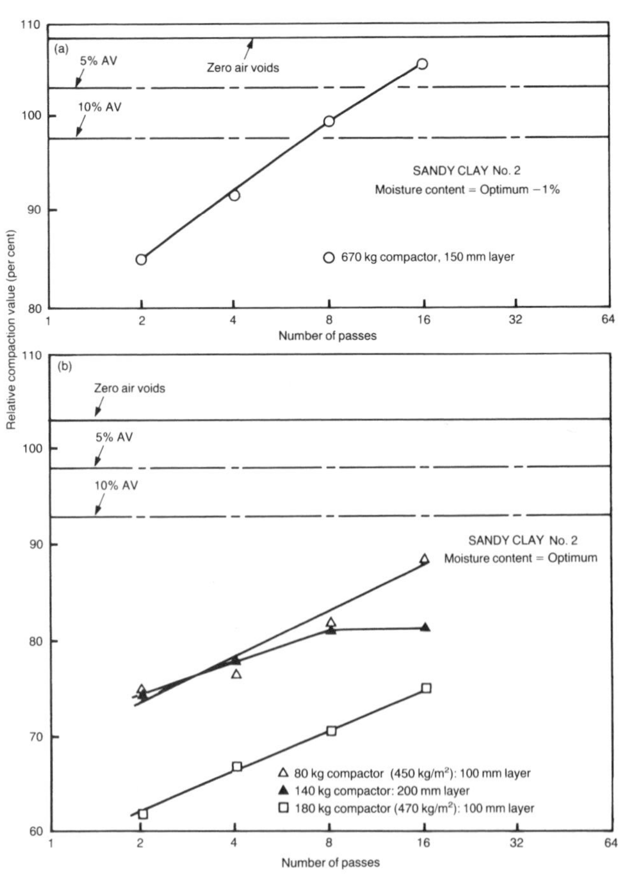

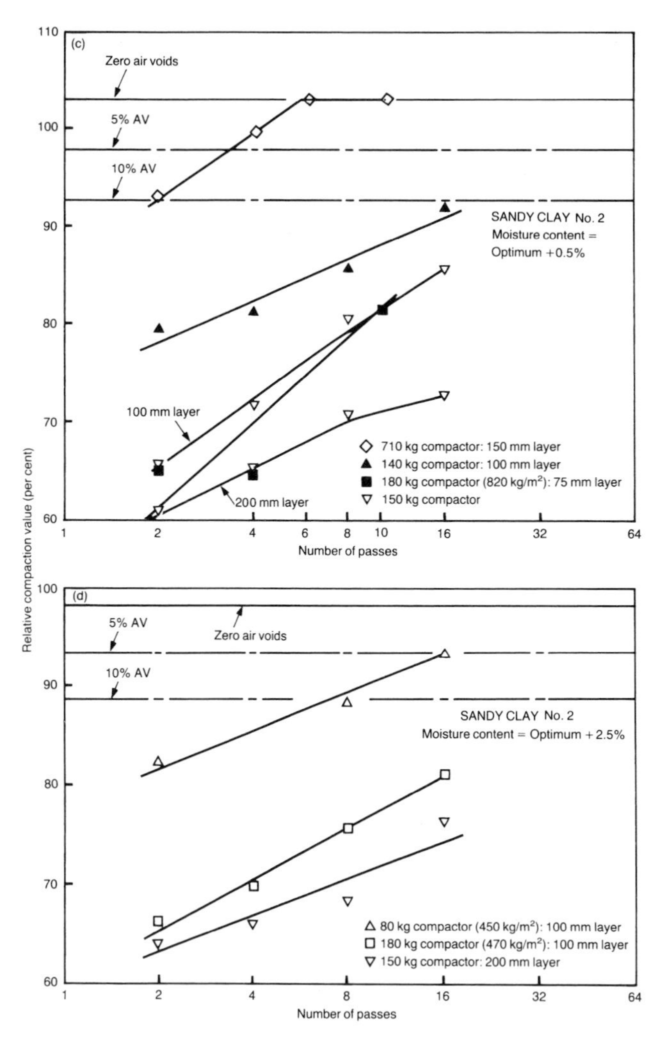

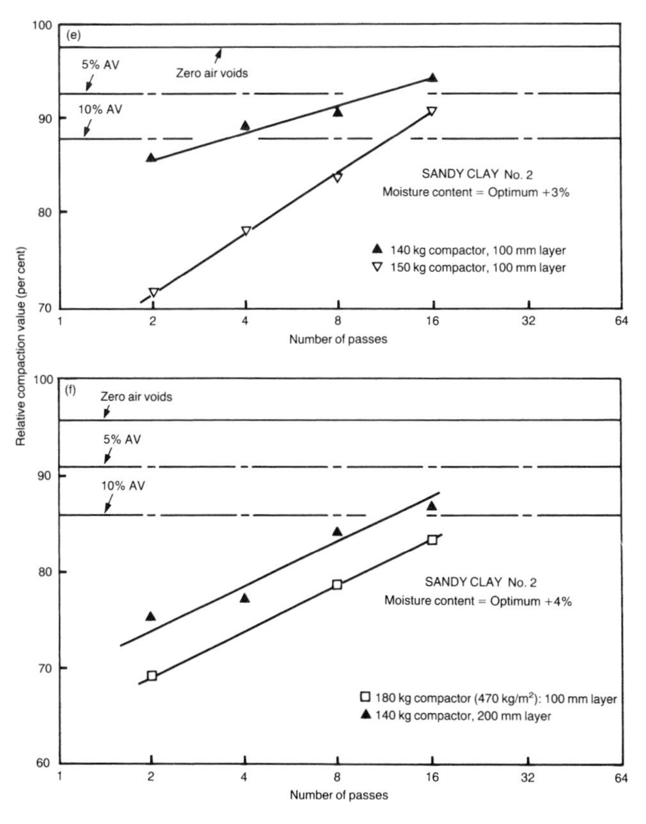

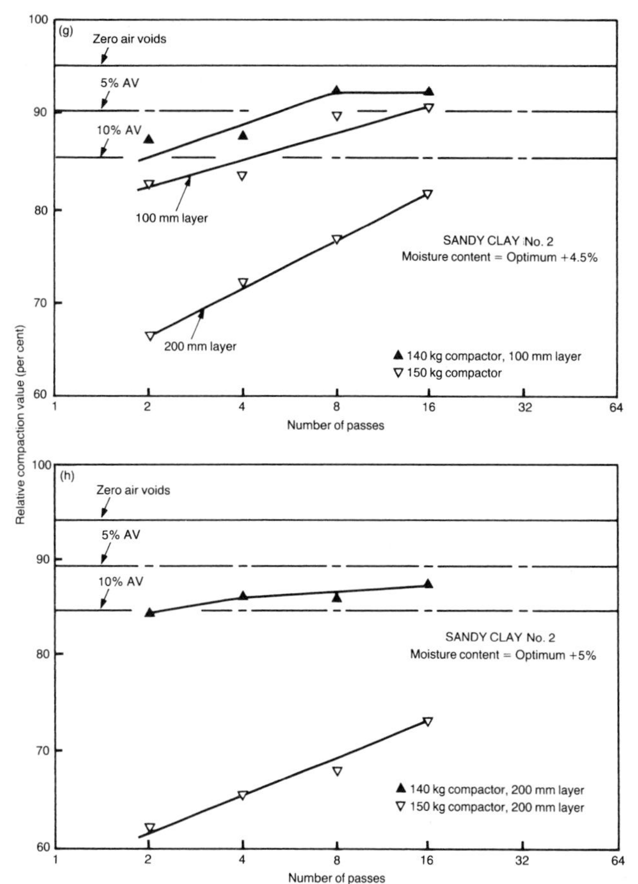

Figure 9.12 Relations between relative compaction and number of passes of vibrating-plate compactors on sandy clay No 2 at eight different values of moisture content. Values are related to the maximum dry density and optimum moisture content obtained in the 2.5 kg rammer compaction test (Figure 3.13)

9.15. Again, the results for each soil are illustrated under a separate figure number, whilst different values of moisture content are in separate graphs within each figure. The results of a comprehensive investigation of a range of lightweight vibrating-plate compactors have been included (Toombs, 1989) and a large number of different graphs, therefore, has been produced.

9.19 Only the 670 kg vibrating-plate compactor was studied on the heavy clay in this series of tests (Figure 9.10) and only the 2.0 t machine was studied on the sandy clay No 1. In general, tests on light cohesive clay soil, or preliminary tests to assess the capability of the machines, revealed that the lighter vibrating-plate compactors were not likely to perform sufficiently well on the heavy clay to justify further investigations on that soil. On the light cohesive soil (sandy clay No 2) and the granular soils tests were carried out with a variety of moisture contents and thicknesses of layer using the

Figure 9.13

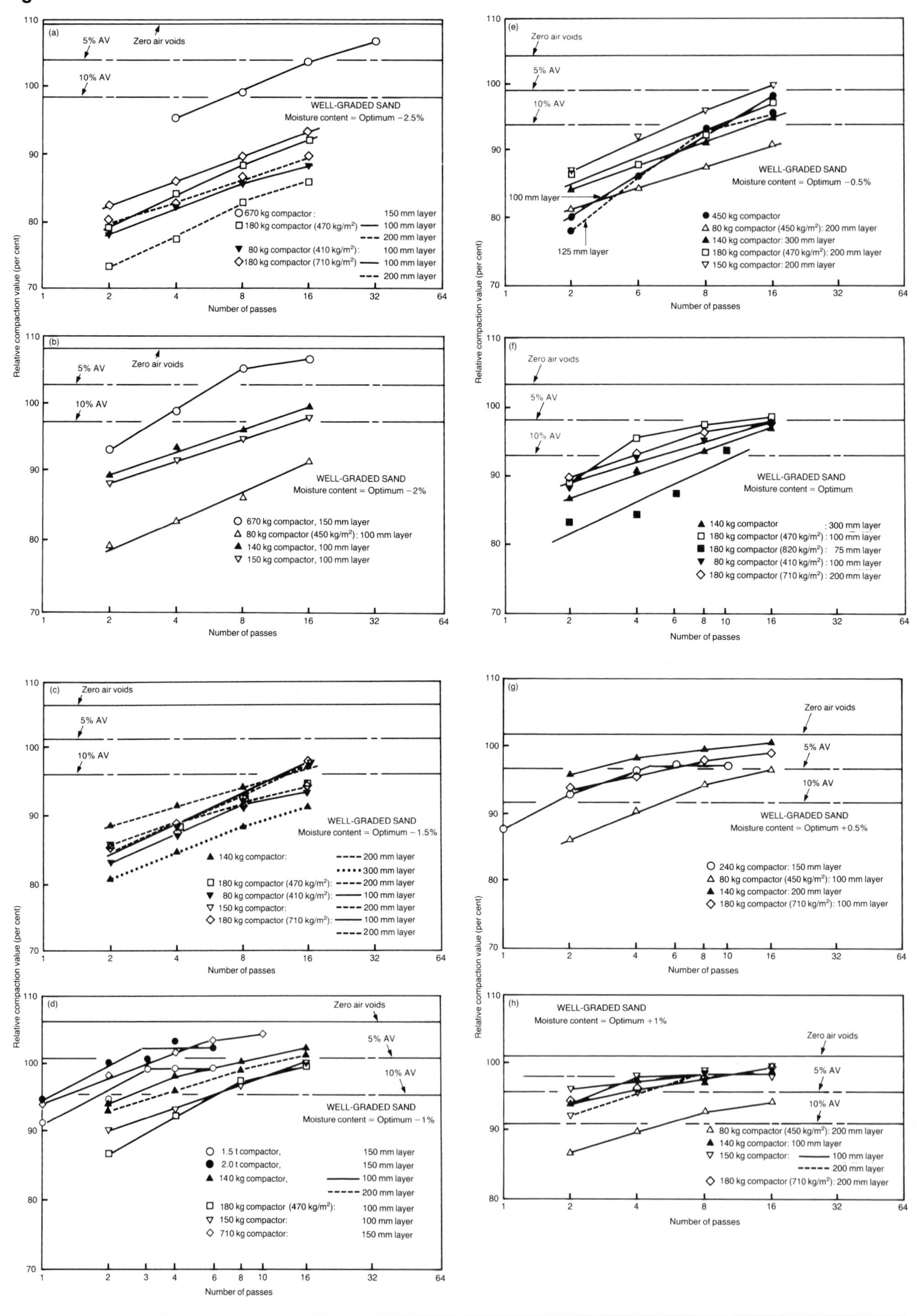

(a)
Zero air voids
5% AV
10% AV
WELL-GRADED SAND
Moisture content = Optimum −2.5%
670 kg compactor: 150 mm layer
180 kg compactor (470 kg/m²) 100 mm layer
200 mm layer
80 kg compactor (410 kg/m²): 100 mm layer
180 kg compactor (710 kg/m²) 100 mm layer
200 mm layer
Number of passes
Relative compaction value (per cent)
(b)
WELL-GRADED SAND
Moisture content = Optimum −2%
670 kg compactor, 150 mm layer
80 kg compactor (450 kg/m²): 100 mm layer
140 kg compactor, 100 mm layer
150 kg compactor, 100 mm layer
(c)
WELL-GRADED SAND
Moisture content = Optimum −1.5%
140 kg compactor: 200 mm layer
300 mm layer
180 kg compactor (470 kg/m²): 200 mm layer
80 kg compactor (410 kg/m²): 100 mm layer
150 kg compactor: 200 mm layer
180 kg compactor (710 kg/m²): 100 mm layer
200 mm layer
(d)
WELL-GRADED SAND
Moisture content = Optimum −1%
1.5 t compactor, 150 mm layer
2.0 t compactor, 150 mm layer
140 kg compactor, 100 mm layer
200 mm layer
180 kg compactor (470 kg/m²): 100 mm layer
150 kg compactor: 100 mm layer
710 kg compactor: 150 mm layer
(e)
WELL-GRADED SAND
Moisture content = Optimum −0.5%
100 mm layer
125 mm layer
450 kg compactor
80 kg compactor (450 kg/m²): 200 mm layer
140 kg compactor: 300 mm layer
180 kg compactor (470 kg/m²): 200 mm layer
150 kg compactor: 200 mm layer
(f)
WELL-GRADED SAND
Moisture content = Optimum
140 kg compactor : 300 mm layer
180 kg compactor (470 kg/m²) : 100 mm layer
180 kg compactor (820 kg/m²) : 75 mm layer
80 kg compactor (410 kg/m²) : 100 mm layer
180 kg compactor (710 kg/m²) : 200 mm layer
(g)
WELL-GRADED SAND
Moisture content = Optimum +0.5%
240 kg compactor: 150 mm layer
80 kg compactor (450 kg/m²): 100 mm layer
140 kg compactor: 200 mm layer
180 kg compactor (710 kg/m²): 100 mm layer
(h)
WELL-GRADED SAND
Moisture content = Optimum +1%
80 kg compactor (450 kg/m²): 200 mm layer
140 kg compactor: 100 mm layer
150 kg compactor: 100 mm layer
200 mm layer
180 kg compactor (710 kg/m²): 200 mm layer

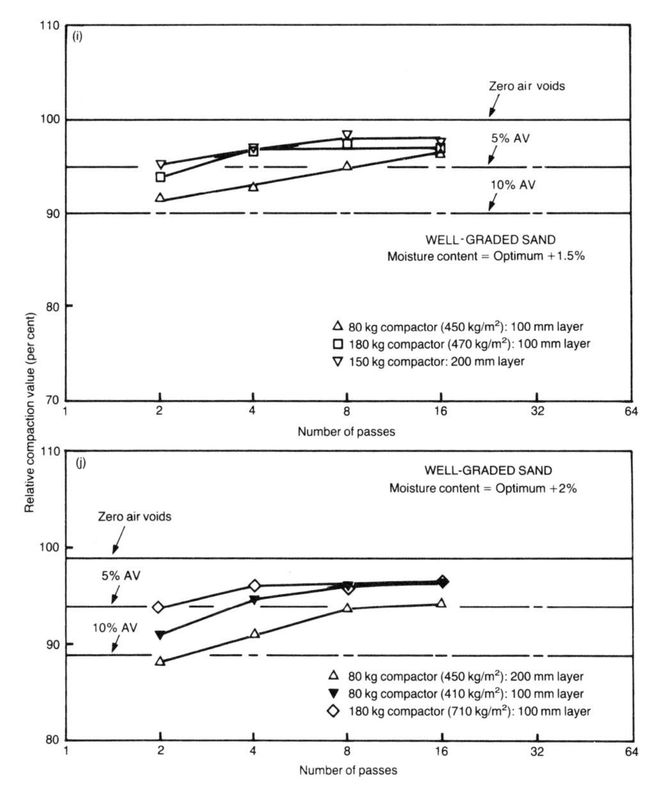

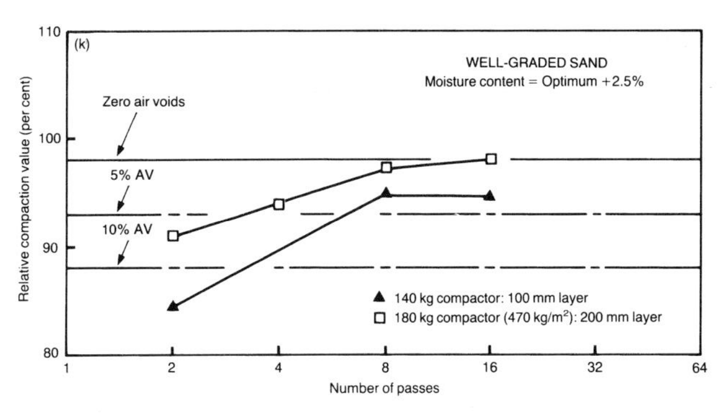

Figure 9.13 Relations between relative compaction and number of passes of vibrating-plate compactors on well-graded sand at eleven different values of moisture content. Values are related to the maximum dry density and optimum moisture content obtained in the 2.5 kg rammer compaction test (Figure 3.13)

various machines listed in Table 9.1. The various relations produced in the figures provide a complex picture of the performance of the machines, and although the results can be selected for individual machines and for particular conditions of use (moisture content,

Figures 9.14 (a) and (b) See caption page 140

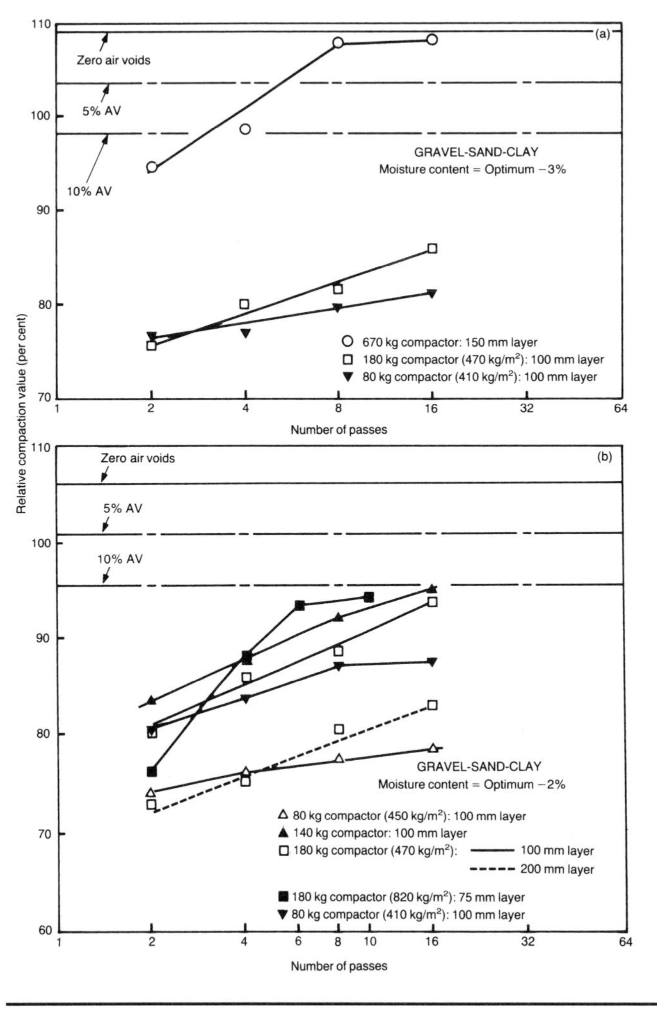

Figures 9.14 (c) and (d) See caption page 140

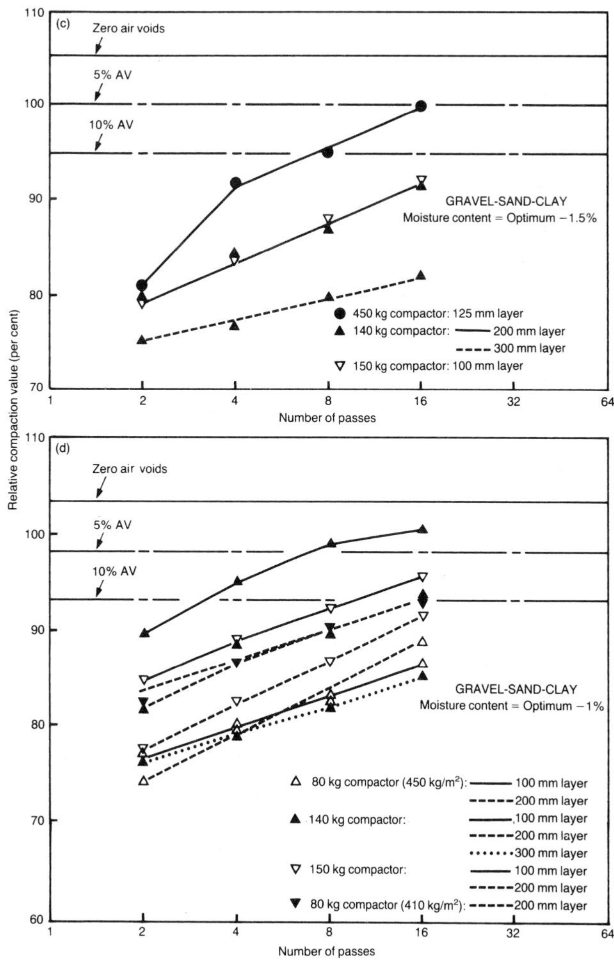

Figures 9.14 (e) and (f) See main caption

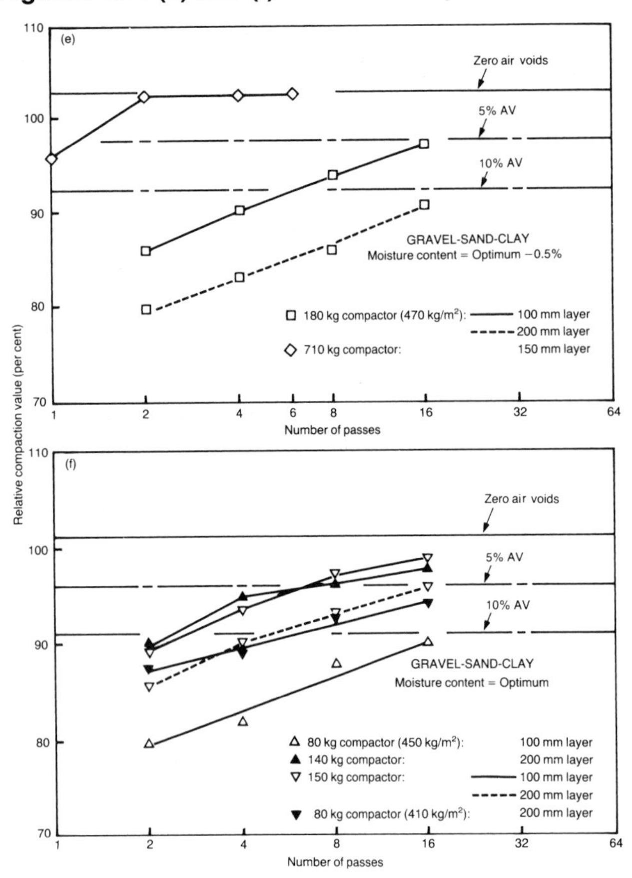

Figures 9.14 (g) and (h) See main caption

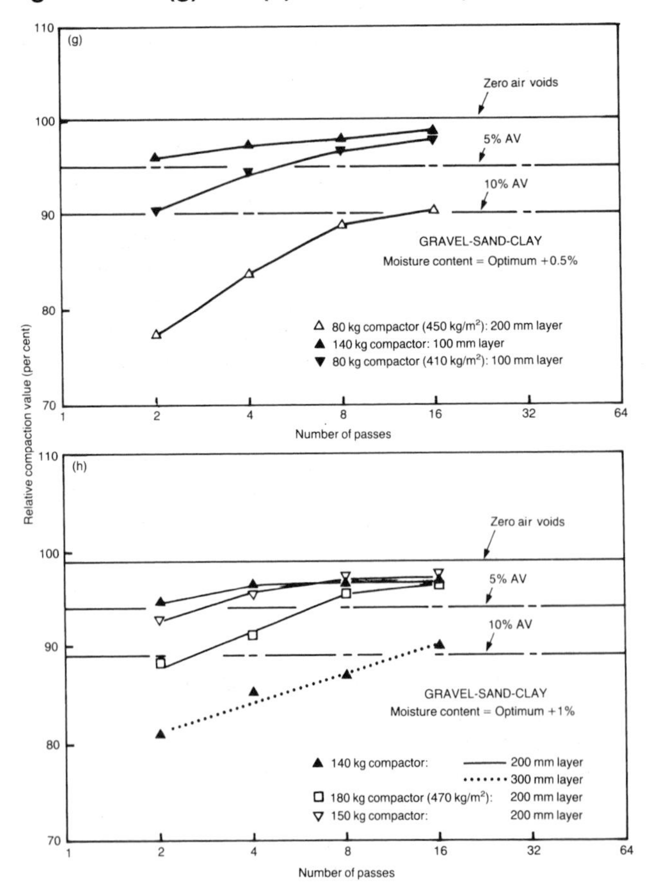

Figures 9.14 (i), (j) and (k) See main caption

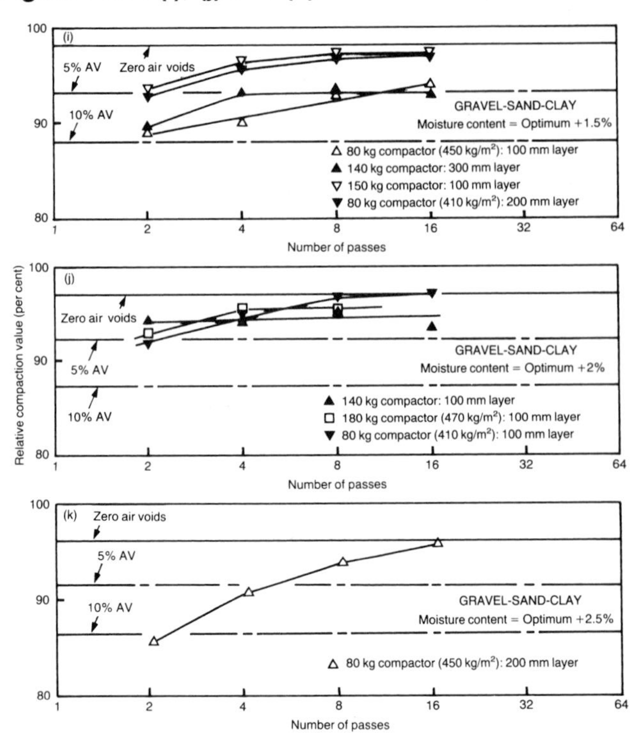

Figure 9.14 Relations between relative compaction and number of passes of vibrating-plate compactors on gravel-sand-clay at eleven different values of moisture content. Values are related to the maximum dry density and optimum moisture content obtained in the 2.5 kg rammer compaction test (Figure 3.13)

thickness of layer) the overall pattern can only be obtained by statistical analyses of the results. As in earlier Chapters the numbers of passes required to achieve specified levels of compaction have been related to relevant parameters. The levels of compaction used were 95 per cent relative compaction (dry density expressed as a percentage of the maximum dry density obtained in the 2.5 kg rammer compaction test at the relevant period) and 10 per cent air voids. The machine parameter chosen was mass per unit area of base-plate (area of contact with the soil surface), this being the factor that is available for all the machines included in the investigations; moisture content and depth of compacted layer were additional variables included in the analyses. The relative effects on performance of mass per unit area, centrifugal force per unit area, and area of base-plate are discussed later in Paragraphs 9.25 to 9.31.

9.20 The results of regression analyses of the results are given in Table 9.2 in the form of regression equations and the correlation coefficients. The effect of depth of layer with

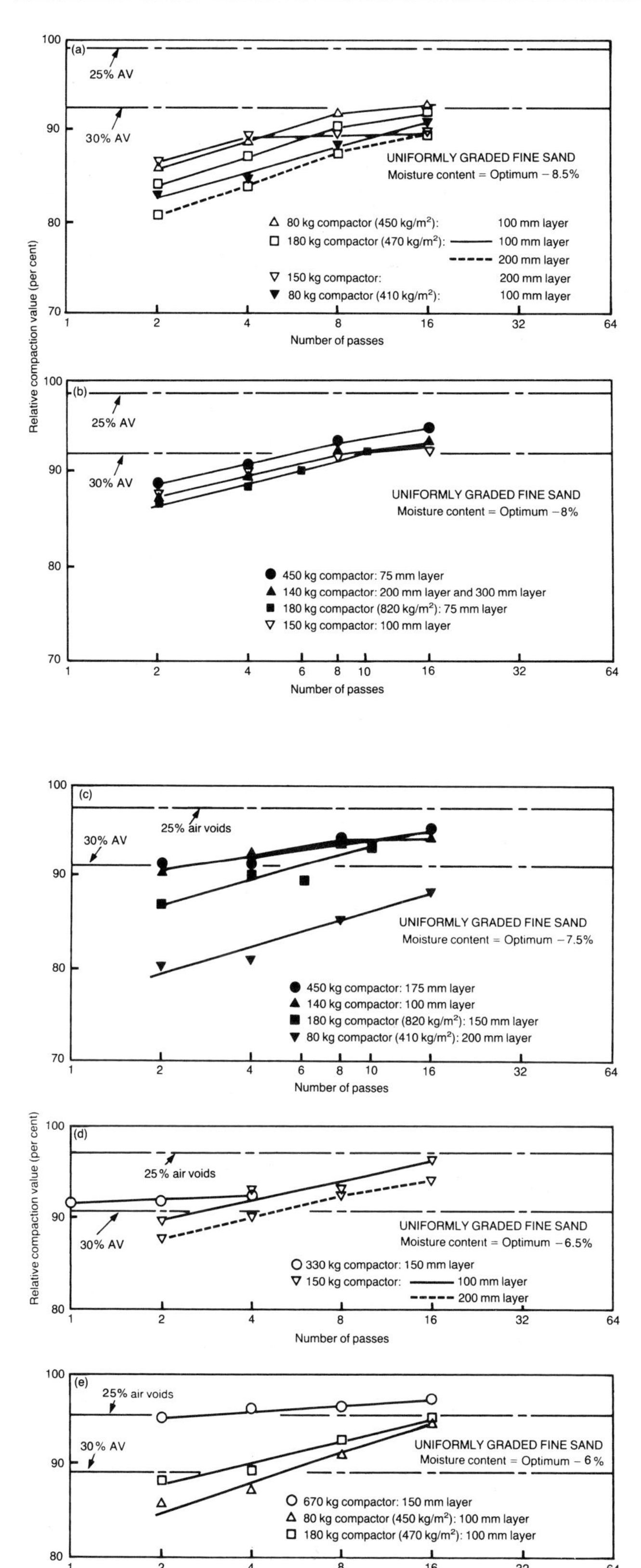

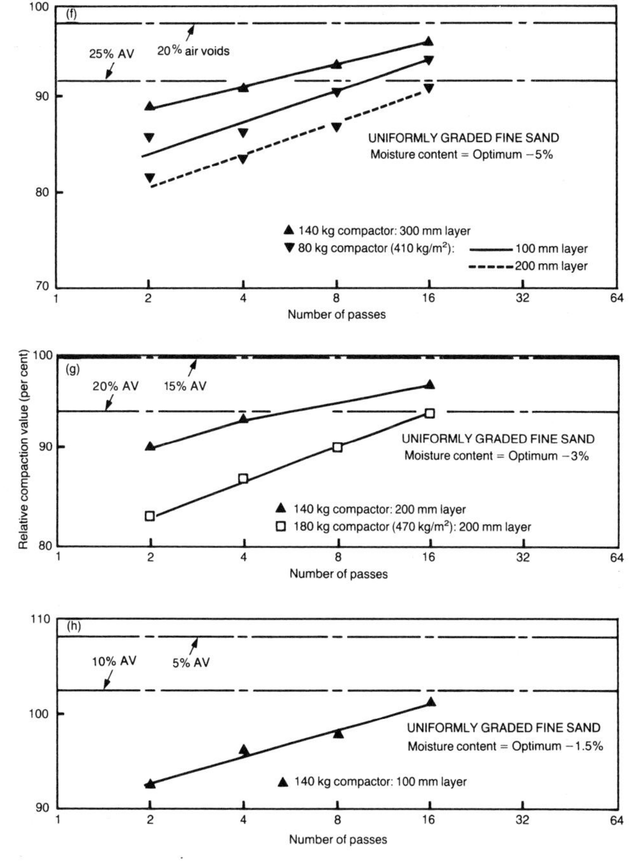

Figure 9.15 Relations between relative compaction and number of passes of vibrating-plate compactors on uniformly graded fine sand at eight different values of moisture content. Values are related to the maximum dry density and optimum moisture content obtained in the 2.5 kg rammer compaction test (Figure 3.13)

sandy clay No 2, level (a) (95 per cent relative compaction) was insignificant within the range of thicknesses of layer for which data were available, ie 75 to 150 mm. The range of depths was restricted in this way as a relative compaction value of 95 per cent was not achieved with any of the machines tested using 200 mm layers (see Figure 9.3). In all the other equations given in Table 9.2 the three independent variables given above were all used. For the determination of the number of passes required to achieve 10 per cent air voids it was found that the well-graded sand and gravel-sand-clay could be considered together to produce a relation with a high correlation coefficient. Only with the gravel-sand-clay at Level (a) (95 per cent relative compaction) was a relatively poor correlation obtained. In all other cases shown in Table 9.2 correlation coefficients in excess of 0.9 were obtained, although a number of particularly erratic results have been omitted to achieve this.

9.21 Examples of relations between the number of passes required and mass per unit area of base-plate are given in Figures 9.16 to 9.19. These are based on the equations with correlation coefficients of 0.9 or more given in Table 9.2. Examples are included for three values of moisture content and, except for Figure

9.16(a), for three depths of compacted layer. It should be remembered that the equations in Table 9.2 and relations given in Figures 9.16 to 9.19 apply only for the ranges of contact pressures, moisture contents and thicknesses of

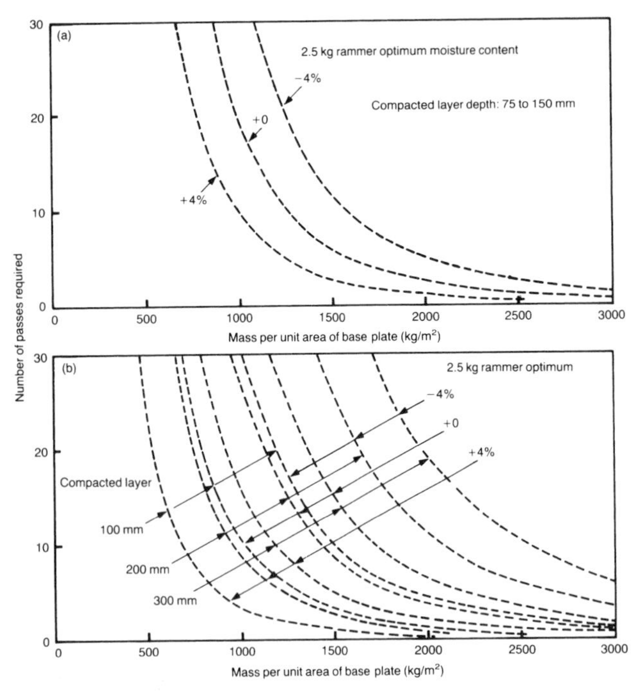

Figure 9.16 Vibrating-plate compactors on sandy clay No 2. Relations between number of passes required and mass per unit area of base plate to achieve (a) 95 per cent of the maximum dry density obtained in the 2.5 kg rammer compaction test and (b) 10 per cent air voids. These graphs were derived by regression analyses of the results given in Figures 9.3 and 9.12

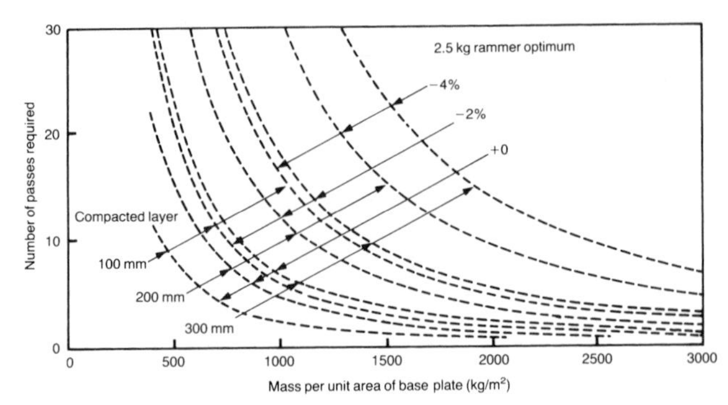

Figure 9.17 Vibrating-plate compactors on well-graded sand. Relations between number of passes required and mass per unit area of base plate to achieve 95 per cent of the maximum dry density obtained in the 2.5 kg rammer compaction test. These relations were derived by regression analyses of the results given in Figures 9.4 and 9.13

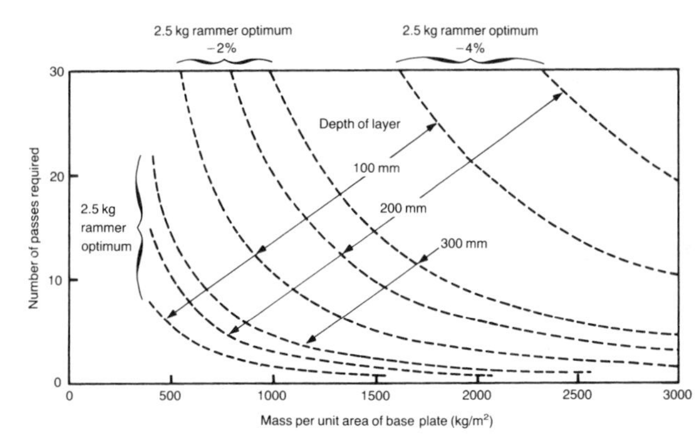

Figure 9.18 Vibrating-plate compactors on well-graded granular soils. Relations between number of passes required and mass per unit area of base plate to achieve 10 per cent air voids. These relations were derived by regression analyses of the results given in Figures 9.4, 9.5, 9.13 and 9.14

Table 9.2 Equations and correlation coefficients to determine the number of passes required to achieve specified levels of compaction with vibrating-plate compactors

Soil	Level*	Regression equation†	No of data points	Correlation coefficient
Sandy clay No 2	(a)	P=10.13–2.95 m–.0769 w‡	8	0.997
	(b)	P=6.78–2.83 m–.120 w+1.36 d	15	0.93
Well-graded sand	(a)	P=3.64–1.75 m–.210 w+.982 d	56	0.90
Gravel-sand-clay	(a)	P=3.87–1.48 m–.117 w+.614 d	40	0.74
Well-graded sand & gravel-sand-clay	(b)	P=3.51–1.73 m–.409 w+.938 d	100	0.91
Uniformly graded fine sand	(a)	P=3.64–1.57 m–.137 w+571 d	23	0.92

*Level (a)=95 per cent of the maximum dry density obtained in the 2.5 kg rammer compaction test
Level (b)=10 per cent air voids
†P=Log_{10} (Number of passes required)
m=Log_{10} (Mass per unit area of base plate in kg/m^2)
w=Moisture content of soil, per cent, minus the optimum moisture content, per cent, obtained in the 2.5 kg rammer compaction test
d=Log_{10} (Depth of compacted layer in mm)
‡Depth of compacted layer in the range 75 to 150 mm

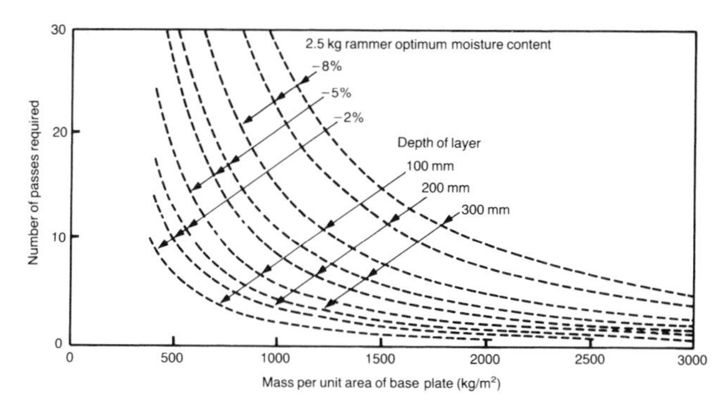

Figure 9.19 Vibrating-plate compactors on uniformly graded fine sand. Relations between number of passes required and mass per unit area of base plate to achieve 95 per cent of the maximum dry density obtained in the 2.5 kg rammer compaction test. These relations were derived by regression analyses of the results given in Figure 9.15

layer in the data from which they have been derived.

Relations between state of compaction and depth within the compacted layer

9.22 Results of investigations of the variation of density with depth in layers of soil compacted by vibrating-plate compactors are given in Figures 9.20 to 9.25. These relations were obtained in the first and second phases of the investigations of the performance of compaction plant and apply only to the larger machines that have been studied. Relations were obtained using the

Figure 9.21 Relation between relative compaction and depth in the compacted layer of sandy clay No 1 obtained with the 2.0 t vibrating-plate compactor. All values are related to the maximum dry density and optimum moisture content obtained in the 2.5 kg rammer compaction test (Figure 3.13)

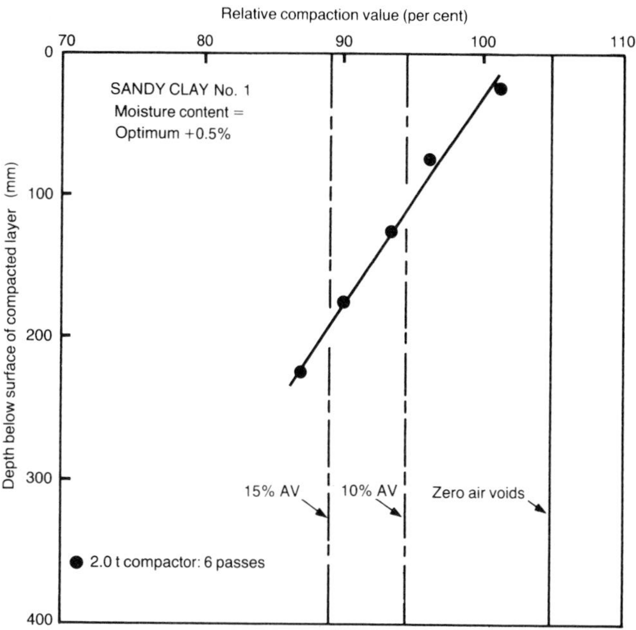

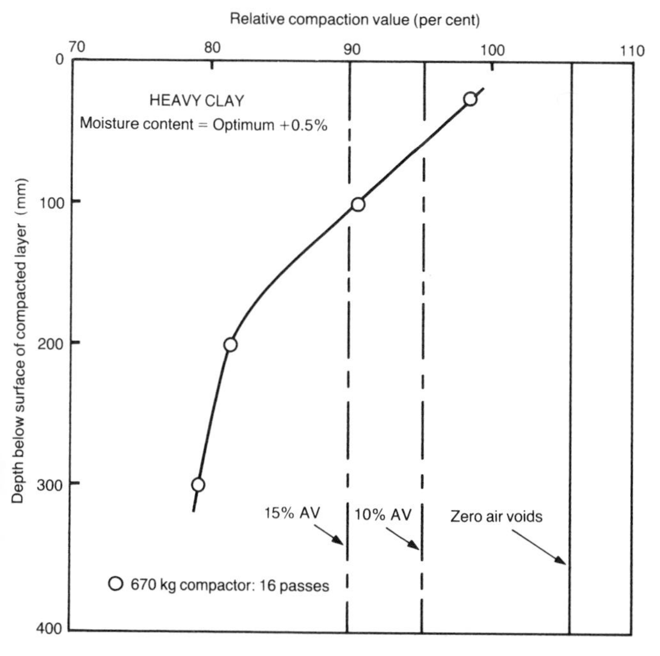

Figure 9.20 Relation between relative compaction and depth in the compacted layer of heavy clay obtained with the 670 kg vibrating-plate compactor. All values are related to the maximum dry density and optimum moisture content obtained in the 2.5 kg rammer compaction test (Figure 3.13)

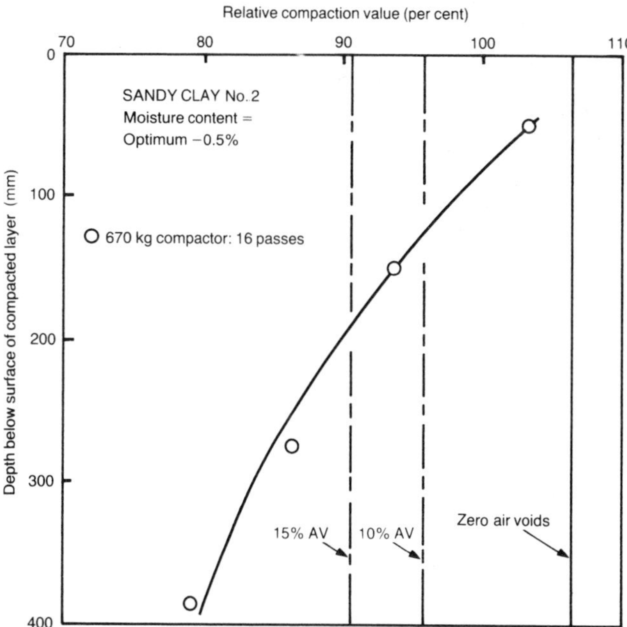

Figure 9.22 Relation between relative compaction and depth in the compacted layer of sandy clay No 2 obtained with the 670 kg vibrating-plate compactor. All values are related to the maximum dry density and optimum moisture content obtained in the 2.5 kg rammer compaction test (Figure 3.13)

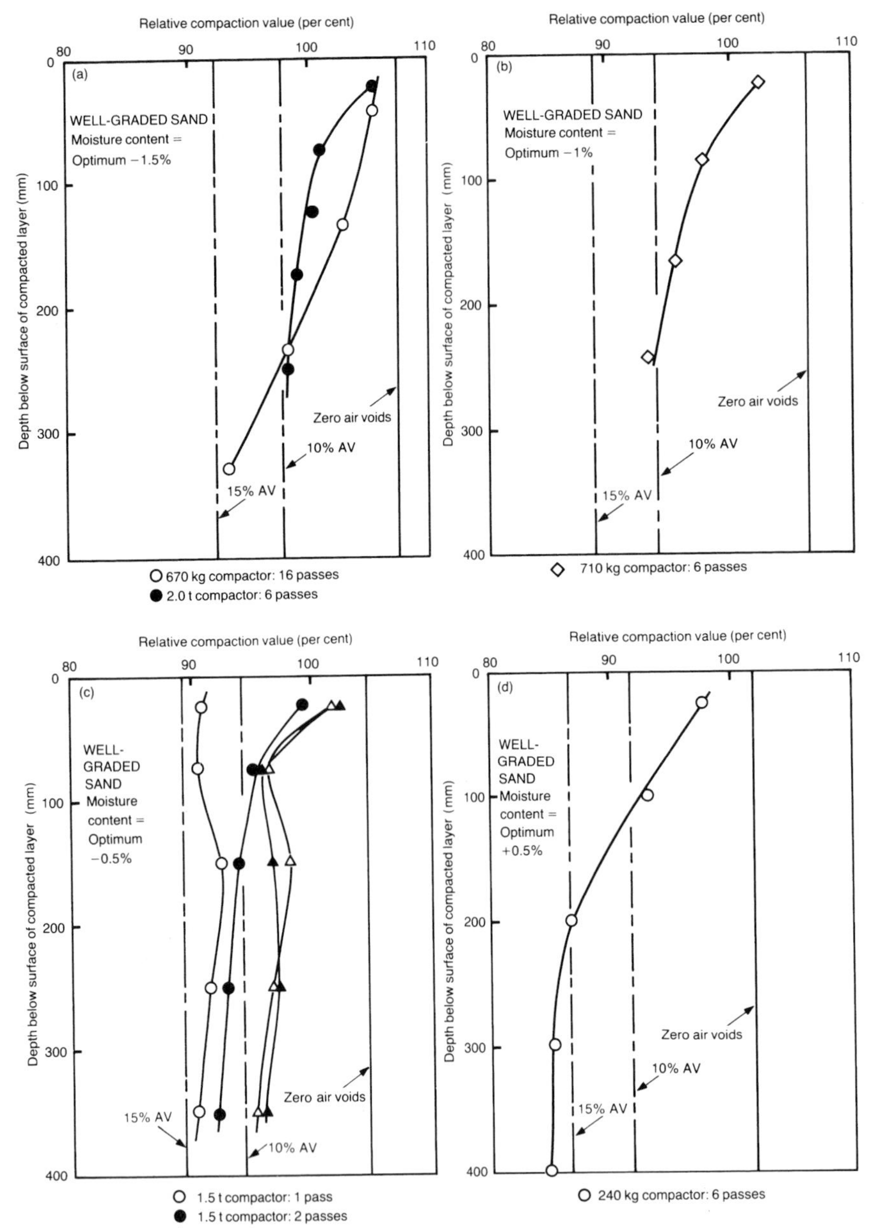

Figure 9.23 Relations between relative compaction and depth in the compacted layer of well-graded sand obtained with vibrating-plate compactors at four different values of moisture content. All values are related to the maximum dry density and optimum moisture content obtained in the 2.5 kg rammer compaction test (Figure 3.13)

670 kg compactor on all soils except sandy clay No 1 and are shown in Figures 9.20, 9.22, 9.23(a), 9.24(a) and 9.25(b). For the particular values of moisture content used, the thicknesses of layer capable of being compacted to achieve 90 per cent relative compaction or 15 per cent air voids at the bottom of the layer varied from about 100 mm on the heavy clay to in excess of 300 mm on well-graded sand; the thickness of the layer of uniformly graded fine sand to achieve 90 per cent relative compaction at the bottom is indeterminate from the results shown but is clearly well in excess of 300 mm (Figure 9.25(b)).

9.23 The 2.0 t vibrating-plate compactor was tested on three soils (Figures 9.21, 9.23(a) and 9.24(b)). At the moisture contents used thicknesses of layer varied from about 180 mm on sandy clay No 1 to over 300 mm with well-graded sand, using the same criteria as given above. The 710 kg and 240 kg compactors were tested on well-graded sand only (Figure 9.23(b) and (d)) and the 450 kg machine on the uniformly graded fine sand only (Figure 9.25(a)). The 1.5 t machine was tested on well-graded sand using various numbers of passes (Figure 9.23(c)). Only with the well-graded sand, therefore, was a sufficient variety of machines and soil conditions used to allow regression analyses to be made. Such analyses have been carried out using mass per unit area of base-plate, moisture content and number of passes as independent variables. It was found that for the range of moisture contents used the effect of moisture content was insignificant. The results

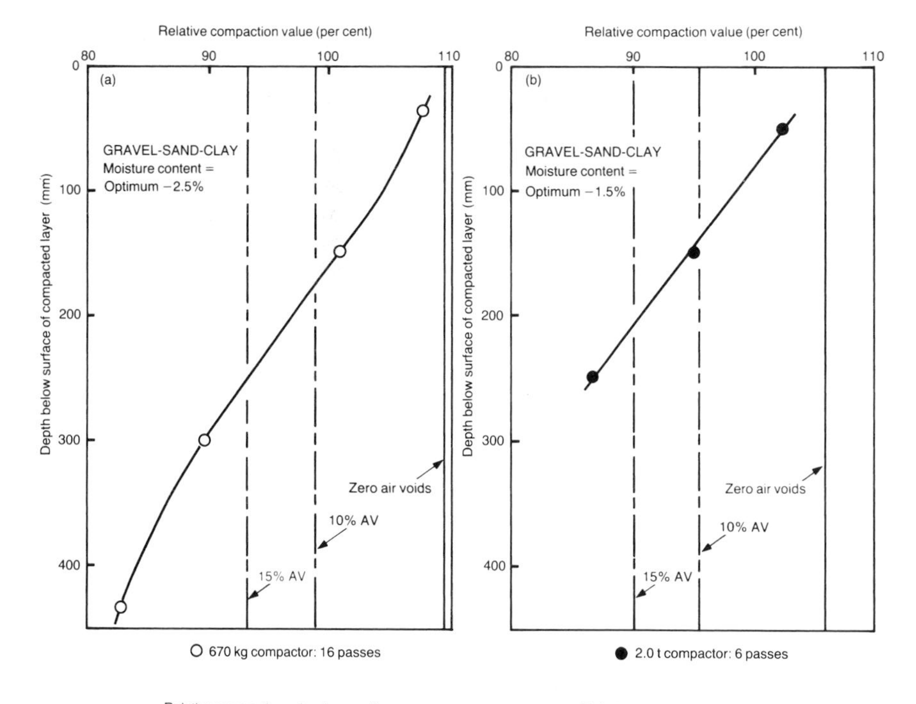

Figure 9.24 Relations between relative compaction and depth in the compacted layer of gravel-sand-clay obtained with vibrating-plate compactors at two different values of moisture content. All values are related to the maximum dry density and optimum moisture content obtained in the 2.5 kg rammer compaction test (Figure 3.13)

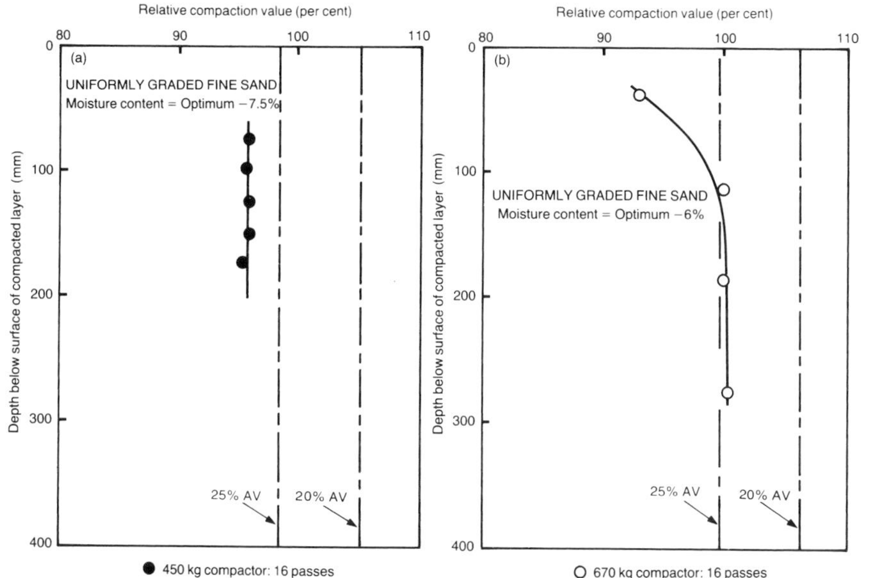

Figure 9.25 Relations between relative compaction and depth in the compacted layer of uniformly graded fine sand obtained with vibrating-plate compactors at two different values of moisture content. All values are related to the maximum dry density and optimum moisture content obtained in the 2.5 kg rammer compaction test (Figure 3.13)

are given in Table 9.3 and relations for 2 and 16 passes, over the range of masses per unit area to which the data applied, are shown in Figure 9.26. The correlations are good with values of 0.95 for the correlation coefficients, but the results should be treated with some reserve as they apply to one soil type only and to a limited range of machines, ie data for compactors with low masses per unit area were not available.

9.24 The depth of layer was, of course, a variable in the determination of the number of passes required to achieve specified states of compaction in the layer (Table 9.2); however, attempts to determine thickness of layer from the relations given in Table 9.2 must be prone to error because of the relatively small range of thicknesses used in those investigations. The relations given in Table 9.2 also indicate that moisture content, even within the range of values in the data used for Table 9.3, did have a significant effect on the performance of the machines. Thus the relative significance of the various factors may depend upon the particular dependent variable to be determined.

Consideration of centrifugal force and area of base-plate as factors influencing the performance of vibrating-plate compactors

9.25 In the review so far in this Chapter of investigations of vibrating-plate compactors, the parameter used to quantify the machine characteristics has been the mass per unit area of base-plate. More recent investigations of the performance of lightweight vibrating-plate compactors were intended specifically to determine the relative effects of the masses, centrifugal forces and areas of base-plate of the machines. The nominal centrifugal force and the centrifugal force per unit area of base-plate are given for each of the small machines listed in Table 9.1, as well as for some of the larger machines. It should be borne in mind that the data relating to dynamic forces, taken from manufacturers' literature, cannot be verified; the values for the larger machines, taken from very early records, must be treated with extreme caution and no attempt has been made here to make use of that early data.

Table 9.3 Equations and correlation coefficients to determine the thickness of layer capable of being compacted by vibrating- plate compactors

Soil: Well-graded sand

Level*	Regression equation†	No of data points	Correlation coefficient
(a)	$D=1.49\,m+.138\,P-2.56$	7	0.95
(b)	$D=1.11\,m+.131\,P-1.22$	8	0.95

*Level (a)=Achievement at the bottom of the layer of 90 per cent of the maximum dry density obtained in the 2.5 kg rammer compaction test

(b)=Achievement at the bottom of the layer of 15 per cent air voids

†$D=\log_{10}$ (Maximum depth of layer to achieve stated level of compaction, mm)

$m=\log_{10}$ (Mass per unit area of base plate, kg/m^2)

$P=\log_{10}$ (Number of passes)

Note The above results apply to a range of moisture contents from 2 per cent below to 1 per cent above the optimum obtained in the 2.5 kg rammer compaction test

9.26 To compare the relative effects of mass, dynamic force and area of base-plate of the smaller machines, regression analyses were

Table 9.4 Effects of various factors on the determination of the number of passes required to achieve given levels of compaction using vibrating-plate compactors. Results of regression and variance analyses

Soil	Layer thickness (mm)	Level*	Variable ‡ m	F	A	w	No of data points
Sandy clay No 2	100	(a)				†	6
		(b)		†	†	†	14
Well-graded sand	100	(a)	†			†	26
		(b)	†			†	24
	200	(a)	†			†	21
		(b)	†			†	20
Gravel-sand-clay	100	(a)	†			†	18
		(b)				†	19
	200	(a)		†		†	18
		(b)				†	19
Uniformly graded fine sand	100	(a)				†	9
	200	(a)	†			†	7

*Level (a)=95 per cent of the maximum dry density obtained in the 2.5 kg rammer compaction test

(b)=10 per cent air voids

‡Variable: $m=\log_{10}$ (Mass per unit area of base plate in kg/m^2)

$F=\log_{10}$ (Centrifugal force per unit area of base plate in kN/m^2)

$A=\log_{10}$ (Area of base plate in m^2)

w=Moisture content of soil, per cent, minus the optimum moisture content, per cent, obtained in the 2.5 kg rammer compaction test

†Indicates the most significant combination of variables to determine the required number of passes

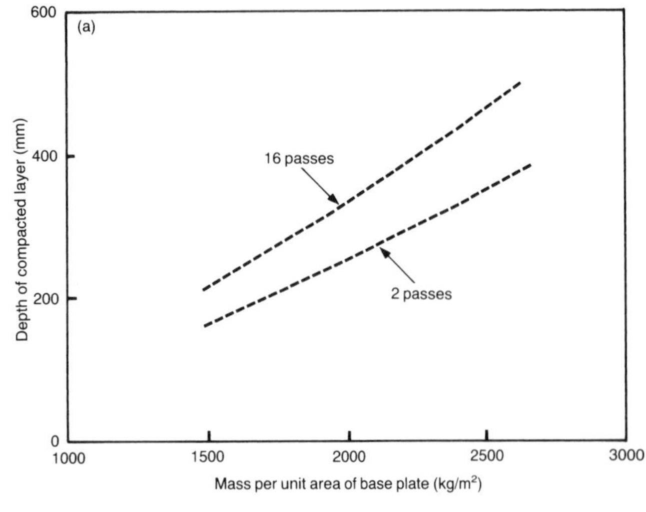

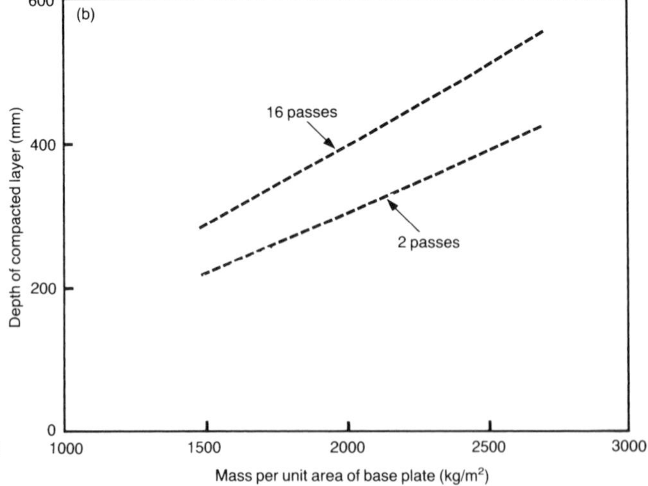

Figure 9.26 Relations between depth of layer and mass per unit area of base plate for well-graded sand at a moisture content between 2 per cent below and 1 per cent above the 2.5 kg rammer optimum:-
(a) 90 per cent relative compaction at the bottom of the layer;
(b) 15 per cent air voids at the bottom of the layer

carried out to produce relations in which the number of passes required to achieve either 95 per cent relative compaction or 10 per cent air voids was the dependent variable, and moisture content, mass per unit area of base-plate, centrifugal force per unit area of base-plate, and area of base-plate were independent variables. Data were analysed for constant thicknesses of layer, and analyses of variance applied to the resultant correlations to determine the relative significance of various combinations of variables. Table 9.4 contains a summary of the findings, in terms of the combinations of variables found to be most significant in the determination of the number of passes.

9.27 In four instances in Table 9.4 it can be seen that only the moisture content had an effect – for sandy clay No 2 at Level (a), with 100 mm layers of uniformly graded fine sand and with both thicknesses of layer with gravel-sand-clay at Level (b). The limited number of data points available for the sandy clay and uniformly graded fine sand, 6 and 9 respectively, may have a bearing on the results, but clearly the relatively small range of values for the machine parameters, in the absence of data for the larger machines, was the main cause of this.

9.28 Where the machine parameters were found to have a significant effect, the most common factor was the mass per unit area of base-plate. Centrifugal force per unit area only featured alone in one set of data (200 mm layers of gravel-sand-clay at level (a)). With the sandy clay the centrifugal force per unit area and area of base-plate were significant variables and this evidence that dynamic force may be the principal factor with clay soils, although very tenuous, must be considered for investigation in the future. For completeness, the regression equations for each of the combinations given in Table 9.4 are given in Table 9.5, using the significant combination of variables in each case. The correlation coefficients, also included in Table 9.5, indicate that, except in one instance (100 mm layers of gravel-sand-clay at level (a)), good correlations were achieved using such combinations.

9.29 The equations given in Table 9.5 are based, as stated earlier, on results of recent investigations using lightweight machines. These are shown in Plates 9.1 to 9.6 and are the first six listed in Table 9.1. The recorded values for the dynamic forces with these machines are likely to be fairly reliable (see Paragraph 9.25),

Table 9.5 Equations and correlation coefficients to determine the number of passes of vibrating-plate compactors to achieve specified levels of compaction, using the most significant combinations of variables

Soil	Layer thickness (mm)	Level*	Regression equation†	No of data points	Correlation coefficient
Sandy clay No 2	100	(a)	P=1.61–.0920 w	6	0.94
		(b)	P=4.10–.855 F+1.24 A–.231 w	14	0.97
Well-graded sand	100	(a)	P=5.82–1.97 m–.206 w	26	0.87
		(b)	P=5.42–1.75 m–.411 w	24	0.94
	200	(a)	P=8.59–2.75 m–.247 w	21	0.88
		(b)	P=8.41–2.75 m–.415 w	20	0.91
Gravel-sand-clay	100	(a)	P=5.91–1.81 m–.289 w	18	0.76
		(b)	P=.641–.418 w	19	0.85
	200	(a)	P=2.91–.981 F–.391 w	18	0.89
		(b)	P=.885–.397 w	19	0.80
Uniformly graded fine sand	100	(a)	P=.256–.164 w	9	0.91
	200	(a)	P=4.89–1.49 m–.127 w	7	0.997

*Level (a)=95 per cent of the maximum dry density obtained in the 2.5 kg rammer compaction test
(b)=10 per cent air voids

†P=Log_{10} (Number of passes required)
m=Log_{10} (Mass per unit area of base plate in kg/m^2)
F=Log_{10} (Centrifugal force per unit area of base plate in kN/m^2)
A=Log_{10} (Area of base plate in m^2)
w=Moisture content of soil, per cent, minus the optimum moisture content, per cent, obtained in the 2.5 kg rammer compaction test

but the range of sizes was small compared with the total range of machines investigated since 1949. To predict the number of passes for any one of the complete range of machines tested the regression analyses summarised in Table 9.2 have to be used; these rely solely on mass per unit area as the variable descriptive of the machine.

9.30 It is not possible to reach general conclusions on the relative effects of the various factors involved in determining the performance of the complete range of vibrating-plate compactors, but for machines less than 200 kg, Tables 9.4 and 9.5 show that mass per unit area is the principal factor with granular soils, but centrifugal force and area of base-plate may be the principal factors with sandy clay soils. For machines with dead weights in excess of 200 kg, the effect of dynamic forces could not be assessed, but fairly good correlations of performance with mass per unit area were achieved (Table 9.2) indicating that this parameter may well provide a reasonable guide to performance. The effect of area of base-plate as an additional variable with the heavier machines was assessed by regression and variance analyses similar to those described above and, for the complete range of vibrating-plate compactors, the additional variable was found to be relatively insignificant in determining the required number of passes. Thus it is concluded that over the range of machines tested the mass per unit area is likely to be a good guide to performance, although it is suggested that further detailed research in the future into the additional effects of dynamic forces with heavier machines would be of considerable interest.

9.31 With regard to the effect of centrifugal force on performance, a limited study was made with the 670 kg vibrating-plate compactor. This machine had the facility for varying the centrifugal force while maintaining other factors, such as frequency of vibration, constant (see Paragraph 9.7). Unfortunately, the machine was tested on uniformly graded fine sand only at a centrifugal force other than the maximum available; on that occasion the reason for reducing the dynamic forces was to reduce the degree of overstressing at the surface of the compacted layer, a common occurrence with this particular type of soil. The states of compaction produced by the two different dynamic forces are shown related to moisture content in Figure 9.6, to number of passes in Figure 9.27 and to depth within the compacted layer in Figure 9.28. The

Figure 9.27 Relations between relative compaction and number of passes of the 670 kg vibrating-plate compactor with two different values of dynamic force on uniformly graded fine sand. Values are related to the maximum dry density and optimum moisture content obtained in the 2.5 kg rammer compaction test (Figure 3.13)

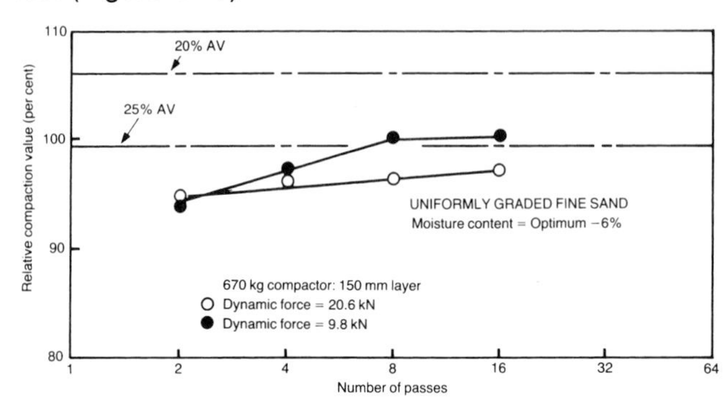

results show that the lower dynamic force produced the higher values of relative compaction over the full range of moisture contents tested (Figure 9.6), and for three or more passes of the machine (Figure 9.27) in 150 mm compacted layers. Figure 9.28 additionally shows that the machine with the reduced dynamic force was capable of achieving levels of compaction equal to or greater than those achieved with the higher dynamic force down to depths in excess of 300 mm in the compacted layer. This particular effect may have been caused by the cushion of loose, overstressed material on the surface of the layer compacted using the higher dynamic force, which might be expected to modulate the vibrational forces reaching the lower parts of the layer. The net result, therefore, was that the effect of centrifugal force on the state of compaction achieved below the overstressed surface layer was fairly small; this result is in keeping with the conclusions reached from the regression analyses described in Paragraphs 9.26 to 9.29.

Effect of frequency of vibration

9.32 One of the lightweight vibrating-plate compactors investigated in the more recent studies, the 80 kg machine with a static contact pressure of 410 kg/m^2, was operated at the recommended frequency (83 Hz) and also at a higher frequency averaging about 105 Hz. The soil used was well-graded sand with a final compacted layer thickness of 100 mm. Results obtained at three different values of moisture content are shown in relations between relative compaction value and number of passes in Figure 9.29. The effect of frequency of vibration

Figure 9.28 Relations between relative compaction and depth in the compacted layer of uniformly graded fine sand obtained with the 670 kg vibrating-plate compactor with two different values of dynamic force. All values are related to the maximum dry density and optimum moisture content obtained in the 2.5 kg rammer compaction test (Figure 3.13)

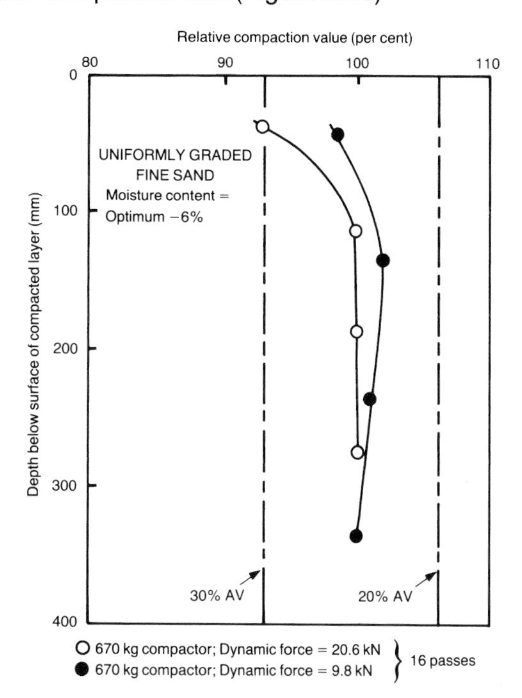

illustrated in the Figure clearly varied with the moisture content, the increased frequency resulting in both higher and lower levels of compaction compared with those produced at the normal frequency of 83 Hz, in the different graphs. The centrifugal force would have been increased in proportion to the square of the frequency, so that the 9.7 kN force in the normal test mode (Table 9.1) would have been increased to a centrifugal force of 15.5 kN at the higher frequency. Thus a 50 per cent increase in centrifugal force did not result in a general improvement in performance, and at the lower moisture content, where an improved performance could be most beneficial, the effect was to produce a reduced level of compaction. This is further evidence, therefore, supporting the conclusion that centrifugal force has only a secondary effect on the performance of vibrating-plate compactors within the range of such forces normally employed in the machines (see Paragraphs 9.25 to 9.31).

Effect of increasing the area of base-plate by the addition of extension plates

9.33 The 450 kg vibrating-plate compactor was provided with extension plates which could be attached to the sides of the existing base-plate to increase its overall area. The machine in its basic form is shown in Plate 9.10 and with extension plates fitted in Plate 9.15. A comparison of the main details of the machine in its two forms is given in Table 9.6. It can be seen that the mass per unit area of base-plate was reduced by about 25 per cent by the addition of the extension plates.

9.34 Tests were carried out with the machine fitted with the extension plates on well-graded sand and uniformly graded fine sand and the results, together with those for the machine in its basic form, are given in Figures 9.30 to 9.32. Relations between relative compaction value and number of passes for well-graded sand are given in Figure 9.30, but because different thicknesses of layer were used for the two modes of use, a direct comparison of the effect was not obtained. However, even with the reduced thickness of layer, the number of passes required to achieve a given level of compaction was increased with

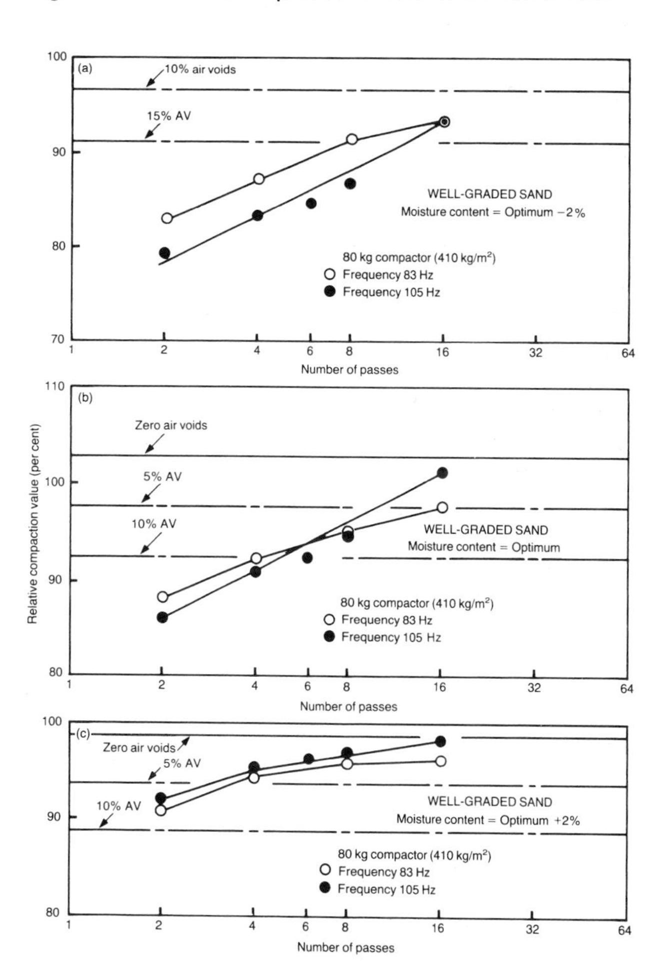

Figure 9.29 Relations between relative compaction and number of passes of the 80 kg vibrating-plate compactor (410 kg/m²) using two different frequencies of vibration, compacting 100 mm layers of well-graded sand

the addition of the extension plates; this effect is compatible with the dominant effect of the mass per unit area of base-plate (see Paragraph 9.28).

Table 9.6 Details of the 450 kg vibrating-plate compactor in its basic form and when fitted with side extensions to the base-plate

Machine: 450 kg vibrating-plate compactor		
	Basic form	With extension plates
Total mass (kg)	448	496
Width of base-plate (m)	0.56	0.84
Contact area of base-plate (m^2)	0.412	0.617
Mass per unit area (kg/m^2)	1090	800
Vibrating mechanism		
No of eccentric shafts	1	1
Frequency (Hz)	53	53
Nominal centrifugal force (kN)	35.5	35.5
Centrifugal force per unit area (kN/m^2)	86	58
Normal speed of travel (km/h)	1.1	0.9

Although the width of strip compacted by the machine fitted with extension plates was increased by a factor of 1.5 over that compacted by the basic machine, the speed of operation was reduced from 1.1 to 0.9 km/h (Table 9.6); the net effect of the results given in Figure 9.30 was a reduction in output of compacted soil (compacted to 95 per cent relative compaction) when using the extension plates to about 70 per cent of that achieved with the machine in its basic mode.

9.35 The relations between relative compaction and number of passes for the uniformly graded fine sand (Figure 9.31) do not provide such a

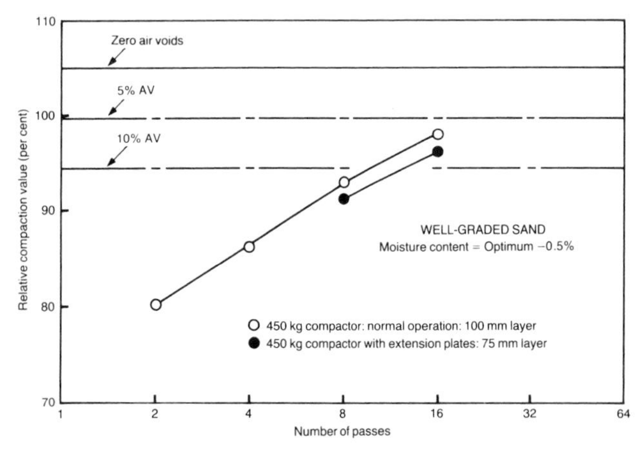

Figure 9.30 Relations between relative compaction and number of passes of the 450 kg vibrating-plate compactor, with and without extension plates, on well-graded sand. Values are related to the maximum dry density and optimum moisture content obtained in the 2.5 kg rammer compaction test (Figure 3.13)

Figure 9.31 Relations between relative compaction and number of passes of the 450 kg vibrating-plate compactor, with and without extension plates, on uniformly graded fine sand. Values are related to the maximum dry density and optimum moisture content obtained in the 2.5 kg rammer compaction test (Figure 3.13)

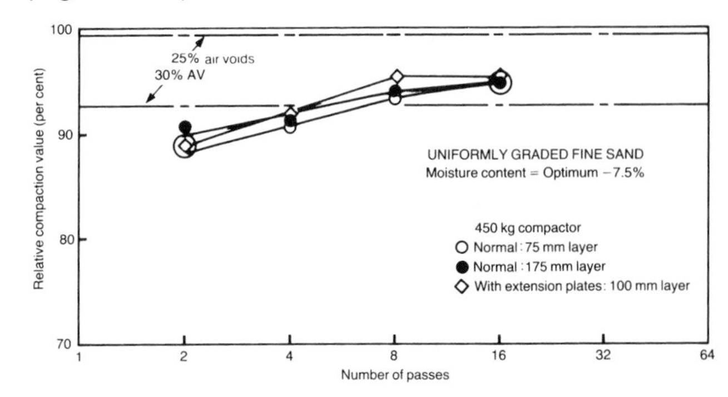

clear picture as those in Figure 9.30, and the improved states of compaction with the thicker, 175 mm, layer over those obtained with the 75 mm layer when using the machine in its basic form indicates that some overstressing of the surface of the layer occurred. Thus the reduced stresses resulting from the addition of the extension plates produced higher states of compaction and a slight reduction in the number of passes to achieve, say, 95 per cent relative compaction, in the 100 mm layer used. The results indicate an increase in output by about 55 per cent by the use of the extension plates with a 100 mm layer, compared with the normal mode of operation on the 175 mm layer; this is further evidence of the benefit of reducing the stresses on this soil as already described in Paragraph 9.31.

9.36 The relations between relative compaction value and depth in the compacted layer of uniformly graded fine sand for the 450 kg machine with and without extension plates (Figure 9.32) were obtained from the results of preliminary tests aimed at determining the appropriate thicknesses of layer to use with the two modes of operation. The addition of extension plates resulted in a tendency to reduce slightly the levels of compaction achieved in the lower parts of the layer, at a depth of about 150 mm, but no distinct difference in results were achieved in the upper parts of the layer. The maximum thickness compacted to achieve a relative compaction value of 90 per cent at the bottom of the layer is judged to be about 180 mm in normal operation and about 150 mm with extension plates, using 16 passes of the machine at the particular moisture content indicated in Figure 9.32. This represents almost identical outputs of compacted soil, allowing for

Figure 9.32 Relations between relative compaction and depth in the compacted layer of uniformly graded fine sand obtained with 16 passes of the 450 kg vibrating-plate compactor with and without extension plates. All values are related to the maximum dry density and optimum moisture content obtained in the 2.5 kg rammer compaction test (Figure 3.13)

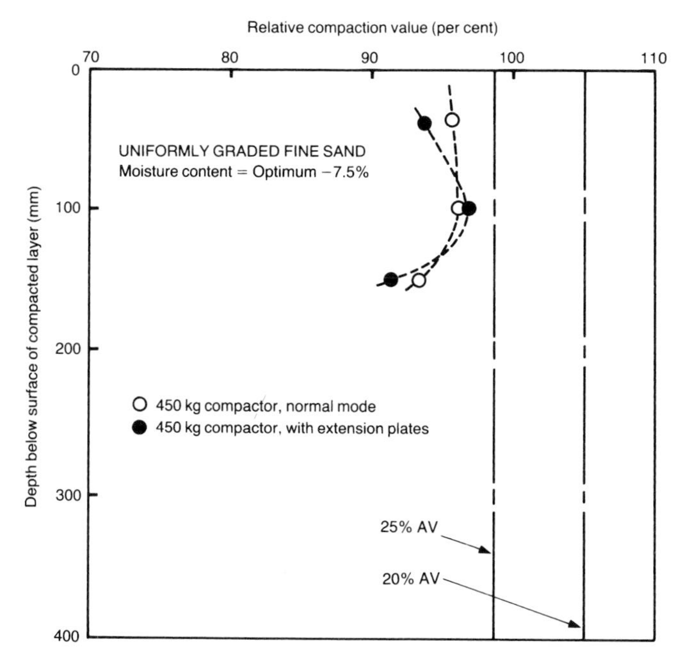

the differences in compacted width and speeds of travel for the two modes of operation.

9.37 It can be concluded from these few results that where a soil is susceptible to overstressing on the surface of the layer the use of extension plates may result in improved outputs of compacted soil in shallow layers. Where the soil is not prone to overstressing on the surface, however, the use of extension plates may have a detrimental effect on the output of soil compacted to a specified level.

Speed of operation of vibrating-plate compactors and methods of control

9.38 The speed of operation of the vibrating-plate compactors was influenced by the characteristics of the vibrating mechanisms and by the soil conditions. Where the characteristics of the vibrating mechanism were fixed, as with machines with single eccentric shafts (see Paragraph 9.3) and those with twin eccentric shafts with a fixed resultant-force angle (see Paragraph 9.8), variations in adhesion between the base-plate and the soil surface was the only factor causing variations in the speed of operation; records show that some of the compactors became immobile on wet soil, irrespective of the soil type. Additionally, the machines usually travelled at lower speeds during the first pass over a loose layer of soil than on subsequent passes. Figure 9.33 illustrates the effect of moisture content on the speed of travel of a number of vibrating-plate compactors for which detailed measurements of speed were obtained during the practical work. The rate of decrease in speed of travel with

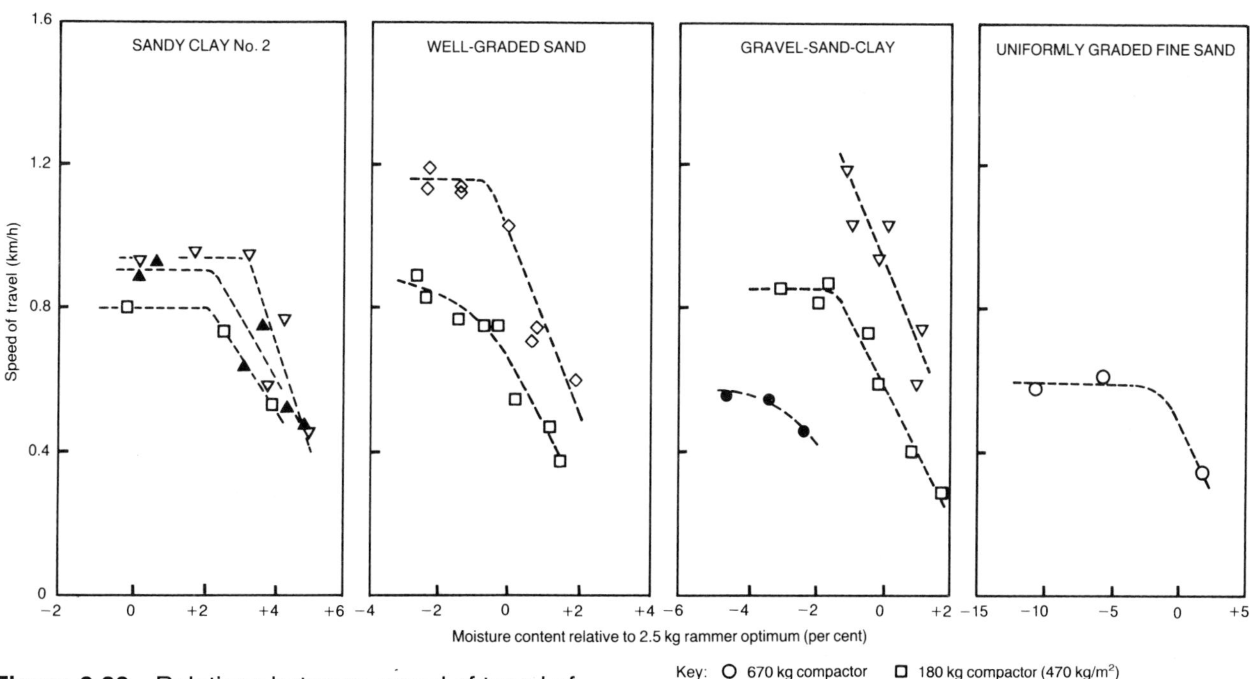

Figure 9.33 Relations between speed of travel of vibrating-plate compactors and the moisture content of soil

increasing moisture content can be seen to be fairly similar for all the machines on all the soils where sufficient results over an adequate range of moisture contents were obtained.

9.39 The speed of travel can be varied independently of the soil conditions with vibrating-plate compactors fitted with twin eccentric shafts and the facility either to vary the phase angle between them or to rotate the housing containing the eccentrics. The speed of travel is varied by varying the angle of the resultant dynamic force, which in turn increases or decreases the horizontal force component. Any increase in speed resulting from an increase in the horizontal component of force has to be compensated, therefore, by a reduction in the vertical component of the force, and hence of the vertical dynamic stress at the contact surface of the base-plate and the soil.

9.40 A vibrating-plate compactor which was studied at more than one speed setting was the 670 kg machine. This had two speeds of travel in both the forward and reverse directions (see Paragraph 9.7) and the normal mode of operation in the tests was to use the fast speed

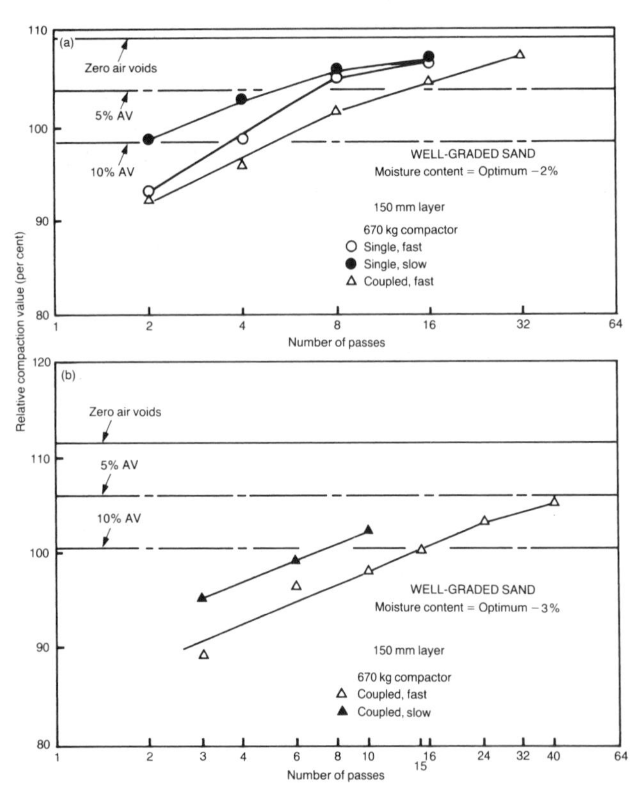

Figure 9.34 Relations between relative compaction and number of passes for various modes of operation of the 670 kg vibrating-plate compactor on well-graded sand. Values are related to the maximum dry density and optimum moisture content obtained in the 2.5 kg rammer compaction test (Figure 3.13)

setting, providing a mean speed of travel of about 1.0 km/h (Table 9.1). The slow speed setting provided a mean speed of travel of about 0.5 km/h, ie half the normal speed, and results obtained using this speed on well-graded sand, together with the equivalent results obtained at the normal speed of travel, are given in Figures 9.34 and 9.35. Some of the results are for coupled machines (two machines linked together transversely) and the effects of such coupling are described later (Paragraphs 9.43 to 9.44); however, the effect of the speed setting would be expected to be fairly similar for both single and coupled machines and the results in this context are discussed together. In Figure 9.34, relating relative compaction to number of passes, results are given for fast and slow speeds of a single machine at a moisture content 2 per cent below the optimum of the 2.5 kg rammer compaction test, and for fast and slow speeds of coupled machines at optimum minus 3 per cent. In both cases the state of compaction produced by a given number of passes was much higher with the slow speed setting, ie the speed at which the larger vertical component of dynamic force was produced. Another factor causing the higher compaction would also have been associated with the number of vibration pulses which were exerted on an area of soil on each pass; this would have been doubled when the speed of travel was halved. It can be easily seen in Figure 9.34 that the number of passes to achieve either 95 per cent relative compaction or 10 per cent air voids was almost exactly doubled when the speed was doubled, thus yielding a constant output of compacted soil to those particular standards. A tentative conclusion from this limited amount of data would be, therefore, that where the speed of travel is varied by changing the angle of the resultant dynamic force, the output of compacted soil to a given standard using a constant layer thickness would remain roughly constant, ie the number of passes must be increased in proportion to any increase in the speed of travel.

9.41 In Figure 9.35 relations between relative compaction and depth within the compacted layer of well-graded sand are given for a single 670 kg vibrating-plate compactor using 16 passes at the fast speed setting (1.0 km/h) and 8 passes at the slow speed setting (0.5 km/h), ie following the principle outlined above to produce a constant level of compaction. There is evidence in these results that the state of compaction at the lower levels of the compacted layer, ie more than 250 mm below the surface, was higher using the slow speed setting,

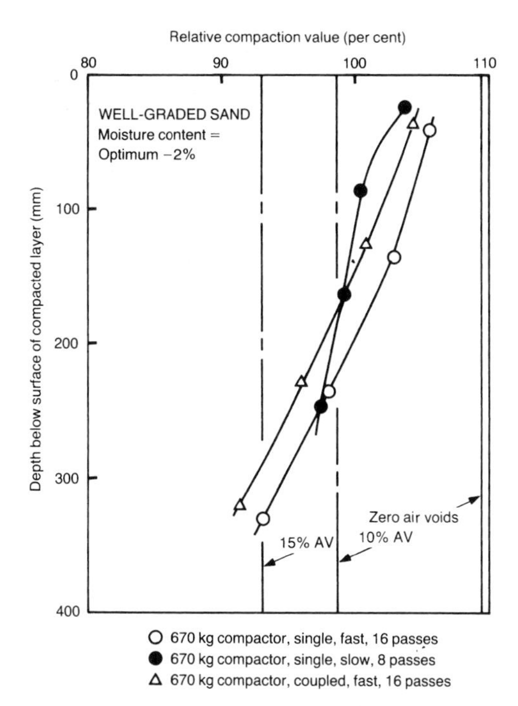

Figure 9.35 Relations between relative compaction and depth in the compacted layer of well-graded sand obtained with various modes of operation of the 670 kg vibrating-plate compactor. All values are related to the maximum dry density and optimum moisture content obtained in the 2.5 kg rammer compaction test (Figure 3.13)

although the limited thickness of layer used in the tests did not show this conclusively. Such a result would, however, be compatible with the increased vertical component of the dynamic force which resulted from the use of the slower speed of travel (note that the centrifugal force of the machine remained constant throughout all these tests). Thus, although output remained constant when using a constant thickness of layer at different speed settings with this machine, it is possible that increased thicknesses of layer could have been compacted using slower speeds of travel, given that such lower speeds were achieved by changing the angle of the resultant dynamic force.

9.42 A limited study was made of the 240 kg vibrating-plate compactor with the vibrator unit locked in both the full-forward (as normally used in the tests) and half-forward positions (see Paragraph 9.6). The tests were carried out on well-graded sand and gravel-sand-clay at one moisture content only in each case. The results are given in Table 9.7. The use of the half-forward position resulted in higher states of compaction than the full-forward position on both soils, albeit at slightly lower values of moisture content. No detailed measurements of speed of travel were made in association with these tests, but the comment was recorded that the machine moved very slowly, at about 1 m/min, with the vibrator unit in the half-forward position, and this was considered too low for economical operation in practice. The contrast between the findings with the two machines shows that a sufficient horizontal component of the dynamic force must be maintained to provide a practical speed of travel, whilst maximising the vertical component within that constraint might provide the opportunity to increase the thickness of layer capable of being compacted, as demonstrated by the results with the 670 kg machine.

The effect of coupling together two vibrating-plate compactors

9.43 Some vibrating-place compactors have the facility for attaching two or more machines side-by-side. The obvious advantage of this is that only one operator is involved in their collective operation and large areas can be covered to provide outputs of compacted soil more compatible with other types of high-production compaction plant. So-called 'gangs' of vibrating-plate compactors, mounted on purpose-built prime movers, also accomplish this objective, although often the individual compactors in such gangs are equivalent to the lightweight types with relatively low performances. The 670 kg vibrating-plate compactor had this facility and a pair of machines were tested in the coupled mode during investigations carried out in 1958. The two machines are shown linked together in Plate 9.16. The combined machines could be manoeuvred by selection of different speeds or even of different directions of travel on the two individual components.

Table 9.7 Results of tests with the 240 kg vibrating-plate compactor using the vibrator unit in two 'speed' positions. Six passes of the machine were used on a 150 mm thick compacted layer

Soil	Position of vibrator unit	Relative compaction (per cent)	Moisture content relative to 2.5 kg rammer optimum (per cent)
Well-graded sand	Full-forward	96	+0.5
	Half-forward	98	0
Gravel-sand-clay	Full-forward	97	+1
	Half-forward	103	0

9.44 Results obtained with coupled machines are directly compared with the results for a single machine, using the fast speeds of travel, in Figures 9.34(a) and 9.35. In the former Figure relations between relative compaction and number of passes are given; the state of compaction produced by the coupled machines was generally lower than that produced by the single machine at any given number of passes. A comparison of the relations between relative compaction and depth in the compacted layer (Figure 9.35) reveals that the coupled machines produced lower states of compaction than the single machine at all levels in the layer. The reason for the loss in performance of the coupled machines was considered to be that each machine partially damped-out the impact effect of the other. The method of linking or 'ganging' the machines must play a considerable part in the damping effect, however, and the results given here would not necessarily apply in all cases of machines operating side-by-side. With the 670 kg machines, although the output of each machine was reduced in a coupled operation the net output of the operator would have been considerably increased; only by comparing the economics of the two modes of operation, taking into account the relative costs of machine and operator, could the most economical solution be determined (see Chapter 15).

Pressures produced in compacted soil by vibrating-plate compactors

9.45 The only attempts to measure the vertical pressure in soil during the passage of a vibrating-plate compactor were associated with the recent studies in 1985 to 1987 of the performance of lightweight machines. Those studies were aimed at determining the relative effects of the mass, centrifugal force and base-plate area on the performance of the compactors. Measurements of pressure were made in two of the soils prepared at a range of moisture contents. The soils were sandy clay No 2 and well-graded sand and the pressure measurements were made at depths of 100 and 200 mm within the compacted layer. The pressure cells used were the type operating on the inductive principle (LVDT type) with the results registered by a high-speed galvanometer ultra-violet recorder. Each pressure cell was individually calibrated in a gas-over-water pressure chamber. Two typical records obtained from pressure cells installed at a depth of 200 mm in well-graded sand are shown in Figure 9.36. The paper speed used was about 100 mm/s and the length of trace corresponding to 1 second duration of time is shown on the Figure. The individual peaks of stress can be very easily counted and records such as those shown in Figure 9.36 were also used to verify the frequency of vibration of each machine. The records shown have been deliberately chosen to illustrate the difference in behaviour of a vibrating-plate compactor on well-graded sand at different moisture contents. In the upper figure, the record was obtained from a pressure cell in the soil at a moisture content about 1.5 per cent above the optimum of the 2.5 kg rammer compaction test. In those particularly wet and soft soil conditions the machine produced a series of pressure peaks of uniform height as it passed over the pressure cell. At a moisture content about 1 per cent below the 2.5 kg rammer optimum (lower trace of Figure 9.36) the

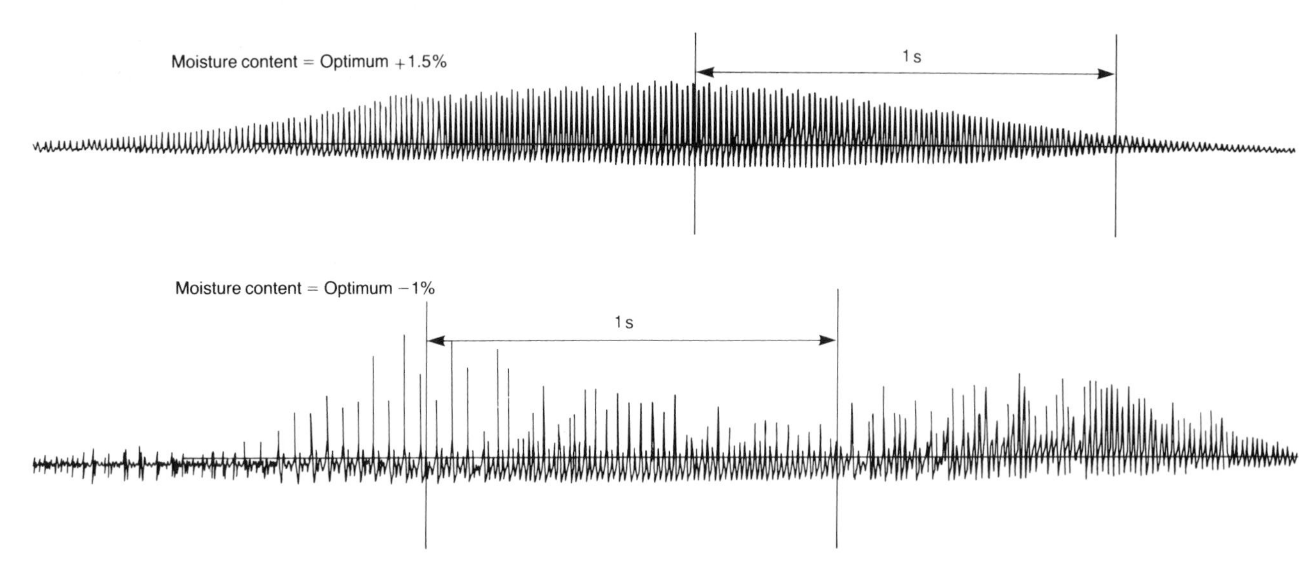

Figure 9.36 Typical records obtained from pressure cells set at 200 mm depth in well-graded sand during the passage of the 180 kg vibrating-plate compactor (470 kg/m² static contact pressure)

machine tended to bounce erratically on the stiff surface of the layer and only intermittent peaks of high pressure were registered. Interpretation of the latter type of pressure trace was extremely difficult.

9.46 An arbitrary decision was made to take the recorded pressure as that corresponding to the height of the 80th percentile peak, ie 20 per cent of the individual peak pressures recorded exceeded that value, taken over that part of the trace recorded when the base-plate of the machine was vertically above the pressure cell. The results are illustrated in Figures 9.37 and 9.38 as relations between recorded pressure and moisture content. With the well-graded sand (Figure 9.38) differences in pressures produced by different machines were considerably greater at the lower values of moisture content, although the results were rather scattered, especially from the pressure cells at a depth of 100 mm. This scatter was clearly associated with the difficulties in interpretation of the pressure traces, as described above. With the sandy clay soil (Figure 9.37) fairly similar relations between pressure and moisture content were obtained with each of the four machines tested.

9.47 To summarise the results, regression analyses were carried out using pressure as the dependent variable and moisture content, mass per unit area of base-plate, centrifugal force per unit area of base-plate, and area of base-plate as independent variables. The results of the analyses are given in Table 9.8, the non-significant variables having been omitted. The Table shows that, for the machines tested, only moisture content was a significant variable affecting the pressure measurements at both depths in the sandy clay soil. With well-graded sand, mass per unit area was also significant with regard to the pressure at 100 mm, and the further addition of the area of base-plate occurred with the pressures recorded at 200 mm. The latter effect might be expected as the bulb of pressure under the larger plates would be more extensive at a constant contact pressure, assuming that the soil behaved elastically. It is noteworthy that the centrifugal force per unit area did not have a significant effect on the recorded pressures. The correlation coefficients given in Table 9.8 are all high except for the results with the 100 mm depth in well-graded sand.

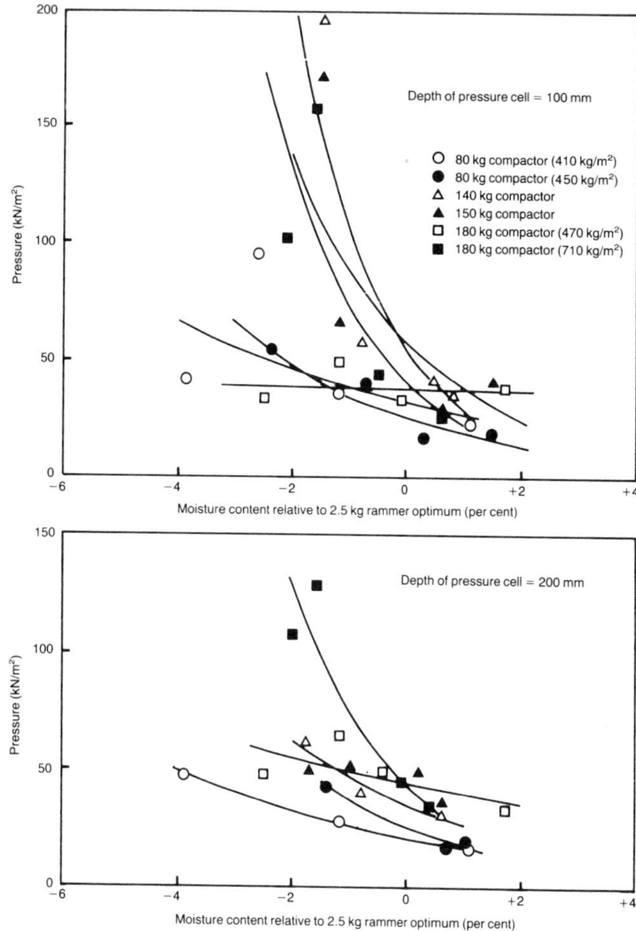

Figure 9.38 Relations between recorded pressure and moisture content at two different depths in well-graded sand using various vibrating-plate compactors

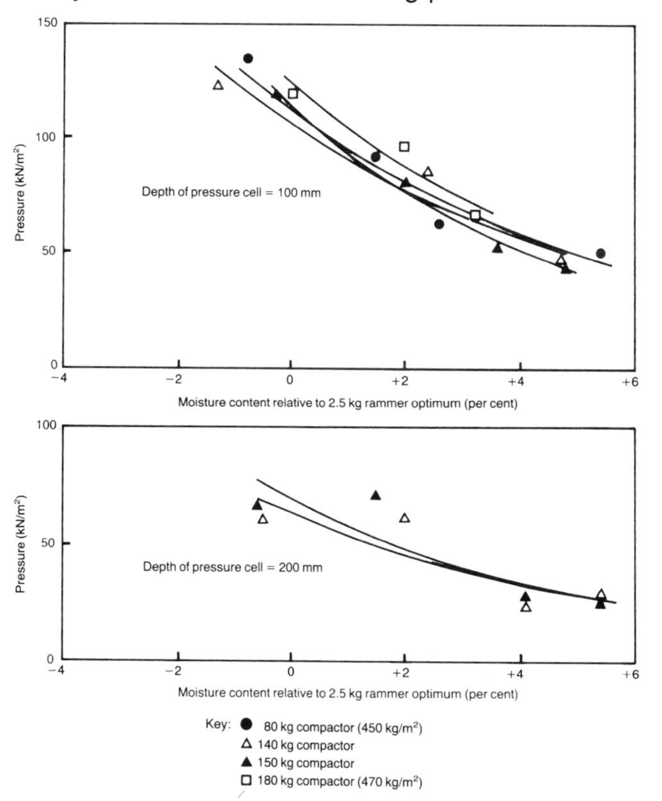

Figure 9.37 Relations between recorded pressure and moisture content at two different depths in sandy clay No 2 using various vibrating-plate compactors

Table 9.8 Results of regression analyses of the data shown in Figures 9.37 and 9.38. Only the significant variables are included in the regression equations

Soil	Depth of pressure cell (mm)	Regression equation*	No of data points	Correlation coefficient
Sandy clay No 2	100	p=2.06–.0769 w	14	0.96
	200	p=1.83–.0752 w	8	0.89
Well-graded sand	100	p=–1.49–.114 w+1.12 m	23	0.75
	200	p=–.656–.0917 w+.975 m +.768 A	20	0.91

*p =Log_{10} (Pressure in kN/m^2)
w=Moisture content of soil, per cent, minus the optimum moisture content, per cent, obtained in the 2.5 kg rammer compaction test
m=Log_{10} (Mass per unit area of base plate in kg/m^2)
A=Log_{10} (Area of base plate in m^2)

9.48 The results of the regression analyses given in Table 9.8 can be compared with the results given in Table 9.5, in which the relative effects of the various machine parameters on the number of passes required to achieve given levels of compaction have been determined. The dominance of mass per unit area of base-plate as the machine factor controlling the number of passes required is confirmed by the analyses of the results of the pressure measurements.

9.49 The inclusion in the soil of a pressure cell with a modulus of elasticity widely dissimilar from that of the soil can be expected to cause an aberration in the original pressure distribution under the compactors. Thus the values of pressure registered may not be strictly accurate and the results in Figures 9.37 and 9.38 and in Table 9.8 should be regarded only as general illustrations of the trends and relative values of the vertical pressures exerted by the machines.

Displacement of the base-plate and body of a vibrating-plate compactor during operation

9.50 It was noted in Paragraph 9.45 that with fairly stiff compacted soils pressure records showed intermittent peaks of high pressure because the machines tended to bounce erratically. This aspect was studied during the earlier investigations, in 1955, of the performance of the 670 kg vibrating-plate compactor. A high-speed ciné film was taken while the machine was compacting a strip of well-graded sand; by examining the film frame-by-frame, measurements were made of the vertical movements with respect to a fixed datum of various parts of the compactor. A small portion of the results obtained with the machine travelling in the forward direction is shown in Figure 9.39. No relation appears to exist between the movements of the various points examined, and the magnitude of the movements indicates that the machine was behaving more as a tamper than as a vibrator. The base-plate oscillated at a frequency of about 20 Hz, ie the frequency of the eccentric shafts, with a maximum amplitude at the front and rear of about 25 mm. Superimposed on this, the main body of the machine, supported above the oscillating base-plate on coil springs (see Plate 9.11) was oscillating at a frequency of about 4 Hz. The maximum amplitude of the base-plate movements only occurred on approximately every second cycle (see trace for the front of the base-plate in Figure 9.39). Erratic movements of this type were clearly the cause of the

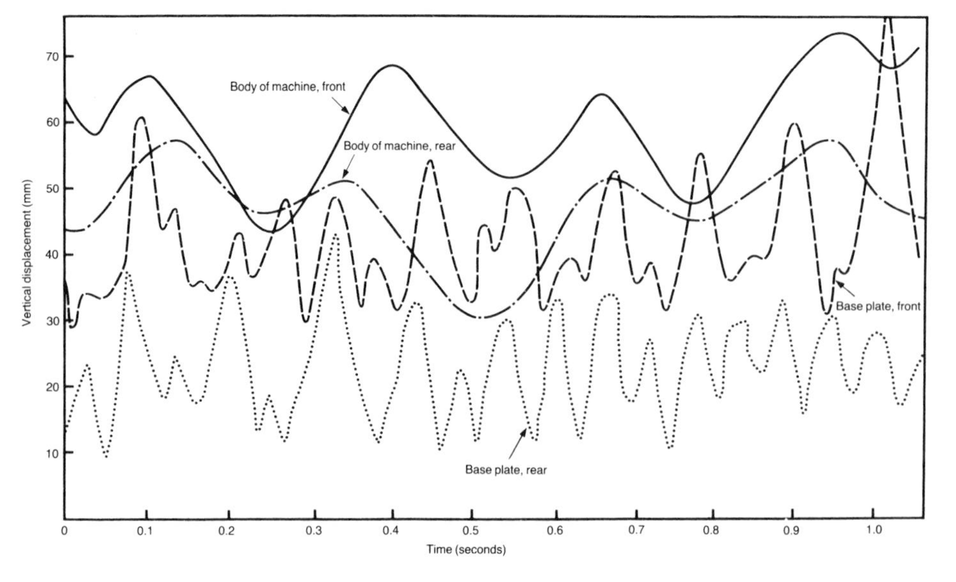

Figure 9.39 Vertical displacement of front and rear of the base plate and the main body of the 670 kg vibrating-plate compactor

intermittent high-pressure peaks obtained in the records of pressures exerted by lightweight compactors at low moisture contents on well-graded sand, as illustrated in the lower part of Figure 9.36. It was considered that the large vertical movements measured with the 670 kg machine were the reason for its efficient performance in the compaction of cohesive soils.

Study of the ability of a vibrating-plate compactor to negotiate an incline

9.51 The 670 kg vibrating-plate compactor was also used in a study to determine the steepest incline on which it was practicable to operate the machine. An embankment of well-graded sand was constructed and at stages during the construction the forward and reverse speeds of the machine were measured while climbing the slope. The results are shown in Figure 9.40.

9.52 The maximum slope of the compacted soil that the machine could climb was about 15° (gradient of 1 in 4). On the assumption that a speed of travel of about 5 m/min (0.3 km/h) was the minimum at which it was practical to operate the machine, it was concluded that a slope of about 12° (gradient of 1 in 5) was the steepest that could be compacted when the vibrating-plate compactor was operated under favourable conditions on well-graded sand. No record appears to have been made of the moisture content of the well-graded sand during these particular tests, but the most favourable moisture content for speed of travel on this soil was probably at least 1 per cent below the optimum of the 2.5 kg rammer compaction test, as indicated by the results, shown in Figure 9.33, of measurements of speeds of travel of other machines.

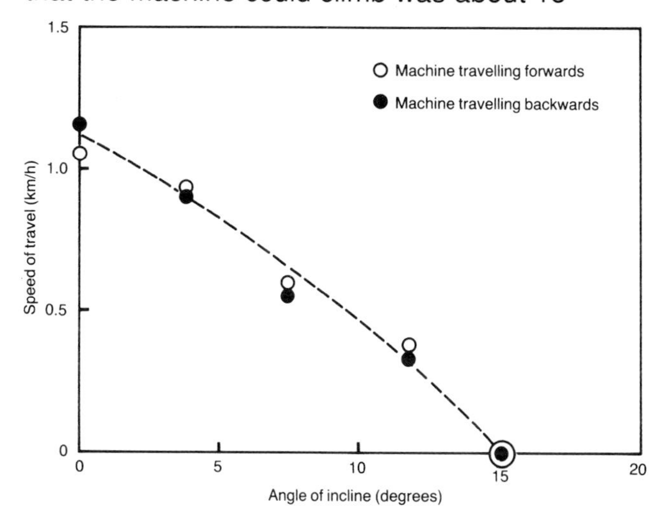

Figure 9.40 Relation between forward and reverse speeds of travel of the 670 kg vibrating-plate compactor and the angle of incline on an embankment of well-graded sand

References

Lewis, W A (1954). Further studies in the compaction of soil and the performance of compaction plant. *Road Research Technical Paper* No 33. HM Stationery Office, London.

Lewis, W A (1961). Recent research into the compaction of soil by vibratory compaction equipment. *Proc. 5th Internat. Conf. Soil Mech*, Vol II. Dunod, Paris.

Toombs, A F (1966a). The performance of a Wacker DVPN-75 3½-cwt vibrating-plate compactor in the compaction of soil. *Ministry of Transport, RRL Laboratory Report* No 26. Road Research Laboratory, Harmondsworth.

Toombs, A F (1966b). The performance of a Delmag SV2 8¾-cwt vibrating-plate compactor in the compaction of soil. *Ministry of Transport, RRL Laboratory Report* No 53. Road Research Laboratory, Crowthorne.

Toombs, A F (1989). The compaction of soil using lightweight vibrating-plate compactors. *Department of Transport, TRRL Research Report* 208. Transport and Road Research Laboratory, Crowthorne.

CHAPTER 10 COMPACTION OF SOIL BY VIBRO-TAMPERS

Description of machines

10.1 Vibro-tampers are also known as vibration rammers or powered rammers, depending on the nomenclature of the individual manufacturer and the country of origin. This type of machine has an engine-driven reciprocating mechanism which acts on a spring system through which vertical oscillations, with an amplitude of 10 to 80 mm depending on the individual machine and the soil conditions, are set up in a base-plate. The most commonly used machines have masses in the range of 50 to 150 kg, and usually operate at a frequency of about 10 Hz. Their main mode of compaction is by impact and they are, therefore, considered suited to the compaction of most types of soil. Because of their potentially low output they are used in confined and small areas where their portability and manoeuvrability are a particular advantage. For situations where fumes from internal combustion engines may be a hazard, electrically driven vibro-tampers are available.

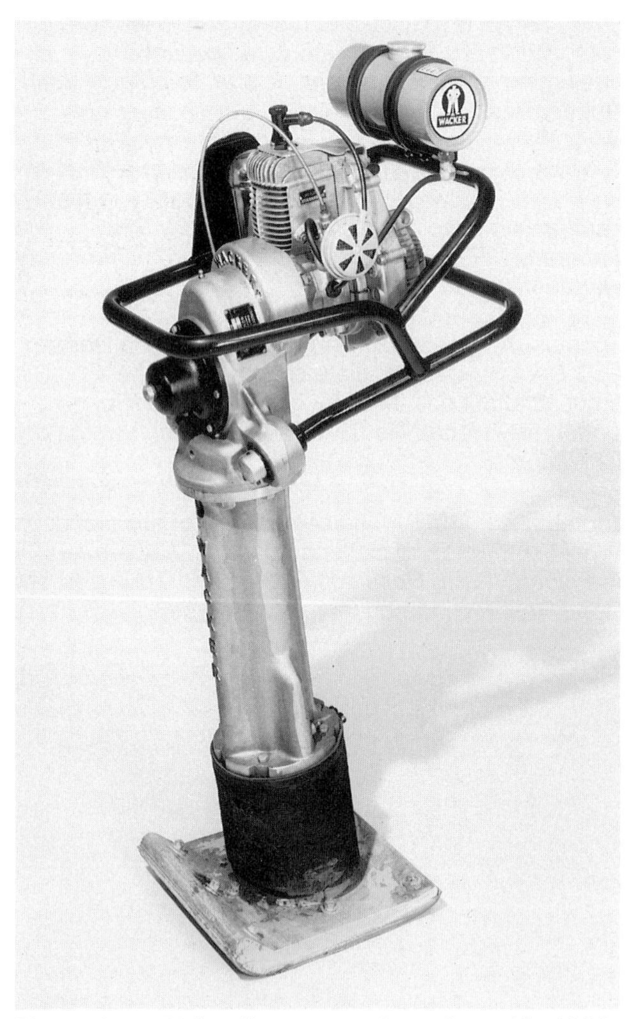

Plate 10.1 56 kg vibro-tamper investigated in 1966

10.2 Vibro-tampers investigated at the Laboratory are illustrated in Plates 10.1 to 10.4 and details are given in Table 10.1. The machines are illustrated and listed in ascending order of total mass. Each consists of a vertical body containing the reciprocating mechanism and spring system with a two-stroke petrol engine mounted at the top. The operator's control in each case comprised a throttle lever mounted on the sprung guide handles, the compacting mechanism being engaged by a centrifugal clutch upon application of full throttle. The front-to-rear profiles of the base-plates were not flat, and the contact areas given in Table 10.1 are best estimates of the areas in contact with the surface of a compacted layer of soil.

Table 10.1 Details of the various machines contributing data on the performance of vibro-tampers

Vibro-tamper	56 kg	59 kg	74 kg	100 kg
Total mass (kg)	56	59	74	103
Plate width (mm)	280	280	280	400
Contact area of plate (m^2)	0.078	0.069	0.084	0.086
Mass per unit area of base-plate (kg/m^2)	720	855	820	1190
Average tamping frequency during tests (Hz)	9.8	9.2	10.8	8.3
Tamping stroke height (mm)	50 (max)	50 (max)	40	25 to 50
Mean speed of travel (km/h)	0.3	0.5	0.7	0.5

10.3 The above descriptions of the machines and the results presented in this Chapter have been taken from Toombs (1966a, b, c) and unpublished records of recent investigations in 1986–87.

Relations between state of compaction produced and moisture content

10.4 Relations between the state of compaction produced by a vibro-tamper and the moisture

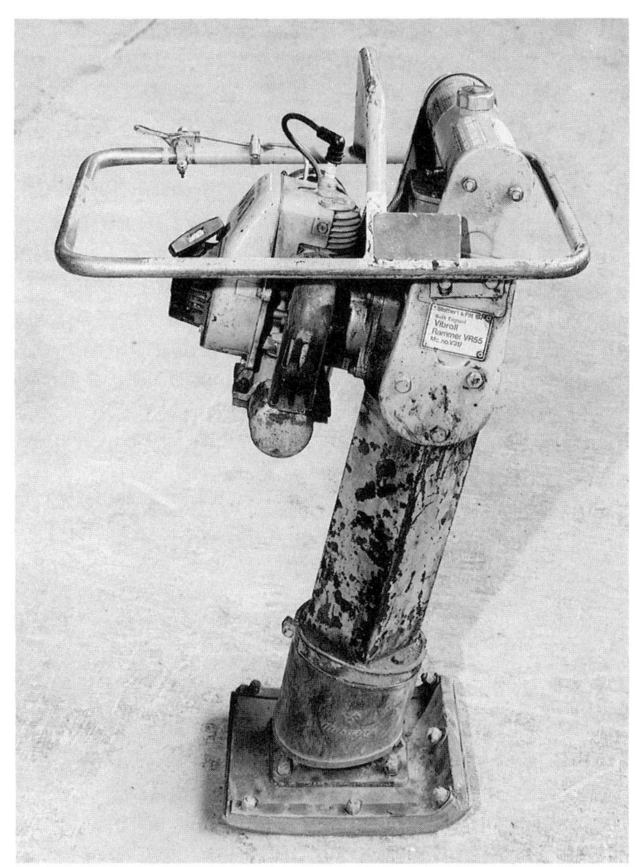

Plate 10.2 59 kg vibro-tamper investigated in 1986–87

Plate 10.4 100 kg vibro-tamper investigated in 1966

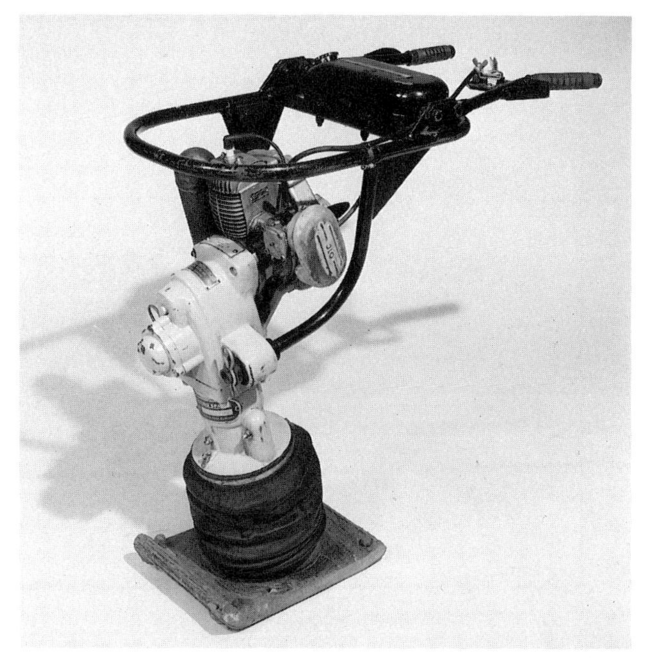

Plate 10.3 74 kg vibro-tamper investigated in 1965

content of the soil are given in Figures 10.1 to 10.4. The results shown were obtained with one machine only, the 59 kg vibro-tamper, in the most recent series of tests. Publications illustrating results obtained with the other machines (Toombs, 1966a, b, c) contain 'calculated' relations between dry density and moisture content, based on the general form of such relations passing through a single co-ordinate in each instance. Such results have been omitted from this section but the data are

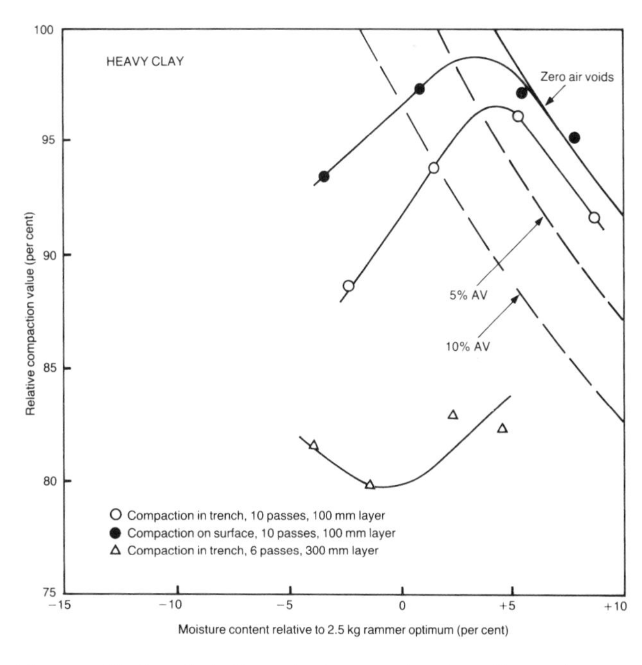

Figure 10.1 Relations between dry density and moisture content for the 59 kg vibro-tamper on heavy clay, expressed in terms of relative compaction with the 2.5 kg rammer test

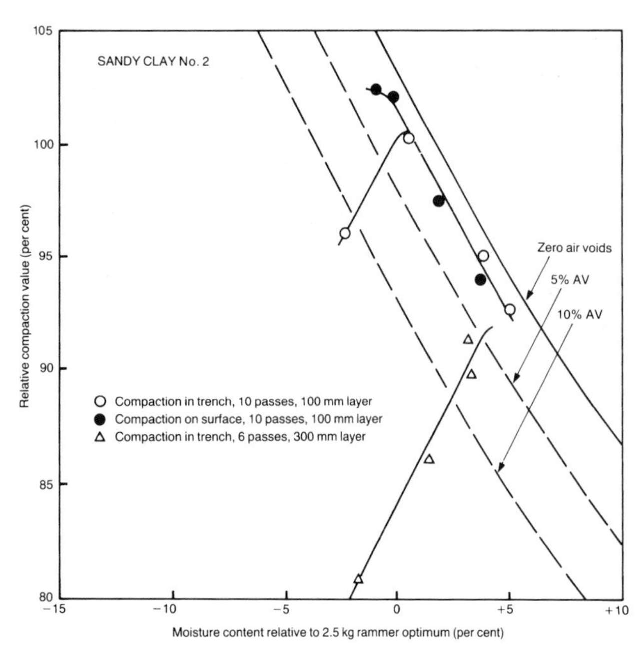

Figure 10.2 Relations between dry density and moisture content for the 59 kg vibro-tamper on sandy clay No 2, expressed in terms of relative compaction with the 2.5 kg rammer test

considered later in Paragraphs 10.9 to 10.12. In Figures 10.1 to 10.4 dry density is expressed in terms of relative compaction value (as a percentage of the maximum dry density obtained in the 2.5 kg rammer compaction test) and the moisture content as the arithmetical difference from the optimum moisture content of the 2.5 kg rammer test; the results obtained in the 2.5 kg rammer compaction test at the relevant period are shown in Figure 3.13.

10.5 The performance of the 59 kg vibro-tamper was studied under two conditions of working. One was in the compaction of an exposed surface layer of soil; the operating procedure in this case was to compact three adjoining base-plate widths of soil and to limit the determination of the in-situ state of compaction to the centre strip. The second condition of working was within a trench excavated in the test soil. The trench was excavated to a width equal to two base-plate widths of the machine, the soil then being spread in the trench to the required thickness and compacted in two adjoining strips; measurements of the in-situ state of compaction were made in both strips. A view of a trench excavated in sandy clay No 2, with an adjacent area of soil prepared for surface compaction, is shown in Plate 10.5.

10.6 The results portrayed in Figures 10.1 to 10.4 are those for the highest number of passes used in each working condition, ie 10 passes on 100 mm layers both on the surface and in the trench and 6 passes on 300 mm layers in the trench only. It is interesting that the levels of compaction achieved on the surface layers of heavy clay, and probably of sandy clay also, were considerably higher than those achieved when working on 100 mm layers in the trench (Figures 10.1 and 10.2), whereas for the granular soils (Figures 10.3 and 10.4) little difference occurred for the 100 mm layers in the two locations. This particular aspect is dealt with in more detail in Paragraphs 10.15 to 10.21. Maximum relative compaction values reached in

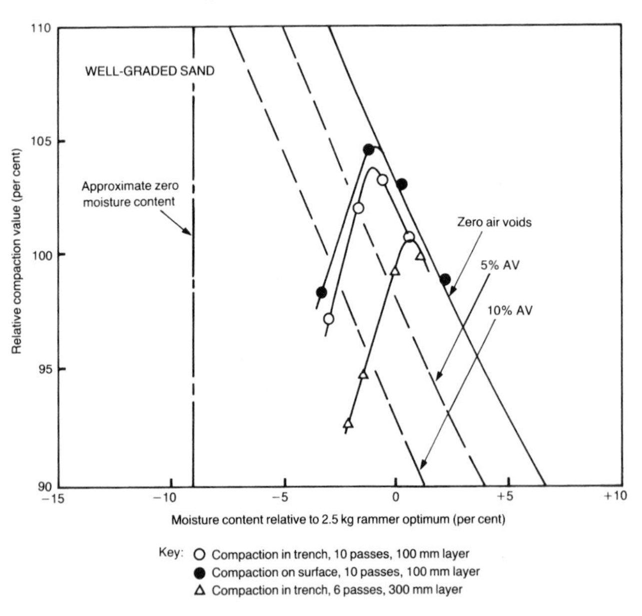

Figure 10.3 Relations between dry density and moisture content for the 59 kg vibro-tamper on well-graded sand, expressed in terms of relative compaction with the 2.5 kg rammer test

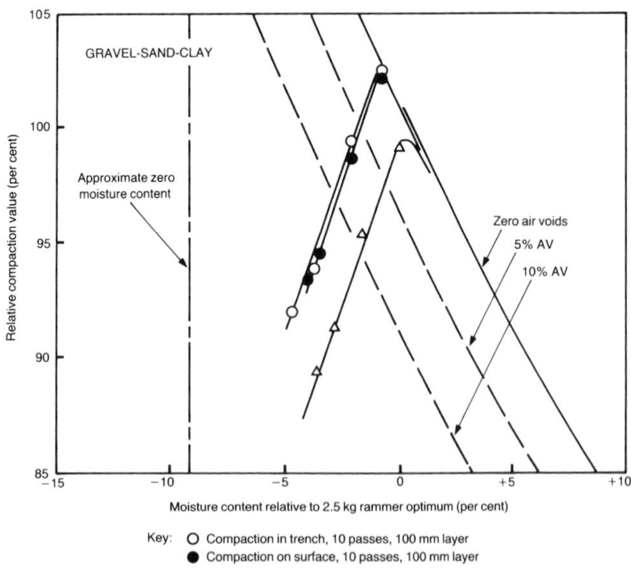

Figure 10.4 Relations between dry density and moisture content for the 59 kg vibro-tamper on gravel-sand-clay, expressed in terms of relative compaction with the 2.5 kg rammer test

Plate 10.5 Trench excavated in sandy clay No 2, with prepared soil alongside for compaction of a surface layer using the 59 kg vibro-tamper

100 mm layers of soil were 96 to 99 per cent with the heavy clay, 100 to 102 per cent with sandy clay No 2, and 102 to 105 per cent with the two granular soils. Optimum moisture content ranged from 3 to 5 per cent above the 2.5 kg rammer optimum with heavy clay to about 1 per cent below that optimum with the three other soils.

10.7 The use of a 300 mm thick layer of soil, with a reduced number of passes (six) resulted, as would be expected, in lower levels of compaction. The reduction in states of compaction was much larger with the clay soils, however, and with heavy clay (Figure 10.1) very poor states of compaction were registered, even at a moisture content of 5 per cent above the 2.5 kg rammer optimum.

10.8 It is clear from these results that, with the 59 kg vibro-tamper, 100 mm was the most appropriate layer thickness with clay soils, but thicker layers, possibly even 300 mm, could be used on the more granular soils provided the moisture content was close to the optimum of the 2.5 kg rammer compaction test.

Relations between state of compaction produced and number of passes

10.9 Results of investigations to determine the effect on the state of compaction achieved of variations in the number of passes of vibro-tampers are given in Figures 10.5 to 10.9. As in previous Chapters, the results for each soil are illustrated under a separate figure number, whilst different values of moisture content are in separate graphs within each figure. All machines listed in Table 10.1 were included in these tests, but only the 59 kg vibro-tamper was tested at more than one value of moisture content with each soil and layer thickness. The thicknesses of layer used depended upon the capabilities of each individual machine, with thicker layers

Figures 10.5 (a) and (b) See caption page 162

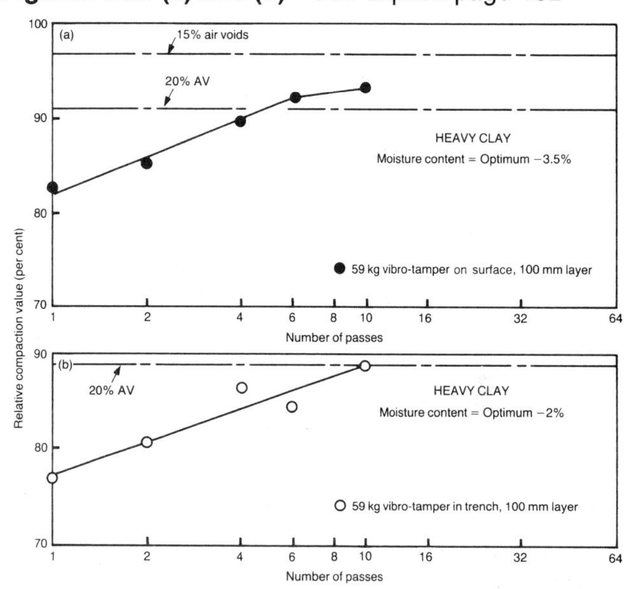

Figures 10.5 (c) and (d) See caption page 162

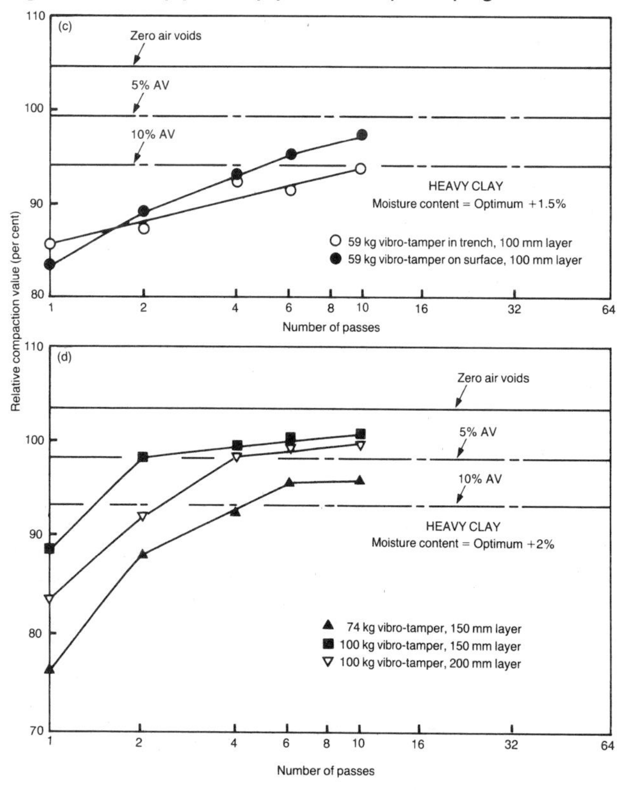

Figures 10.5 (e) and (f) See main caption

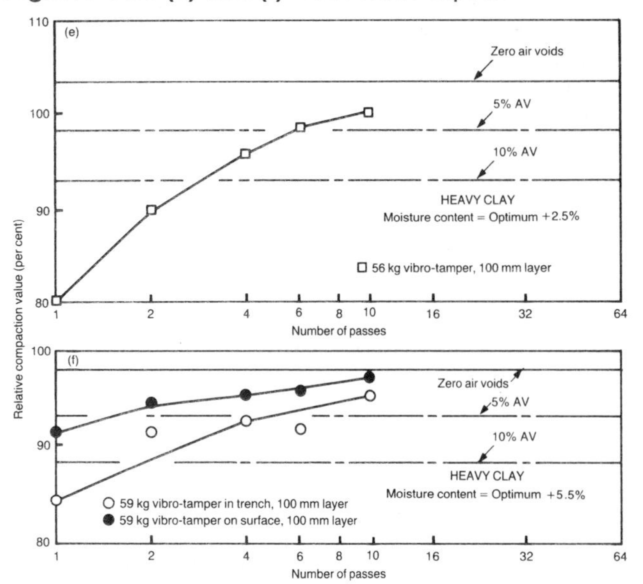

Figures 10.5 (g) See main caption

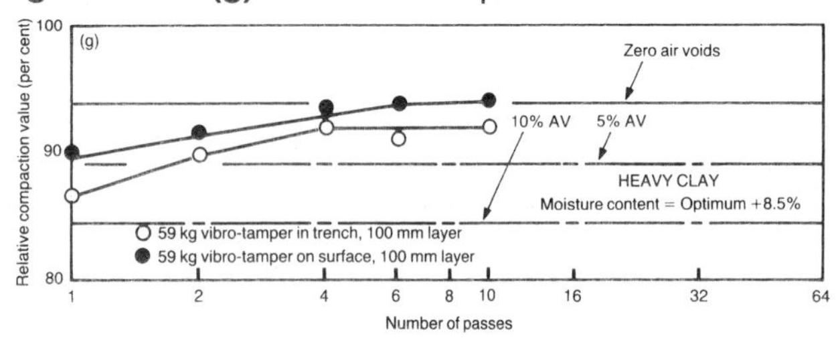

Figure 10.5 Relations between relative compaction and number of passes for vibro-tampers on heavy clay at seven different values of moisture content. Values are related to the maximum dry density and optimum moisture content obtained in the 2.5 kg rammer compaction test (Figure 3.13)

being used with the larger machines; on some soils two thicknesses of layer were tested with particular machines (Figures 10.5(d), 10.6(b) and (d), and 10.9(b)).

10.10 Regression analyses have been carried out on the results portrayed in Figures 10.5 to 10.9, and those given in Figures 10.1 to 10.4 have also been incorporated. With the 59 kg vibro-tamper only the surface compaction results were included. The number of passes required to achieve the specified levels of compaction of 95 per cent relative compaction and 10 per cent air voids were related to relevant parameters. Machine parameters tried were the mass of the machine and the mass per unit contact area; both these parameters are given in Table 10.1. Moisture content and depth of compacted layer were additional variables included in the analyses, although the high level of correlation between the mass of the machine and the thickness of layer (see Paragraph 10.9) might be expected to obscure the full effect of the latter factor.

10.11 The results of the regression analyses are given in Table 10.2 in the form of regression equations and correlation coefficients. It was found that there was little difference in the correlation coefficients when using either of the machine factors (mass or mass per unit area) but the mass per unit area was found to be marginally better in five out of eight comparisons and the results incorporating that factor, therefore, are given in the Table. The effects of variations in the thickness of layer with heavy clay and gravel-sand-clay, within the range used, were of very low significance and the factor has been omitted from the equations for those soils. As can be seen by inspection of Figure 10.9, the relations between relative compaction and number of passes obtained on uniformly graded

Figure 10.6

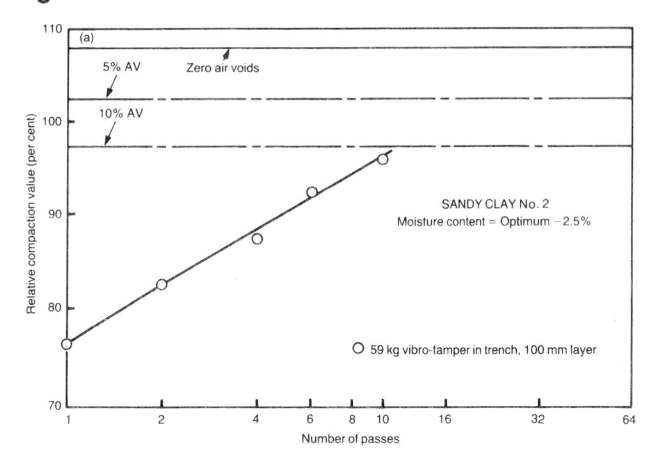

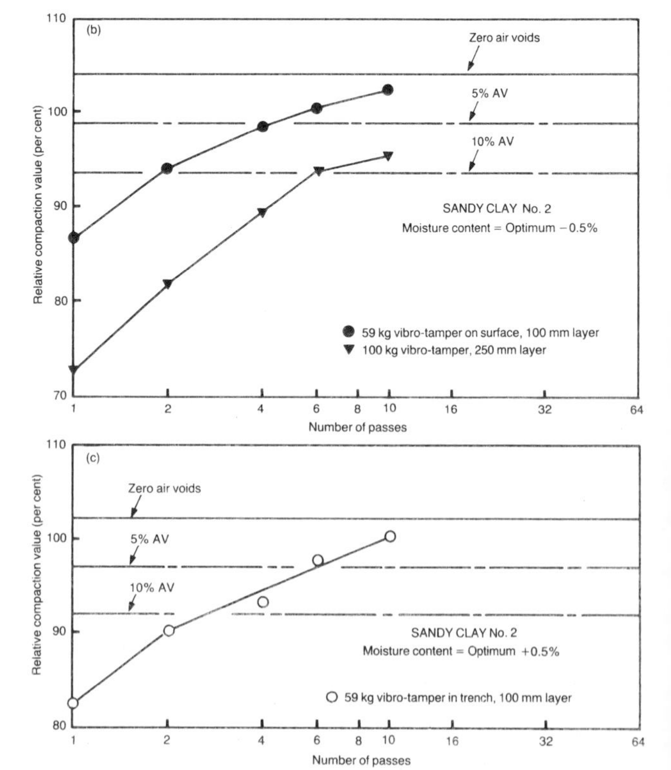

fine sand were very similar for all the machines; very poor correlations were obtained in the regression analyses with that soil, therefore, and no results have been given. It was found that combining the results for two or more of the soils produced relatively poor correlations and it has

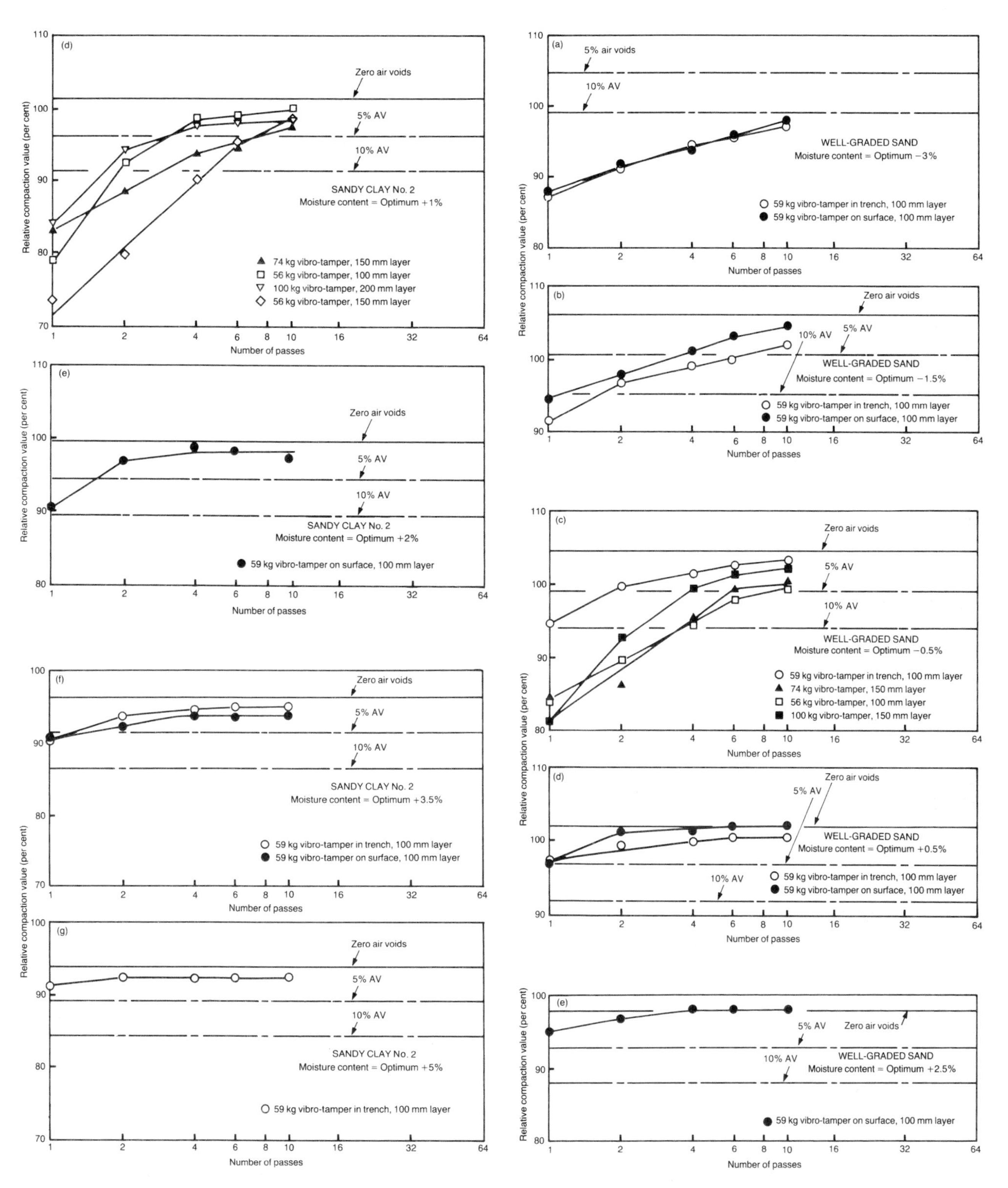

Figure 10.6 Relations between relative compaction and number of passes for vibro-tampers on sandy clay No 2 at seven different values of moisture content. Values are related to the maximum dry density and optimum moisture content obtained in the 2.5 kg rammer compaction test (Figure 3.13)

Figure 10.7 Relations between relative compaction and number of passes for vibro-tampers on well-graded sand at five different values of moisture content. Values are related to the maximum dry density and optimum moisture content obtained in the 2.5 kg rammer compaction test (Figure 3.13)

been concluded that, in general, the performances of the vibro-tampers varied between each of the test soils.

10.12 Examples of relations between the number of passes required and mass per unit area of base-plate are given in Figures 10.10 to 10.13. Relations have been included for three

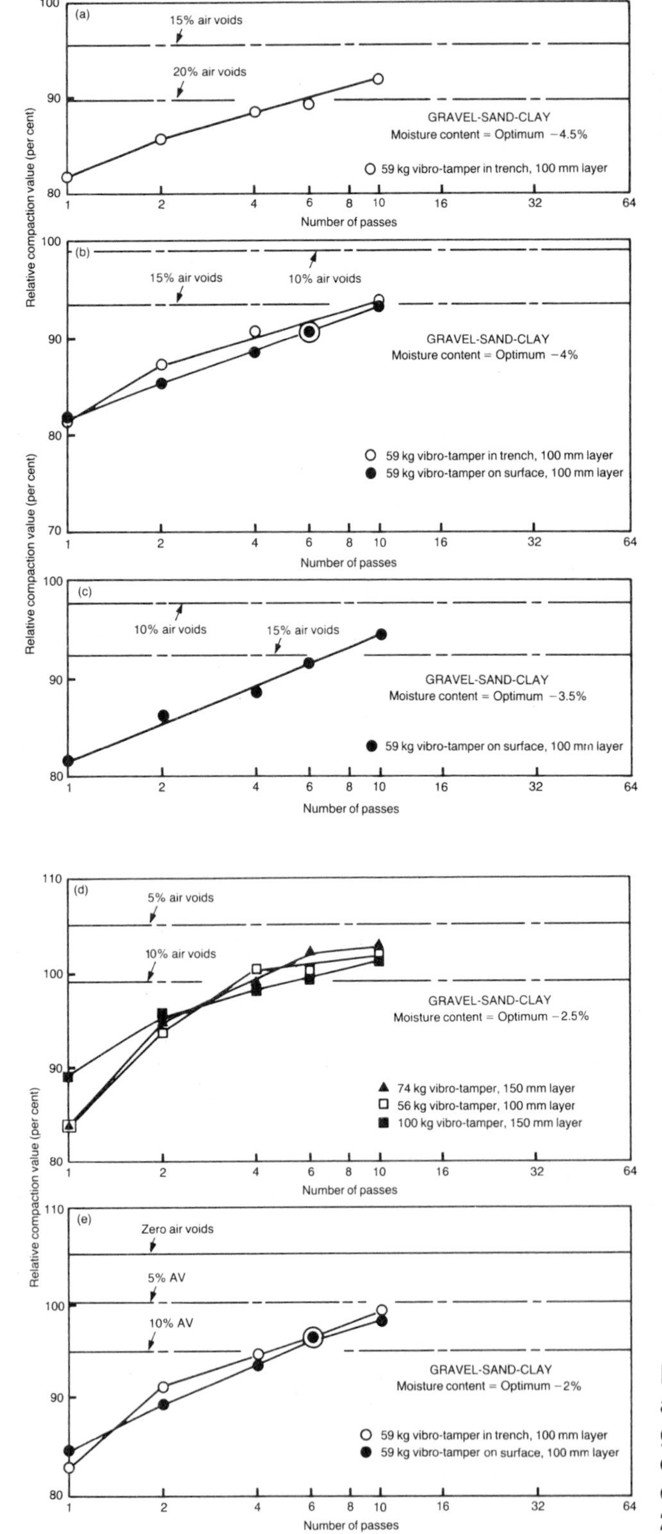

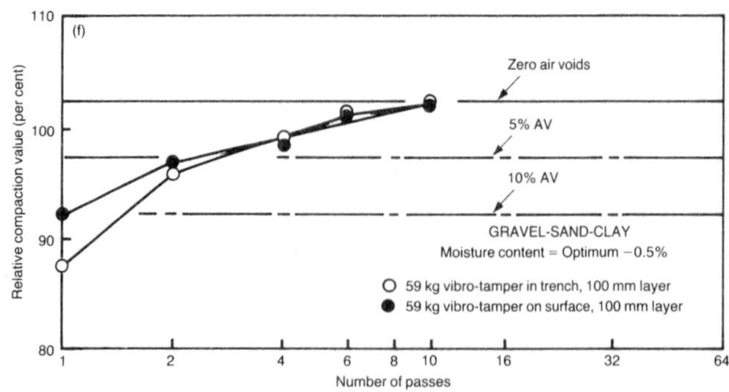

Figure 10.8 Relations between relative compaction and number of passes for vibro-tampers on gravel-sand-clay at six different values of moisture content. Values are related to the maximum dry density and optimum moisture content obtained in the 2.5 kg rammer compaction test (Figure 3.13)

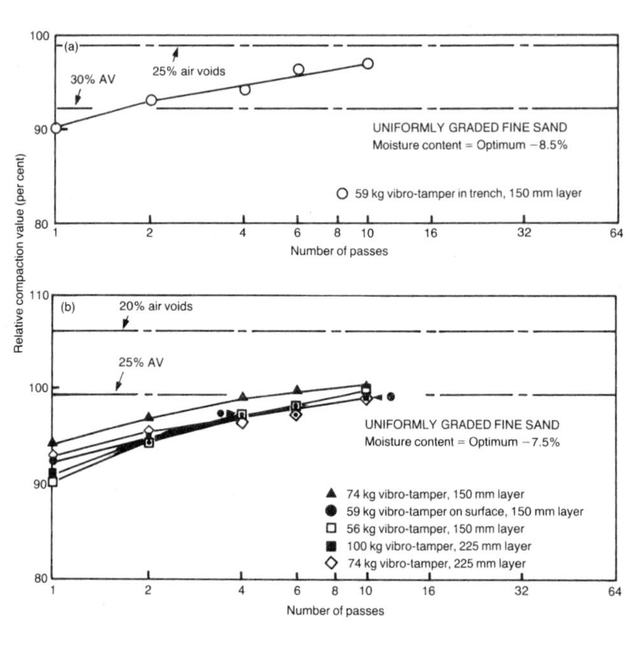

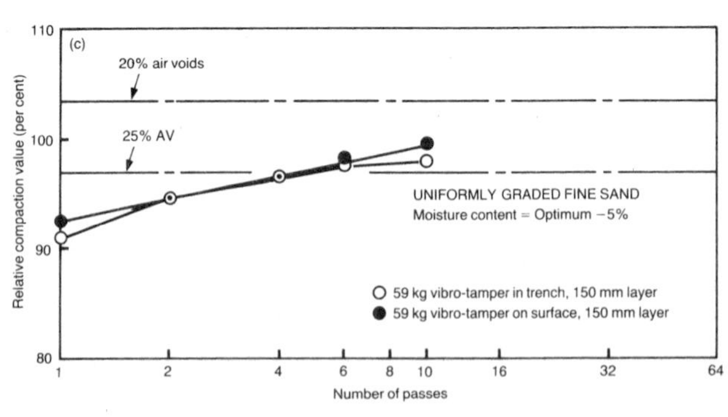

Figure 10.9 Relations between relative compaction and number of passes for vibro-tampers on uniformly graded fine sand at three different values of moisture content. Values are related to the maximum dry density and optimum moisture content obtained in the 2.5 kg rammer compaction test (Figure 3.13)

values of moisture content and, except for Figures 10.10 and 10.13, three depths of compacted layer. The omission of the depth of layer in the variables used with heavy clay and gravel-sand-clay has been discussed in Paragraph 10.11.

Table 10.2 Equations and correlation coefficients to determine the number of passes required to achieve specified levels of compaction with vibro-tampers

Soil	Level*	Regression equation†	No of data points	Correlation coefficient
Heavy clay	(a)	P=6.71–1.99 m–.0941 w‡	8	0.92
	(b)	P=5.89–1.67 m–.228 w‡	7	0.99
Sandy clay No 2	(a)	P=4.04–2.78 m–.111 w +2.22 d	7	0.96
	(b)	P=4.35–2.87 m–.160 w +2.13 d	7	0.99
Well-graded sand	(a)	P=1.63–2.65 m–.212 w +3.08 d	7	0.90
	(b)	P=1.66–2.13 m–.327 w +2.31 d	6	0.90
Gravel-sand-clay	(a)	P=7.24–2.51 m–.325 w‡	7	0.90
	(b)	P=2.05–.826 m–.488 w‡	7	0.99

*Level (a)=95 per cent of the maximum dry density obtained in the 2.5 kg rammer compaction test
Level (b)=10 per cent air voids
†P=Log_{10} (Number of passes required)
m=Log_{10} (Mass per unit area of base plate in kg/m^2)
w=Moisture content of soil, per cent, minus the optimum moisture content, per cent, obtained in the 2.5 kg rammer compaction test
d=Log_{10} (Depth of compacted layer in mm)
‡Depth of compacted layer in the range 100 to 150 mm

Relations between state of compaction and thickness of the compacted layer

10.13 In tests with the 59 kg vibro-tamper, using 3 and 6 passes to compact a layer of soil 300 mm thick in a trench, measurements of the in-situ state of compaction were made in the upper 100 mm and 200 mm of the compacted layer as well as in the complete 300 mm thickness of the layer. This was achieved using a nuclear density-moisture gauge, suitably calibrated, with moisture content determinations using the normal British Standard oven-drying

Figure 10.10 Vibro-tampers on heavy clay. Relations between number of passes required and mass per unit area of base plate to achieve (a) 95 per cent of the maximum dry density obtained in the 2.5 kg rammer compaction test and (b) 10 per cent air voids. These graphs were derived by regression analyses of the results given in Figures 10.1 and 10.5

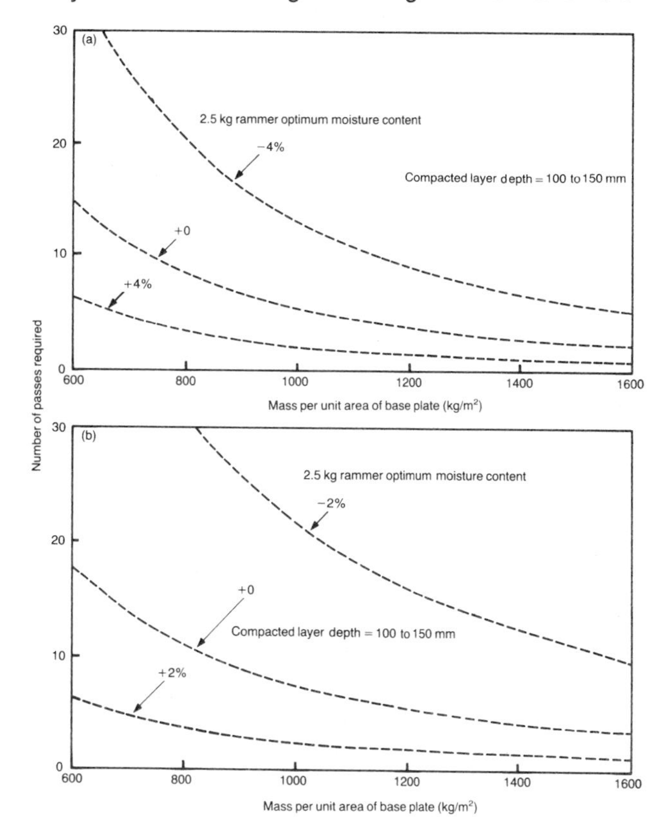

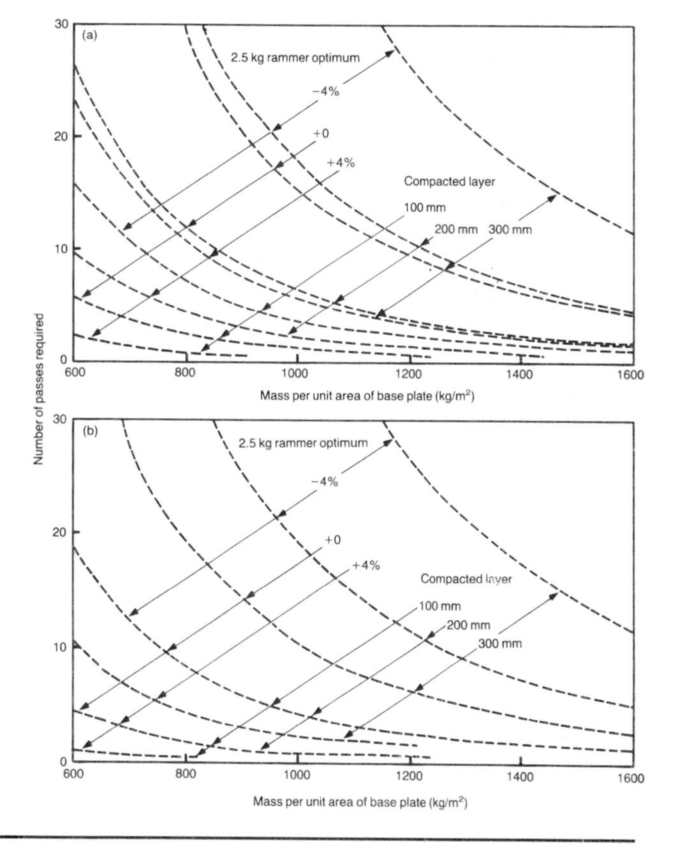

Figure 10.11 Vibro-tampers on sandy clay No 2. Relations between number of passes required and mass per unit area of base plate to achieve (a) 95 per cent of the maximum dry density obtained in the 2.5 kg rammer compaction test and (b) 10 per cent air voids. These graphs were derived by regression analyses of the results given in Figure 10.6

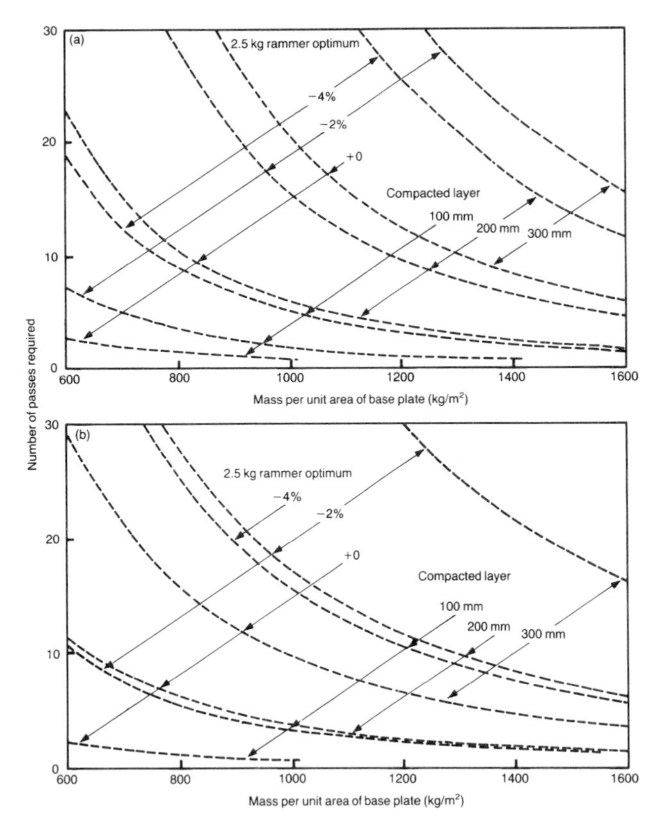

Figure 10.12 Vibro-tampers on well-graded sand. Relations between number of passes required and mass per unit area of base plate to achieve (a) 95 per cent of the maximum dry density obtained in the 2.5 kg rammer compaction test and (b) 10 per cent air voids. These graphs were derived by regression analyses of the results given in Figures 10.3 and 10.7

Figure 10.13 Vibro-tampers on gravel-sand-clay. Relations between number of passes required and mass per unit area of base plate to achieve (a) 95 per cent of the maximum dry density obtained in the 2.5 kg rammer compaction test and (b) 10 per cent air voids. These graphs were derived by regression analyses of the results given in Figures 10.4 and 10.8

procedure (British Standards Institution, 1990) (see Chapter 18 for information on the use of nuclear gauges). Although the results for the upper parts of the layer would be slightly poorer than might be expected if the same thickness was compacted as a single layer on a well-compacted foundation, a good (but slightly pessimistic) indication of the potential thickness of layer that the machine was capable of compacting at a particular moisture content could be obtained. The results are given in Figures 10.14 to 10.18.

10.14 To summarise the results, relations between thickness of layer and moisture content are given in Figures 10.19 and 10.20. These show respectively the maximum thickness of layer capable of being compacted to 95 per cent relative compaction and 10 per cent air voids, estimated from the relations given in Figures 10.14 to 10.18; in some instances the values used were obtained by extrapolating the general trends of the relations. As mentioned in Paragraph 10.13, it is likely that slightly thicker

Figure 10.14

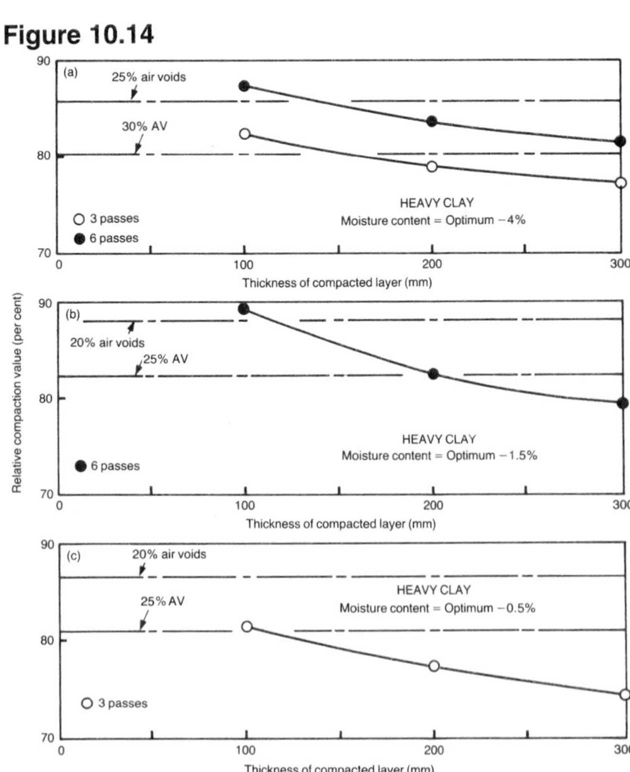

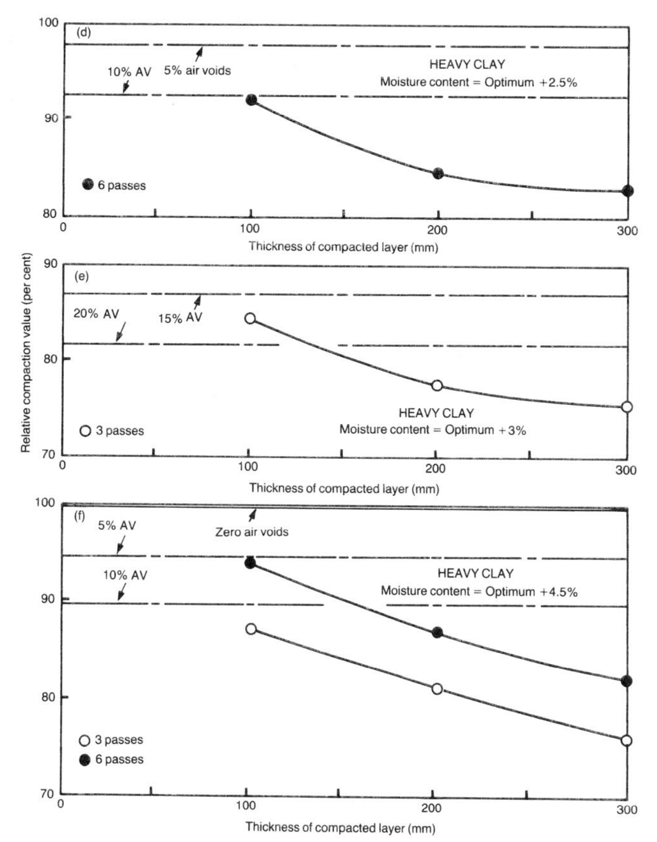

Figure 10.14 Relations between the average relative compaction and the total thickness of the compacted layer using 3 and 6 passes of the 59 kg vibro-tamper on heavy clay

layers could be compacted to the same criteria if they were compacted as individual layers on a firm (well-compacted) foundation. These results, of course, apply only to the 59 kg vibro-tamper and only for the numbers of passes tested.

Figure 10.15

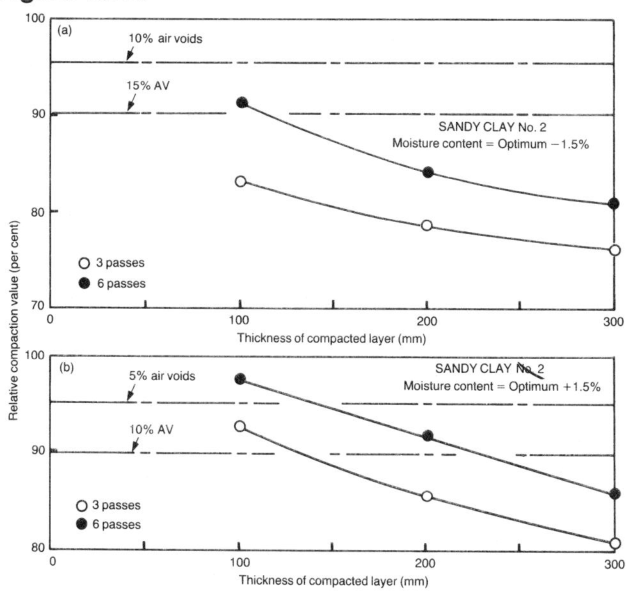

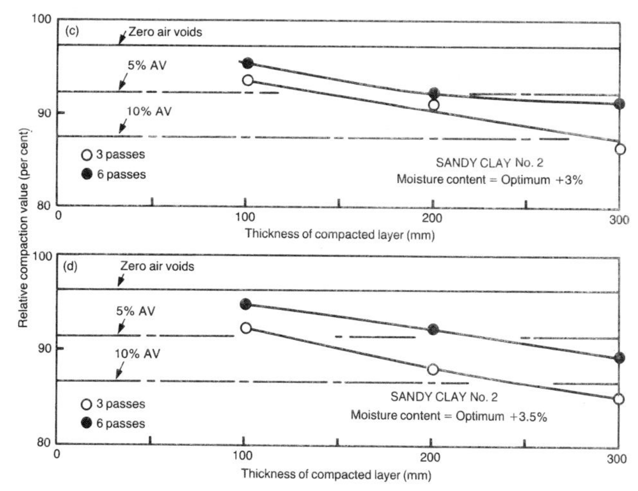

Figure 10.15 Relations between the average relative compaction and the total thickness of the compacted layer using 3 and 6 passes of the 59 kg vibro-tamper on sandy clay No 2

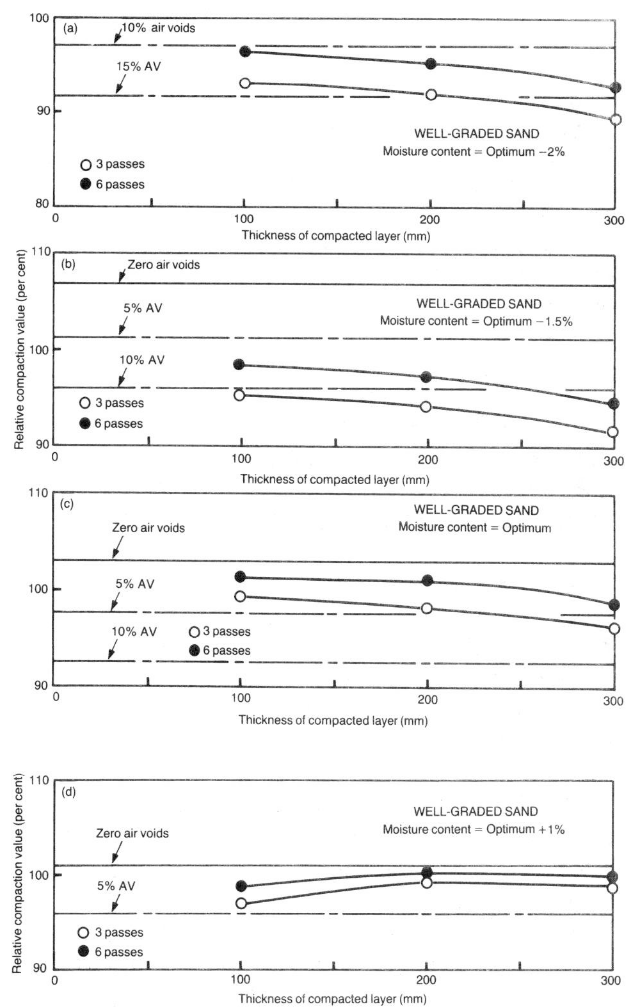

Figure 10.16 Relations between the average relative compaction and the total thickness of the compacted layer using 3 and 6 passes of the 59 kg vibro-tamper on well-graded sand

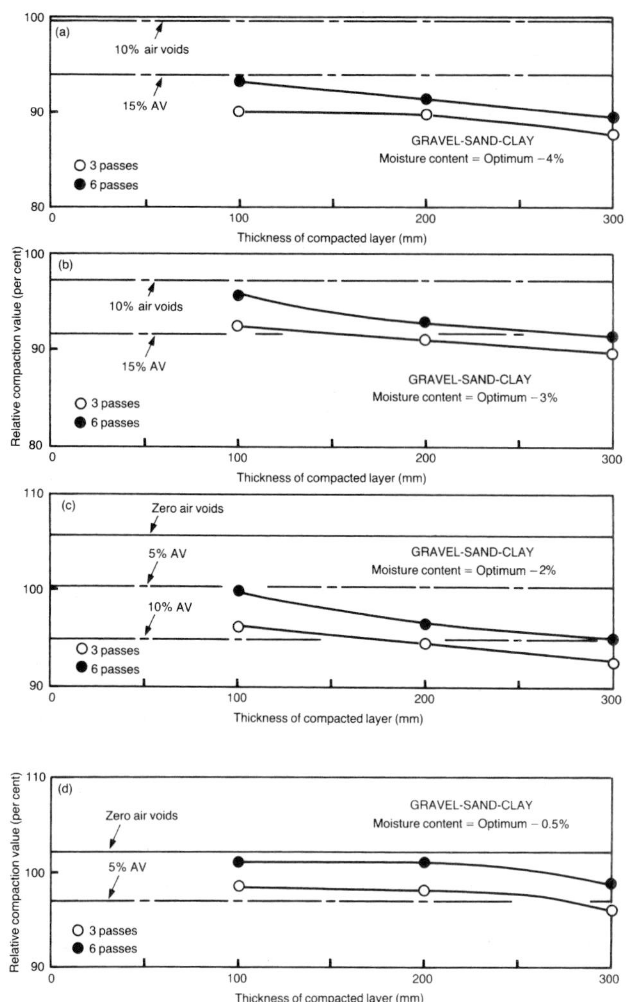

Figure 10.17 Relations between the average relative compaction and the total thickness of the compacted layer using 3 and 6 passes of the 59 kg vibro-tamper on gravel-sand-clay

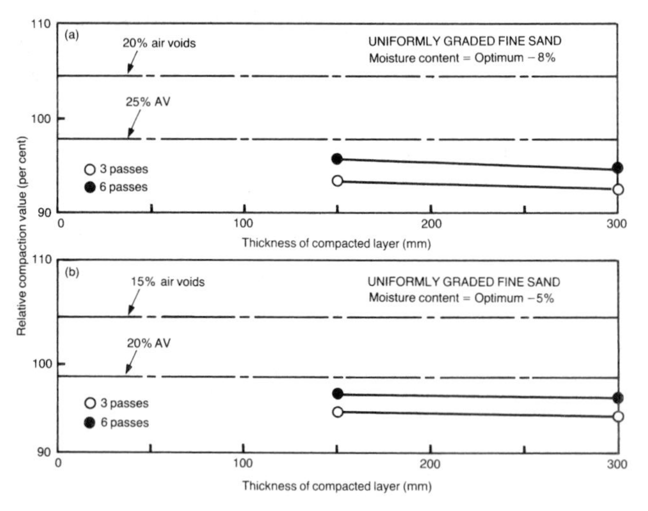

Figure 10.18 Relations between the average relative compaction and the total thickness of the compacted layer using 3 and 6 passes of the 59 kg vibro-tamper on uniformly graded fine sand

Figure 10.19 Relations between depth of layer and moisture content for 3 and 6 passes of the 59 kg vibro-tamper. Values given are the maximum thicknesses of layer capable of being compacted to a relative compaction value of 95 per cent, estimated from Figures 10.14 to 10.18

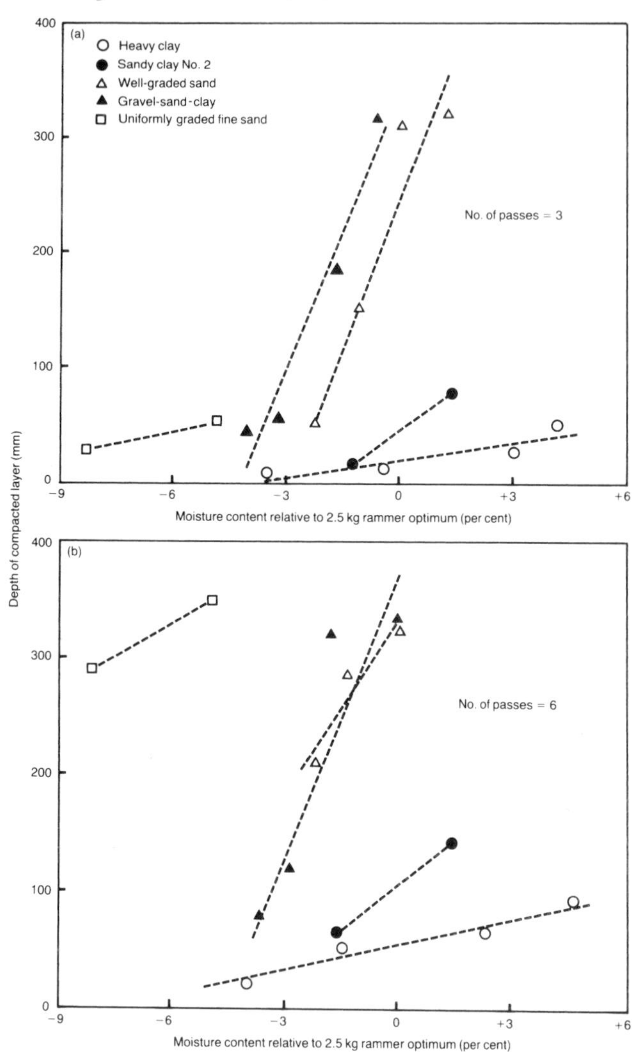

A comparison of the results produced in the compaction of surface layers and layers of soil confined within a trench

10.15 In all the investigations of compaction plant prior to 1985 the procedures involved the compaction of a single layer of controlled thickness prepared at the surface of the test soil. The soil was contained within a concrete-sided pit having dimensions which were large in relation to those of the equipment being tested. This procedure was used even for small-scale equipment which would normally operate in trenches. The normal procedure with small plant was to compact three adjoining strips of soil, each strip equal to the width compacted by the machine, and to carry out all measurements of

Figure 10.20 Relations between depth of layer and moisture content for 3 and 6 passes of the 59 kg vibro-tamper. Values given are the maximum thicknesses of layer capable of being compacted to an air void content of 10 per cent, estimated from Figures 10.14 to 10.18

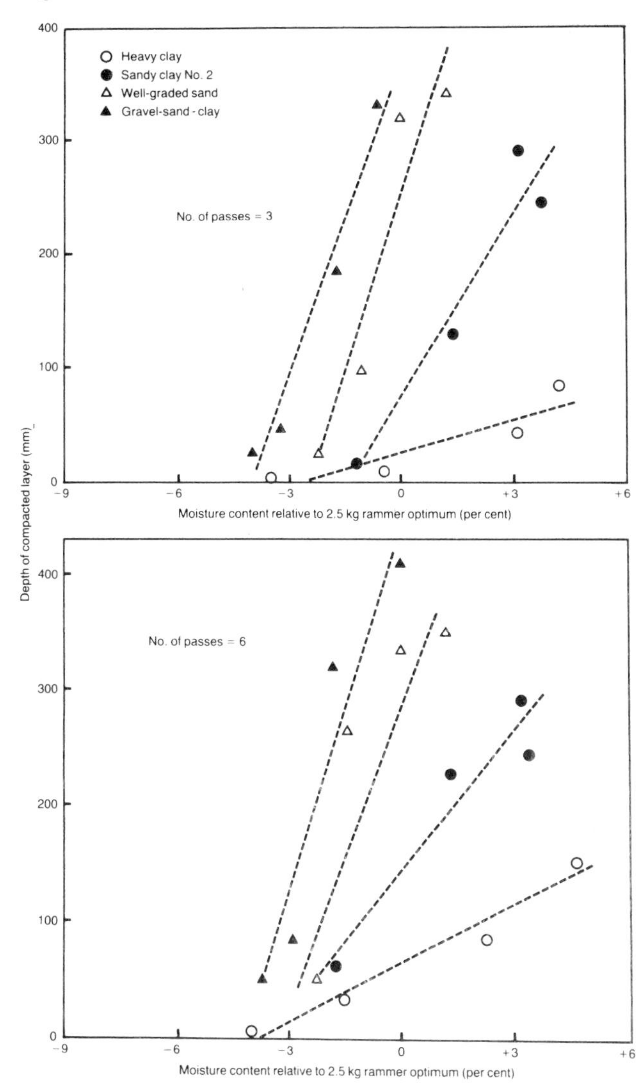

the in-situ state of compaction in the central strip. It was considered that the outer strips provided sufficient confinement of the soil in the central strip to simulate the effect of compacting the soil within a trench.

10.16 When compacting soil in a trench the side-walls might be expected to provide some additional confinement which could enhance the compactive effect of a machine; on the other hand, friction or adhesion between the layer being compacted and the side-wall might result in a reduction in the compactive effect.

10.17 To assess if either of these effects predominated, an investigation was carried out in 1986–87, with the 59 kg vibro-tamper, in which layers of soil, 100 mm thick, were compacted on the surface of the test soil and also in a trench excavated in the compacted test soil. Details of the method of investigation have already been given in Paragraph 10.5 and a photograph of the sandy clay No 2 prepared for test is shown in Plate 10.5. The results have been included in Figures 10.1 to 10.9. It should be noted that in the analyses of the effect of the number of passes of vibro-tampers and the discussions of those effects in Paragraphs 10.9 to 10.12, only the results for surface compaction were taken for the 59 kg vibro-tamper; all other vibro-tampers investigated were tested, as per normal practice, on surface layers.

10.18 Relations between relative compaction and moisture content (Figures 10.1 to 10.4) compare directly the results of compaction in the two locations using 10 passes of the machine on 100 mm thick compacted layers. With heavy clay (Figure 10.1) there is a discernible difference between the results (see Paragraph 10.6), with the states of compaction produced in the trench being almost 5 per cent lower in relative compaction value at moisture contents less than the 2.5 kg rammer optimum plus 5 per cent. The same effect could also have occurred with the sandy clay No 2 (Figure 10.2) at moisture contents less than the optimum of the 2.5 kg rammer compaction test, although an insufficient range of moisture contents was used in the surface layers to provide conclusive evidence. Direct comparisons of the relations between relative compaction and number of passes can only be made where the moisture contents tested in the two locations were similar. Figures 10.5(c), (f) and (g) all show higher states of compaction with heavy clay in the surface layers. Figures 10.6(f), 10.7(a), (b) and (d), 10.8(b), (e) and (f) and 10.9(c) all show little difference in the relations produced in the two locations when the soil type and the moisture content were the same.

10.19 The results produced by the 59 kg vibro-tamper on 100 mm layers of soil in both surface and trench locations have been further analysed to determine the number of passes necessary to produce the criteria adopted earlier in this Chapter, ie 95 per cent relative compaction and 10 per cent air voids. Relations between the required number of passes and moisture content are given in Figures 10.21 to 10.23. These confirm the differences in the results for the heavy clay (Figure 10.21) and provide evidence of a similar effect with the sandy clay No 2 (Figure 10.22); the similarity of the results in the two locations with the granular soils is also

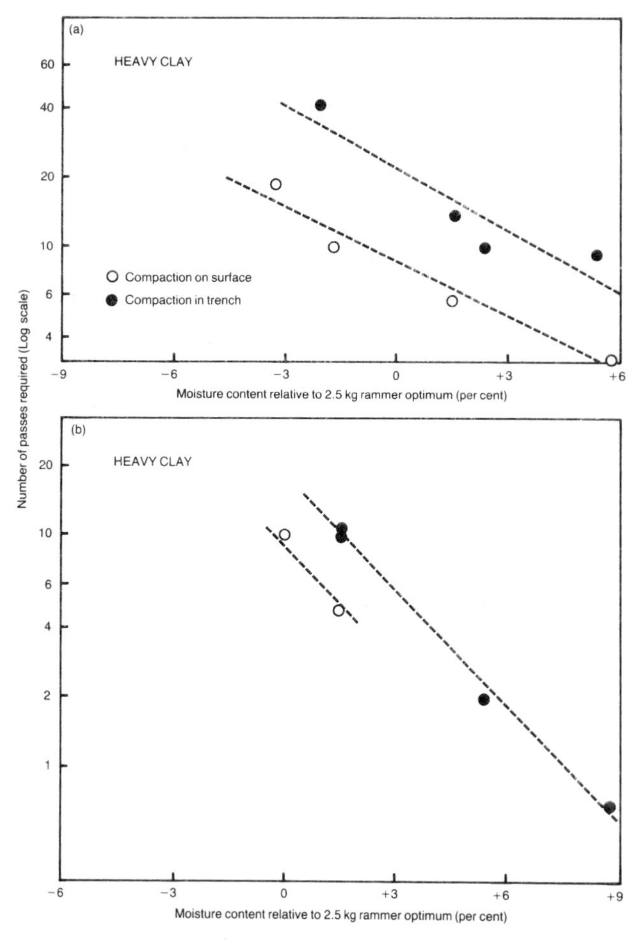

Figure 10.21 Relations between number of passes of the 59 kg vibro-tamper and moisture content of 100 mm compacted layers of heavy clay to achieve (a) 95 per cent of the maximum dry density obtained in the 2.5 kg rammer compaction test and (b) 10 per cent air voids

confirmed (Figure 10.23). In addition, Figure 10.23(b) shows that a single relation can be used for gravel-sand-clay and well-graded sand when operating to the 10 per cent air void criterion with this particular machine (the 59 kg vibro-tamper).

10.20 It can be concluded from this investigation that with plastic materials, similar in character to the heavy clay and sandy clay No 2 soils used in the Laboratory's investigations, significant reductions in performance of vibro-tampers, and probably other lightweight compactors, may occur when working in trenches compared with that achieved in surface layers. The reason for the reductions in the states of compaction is not readily apparent, but if it is caused by side-wall friction or adhesion, one of the mechanisms suggested in Paragraph 10.16, then all the soils would be expected to exhibit the same effect. A possible explanation is that arching occurs between the individual lumps of clay confined by the walls of the trench; a lightweight machine such as the 59 kg vibro-tamper may then require an increased number of passes to deform the lumps so as to increase the dry density at the bottom of a 100 mm layer under such conditions of confinement.

10.21 As all other investigations of lightweight plant have been carried out on surface layers, the number of passes derived for a specified state of compaction may have to be increased for clay soil when working within a trench; the results given in Figures 10.21 and 10.22 indicate that the number of passes should be increased (multiplied) by a factor of 2 to 2.5. This factor can be assumed to be constant over the ranges of moisture content tested with the heavy clay and sandy clay No 2 and for which results are plotted in Figures 10.21 and 10.22.

Mobility of vibro-tampers

10.22 For test purposes the normal practice of preparing a loose layer of soil, by mixing to a fine tilth with a rotary cultivator, was adopted. In most cases the vibro-tampers suffered from some loss of mobility during the first pass over

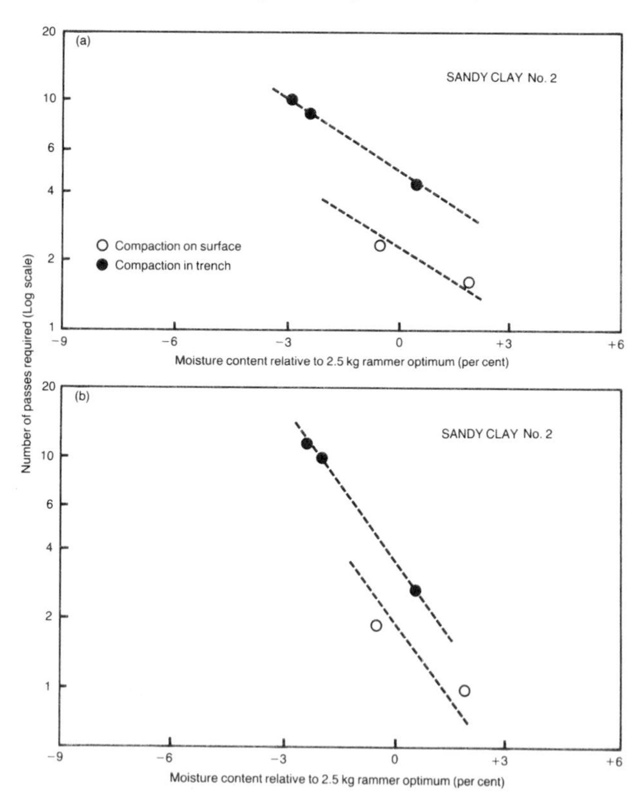

Figure 10.22 Relations between number of passes of the 59 kg vibro-tamper and moisture content of 100 mm compacted layers of sandy clay No 2 to achieve (a) 95 per cent of the maximum dry density obtained in the 2.5 kg rammer compaction test and (b) 10 per cent air voids

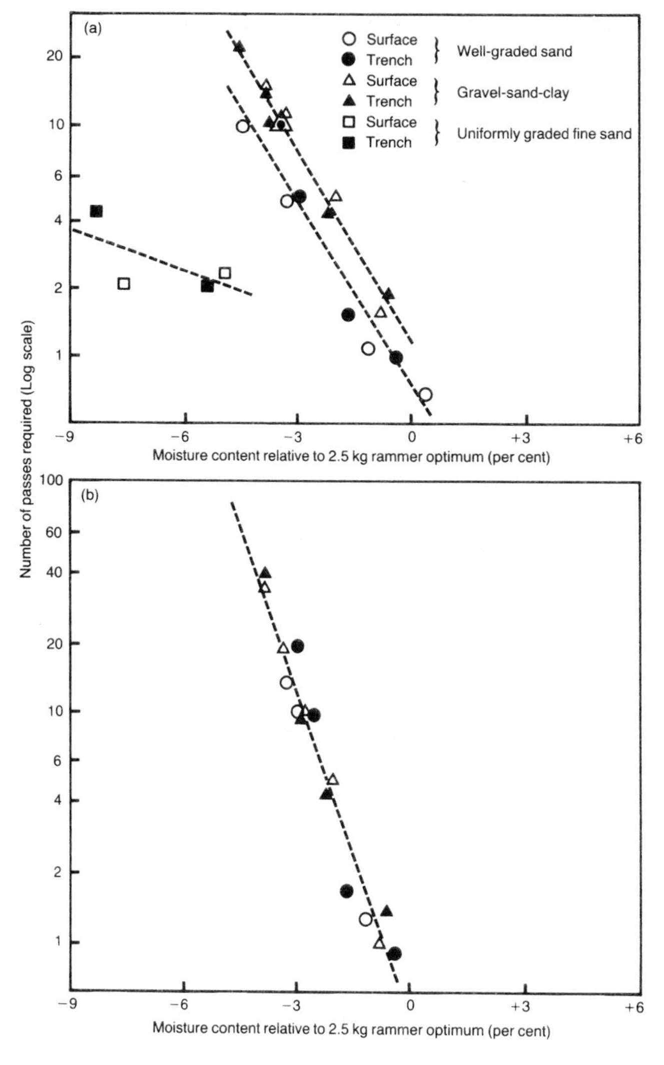

Figure 10.23 Relations between number of passes of the 59 kg vibro-tamper and moisture content of 100 mm compacted layers of granular soils to achieve (a) 95 per cent of the maximum dry density obtained in the 2.5 kg rammer compaction test and (b) 10 per cent air voids

loose soil. The lighter machines (56 kg and 59 kg vibro-tampers) were assisted during the first pass by being pushed by the operator to maintain forward momentum. The 100 kg machine was occasionally pulled during the first pass by a second operator using a rope, a procedure recommended by the manufacturer for use when operating under difficult conditions. It was considered at the time that the very loose soil conditions created for these tests added greatly to the loss of mobility and are unlikely to be encountered on site. The 74 kg machine did not exhibit this problem, and the speeds of travel given in Table 10.1 appear to reflect the relative difficulties in mobility on the first pass under the particular test conditions.

References

British Standards Institution (1990). British Standard methods of test for soils for civil engineering purposes: BS 1377: Part 2 Classification tests. BSI, London.

Toombs, A F (1966a). An investigation into the performance of a Winget Dynapac CO.10 1¼-cwt vibro-tamper in the compaction of soil. *Ministry of Transport, RRL Laboratory Report* No 10. Road Research Laboratory, Harmondsworth.

Toombs, A F (1966b). The performance of a Wacker BS 50K 1-cwt vibro-tamper in the compaction of soil. *Ministry of Transport, RRL Laboratory Report* No 27. Road Research Laboratory, Harmondsworth.

Toombs, A F (1966c). The performance of a Wacker BS 100 2-cwt vibro-tamper in the compaction of soil. *Ministry of Transport, RRL Laboratory Report* No 28. Road Research Laboratory, Harmondsworth.

CHAPTER 11 COMPACTION OF SOIL BY POWER RAMMERS

Description of machines

11.1 Power rammers rely on an internal combustion cylinder in which firings of a petrol-air mixture are individually actuated by the operator to provide a form of impact compaction. The piston in the cylinder acts on a base plate, which supports the main body of the machine on springs. The normal type of power rammer has a mass of about 100 kg, with a circular base of about 250 mm diameter. Machines operating on the same principle but weighing about 600 kg and with a base diameter of about 750 mm have also been employed, with the name 'frog rammer', and have been included in investigations at the Laboratory (Lewis, 1954).

Plate 11.1 600 kg power rammer (frog rammer) investigated in 1948

11.2 With all these machines the firing of the petrol-air mixture causes the machine to leap upwards about 300 mm. In the case of the smaller type of machine the forward movement is controlled by the angle at which it is held by the operator. With the larger machine the forward movement is controlled by the particular inclination of the machine on the angled base.

Plate 11.2 100 kg power rammer (A) investigated in 1949

Plate 11.3 100 kg power rammer (B) investigated in 1949

Plate 11.4 100 kg power rammer (C) investigated in 1949

Plate 11.5 100 kg power rammer (D) investigated in 1949

Plate 11.7 100 kg power rammer with experimental bases of various sizes investigated in 1950

Plate 11.6 100 kg power rammer (D) with standard base compacting a test soil

Power rammers are generally regarded as capable of compacting most soils, but because of their relatively low output they are restricted to the compaction of soil in confined areas or spaces such as in trenches and adjacent to structures.

11.3 The machines in this category were tested at the Laboratory during the early series of investigations, from 1946 to 1953 (see Chapter 3). The 600 kg power rammer, otherwise known as a frog rammer, is illustrated in Plate 11.1 and a sectional diagram showing its main components is given in Figure 11.1. Four power rammers of the standard type were investigated in one combined study and are illustrated in

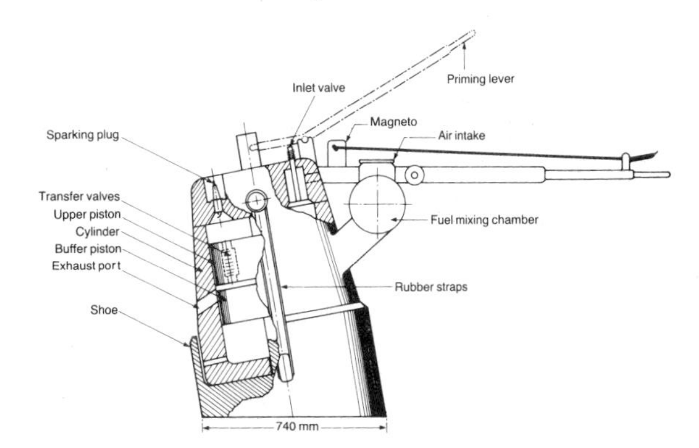

Figure 11.1 Sectional diagram of the 600 kg power rammer (frog rammer)

Plates 11.2 to 11.5; these have been identified as Machines A, B, C and D. An illustration of Machine D in use is shown in Plate 11.6. A special investigation of the effect of the size of the base of the power rammer was also carried out using a 100 kg machine similar to Machine D mentioned above. The machine with its various bolt-on bases is shown in Plate 11.7.

11.4 Details of the various machines are given in Table 11.1. Included in the details are the energy per blow, based on the mean height of jump and the total mass of the power rammer, and the specific energy (the energy per blow per unit area of base). Table 11.1 shows that a wide range of values of specific energy was achieved by the use of the various experimental sizes of base, listed as (i) to (iv) in the Table.

11.5 The above descriptions of the machines and the results presented in this Chapter have been taken largely from Lewis (1954, 1951), but the information on pressure measurements was taken from Tanner and Morris (1950).

Relations between state of compaction produced and moisture content

11.6 Relations between the state of compaction produced by power rammers and the moisture content of the soil are given in Figures 11.2 to 11.6. Dry density has been expressed in terms of relative compaction value (as a percentage of the maximum dry density obtained in the 2.5 kg rammer compaction test) and the moisture content as the arithmetical difference from the optimum moisture content of the 2.5 kg rammer test; the results obtained in the 2.5 kg rammer compaction test at the relevant period are shown in Figure 3.13.

11.7 Although four different standard 100 kg power rammers (Machines A to D in Table 11.1) were investigated, only one relation is shown in each Figure. It was found that the results for the four machines were so similar that they could be averaged to produce the results shown. Results for the 100 kg power rammer with each of the experimental bases have also been included for the soils on which they were tested, ie heavy clay (Figure 11.2) and sandy clay No 1 (Figure 11.4).

11.8 Previous published relations between dry density and moisture content obtained using power rammers (Lewis, 1954) refer to the soil being fully compacted, although reference is also made to the application of 10 passes of the 100 kg power rammer fitted with the experimental bases. As far as can be ascertained from records made at the time of the investigations, the 600 kg power rammer gave 4 'coverages' of the soil to produce the relations

Table 11.1 Details of the various machines contributing data on the performance of power rammers

Power rammer		Total mass (kg)	Height of jump (mean) (mm)	Energy per blow (J)	Diameter of base (mm)	Area of base (m^2)	Specific energy (Energy per blow per unit area of base) (kJ/m^2)
100 kg Standard	A	97.5	300–360 (330)	315	240	.0457	6.9
100 kg Standard	B	98.4	300–360 (330)	320	240	.0457	7.0
100 kg Standard	C	98.4	300–360 (330)	320	250	.0507	6.3
100 kg Standard	D	114.3	300–360 (330)	370	240	.0457	8.1
100 kg Standard	(Mean)						7.1
600 kg		613	305 approx.	1835	740	.4261	4.3
100 kg Experimental	(i)	104	300–360 (330)	335	190	.0285	11.8
100 kg Experimental	(ii)	111	300–360 (330)	360	300	.0700	5.1
100 kg Experimental	(iii)	117	290–340 (320)	365	360	.1029	3.6
100 kg Experimental	(iv)	120	280–330 (305)	360	410	.1338	2.7

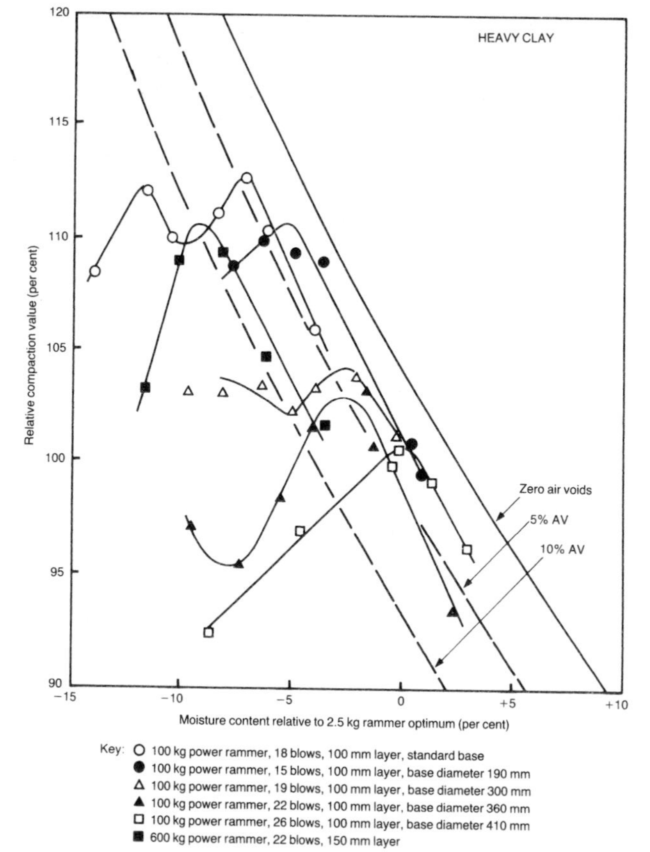

Figure 11.2 Relations between dry density and moisture content for power rammers on heavy clay, expressed in terms of relative compaction with the 2.5 kg rammer test

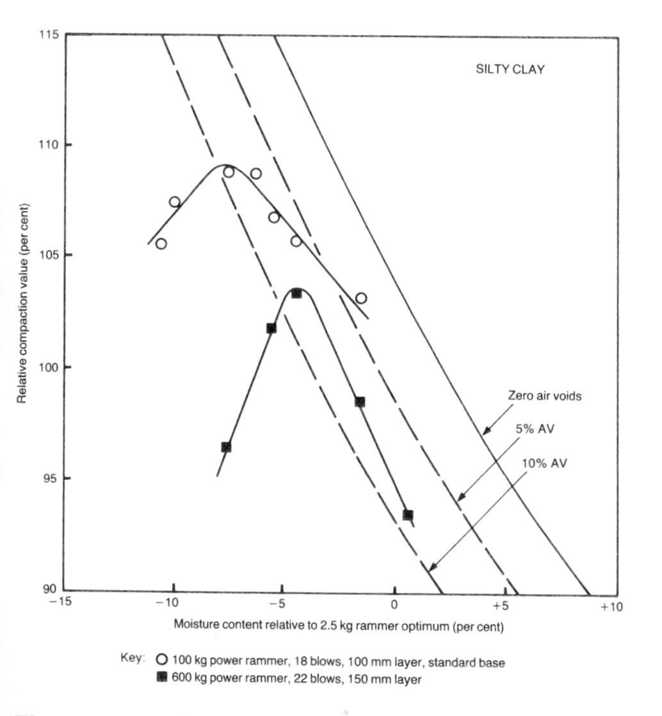

Figure 11.3 Relations between dry density and moisture content for power rammers on silty clay, expressed in terms of relative compaction with the 2.5 kg rammer test

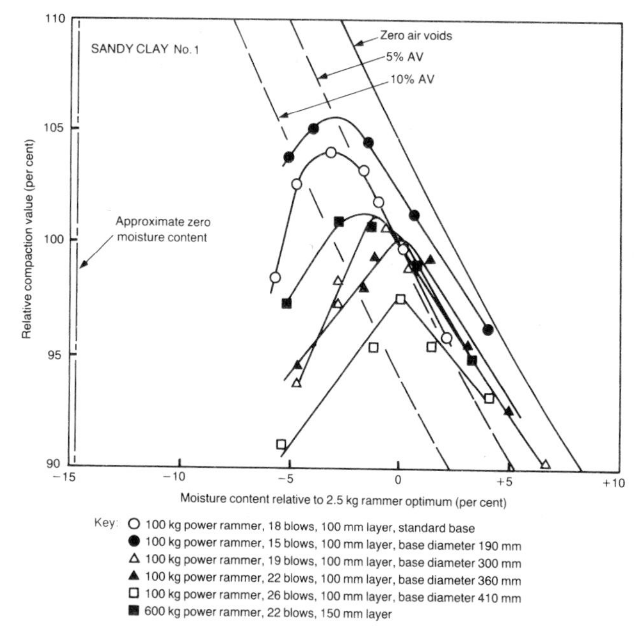

Figure 11.4 Relations between dry density and moisture content for power rammers on sandy clay No 1, expressed in terms of relative compaction with the 2.5 kg rammer test

shown in Figures 11.2 to 11.6, while the 100 kg machines gave 10 'passes'. Because of the different sizes of base and the individual operating characteristics of the machines, the number of blows received by each individual area of soil varied from one machine to another; the numbers of blows given in the keys to the Figures are the best estimates that can be made. Table 11.2 gives a summary of the information on which these values have been based.

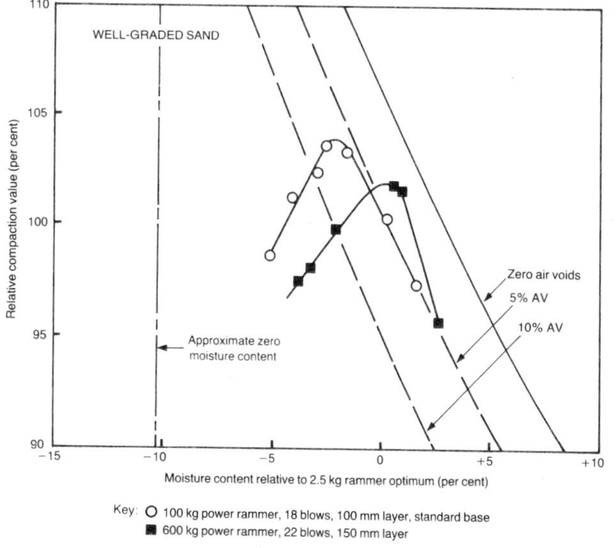

Figure 11.5 Relations between dry density and moisture content for power rammers on well-graded sand, expressed in terms of relative compaction with the 2.5 kg rammer test

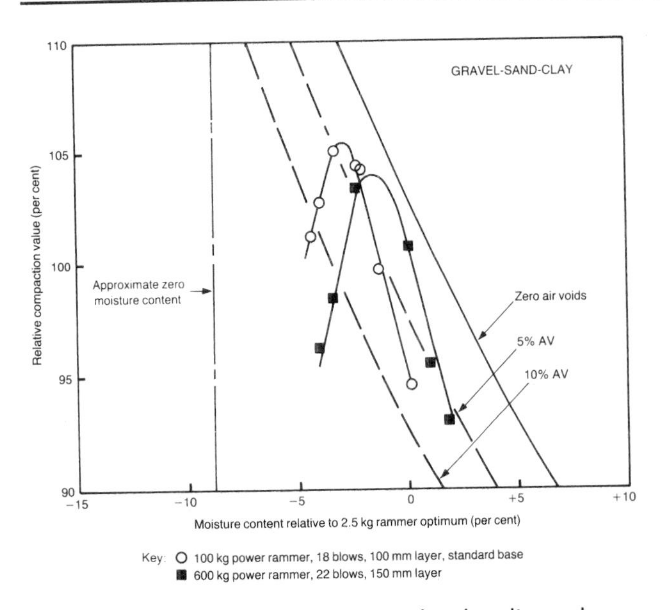

Figure 11.6 Relations between dry density and moisture content for power rammers on gravel-sand-clay, expressed in terms of relative compaction with the 2.5 kg rammer test

Table 11.2 Derivation of the average number of blows applied to each individual area of soil on one pass of each of the power rammers

(1) Power rammer	(2) Area of base (m^2)	(3) Observed number of blows per m^2 per pass	(4) Calculated average number of blows on each individual area of soil per pass (2)×(3)
100 kg standard			
A, B, D	.0457	40	1.8
C	.0507	40	2.0
600 kg	.4261	13	5.5
100 kg experimental			
(i)	.0285	54	1.5
(ii)	.0700	27	1.9
(iii)	.1029	21	2.2
(iv)	.1338	19	2.6

11.9 Although no specific reference has been made to the actual final thickness of the compacted layer in published documents, early records indicate that the thickness of layer tested with the 100 kg power rammers was 100 mm, whilst that tested with the 600 kg machine was 150 mm.

11.10 The results portrayed in Figures 11.2 to 11.6 have been summarised in Figure 11.7 in relations between the maximum relative compaction value and specific energy and between optimum moisture content and the same energy factor. In these relations the variation in thickness of layer between 100 and 150 mm and the variations in the number of applied blows between 15 and 26 have been ignored, the conditions for 'full compaction' being assumed to apply in all cases. The general trends of the relations indicate that the maximum relative compaction value increased with increase in the specific energy, whilst the optimum moisture content decreased. The maximum relative compaction values were highest for any particular value of specific energy with the heavy and silty clays and lowest with the sandy clay No 1; the granular soils, on which only the standard 100 kg and 600 kg power rammers were tested, occupied an intermediate level. The optimum relative moisture content values were very similar for the sandy clay and the granular soils, with those for heavy and silty clays very much lower. Maximum relative compaction values ranged from 97 to 101 per cent at the lower values of specific energy to 105 to 111 per cent at the highest energy value. Optimum relative moisture contents were more scattered about the relations given, but the trends indicate that the values ranged from about optimum of the 2.5 kg rammer compaction test at low specific energy levels to about 4 to 7 per cent below that optimum, depending on soil type, at the high specific energy levels. The results which apply to the standard 100 and

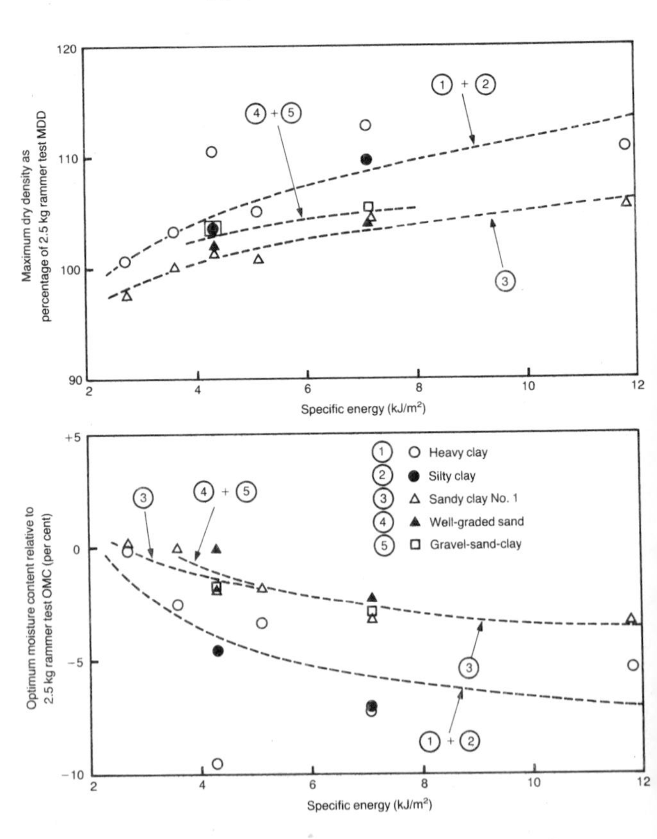

Figure 11.7 Maximum dry densities and optimum moisture contents related to the specific energy of power rammers. Results for 100 to 150 mm compacted layers from Figures 11.2 to 11.6

600 kg power rammers may be identified in Figure 11.7 as those plotted at specific energy values of 4.3 kJ/m² (600 kg machine) and 7.1 kJ/m² (100 kg machines).

Relations between state of compaction produced and number of blows

11.11 Results of investigations to determine the effect on the state of compaction achieved of variations in the number of blows of power rammers are given in Figures 11.8 to 11.12. The results for each soil are illustrated under a

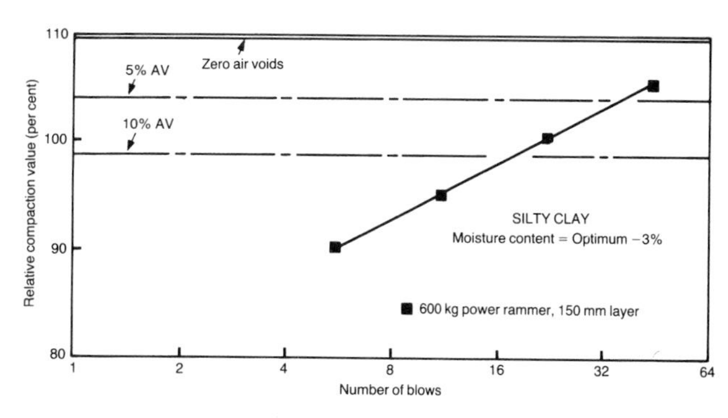

Figure 11.9 Relation between relative compaction and number of blows for the 600 kg power rammer on silty clay. Values are related to the maximum dry density and optimum moisture content obtained in the 2.5 kg rammer compaction test (Figure 3.13)

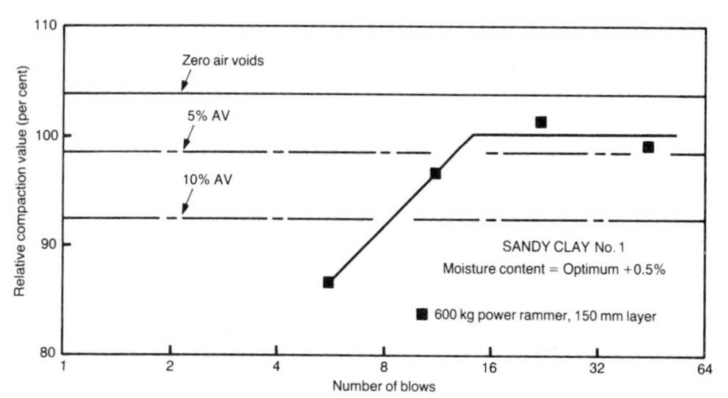

Figure 11.10 Relation between relative compaction and number of blows for the 600 kg power rammer on sandy clay No 1. Values are related to the maximum dry density and optimum moisture content obtained in the 2.5 kg rammer compaction test (Figure 3.13)

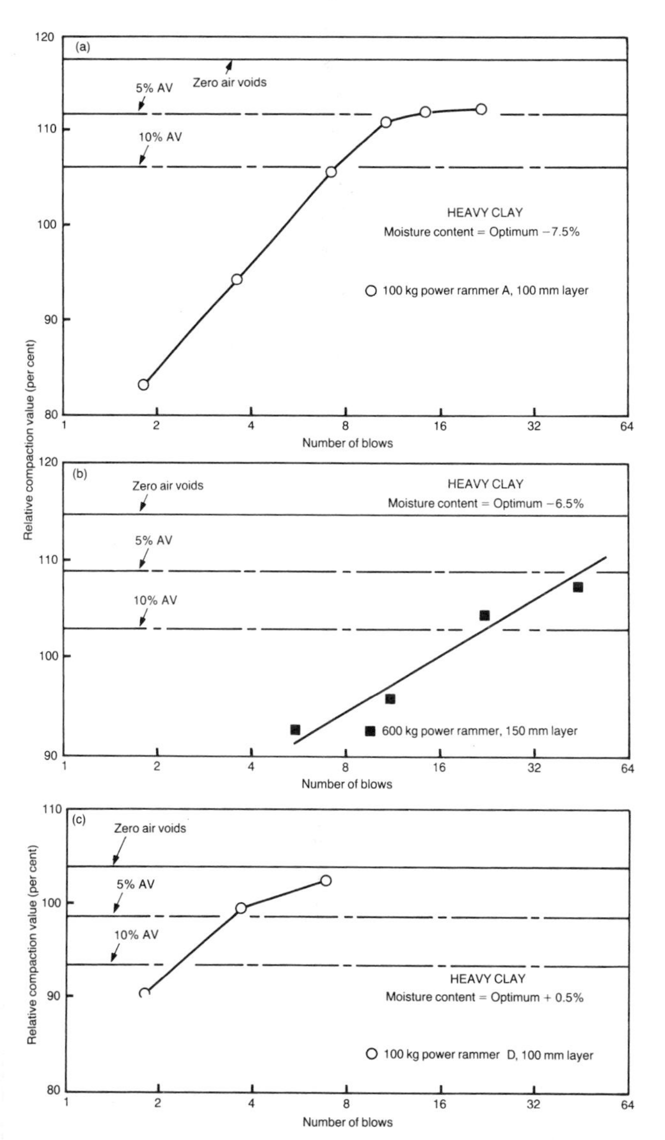

Figure 11.8 Relations between relative compaction and number of blows for power rammers on heavy clay at three different values of moisture content. Values are related to the maximum dry density and optimum moisture content obtained in the 2.5 kg rammer compaction test (Figure 3.13)

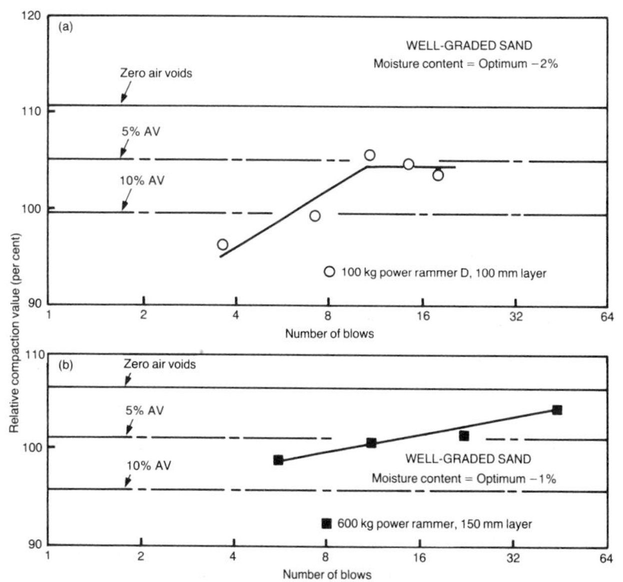

Figure 11.11 Relations between relative compaction and number of blows for power rammers on well-graded sand at two different values of moisture content. Values are related to the maximum dry density and optimum moisture content obtained in the 2.5 kg rammer compaction test (Figure 3.13)

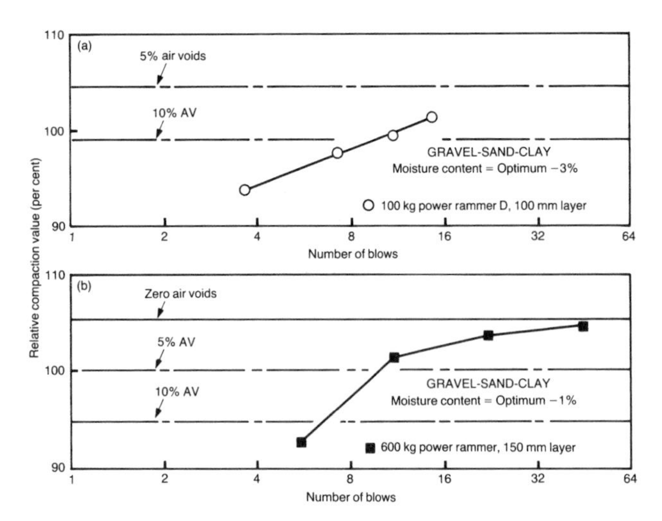

Figure 11.12 Relations between relative compaction and number of blows for power rammers on gravel-sand-clay at two different values of moisture content. Values are related to the maximum dry density and optimum moisture content obtained in the 2.5 kg rammer compaction test (Figure 3.13)

separate figure number, whilst different values of moisture content are in separate graphs within each figure. Only the 600 kg power rammer and the 100 kg standard machines A and D (Table 11.1) contributed to the results given in these particular relations. The thicknesses of layer used, 100 mm and 150 mm, were similar for each particular machine to those used in the determination of the effects of variations in moisture content (Paragraphs 11.6 to 11.10). The numbers of blows plotted in the figures have been determined from the originally recorded numbers of passes and the information given in Table 11.2. With the particular machines tested, the variation in specific energy values was rather limited, but by incorporating the results given in Figures 11.2 to 11.6, especially those for the complete range of specific energy values involved (Figures 11.2 and 11.4), sufficient data have been obtained to allow regression analyses to be made. The variation in thickness of layer (100 to 150 mm) has been ignored and the numbers of blows required to achieve both 95 per cent relative compaction (dry density as a percentage of the maximum dry density obtained in the 2.5 kg rammer compaction test) and 10 per cent air voids have been related to specific energy and moisture content.

11.12 The results of the regression analyses are given in Table 11.3 in the form of regression equations and correlation coefficients. Combinations of different soils have also been tested in the regression analyses and, where good correlations were obtained, those combinations have been given. Thus, heavy clay and silty clay were found to provide good correlations when all the results were taken together, as also were well-graded sand and gravel-sand-clay, although in this instance only when using the 95 per cent relative compaction criterion. The correlation coefficient for well-graded sand to level (b) was considered to be too low, with the small number of data available, for the results of the analysis to be included in the Table.

Table 11.3 Equations and correlation coefficients to determine the number of blows of power rammers required to achieve specified levels of compaction. Layer thickness 100 to 150 mm

Soil	Level*	Regression equation†	No of data points	Correlation coefficient
Heavy clay & silty clay	(a)	P=2.22–2.03 E –.0231 w	7	0.95
	(b)	P=1.97–1.61 E –.0623 w	11	0.91
Sandy clay No 1	(a)	P=1.62–.855 E –.0516 w	6	0.93
	(b)	P=1.69–1.03 E –.112 w	7	0.95
Gravel-sand-clay	(b)	P=1.36–1.29 E –.247 w	4	0.99
Well-graded sand & gravel-sand-clay	(a)	P=1.69–1.57 E –.128 w	6	0.99

*Level (a)=95 per cent of the maximum dry density obtained in the 2.5 kg rammer compaction test
Level (b)=10 per cent air voids
†P=Log_{10} (Number of blows required)
E=Log_{10} (Specific energy in kJ/m^2)
w=Moisture content of soil, per cent, minus the optimum moisture content, per cent, obtained in the 2.5 kg rammer compaction test

11.13 Although the results of the regression analyses provide a useful guide on the effects of the specific energy of a power rammer and of the moisture content of the soil on the number of blows required, many of the data were determined by means of the 100 kg machine fitted with experimental bases of various sizes. These results increased the ranges of the variables used and so added to the accuracy of the regression equations, but the results for standard rammers, which are still in use at the present time, are of most interest. Relations between the number of blows required to achieve 95 per cent relative compaction and 10 per cent air voids and moisture content are given, therefore, in Figure 11.13 for the 600 kg

Figure 11.13 Relations between number of blows of the 600 kg power rammer and moisture content to achieve (a) 95 per cent of the maximum dry density obtained in the 2.5 kg rammer compaction test and (b) 10 per cent air voids. The relations drawn through the points are based on those given in Table 11.3 for 100 to 150 mm layers

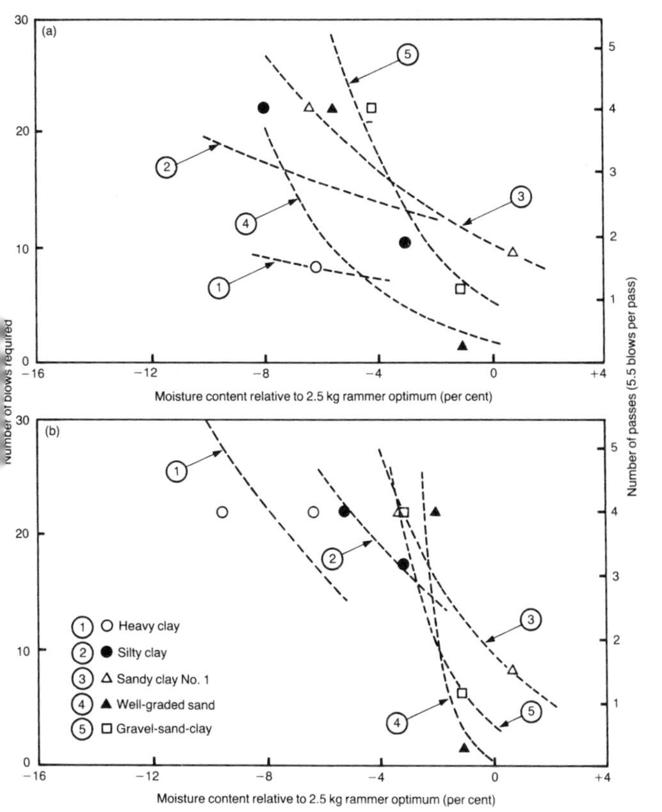

power rammer and in Figure 11.14 for the 100 kg standard power rammers. The individual points shown have been determined from Figures 11.2 to 11.6 and 11.8 to 11.12, in some cases by extrapolating the existing information where it has been considered reasonable to do so. The relations drawn through the points are

Figure 11.15 (a) See caption page 180

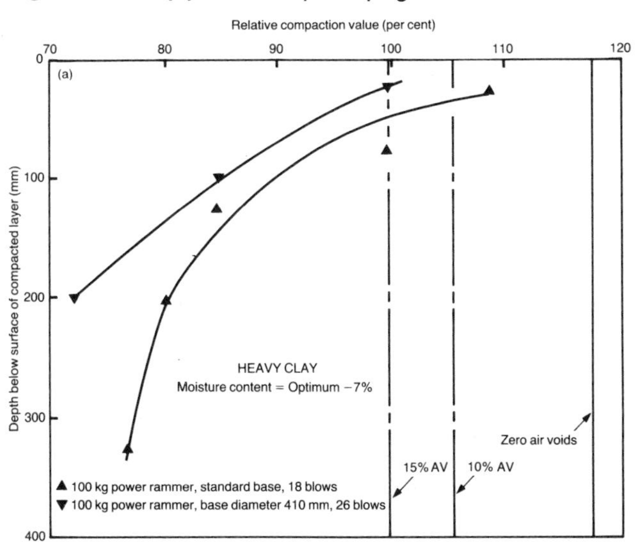

Figure 11.14 Relations between number of blows of a standard 100 kg power rammer and moisture content to achieve (a) 95 per cent of the maximum dry density obtained in the 2.5 kg rammer compaction test and (b) 10 per cent air voids. The relations drawn through the points are based on those given in Table 11.3 for 100 to 150 mm layers

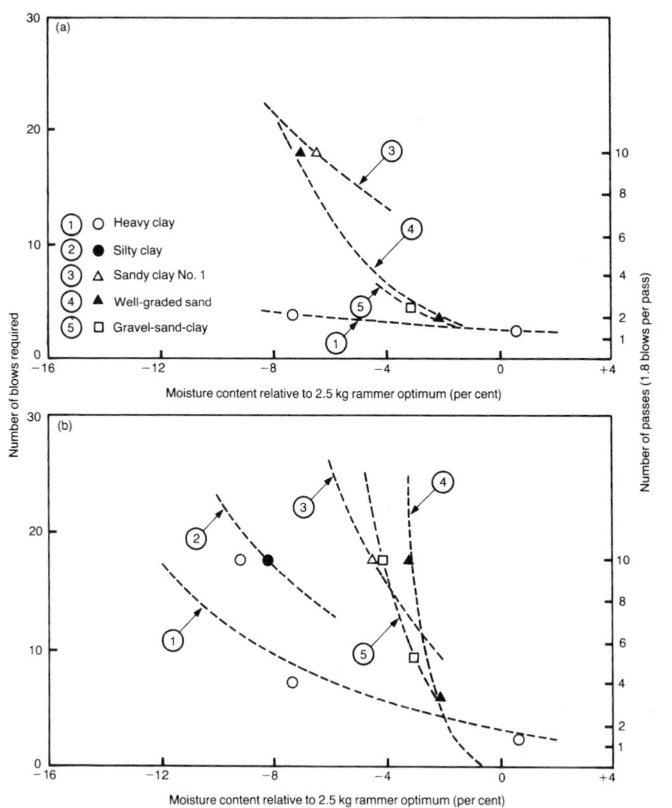

based on the rates of change of 'P' with change in 'w' in Table 11.3 for each soil type and compaction criterion. The equivalent number of passes required, based on the information given in Table 11.2, is shown on the right of the figures. The different relations for the different types of soil indicate very clearly that a general

Figure 11.15 (b) See caption page 180

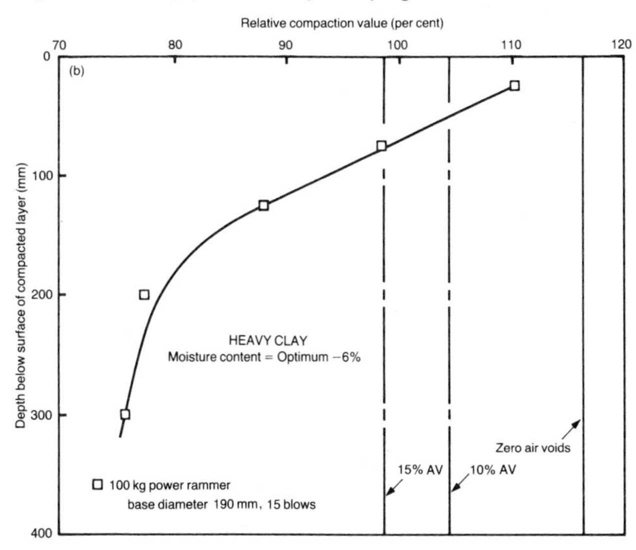

Figures 11.15 (c) and (d) See main caption

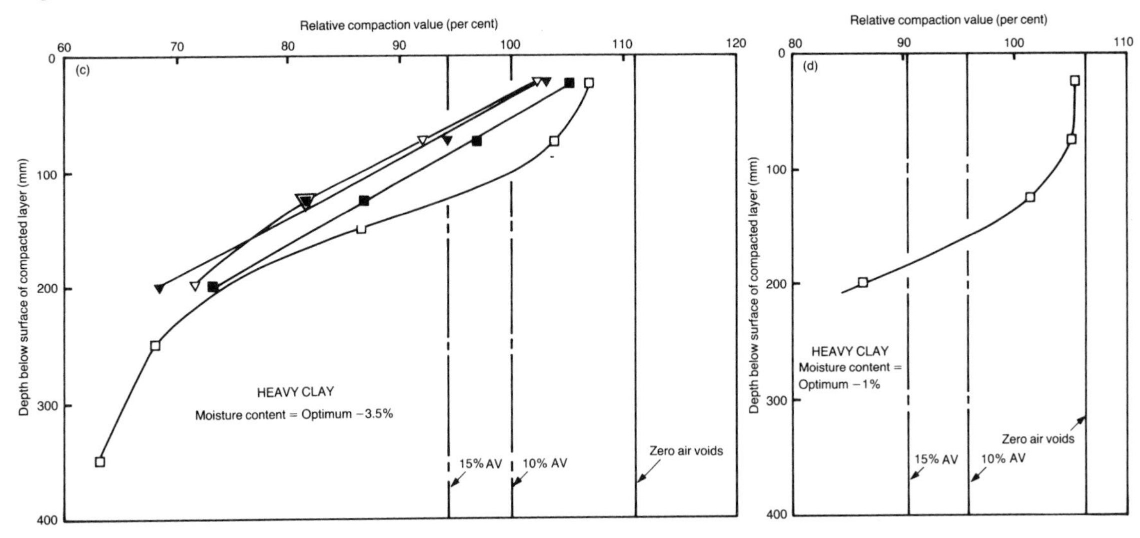

Figure 11.15 Relations between relative compaction and depth in the compacted layer of heavy clay obtained with power rammers at four different values of moisture content. All values are related to the maximum dry density and optimum moisture content obtained in the 2.5 kg rammer compaction test (Figure 3.13)

statement cannot be made regarding the appropriate number of blows to be applied for a given relative moisture content, although reference to the figures will provide some guidance. In general, it appears that sandy clay No 1, which had a very low plasticity (Figure 3.2), behaved more similarly to the well-graded sand and gravel-sand-clay than it did to the heavy clay and silty clay, a conclusion which is also supported by the relations in Figure 11.7.

Relations between state of compaction and depth within the compacted layer

11.14 Results of investigations of the variation of density with depth in layers of soil compacted by power rammers are given in Figures 11.15 to 11.19. The 600 kg power rammer was tested on the silty clay (Figure 11.16 (c)), well-graded sand (Figure 11.18 (b)) and gravel-sand-clay (Figure 11.19 (b)). The 100 kg power rammer with the standard base was tested on heavy clay (Figure 11.15(a)), well-graded sand (Figure 11.18(a)) and gravel-sand-clay (Figure 11.19(a)). The 100 kg power rammer with experimental bases of various sizes was tested on the three clay soils (Figures 11.15 to 11.17). In all cases the final thickness of layer used was equal to or in excess of 200 mm and the numbers of blows used were such that maximum possible compaction can be assumed to have been achieved.

11.15 To summarise the results regression analyses have been carried out to determine the effects of specific energy, diameter of base, and moisture content on the maximum thickness of layer capable of compaction by power rammers. The criteria used were similar to those in previous chapters, ie the achievement of 90 per

Figure 11.16 (a) See caption below

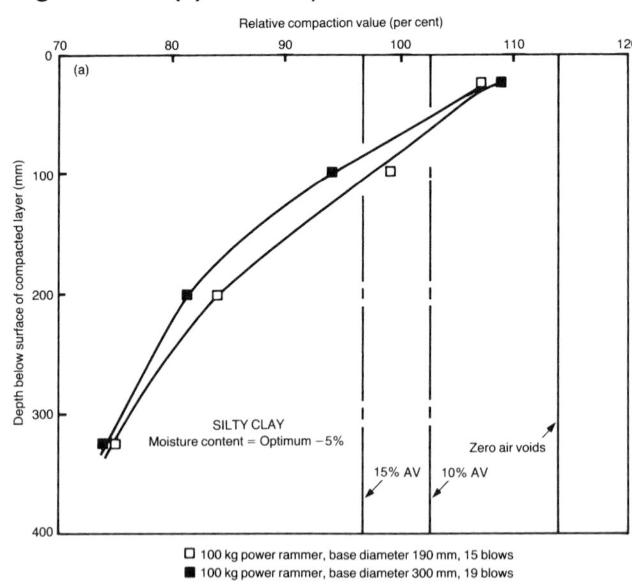

Figure 11.16 Relations between relative compaction and depth in the compacted layer of silty clay obtained with power rammers at three different values of moisture content. All values are related to the maximum dry density and optimum moisture content obtained in the 2.5 kg rammer compaction test (Figure 3.13)

Figures 11.16 (b) and (c) See main caption p. 180

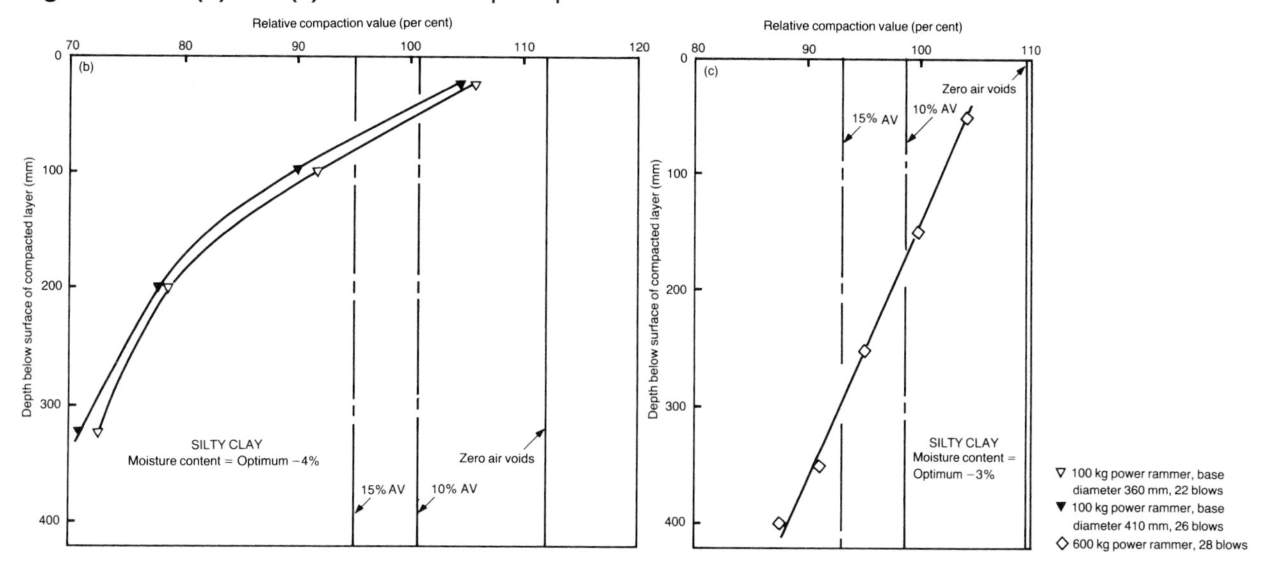

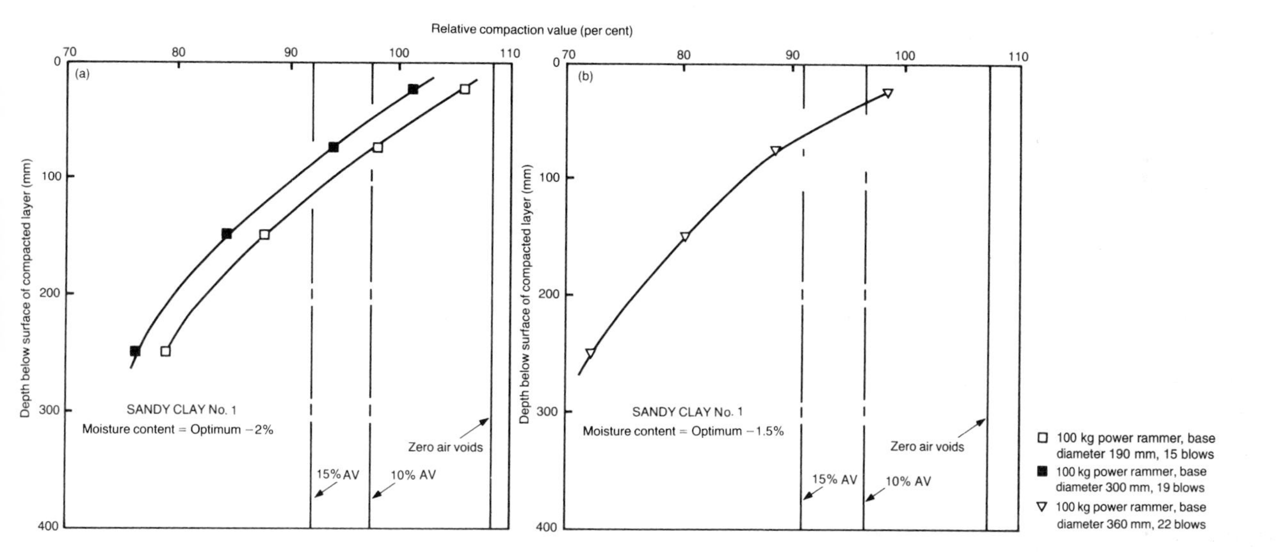

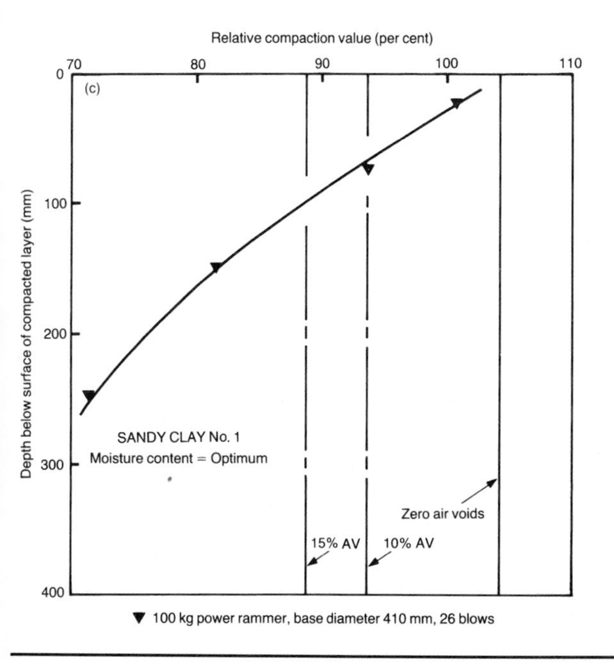

cent relative compaction and of 15 per cent air voids at the bottom of the layer. The results are given in Table 11.4.

11.16 For those soils on which the 600 kg power rammer was not used there was a high level of correlation between specific energy and diameter of base (as would be expected as the power rammers used had very similar total masses, see Table 11.1). Thus the maximum depths of layer capable of being compacted to the specified criteria are related solely to specific

Figure 11.17 Relations between relative compaction and depth in the compacted layer of sandy clay No 1 obtained with power rammers at three different values of moisture content. All values are related to the maximum dry density and optimum moisture content obtained in the 2.5 kg rammer compaction test (Figure 3.13)

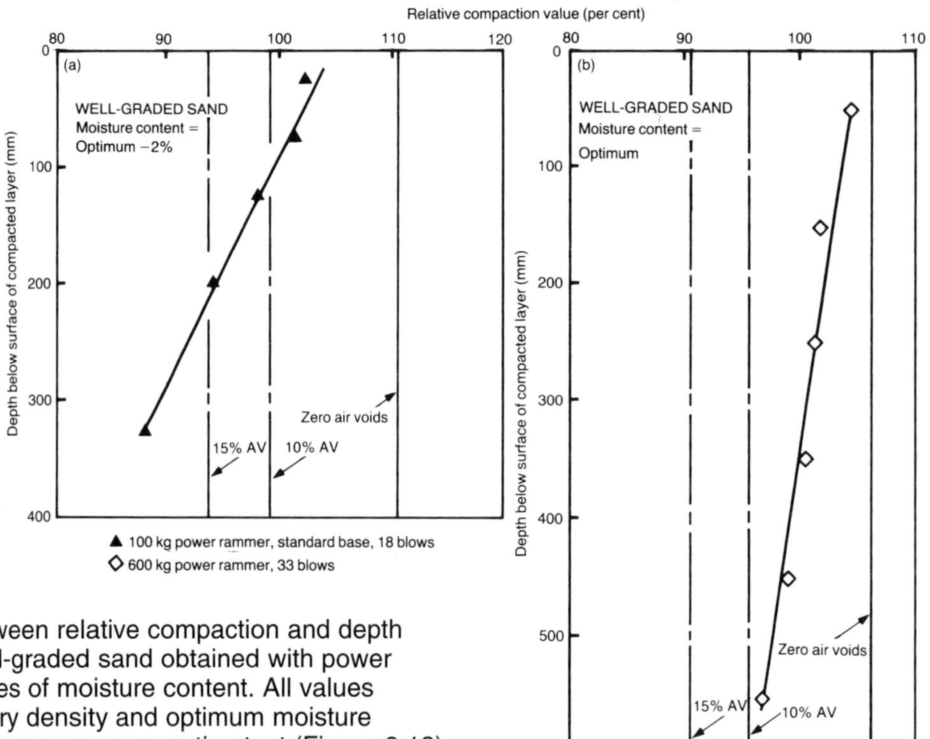

Figure 11.18 Relations between relative compaction and depth in the compacted layer of well-graded sand obtained with power rammers at two different values of moisture content. All values are related to the maximum dry density and optimum moisture content obtained in the 2.5 kg rammer compaction test (Figure 3.13)

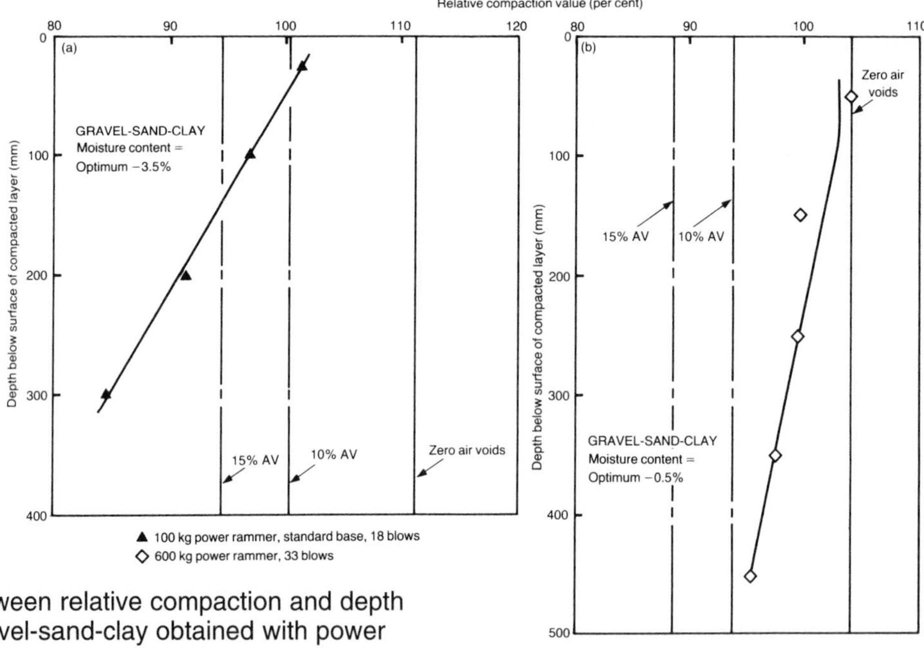

Figure 11.19 Relations between relative compaction and depth in the compacted layer of gravel-sand-clay obtained with power rammers at two different values of moisture content. All values are related to the maximum dry density and optimum moisture content obtained in the 2.5 kg rammer compaction test (Figure 3.13)

energy and moisture content (Table 11.4), with the qualification that the results apply only to the 100 kg power rammer. Good correlations were obtained with both the heavy clay and the sandy clay No 1. With the silty clay, on which the 100 kg machine with various sizes of base and the 600 kg machine were tested, the range of moisture content values used was fairly small (from 5 to 3 per cent below the 2.5 kg rammer optimum) and the moisture content factor was found to be of low significance; Table 11.4 shows that very high levels of correlation were achieved using specific energy and diameter of base as independent variables. As only the standard 100 kg and 600 kg power rammers were used in tests on well-graded sand and gravel-sand-clay an insufficient range of factors was available to provide meaningful correlations.

11.17 Relations between depth of compacted layer and specific energy for a range of moisture contents, based on the results given in Table 11.4, are given for heavy clay in Figure 11.20 and for sandy clay No 1 in Figure 11.21. These represent the effect of varying the size of base of a 100 kg power rammer assuming that the soil is fully compacted by at least 15 blows of the rammer. Possible relations between the same two factors for a range of base diameters are also given for silty clay in Figure 11.22; these have been obtained using both the 100 kg and 600 kg machines but apply only to the range of moisture contents from 5 to 3 per cent below the 2.5 kg rammer optimum.

11.18 The maximum depths of layer for the standard power rammers, under the conditions in which they were tested on all the soils, are given in Table 11.5. The depths of layer satisfactorily compacted by the 600 kg machine were always considerably greater than those of the 100 kg power rammer; however, the moisture contents used were always higher with the larger machine and this factor would have been beneficial to its performance. The regression lines, plotted in Figure 11.22, for silty clay within a limited range of moisture contents, confirm that at the specific energies for the two machines, ie 4.3 and 8.1 kJ/m² for the 600 kg and 100 kg power rammers respectively, the larger base diameter (740 mm) with the 4.3 kJ/m² of the 600 kg machine could compact a layer of more than 300 mm, whereas the standard base diameter (240 mm) with the 8.1 kJ/m² of the 100 kg machine only achieved a 100 to 150 mm thickness. Thus the larger mass allied with a larger diameter of base, even though they combined to produce the lower

Table 11.4 Equations and correlation coefficients to determine the thickness of layer capable of being compacted by power rammers. These results are for full compaction utilising at least 15 blows of the power rammer

Soil	Level*	Regression equation†	No of data points	Correlation coefficient
Results obtained with the 100 kg machine with various sizes of base				
Heavy clay	(a)	D=1.87+.373 E+.0299 w	8	0.97
	(b)	D=1.82+.539 E+.0868 w	8	0.99
Sandy clay No 1	(a)	D=1.69+.579 E+.0773 w	4	0.93
	(b)	D=1.70+.650 E+.138 w	4	0.94
Results obtained with the 100 kg and 600 kg machines‡				
Silty clay	(a)	D=–2.13+1.03 E+1.41 d	5	0.999
	(b)	D=–2.78+1.08 E+1.59 d	5	≈1.0

*Level (a)=Achievement at the bottom of the layer of 90 per cent of the maximum dry density obtained in the 2.5 kg rammer compaction test
(b)=Achievement at the bottom of the layer of 15 per cent air voids

†D=Log_{10} (Maximum depth of layer to achieve stated level of compaction, mm)
E=Log_{10} (Specific energy in kJ/m²)
w=Moisture content of soil, per cent, minus the optimum moisture content, per cent, obtained in the 2.5 kg rammer compaction test
d=Log_{10} (Diameter of base of rammer in mm)

‡Moisture content in the range from 5 to 3 per cent below the optimum obtained in the 2.5 kg rammer compaction test

Table 11.5 Maximum depths of layer capable of being compacted by standard power rammers for the soil conditions tested

Power rammer:–		100 kg		600 kg	
State of compaction at the bottom of the layer:–		90% relative compaction	15% air voids	90% relative compaction	15% air voids
Soil	Moisture content relative to 2.5 kg rammer optimum (per cent)	Depth of compacted layer (mm)			
Heavy clay	–7	100	50	–	–
Silty clay	–3	–	–	370	300
Well-graded sand	–2	290	210	–	–
	0	–	–	More than 600	More than 600
Gravel-sand-clay	–3.5	210	140	–	–
	–0.5	–	–	More than 500	More than 500

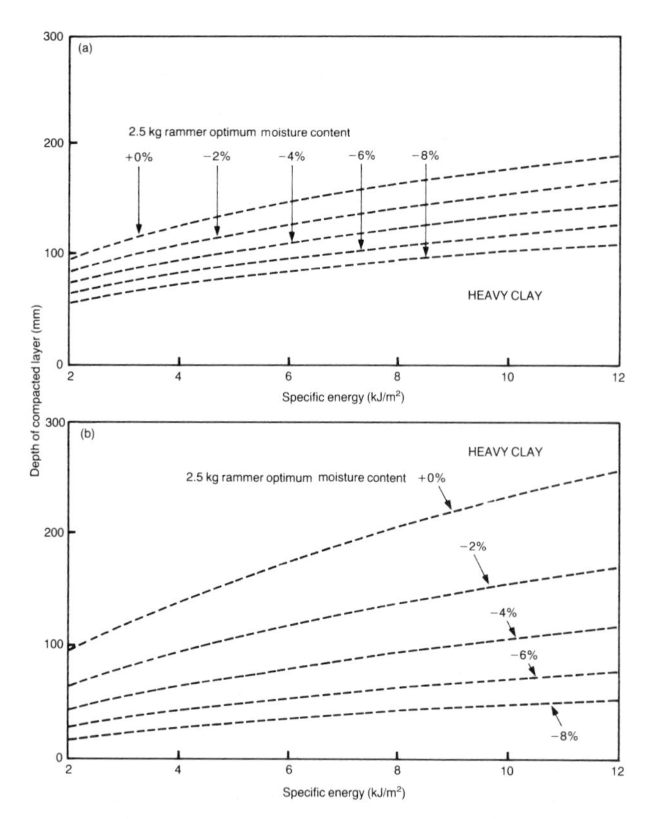

Figure 11.20 Relations between depth of layer and specific energy for compaction to refusal by the 100 kg power rammer (at least 15 blows) to achieve (a) 90 per cent relative compaction and (b) 15 per cent air voids at the bottom of the layer. Results obtained using the power rammer with various sizes of base on heavy clay

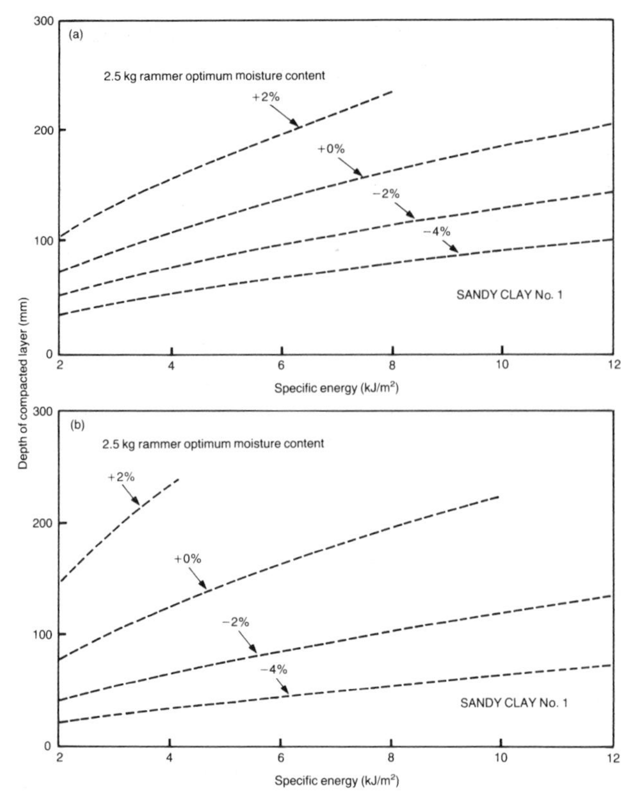

Figure 11.21 Relations between depth of layer and specific energy for compaction to refusal by the 100 kg power rammer (at least 15 blows) to achieve (a) 90 per cent relative compaction and (b) 15 per cent air voids at the bottom of the layer. Results obtained using the power rammer with various sizes of base on sandy clay No 1

specific energy, were the most effective combination for the production of the greatest possible thickness of compacted layer.

11.19 It is considered that the good correlations, given in Table 11.3 for the regression equations for the determination of the number of blows required, were obtained by virtue of the relatively thin layers used in the tests. For such thin layers, 100 mm for the 100 kg machines and 150 mm for the 600 kg rammer, the effects of the variations in base diameter, other than the direct influence on specific energy, would have been largely nullified (see also the discussion on pressures generated in soil in Paragraphs 11.23 to 11.26).

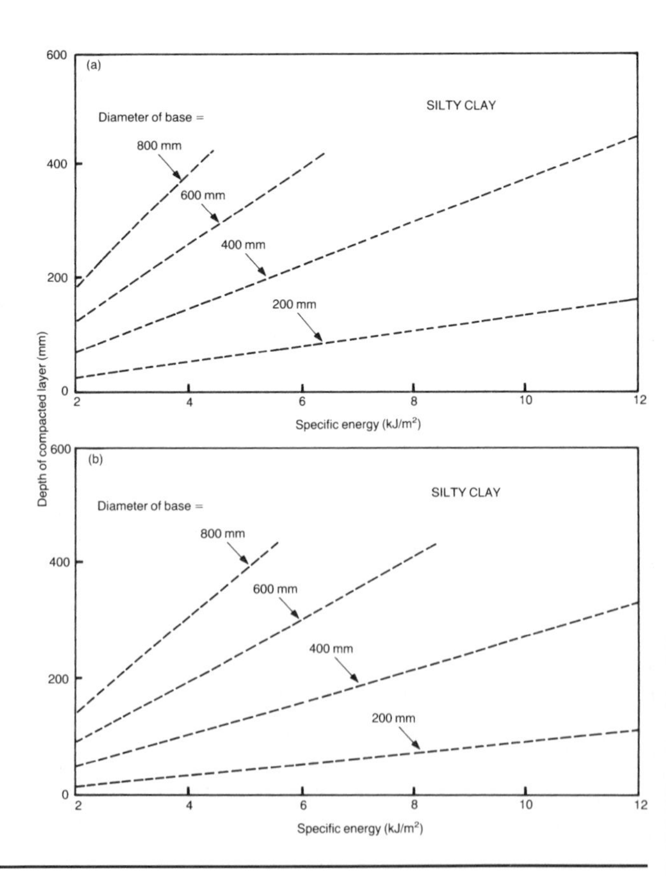

Figure 11.22 Relations between depth of layer and specific energy for compaction to refusal by power rammers (at least 15 blows) to achieve (a) 90 per cent relative compaction and (b) 15 per cent air voids at the bottom of the layer. Results obtained with the 600 kg and 100 kg power rammers, the latter with various sizes of base, on silty clay with moisture contents in the range 5 to 3 per cent below the 2.5 kg rammer optimum

Effect of varying the diameter of the base of the 100 kg power rammer

11.20 The results produced by the use of experimental sizes of base on a 100 kg power rammer have been incorporated in Figures 11.2, 11.4, and 11.15 to 11.17. These show relations between state of compaction and moisture content and between state of compaction and depth in the compacted layer, both for large numbers of blows, assumed to represent the fully compacted condition. The results obtained by use of the experimental bases have provided valuable additional data in statistical analyses; in particular, they contributed to a significant range of specific energies (the energy per blow per unit area of base). The influences of change in specific energy on the maximum relative compaction value and optimum moisture content are shown in Figure 11.7 and its influence on thickness of layer capable of being compacted is shown in Figures 11.20 to 11.22.

11.21 It can be concluded from Figure 11.7 that by increasing the base diameter on a 100 kg power rammer (thereby reducing its specific energy) the maximum relative compaction value was reduced and the optimum moisture content was increased. It was concluded at the time of the tests that there was no significant difference made to the state of compaction obtained at the higher moisture contents at which the soils are normally encountered in the British Isles (Lewis, 1951, 1954). Figures 11.20 and 11.21 show that the increase in diameter of base, resulting in a decrease in specific energy, reduced the depth of layer capable of being compacted, although the effect was fairly small except at higher moisture contents. A study of Figure 11.22 in conjunction with Table 11.1 (the latter gives the values of specific energy and of base diameter for each of the experimental bases) indicates that very similar thicknesses result for the small range of moisture contents used with the silty clay soil. It was concluded at the time of the tests that the increase in the depth of penetration of the stresses produced by the rammer with bases of larger diameter was counter-balanced by the reduced levels of stress.

11.22 Apart from its effect on the state of compaction obtained, increasing the size of the base of the power rammer resulted in an increase in the area of soil compacted by each blow and an increase in the number of blows per pass (Table 11.2); the potential output of the machine, therefore, was improved. On the other hand, the power rammer was found to be more difficult to handle with the larger bases, which showed a tendency to adhere to the soil. No difficulty was experienced in operating the power rammer with the base of 300 mm diameter, and it was concluded that for use in the British Isles it would be an advantage if the standard base of the 100 kg power rammer (240 to 250 mm diameter) was increased in size to a diameter of 300 mm. The effect of base diameter on pressures generated in the soil is discussed in the following section.

Measurement of pressures produced in soil by a 100 kg power rammer

11.23 During the course of the tests to produce the results shown in Figure 11.16, piezoelectric pressure gauges were buried in the silty clay soil at various depths, varying from about 100 mm to in excess of 400 mm, below the surface of the final compacted layer. Piezoelectric pressure gauges each contain a quartz crystal which produces an electrical charge in proportion to the magnitude of the applied pressure. The outputs from the pressure gauges were registered on a cathode-ray tube and recorded by a rotating-drum camera; rather sophisticated equipment for 1950.

11.24 The 100 kg power rammer, equipped in turn with each of the experimental bases, was operated normally across the fully compacted soil along the centre-line of the gauge installation. A typical recorded gauge response is shown in Figure 11.23. The high-speed record in Figure 11.23(b) shows the various pressure pulses registered during a single cycle of the power rammer operation. These were:–

(1) The pressure pulse A, caused by the firing of the petrol-air mixture in the internal combustion cylinder and the resulting downward thrust of the base causing the rammer to leap in the air.

(2) The pressure pulse B as the base of the descending rammer struck the surface of the soil.

(3) The pressure pulse C as the main body of the rammer reached the limit of downward travel, striking the base after compressing the springs between the main body and the base. The associated action of the piston in the internal combustion cylinder primed the machine for the next firing.

(4) The reflected pressure D which was always small and was ignored.

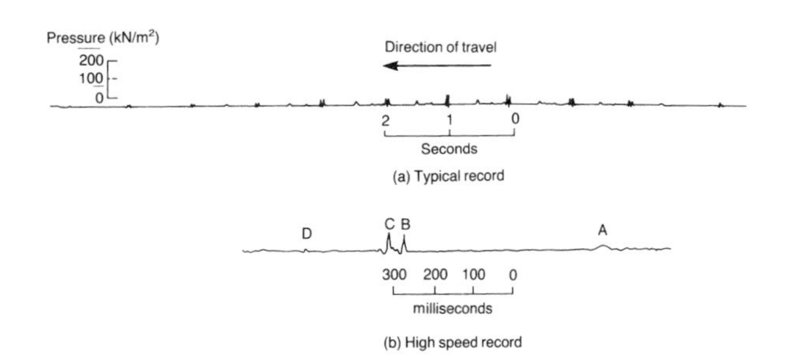

Figure 11.23 Typical records from pressure cells buried in silty clay during compaction by the 100 kg power rammer

11.25 The results shown in Figure 11.23 indicate that the pressure at C was slightly higher than that at B, reflecting the relative masses of the base and piston (pulse B) and the main body of the machine (pulse C).

11.26 Relations between the maximum recorded pressures and the depth below the surface of the compacted layer, for the various sizes of base, are shown in Figure 11.24. The accuracy of the values of pressure given cannot be verified, but the relative effects of the different sizes of base and the different depths in the soil can be expected to be valid. The variations in recorded pressure with depth follow the same trends as the variation in relative compaction with depth shown in Figure 11.16 for the same soil. As also indicated in Figure 11.16, the net effect of varying the base diameter was fairly small. As mentioned in Paragraph 11.21, this was considered to be caused by the increase in the depth of penetration of the stresses produced by the larger bases being counter-balanced by the reduced levels of stress associated with the reduced values of specific energy. The thicknesses of layer capable of being compacted at the moisture contents tested (4 to 5 per cent below the 2.5 kg rammer optimum) to the criteria of 90 per cent relative compaction or 15 per cent air voids at the bottom of the layer were in the range of 70 to 150 mm, depending on the size of experimental base and the criterion used. At these depths the pressures shown in Figure 11.24 were generally in excess of 200 kN/m^2 and were largely directly related to the specific energy of the machine, ie the pressure was highest for the base with the smallest diameter. The influence of the larger bases on the pressure distribution at greater depths in the compacted layer, although discussed by Tanner and Morris (1950), can only be of minor interest. Assuming that the pressures recorded were an accurate representation of the pressure distribution in the soil, the average pressure indicated to be necessary to achieve 90 per cent relative compaction in the silty clay at the moisture content of the tests was about 250 kN/m^2.

General comments

11.27 The analyses presented in this Chapter relating to the performance of power rammers have been based to a considerable extent on the results obtained with the 100 kg machine fitted with experimental bases of various sizes. The range of specific energies was considerably enhanced by this and the validity of the statistical analyses was improved. However, the results for the extremes of the range of specific energy (Table 11.1) are only of interest to a designer contemplating possible new forms of power rammer. The standard 100 kg power rammer had a specific energy in the range of about 6 to 8 kJ/m^2 and the 600 kg machine had a lower value of just over 4 kJ/m^2 (Table 11.1). The results pertaining to these particular machines and to their values of specific energy would be of primary interest to the practising engineer.

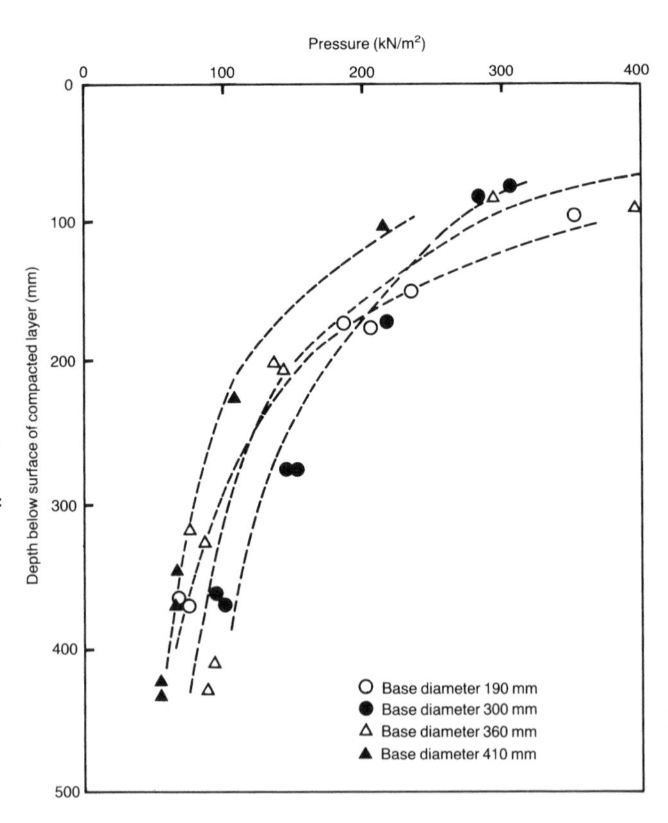

Figure 11.24 Relations between maximum recorded pressure and depth in a layer of silty clay compacted by the 100 kg power rammer fitted with various sizes of base. See Figure 11.16 for details of moisture contents

References

Lewis, W A (1951). Performance of power rammers and their use in the reinstatement of trenches. The Surveyor and Municipal and County Engineer, CX, 3082, 187–8.

Lewis, W A (1954). Further studies in the compaction of soil and the performance of compaction plant. *Road Research Technical Paper* No 33. HM Stationery Office, London.

Tanner, J S and Morris, S A H (1950). Investigation of the effects on compaction produced by increasing the size of the base of a 2-cwt power rammer. *D.S.I.R., RRL Research Note* No RN/1407/JST.SAHM (Unpublished).

CHAPTER 12 COMPACTION OF SOIL BY DROPPING-WEIGHT COMPACTORS

Description of machines

12.1 The simple dropping of a dead weight from a controlled height, using some form of hoist mechanism, is attractive as a method of compaction. Very high energy can be generated by using large weights and large heights of drop (Thompson and Herbert, 1978). For compaction of relatively shallow layers in cut-and-fill operations the mass is usually restricted to about 500 kg with heights of drop variable up to a maximum of about 2 m. For compaction of soil in trenches and close to structures, self-propelled machines with mechanical traversing mechanisms are available. With different attachments these machines can also be used as breakers, post drivers, etc.

12.2 Three machines in the dropping-weight compactor class have been investigated in detail at the Laboratory to determine their performance or to investigate the effects of various factors. The machines are illustrated in Plates 12.1 to 12.5 and details are given in Table 12.1.

12.3 The first machine (Plates 12.1 and 12.2), tested in late 1953 and early 1954, was designed to form the compaction unit in a soil-stabilisation train; as such, it was intended to provide adequate compaction in one pass. It was a towed machine, therefore, with a compacting mechanism consisting of six 200 kg cast-iron blocks arranged so that they were individually lifted by a set of cams mounted on a transverse shaft and allowed a drop 'freely' on to the surface of the soil. The machine will be referred to throughout this Chapter as the multi dropping-weight compactor or multi-weight machine. The total lift of the cams was about 190 mm and by means of a lever system each block was lifted about 330 mm (Table 12.1). Each weight was 305 mm wide with a horizontal surface 150 mm long and a front face about 200 mm long inclined at an angle of about 30 degrees to the horizontal; the shape of the dropping weight is just apparent in Plate 12.2. It has been assumed in this Chapter that the soil deformed sufficiently during compaction so that the inclined portion of the base formed part of the contact area in

Plate 12.1 Multi dropping-weight compactor investigated in 1953–54

Table 12.1 Details of the machines and their various modes of operation contributing data on the performance of dropping-weight compactors

Dropping-weight compactor	Mass of rammer (kg)	Height of drop (m)	Efficiency*	Energy per blow† (J)	Dimensions of base of rammer (m×m)	Area of base of rammer (m^2)	Specific energy‡ (kJ/m^2)	Blows per pass
Multi-weight machine	204	0.330	0.78	515	.305×.330	0.101	5.1	10
Experimental rig	51.3	1.77	0.87 (estimated)	775	.305×.305	0.0930	8.3	1
	51.3	2.44	0.92	1130	.305×.305	0.0930	12.1	1
	65.3	2.44	0.92	1438	.305×.305	0.0930	15.5	1
	102.5	1.22	0.82	1006	.305×.305	0.0930	10.8	1
	102.5	2.44	0.92	2257	.305×.610	0.186	12.1	1
	115.2	2.44	0.92	2537	.457×.457	0.209	12.1	1
	172.8	2.44	0.92	3805	.457×.457	0.209	18.2	1
	205.0	0.61	0.66	810	.305×.305	0.0930	8.7	1
Mobile machine	588	0.9	0.96	4984	.305×.305	0.0930	53.6	2
	588	2.0	0.88	10152	.305×.305	0.0930	109	1
	588	2.2	0.88	11167	.305×.305	0.0930	120	1

* Efficiency=Actual acceleration/Gravitational acceleration
†Energy per blow=Mass×Height of drop×Efficiency×Gravitational acceleration
‡Specific energy=Energy per blow/Area of base of rammer

Plate 12.2 Rear view of the multi dropping-weight compactor showing arrangement of weights

addition to the horizontal section. The 36 kW diesel engine mounted on the machine drove the transverse cam shaft to operate the compacting mechanism. The power unit also drove a levelling rotor carrying 26 blades, which fed soil under an adjustable screed board before compaction by the dropping weights. Where the effect of using more than one pass of the machine was studied the blades of the levelling rotor had to be removed. The towing tractor was equipped with an auxiliary 'creeper' gearbox which allowed the speed of travel of the unit to be controlled to the recommended 1.8 m/min

Plate 12.3 Experimental dropping-weight compactor investigated in 1954. The rammer has been raised to an indicated height of about 5 ft (1.5 m) and the operator is about to trip the quick-release mechanism

Plate 12.4 Close-up of the rammer of the experimental dropping-weight compactor showing alternative bolt-on base-plates and additional ballast weights

(6 ft/min). The speed of the engine on the compactor was such that each dropping weight gave 54 blows per minute, and this resulted in each area of soil receiving 10 blows on a single pass of the machine (Table 12.1).

12.4 The experimental dropping-weight compactor (also referred to as the experimental rig) is illustrated in Plates 12.3 and 12.4. The equipment comprised a special rammer working in conjunction with a pile-driving frame suspended from the jib of an excavator. The rammer consisted of a light-alloy cage, the mass of which could be adjusted between about 50 kg and 210 kg by bolting steel plates inside the cage. The area and shape of the rammer base could also be altered from a minimum size of about 0.09 m^2 to a maximum size of about 0.2 m^2 by bolting special steel plates to the base of the rammer (Plate 12.4). The rammer could be lifted to any height up to about 2.7 m above the ground and, with the aid of a quick-release mechanism, could be allowed to drop freely on to the surface of the soil. A scale attached to the side of the pile-driving frame assisted the operator of the release mechanism to maintain a constant height of drop.

Plate 12.5 The mobile dropping-weight compactor investigated in 1968

12.5 The third of the dropping-weight compactors investigated was tested in the most recent series of investigations. It is illustrated in Plate 12.5 and is referred to in this Chapter as the mobile dropping-weight compactor or the mobile machine. It consisted of a four-wheeled chassis, with the dropping-weight assembly mounted at the front and the engine at the rear. The 36 kw diesel engine drove a pump supplying oil to the hydraulic mechanisms for lifting the dropping weight, for traversing the dropping-weight assembly and tilting it about the longitudinal and transverse axes of the machine, and for the 'creeper' drive used to move the machine slowly when working. For normal travel between sites the engine drove the front wheels through a conventional clutch and gearbox. The rammer rose and fell in a welded frame guide and was raised to a pre-selected height for each blow by a cable running over a pulley at the top of the guide, and thence through a series of pulleys to a hydraulic piston. The raising and dropping of the rammer could be cycled automatically leaving the operator free to control the point of application of the blow by use of the traversing mechanism or the 'creeper' drive to the wheels. The normal method of operation of the machine during the investigations was to use the automatic ramming cycle and keep the traversing mechanism operating continuously. The frequency of rammer blows increased as the height of drop was reduced and this resulted in twice the number of blows per pass at half the height of drop (Table 12.1). The facility for tilting the guide frame, used in practice to compact areas with access difficulties, was not utilised in the Laboratory's investigations.

12.6 The descriptions of the machines given above and the results presented in this Chapter have been taken from Lewis and Parsons (1958), Lewis (1957) and Parsons and Toombs (1968).

Efficiency of the dropping-weight compactors

12.7 In all three machines free-fall of the dropping weight or weights was inhibited by numerous sources of friction. In the multi-weight machine the pivots in the lever system, and with the experimental rig the pulleys running down the pile-driving frame, were potential sources of friction. With the mobile machine friction in the rammer guides and pulleys and the drag of the hydraulic piston used for lifting the rammer would have contributed to loss of efficiency of the dropping weight.

12.8 With the machines tested in 1953 and 1954 the efficiency of the free fall conditions was determined by filming the falling weight using a high-speed ciné camera, with a vertical scale and a rotating disc also in the frame to provide distance and time scales. Only the last 760 mm of travel of the rammer on the experimental rig could be photographed in this way. The terminal velocities and overall accelerations were calculated from a frame-by-frame analysis of the film.

12.9 In the more recent tests with the mobile dropping-weight compactor the method of measuring the efficiency of the falling weight was to attach a flexible steel strip to the rammer with a loaded paint brush attached to its free end. The strip was set oscillating and on release of the rammer a curve was traced on a strip of paper held vertically in front of the falling rammer. The frequency of the oscillating steel strip was easily determined to provide a time base and an analysis of the traced curve allowed a linear relation between velocity and time to be plotted; the slope of this line was a measure of the acceleration of the dropping weight.

12.10 The results of the determinations of acceleration of the dropping weights, expressed in terms of efficiency, are given in Table 12.1. The values of energy per blow and specific energy, also given in the Table, all take account of the efficiency as measured by the methods described above.

Relations between state of compaction produced and moisture content

12.11 Relations between the state of compaction produced by dropping-weight compactors and the moisture content of the soil are given in Figures 12.1 to 12.4. Dry density has again been expressed in terms of relative

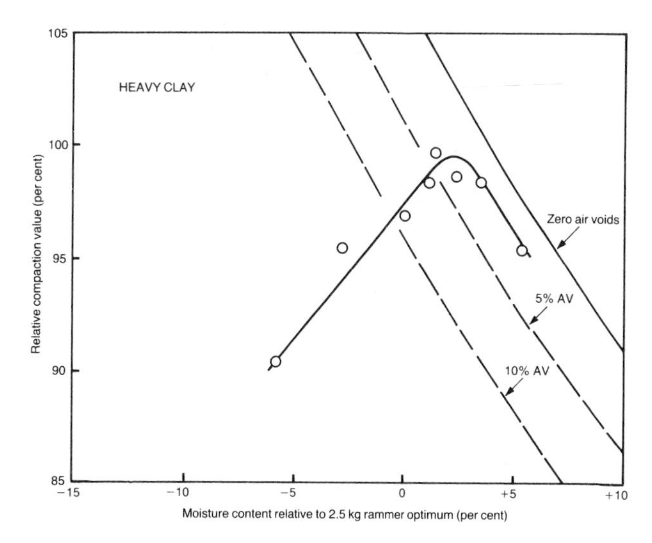

Figure 12.1 Relation between dry density and moisture content for 1 pass of the multi-dropping-weight compactor on heavy clay, expressed in terms of relative compaction with the 2.5 kg rammer test. Layer thickness = 150 mm

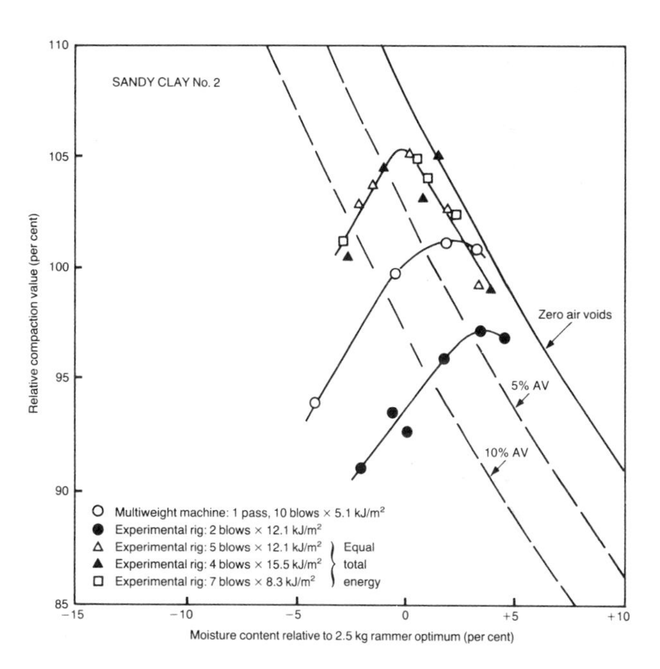

Figure 12.2 Relations between dry density and moisture content for dropping-weight compactors on sandy clay No 2, expressed in terms of relative compaction with the 2.5 kg rammer test. Layer thickness = 150 mm

compaction value (as a percentage of the maximum dry density obtained in the 2.5 kg rammer compaction test) and the moisture content as the arithmetical difference from the optimum moisture content of the 2.5 kg rammer test; the results obtained in the 2.5 kg rammer compaction test at the relevant period are shown in Figure 3.13.

12.12 The multi dropping-weight compactor was tested on all four of the soils used, with one pass

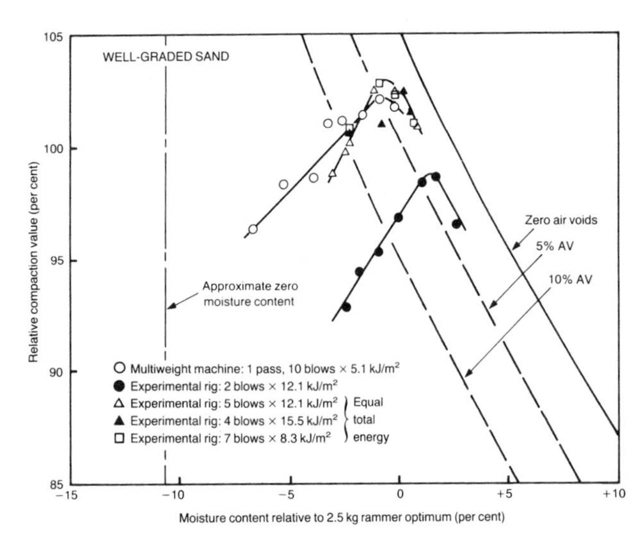

Figure 12.3 Relations between dry density and moisture content for dropping-weight compactors on well-graded sand, expressed in terms of relative compaction with the 2.5 kg rammer test. Layer thickness = 150 mm

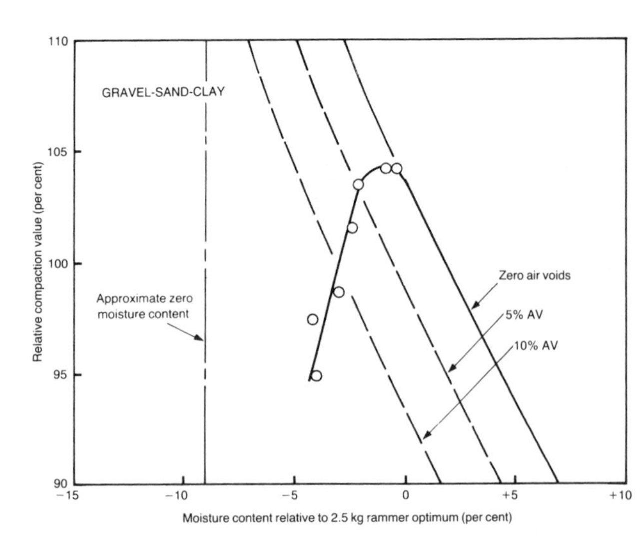

Figure 12.4 Relations between dry density and moisture content for 1 pass of the multi-dropping-weight compactor on gravel-sand-clay, expressed in terms of relative compaction with the 2.5 kg rammer test. Layer thickness = 150 mm

of the machine (10 blows) on 150 mm compacted layers (Figures 12.1 to 12.4). The experimental rig was tested on the sandy clay No 2 (Figure 12.2) and well-graded sand (Figure 12.3) only, using two levels of energy on 150 mm layers. All the results for the experimental rig were obtained using the 305 mm square rammer base and, as Table 12.1 shows, the different levels of specific energy given in the key to each Figure were achieved by using different combinations of height of drop and mass of the rammer. The relations produced by the three combinations of numbers of blows and specific energies, which were designed to produce approximately the same total energy per unit area of base (ie number of blows×specific energy = constant) were identical and only one curve has been drawn through the points. It should be noted that the results given are for a comparatively small number of blows (compare Figures 11.2 to 11.6 for power rammers) and the maximum achievable levels of compaction would not have been reached.

12.13 The results given in Figures 12.1 to 12.4 have been analysed and found to be directly related to the total energy per unit area of base (specific energy×number of blows). These findings have been summarised in Figure 12.5, which shows relations between the maximum relative compaction value and total energy per unit area of base and between optimum moisture content and the same energy factor. The maximum relative compaction value increased with increase in total energy per unit area of base, whilst the optimum moisture content decreased. For the range of total energy per unit area of base used in the investigations, maximum relative compaction values of 97 per cent to 105 per cent were obtained, with

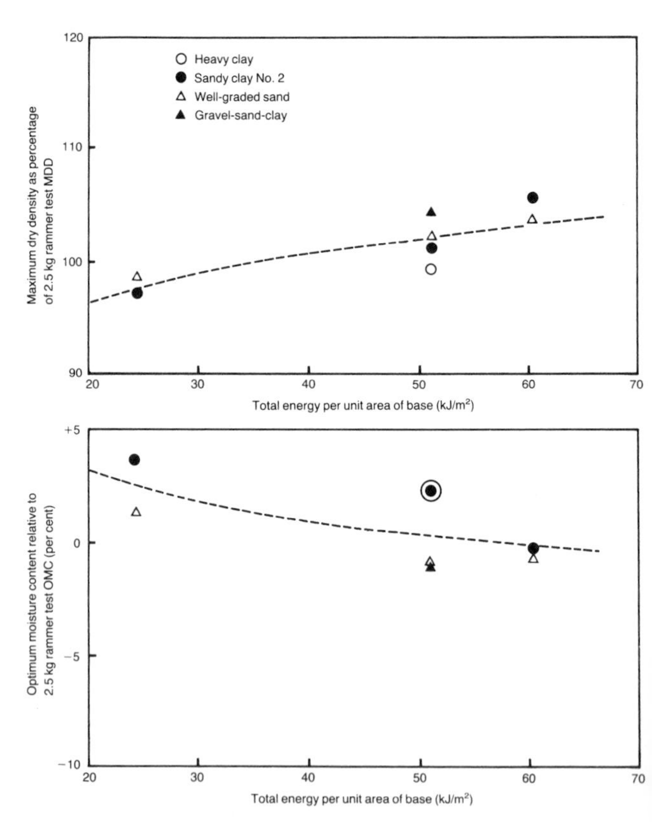

Figure 12.5 Maximum dry densities and optimum moisture contents related to the total energy per unit area of base (specific energy x number of blows) of dropping-weight compactors. Results from Figures 12.1 to 12.4 for 150 mm compacted layers and rammer bases approximately 300 mm square

optimum moisture contents from 4 per cent above to 1 per cent below the 2.5 kg rammer optimum moisture content.

12.14 A further interesting feature is that soil type appears to have only a small effect on the relations in Figure 12.5 and this may be a reflection of the similarity between the mode of compaction for the full-scale machines and the laboratory test with which they have been compared in the relative compaction approach. A simple calculation reveals that, in the 2.5 kg rammer compaction test (British Standards Institution, 1990), the total energy per unit area of the rammer for each of the three layers compacted in the mould is about 23 kJ/m^2. At this level of total energy per unit area of base, Figure 12.5 indicates that the full-scale dropping-weight compactors produced maximum relative compaction values of about 97 per cent and optimum moisture contents about 3 per cent above the 2.5 kg rammer optimum. The small differences of these values from 100 per cent and zero respectively could easily be the result of the beneficial confining effect of the mould walls in the laboratory test.

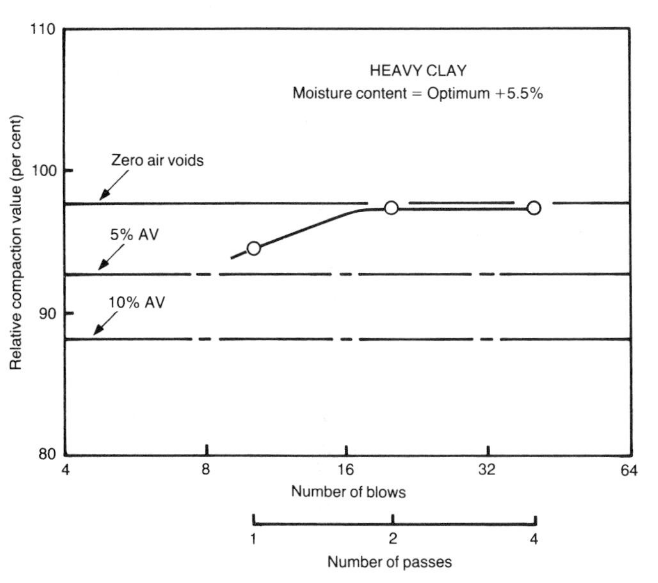

Figure 12.6 Relation between relative compaction and number of blows or passes of the multi-dropping-weight compactor on 150 mm layers of heavy clay. Values are related to the maximum dry density and optimum moisture content obtained in the 2.5 kg rammer compaction test (Figure 3.13)

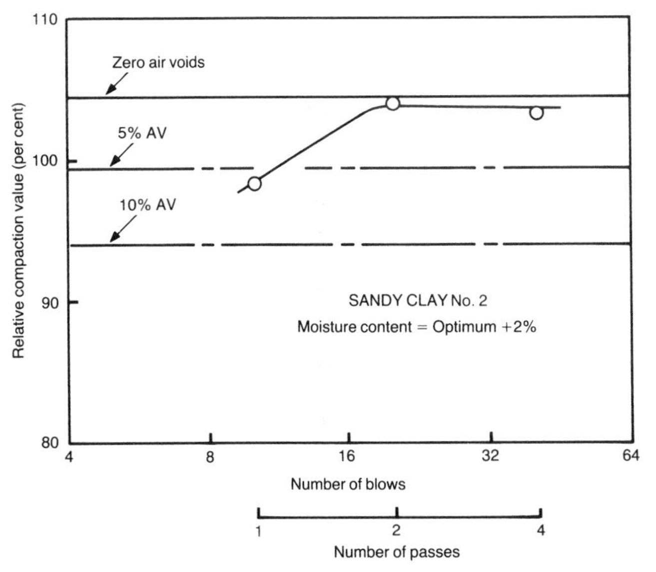

Figure 12.7 Relation between relative compaction and number of blows or passes of the multi-dropping-weight compactor on 150 mm layers of sandy clay No 2. Values are related to the maximum dry density and optimum moisture content obtained in the 2.5 kg rammer compaction test (Figure 3.13)

Relations between state of compaction produced and number of blows

12.15 Investigations specifically of the effect of numbers of blows of dropping-weight compactors on the state of compaction produced were only carried out with the multi-weight machine and, even then, only on the heavy clay and sandy clay No 2. The results are given in Figures 12.6 and 12.7. These indicate that after 1 pass (10 blows) a further increase in relative compaction value was achieved with a second pass, although at the moisture contents of the tests the increase was fairly small. As mentioned in Paragraph 12.3, the machine was designed as a single-pass compactor and the results in Figures 12.1 to 12.4 give a better indication of the likely performance for a range of moisture content conditions. However, the facility for varying the forward speed of travel would have allowed an increase (or decrease) in the number of blows delivered to the soil if that had been considered appropriate.

12.16 Regression analyses have been made of the results given in Figures 12.1, 12.2, 12.6 and 12.7 to determine relations, given in Figure 12.8, between the total energy required per unit area of base and moisture content of the clay soils for the achievement of both 95 per cent relative compaction and 10 per cent air voids in 150 mm compacted layers. The regression equations and correlation coefficients (r) are given in the Figure. To determine the number of blows required to achieve either of these criteria with other dropping-weight compactors on the same thickness of layer, the total energy determined from Figure 12.8 should be divided by the specific energy of the dropping weight. Results of further investigations described in Paragraphs 12.30 to 12.32 indicate that, within practical limits, the size and shape of the rammer base should not affect this determination for a layer thickness of 150 mm.

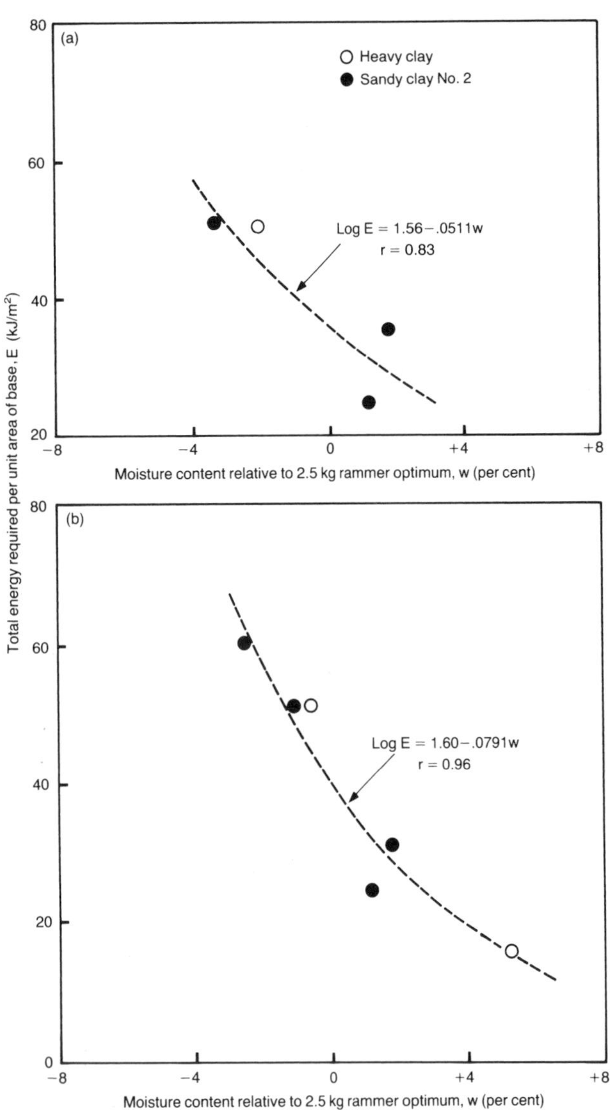

Figure 12.8 Relations between total energy required per unit area of base of dropping-weight compactors and moisture content of clay soils to achieve (a) 95 per cent of the maximum dry density obtained in the 2.5 kg rammer compaction test and (b) 10 per cent air voids. Layer thickness = 150 mm

Relations between state of compaction and depth within the compacted layer

12.17 Results of investigations of the variation of state of compaction with depth in layers of soil compacted by dropping-weight compactors are given in Figures 12.9 to 12.13. The mobile machine was tested on five soils and relations between relative compaction value and depth below the surface of the compacted layer are given for various numbers of blows; the level of specific energy was also varied in some instances. In addition, the multi dropping-weight compactor was tested on heavy clay and sandy clay No 2, using 1 pass of the machine, resulting in each area of soil receiving 10 blows of the dropping weights.

12.18 A variation from the normal procedure was used in the tests with the mobile dropping-weight compactor. With this machine a trench about 1.2 m wide and at least 600 mm deep was excavated in each of the test soils and the excavated soil was placed alongside the trench and prepared for test at the required moisture content by mixing with a rotary cultivator. (This was a similar process to that used for the tests, described in Chapter 10, using a 59 kg vibro-tamper; an excavated trench for those tests is shown in Plate 10.5.) The whole of the prepared soil was replaced in the trench in a single loose layer and each one-third of the length of the trench was compacted by two, four and eight passes respectively. Compaction was carried out using the automatic ramming cycles (see Paragraph 12.5) with the machine straddling the trench. Employing the traversing mechanism and the forward/reverse 'creeper' drive, one pass over a section was completed before the subsequent pass was commenced. The number of blows delivered to each area of soil was found to be one per pass for the larger heights of drop and two blows per pass were given for the 0.9 m height of drop (Table 12.1).

12.19 The results obtained on the heavy clay and sandy clay (Figures 12.9 and 12.10) were of the usual type with the state of compaction generally decreasing as the depth in the layer increased. With the well-graded sand, gravel-sand-clay and uniformly graded fine sand (Figures 12.11 to 12.13), however, the highest states of compaction occurred in the middle or lower parts of the layer, indicating that a considerable amount of overstressing occurred in the upper parts of the compacted layer under the action of the 588 kg rammer with its 305 mm square contact area.

12.20 In analysing the results to determine the factors affecting the depth of layer capable of being compacted, one important factor which could not be included was the area of the base of the rammer, which was constant throughout these particular tests. All findings apply, therefore, only to rammers with bases approximately 300 mm square. Independent

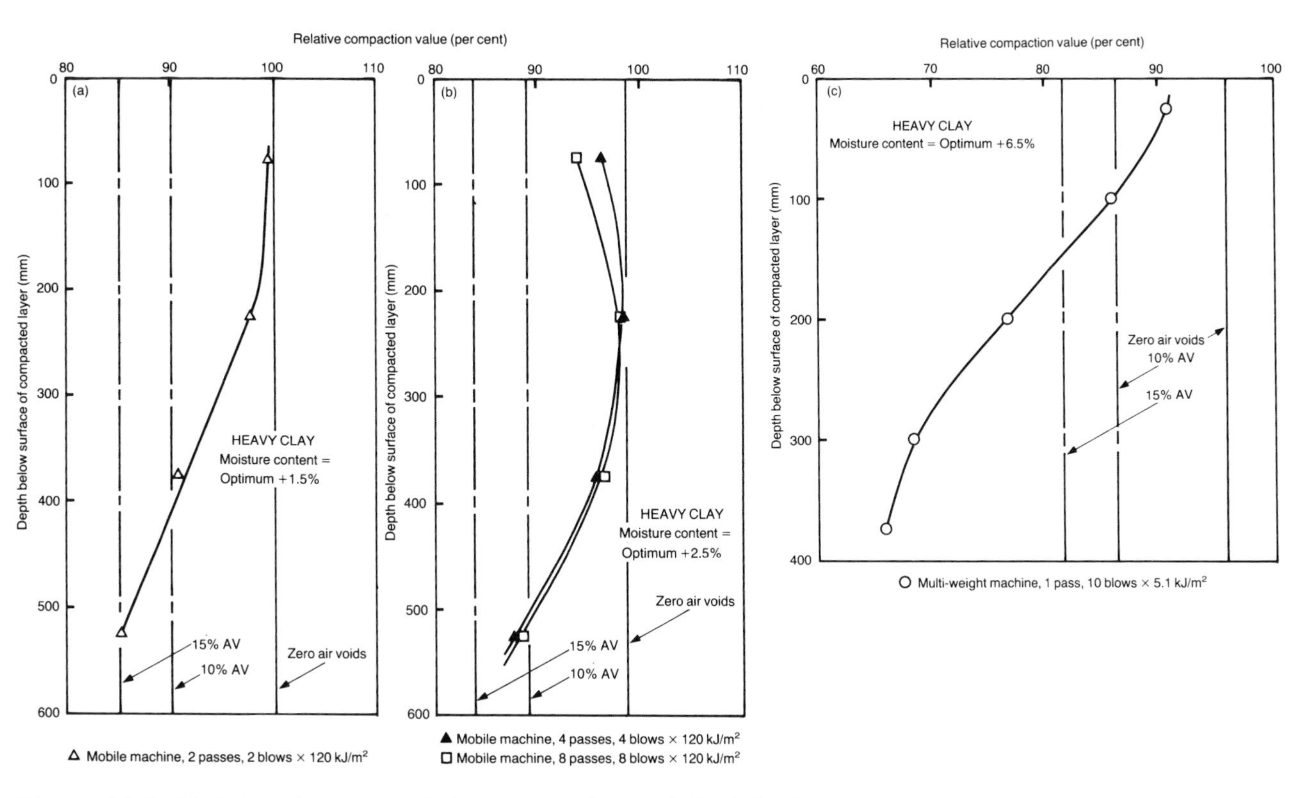

Figure 12.9 Relations between relative compaction and depth in the compacted layer of heavy clay obtained with dropping-weight compactors. All values are related to the maximum dry density and optimum moisture content obtained in the 2.5 kg rammer compaction test (Figure 3.13)

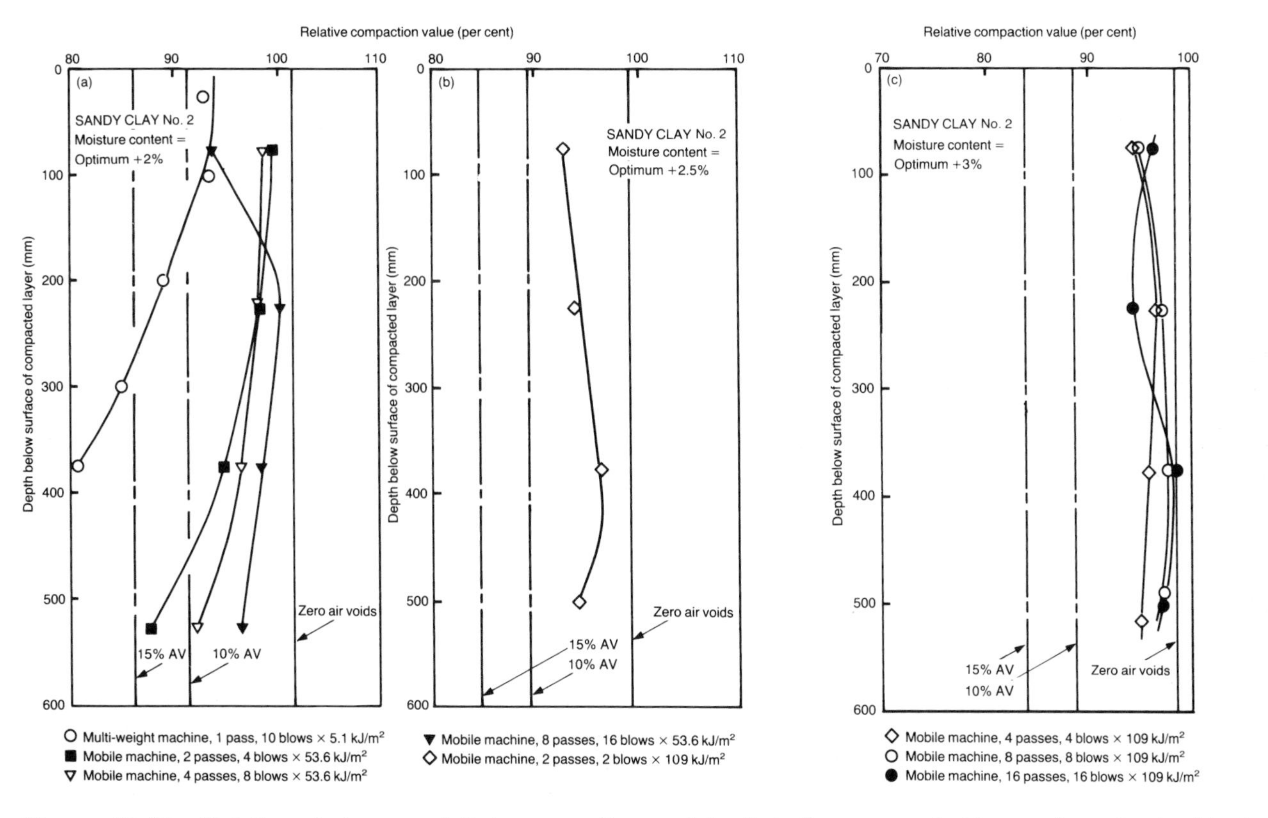

Figure 12.10 Relations between relative compaction and depth in the compacted layer of sandy clay No 2 obtained with dropping-weight compactors. All values are related to the maximum dry density and optimum moisture content obtained in the 2.5 kg rammer compaction test (Figure 3.13)

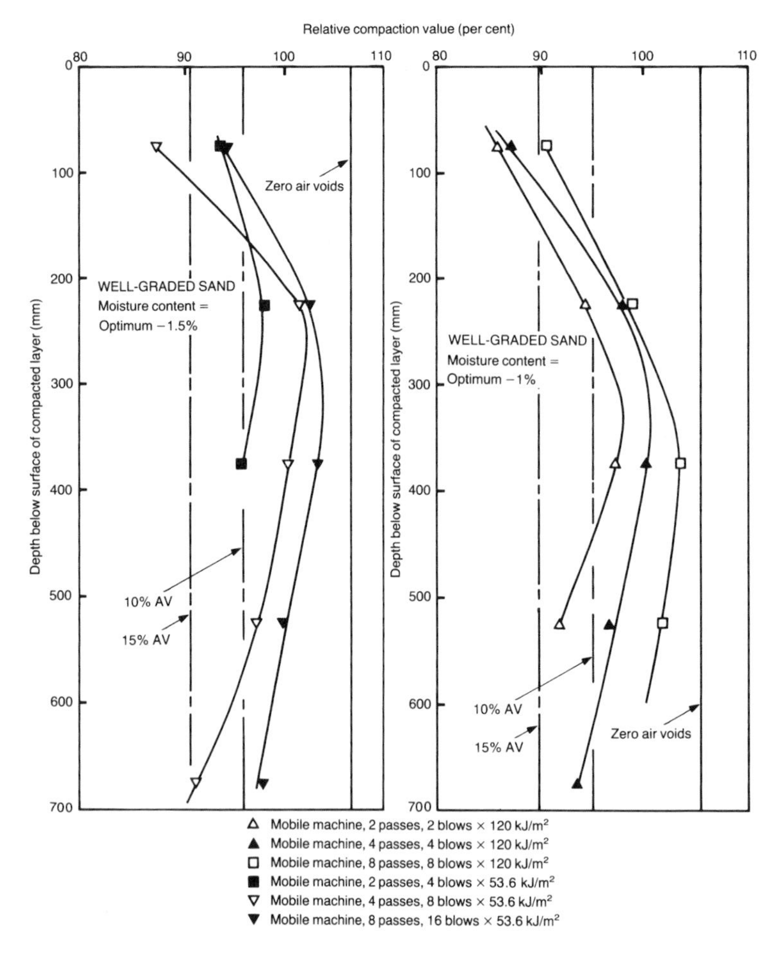

Figure 12.11 Relations between relative compaction and depth in the compacted layer of well-graded sand obtained with a dropping-weight compactor. All values are related to the maximum dry density and optimum moisture content obtained in the 2.5 kg rammer compaction test (Figure 3.13)

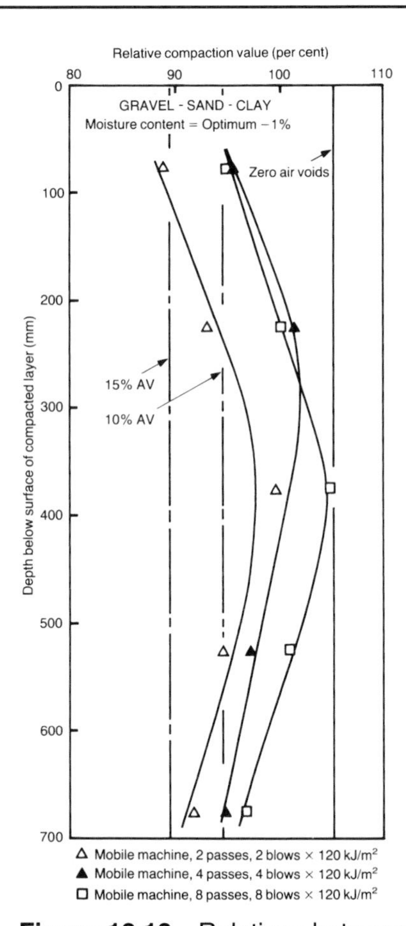

Figure 12.12 Relations between relative compaction and depth in the compacted layer of gravel-sand-clay obtained with a dropping-weight compactor. All values are related to the maximum dry density and optimum moisture content obtained in the 2.5 kg rammer compaction test (Figure 3.13)

variables used in the analyses were the specific energy, number of blows and moisture content (relative to the 2.5 kg rammer optimum). Regression analyses were made with the compacted thickness of layer as the dependent variable. This was determined from Figures 12.9 to 12.13 as the thickness capable of compaction to achieve either a dry density equal to 90 per cent of the maximum dry density obtained in the 2.5 kg rammer compaction test (90 per cent relative compaction) or 15 per cent air voids at the bottom of the layer. Where overstressing of the surface occurred, as with the three granular soils, the thickness of material at the surface with a relative compaction value less than 95 per cent or an air content more than 10 per cent was arbitrarily deducted for the relative compaction and air voids criteria respectively.

12.21 With heavy clay and sandy clay No 2 very little variation occurred in the relations between relative compaction and depth in the layer for the various modes of use of the mobile dropping-weight compactor. In all instances with that machine the thickness of layer capable of being compacted to the set criteria was 500 to 600 mm. The multi dropping-weight compactor appears to have been capable of compacting layers of 150 mm to 200 mm to the same criteria. These depths apply only to the moisture contents used in the tests. No significant correlations between depth of layer and any of the other variables were obtained.

12.22 The results for well-graded sand and gravel-sand-clay, obtained with the mobile machine only, have been included together in Figure 12.14. It was found in the regression analyses for these well-graded granular soils that the specific energy was not a significant factor, but that the number of blows and moisture content were both highly significant. It is probable that with the granular soils any gain in the depth at which 90 per cent relative

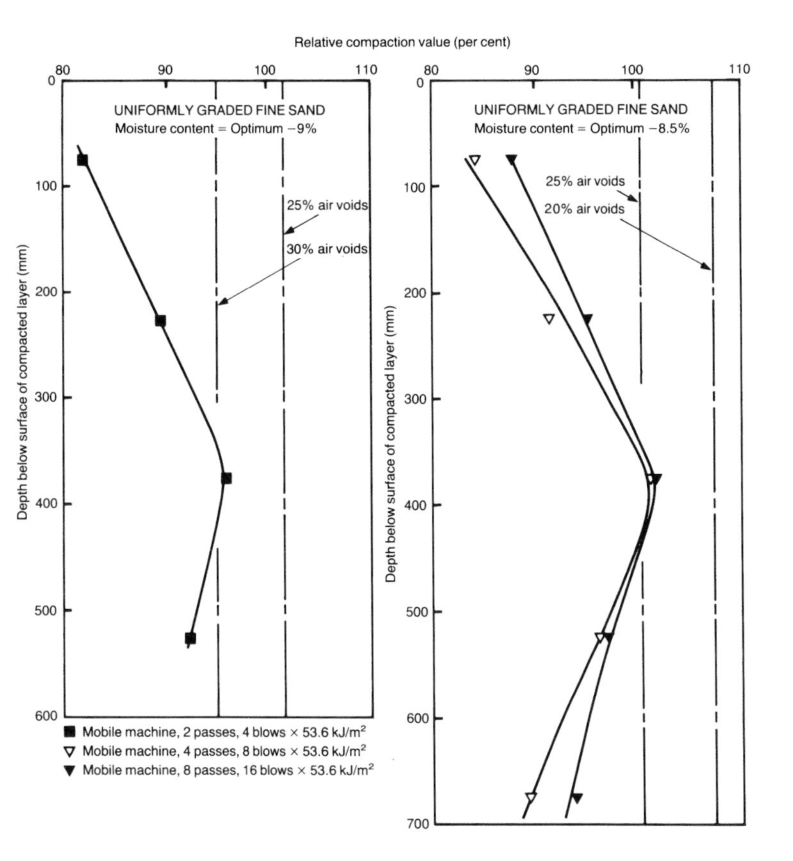

Figure 12.13 Relations between relative compaction and depth in the compacted layer of uniformly graded fine sand obtained with a dropping-weight compactor. All values are related to the maximum dry density and optimum moisture content obtained in the 2.5 kg rammer compaction test (Figure 3.13)

compaction or 15 per cent air voids were obtained, achieved by increasing specific energy, would have been lost by an increase in the depth of overstressed soil at the surface. As well as indicating the individual results obtained with the compactor, Figure 12.14 also shows the results of the regression analyses for four values of moisture content. The regression equations and correlation coefficients are given in Table 12.2. The relations show that the depth of the granular soils capable of being compacted was very sensitive to the moisture content.

12.23 With the uniformly graded fine sand only the number of blows was varied and the results and the regression line are shown in Figure 12.15; the regression equation and correlation coefficient have been included in Table 12.2.

Figure 12.14 Relations between depth of layer and number of blows of the mobile dropping-weight compactor on granular soils to achieve (a) 90 per cent relative compaction and (b) 15 per cent air voids at the bottom of the layer. The depths plotted have been reduced by the depth of overstressed soil at the surface of the layer (see Figures 12.11 and 12.12)

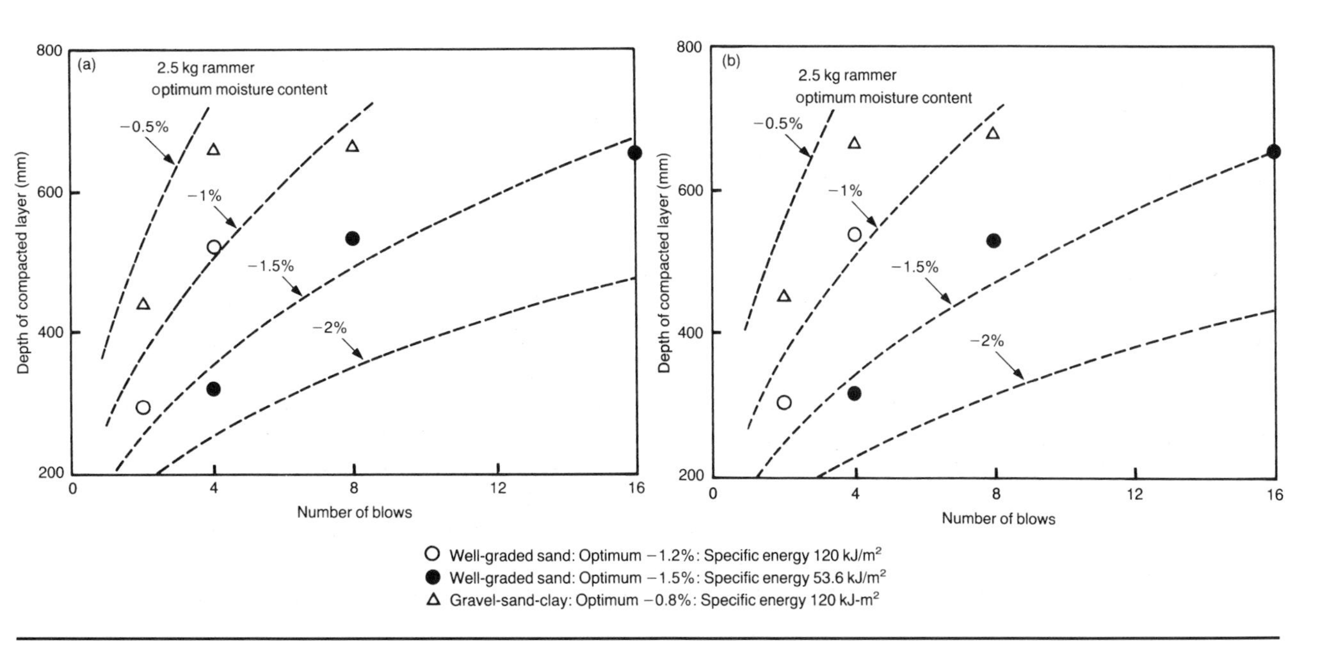

Again, a considerable depth of overstressed soil was produced at the surface of the layer (see Figure 12.13) but this was reduced at the higher numbers of blows of the compactor. The total depth of soil conforming to the criteria of adequate compaction ranged from about 200 mm after 4 blows to about 500 mm after 16 blows, using the low specific energy associated with a height of drop of 0.9 m.

Table 12.2 Equations and correlation coefficients to determine the thickness of layer of granular soil capable of being compacted by dropping-weight compactors

Soil	Level*	Regression equation†	No of data points	Correlation coefficient
Well-graded sand & gravel-sand-clay	(a)	D=2.73+.464 B +.305 w‡	8	0.95
	(b)	D=2.78+.464 B +.349 w‡	8	0.94
Uniformly graded fine sand	(a)	D=1.98+.613 B#	3	0.95

*Level (a)=Achievement at the bottom of the layer of 90 per cent of the maximum dry density obtained in the 2.5 kg rammer compaction test, and at the top of the layer of at least 95 per cent of the same maximum dry density

(b)=Achievement at the bottom of the layer of 15 per cent air voids, and at the top of the layer of no more than 10 per cent air voids

†D=Log_{10} (Maximum depth of layer to achieve stated level of compaction, mm)

B=Log_{10} (Number of blows applied)

w=Moisture content of soil, per cent, minus the optimum moisture content, per cent, obtained in the 2.5 kg rammer compaction test

‡Specific energy in the range from 50 to 120 kJ/m^2; base of rammer 305 mm square

#Specific energy 54 kJ/m^2; base of rammer 305 mm square; moisture content in the range of 9.5 to 8.5 per cent below the optimum obtained in the 2.5 kg rammer compaction test

12.24 The relations between state of compaction and depth of layer were determined under a very limited range of conditions, particularly with regards to moisture content. The relations given in Figures 12.9 to 12.15 do not allow the prediction of the performance of dropping-weight compactors outside the range of moisture contents used for each particular soil type.

The effect of the levelling rotor on the state of compaction produced by the multi dropping-weight compactor

12.25 As mentioned in Paragraph 12.3, the multi dropping-weight compactor was equipped with a

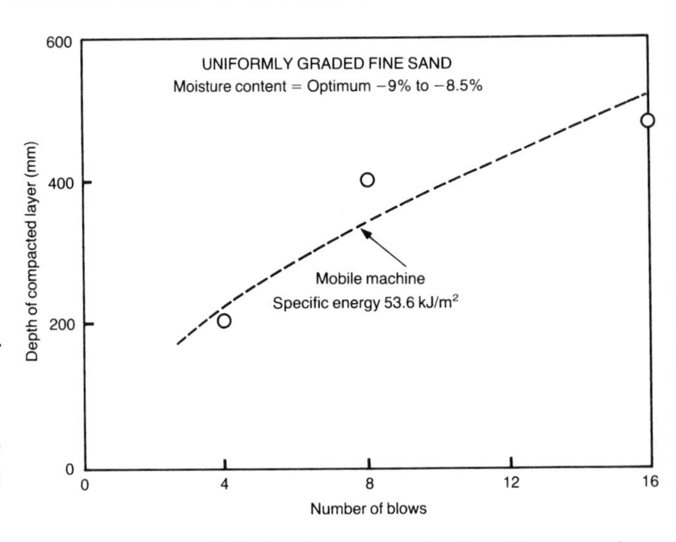

Figure 12.15 Relation between depth of layer and number of blows using the mobile dropping-weight compactor (specific energy 53.6 kJ/m^2) on uniformly graded fine sand to achieve 90 per cent relative compaction at the bottom of the layer. The depths plotted have been reduced by the depth of overstressed soil at the surface of the layer (see Figure 12.13)

levelling rotor with 26 blades, which fed soil under an adjustable screed board to be compacted by the dropping weights. As also mentioned, the blades of the levelling rotor had to be removed when studies were made of the effects of more than one pass of the machine. To provide information on the effect of the rotor blades on the state of compaction produced in one pass of the machine, tests were carried out on the heavy clay and sandy clay No 2 with the compactor fitted with the rotor blades and with the blades removed. Measurements of dry density, using the sand-replacement method, were made in the top 50 mm and the top 150 mm of the compacted layer in each case; the dry density of the soil in the layer between 50 and 150 mm was calculated from the results. The relative compaction values obtained are given in Table 12.3 and show that the rotor blades produced a very slight increase in the state of compaction of the top 50 mm, but below this depth the compaction did not appear to be affected to any significant extent by the presence of the blades.

12.26 The conclusion reached at the time of the investigation was that the effect of the blades was too small to justify the large amount of power required to drive the rotor; only about 4 kW was required to operate the dropping weights and the remainder of the 36 kW output of the power unit was required to drive the rotor. It was suggested (Lewis and Parsons, 1958) that consideration should be given to the provision of

Table 12.3 Results of measurements of states of compaction at various depths in clay soils compacted by the multi dropping-weight compactor fitted with rotor blades and with rotor blades removed

Depth in layer (mm)	Relative compaction value (per cent)			
	Heavy clay Moisture content=Optimum+5%		Sandy clay No 2 Moisture content=Optimum+1.5%	
	Blades fitted	Blades removed	Blades fitted	Blades removed
0–50	98	95	100	99
0–150	96	94	99	98
50–150 (deduced)	95	94	99	98

a much simpler type of levelling device. It was admitted, however, that the rotor might have been of value in providing some additional mixing of the stabilised soil.

Theoretical analysis of the factors influencing the performance of dropping-weight compactors

12.27 To give an indication of the important factors to be considered in the design of impact compactors in general, and dropping-weight compactors in particular, Lewis (1957) produced a simplified theoretical analysis of the impact pressures produced on the surface of soil by a rammer. The experimental dropping-weight compactor shown in Plates 12.3 and 12.4 was used to verify the theoretical analysis.

12.28 From the well known equations of motion:–

$$V^2 = 2\,f\,x \quad \ldots\ldots (1)$$

$$\text{and } p\,A = M\,f \quad \ldots\ldots (2)$$

where V =velocity of rammer on impact
f =deceleration of rammer on striking soil
x =deformation of soil during impact
p =pressure generated on surface of soil by the impact
A =area of rammer base
and M =mass of rammer

$$\text{Hence:– } p = \sqrt{\left(\frac{M\,k_s\,V^2}{2\,A}\right)} \quad \ldots\ldots (3)$$

where $k_s = \frac{p}{x}$
=dynamic modulus of deformation of the soil

In the case of a rammer falling freely from a height h:–

$$p = \sqrt{\left(\frac{M\,h\,g\,k_s}{A}\right)} \quad \ldots\ldots (4)$$

If the acceleration of the falling weight is less than g as a result of frictional losses:–

$$p = \sqrt{\left(\frac{M\,h\,g'\,k_s}{A}\right)} \quad \ldots\ldots (5)$$

where g'=actual acceleration of the falling rammer.

$$p = \sqrt{(E_s\,k_s)} \quad \ldots\ldots (6)$$

where E_s=specific energy

12.29 These relations indicate that the impact pressure is a function of the energy per unit area of the rammer base (specific energy) and the deformation properties of the soil under dynamic conditions of loading. The latter factor is also likely to be a function to some extent of the area and shape of the rammer base, but little information was available on that aspect at the time that the analysis was made. If it is assumed that the dynamic modulus of deformation behaves similarly to the static modulus of deformation in that the modulus is often found to be inversely proportional to the square root of the loaded area, then:–

$$k_s = \frac{C}{\sqrt{A}} \quad \ldots\ldots (7)$$

where C is a constant

The expression for the impact pressure developed can then be written:–

$$p = \sqrt{\left(\frac{M\,V^2\,C}{A\,2\sqrt{A}}\right)} \quad \ldots\ldots (8)$$

where C = constant for the particular soil conditions.

Thus, if the rammer area is changed, the compaction energy provided by each blow per unit area of rammer base (specific energy) ($\frac{1}{2} M \frac{V^2}{A}$) would have to be kept proportional to the square root of the area of the rammer base ($\sqrt{A}$) for a constant pressure to be developed.

Tests of the validity of the theoretical analysis using the experimental dropping-weight compactor

12.30 The effects of variations in height of drop and mass of the rammer were explored using the 305 mm square rammer base. With theoretically constant specific energy, results for two and five blows on 150 mm layers of sandy clay No 2 and well-graded sand are given in Table 12.4. The results supported the conclusions reached in the theoretical analysis that the state of compaction produced by impact compaction is a function of the kinetic energy per unit area of the rammer for a given size of rammer base; the losses of efficiency at the lower heights of drop (see Table 12.1) were not, however, reflected in the results. The small variations that were obtained in the relative compaction values for a constant amount of compactive energy were explained at the time as being either within limits of experimental error or accounted for by differences in the moisture content of the soil.

12.31 A further study was made into the effect of size of rammer base, with tests involving a constant compactive energy per unit area of the rammer base (theoretical values of 26.4 and 66 kJ/m^2) and also with the compactive energy proportional to the square root of the area of the rammer base (26.4 and 66 kJ/m^2 for the 305 mm square base and 39.6 and 99 kJ/m^2 for the 457 mm square base). The latter tests were carried out to maintain constant impact pressures, assuming that the dynamic modulus of deformation of the soil is inversely proportional to the square root of the loaded area (see Paragraph 12.29). The results of this further study are given in Table 12.5. A comparison of the results for the 305 mm and 457 mm square rammer bases shows that constant states of compaction in the top 150 mm of compacted soil were obtained only when the specific energy (compactive energy per unit area of the rammer base) was kept constant. When the compactive energy provided by the 457 mm square base was increased in proportion to the square root of the plate area slightly higher states of compaction were obtained. Lewis (1957) suggested that the reason for this was the more uniform pressure distribution with depth under the larger plate, resulting in a smaller reduction of density with depth through the 150 mm thick layer, even though the pressure, and hence density, at the surface may have been equal.

12.32 It was concluded from these results that as far as the design of impact compactors for the compaction of 150 mm to 200 mm thick layers was concerned, the most important, and probably only, design factor to be considered is the specific energy (energy per unit area of base). The results of the statistical analyses of results for clay soils, given in Figure 12.8, can be used, therefore, for sizes and shapes of rammer base other than those tested, as mentioned in Paragraph 12.16. With thicker layers, however, the increase of energy per unit area in proportion to the square root of the area of the rammer, as indicated by the theory, would be valid. The results illustrated earlier in this Chapter, shown

Table 12.4 Effect on state of compaction produced by the experimental rig of varying height of drop and mass of rammer to maintain a near-constant total compactive energy per unit area of rammer base

1 Size of rammer base plate (mm)	2 Height of drop of rammer (m)	3 Theoretical impact velocity of rammer* (m/s)	4 Total mass of rammer (kg)	5 Number of blows	6 Total compactive energy per unit area of rammer: Theoretical (kJ/m^2)	7 Actual (kJ/m^2)	8 Sandy clay No 2: Moisture content relative to 2.5 kg rammer optimum (%)	9 Relative compaction value (%)	10 Well-graded sand: Moisture content relative to 2.5 kg rammer optimum (%)	11 Relative compaction value (%)
	2.44	6.9	51.3	2	26.4	24.2	+0	93	−2	95
305×305	1.22	4.9	102.5	2	26.4	21.6	+0.5	94	−1.5	96
	0.61	3.5	205.0	2	26.4	17.4	+1	98	−2	95
	2.44	6.9	51.3	5	66.0	60.5	+0	104	−2.5	100
305×305	1.22	4.9	102.5	5	66.0	54.0	+0	105	−2	101
	0.61	3.5	205.0	5	66.0	43.5	+0.5	107	−2	101

*See Table 12.1 for efficiency of the system; frictional losses caused reductions in impact velocity, resulting in the actual values of energy in Column 7.

Table 12.5 Effect on state of compaction produced by the experimental rig of varying the size and shape of the rammer base plate

1 Size of rammer base plate (mm)	2 Height of drop of rammer (m)	3 Total mass of rammer (kg)	4 Number of blows	5 Total compactive energy per unit area of rammer Theoretical (kJ/m²)	6 Actual (kJ/m²)	7 Sandy clay No 2 Moisture content relative to 2.5 kg rammer optimum (%)	8 Relative compaction value (%)	9 Well-graded sand Moisture content relative to 2.5 kg rammer optimum (%)	10 Relative compaction value (%)
305×305	2.44	51.3	2	26.4	24.2	+0	93	–2	95
457×457	2.44	115.2	2	26.4	24.2	+0	93	–2	95
305×610	2.44	102.5	2	26.4	24.2	+0	94	–2	94
457×457	2.44	172.8	2	39.6	36.4	+0.5	96	–2	97
305×305	2.44	51.3	5	66.0	60.5	+0	104	–2.5	100
457×457	2.44	115.2	5	66.0	60.5	–0.5	104	–1.5	101
305×610	2.44	102.5	5	66.0	60.5	+1	104	–2	100
457×457	2.44	172.8	5	99.0	91.0	+0	106	–2	103

in Figures 12.2, 12.3, 12.5 and 12.8, and those for power rammers in Figure 11.7 and Table 11.3, all show that the dominating factor in impact compaction of layers with a maximum thickness of 150 mm was the specific energy or total energy per unit area of rammer base.

Measurement of pressures produced in soil by the experimental dropping-weight compactor

12.33 Measurements were made of pressures developed in the sandy clay No 2 by the experimental rig using the 305 mm square rammer base. Height of drop and mass of rammer were varied in inverse proportion to maintain a theoretically constant compactive energy per blow. Piezoelectric gauges, similar in type to those described in Paragraph 11.23, were installed at depths of 150 mm and 300 mm in the soil. The soil was at a moisture content approximately equal to the optimum of the 2.5 kg rammer compaction test and had been compacted to a relative compaction value of about 97 per cent.

12.34 The recorded impact pressures, taken from Lewis (1957), are given in Table 12.6. Each value is a mean of at least five observations and Lewis included an empirical allowance of 20 per cent for gauge error in over-recording the actual pressure; such errors were considered to arise from the differences between the elastic properties of the gauge and the soil in which it was embedded. Values of impact pressure at the depths of the gauges were also calculated and are given in Table 12.6. For these values the surface impact pressures were first calculated using the values of specific energy from Table 12.1 and a measured static modulus of deformation of 163 MN/m²/m, and the reduced pressures at the depths of the gauges were then calculated using the Boussinesq distribution of stress with depth (Jurgenson, 1934).

Table 12.6 Measured and calculated impact pressures developed in sandy clay No 2 beneath the 305 mm square rammer of the experimental dropping-weight compactor

Height of drop of rammer (m)	Mass of rammer (kg)	Initial depth of pressure gauge (mm)	Recorded pressure (kN/m²)	Calculated pressure (kN/m²)
0.61	205.0	150	630	830
1.22	102.5	150	700	930
2.44	51.3	150	690	980
0.61	205.0	300	420	410
1.22	102.5	300	470	450
2.44	51.3	300	530	480

12.35 The recorded pressures show that, for a given rammer base, the use of constant compactive energy (specific energy) resulted in constant impact pressures, irrespective of the height of drop or impact velocity, thus confirming the conclusion reached in the theoretical analysis (see Paragraph 12.29). The calculated impact pressures also show good agreement with the measured values.

References

British Standards Institution (1990). British Standard methods of test for soils for civil engineering purposes: BS 1377 : Part 4 Compaction related tests. BSI, London.

Jurgenson, L (1934). The application of theories of elasticity and plasticity to foundation problems. J Boston Soc. Civ. Engrs, 21 (3), 206–41.

Lewis, W A (1957). A study of some of the factors likely to affect the performance of impact compactors on soil. *Proc. 4th Int. Conf. Soil Mech. and Foundation Engrg.* Butterworths Scientific Publications, London, Vol 2, 145–150.

Lewis, W A and Parsons, A W (1958). An investigation of the performance of a prototype dropping weight soil compactor. *D.S.I.R., Road Research Laboratory, Research Note* No RN/3280/WAL.AWP (Unpublished).

Parsons, A W and Toombs, A F (1968). The performance of an Arrow D.500 dropping-weight compactor in the compaction of soil. *Ministry of Transport, RRL Laboratory Report* LR 229, Road Research Laboratory, Crowthorne.

Thompson, G H and Herbert, A (1978). Compaction of clay fills in-situ by dynamic consolidation. *Clay Fills,* Institution of Civil Engineers, London, 197–204.

CHAPTER 13 COMPACTION OF SOIL BY TRACK-LAYING TRACTORS

Introduction

13.1 Although tractors are not, strictly speaking, compaction equipment, they are often used to tow compaction plant and might be expected to contribute to the state of compaction produced. Track-laying tractors can also be used for light compaction of difficult areas such as side slopes of cuttings and embankments or where difficult soil conditions prevail such as in deposition areas for soils considered too wet for use in engineering works. In compaction specifications for major works track-laying tractors are not considered appropriate for use as compaction equipment because of their relatively low contact pressures.

Plate 13.1 6 t track-laying tractor investigated in 1950

13.2 The potential compaction performances of two tractors were studied in the early series of investigations at the Laboratory. These were the 6 t and 11 t machines illustrated in Plates 13.1 and 13.2. Details of relevance to their performance as compactors are given in Table 13.1.

Table 13.1 Details of the track-laying tractors used in investigations to determine their performance as compaction equipment

	6 t track-laying tractor	11 t track-laying tractor
Total mass (kg)	5820	10960
Width of each track (mm)	360	510
Length of track in contact with ground (m)	1.58	2.13
Area of contact (both tracks) (m^2)	1.13	2.17
Mean mass per unit area of track (kg/m^2)	5130	5060
Mean contact pressure (kN/m^2)	50.3	49.6

13.3 The results presented in this Chapter have been taken from Lewis (1954) and West (1951).

Relations between state of compaction produced and moisture content

13.4 Relations between the state of compaction produced by track-laying tractors and the moisture content of the soil are given in Figures 13.1 to 13.4. Relative compaction values are shown plotted against the difference between the actual moisture content of the soil and the optimum moisture content obtained in the 2.5 kg rammer compaction test. Relations for 32 passes of the 6 t tractor were obtained on four soils, ie heavy clay, silty clay, well-graded sand and gravel-sand-clay (Figures 13.1 to 13.4); those for 32 passes of the 11 t tractor were determined on heavy clay and gravel-sand-clay only (Figures 13.1 and 13.4). A compacted thickness of 150 mm was used throughout these series of tests.

13.5 Where both machines were tested, fairly similar results were obtained with the 11 t and 6 t track-laying tractors (Figures 13.1 and 13.4). This might be expected from the similarity of the

Plate 13.2 11 t track-laying tractor investigated in 1951

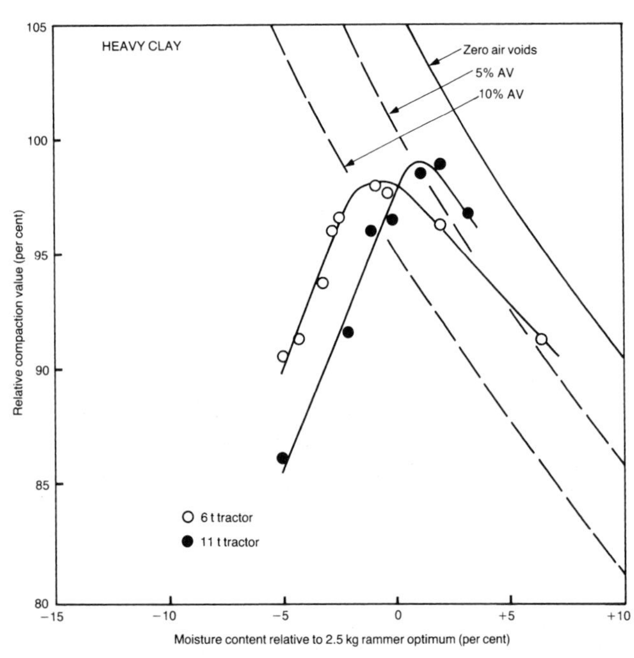

Figure 13.1 Relations between dry density and moisture content for 32 passes of track-laying tractors on 150 mm layers of heavy clay, expressed in terms of relative compaction with the 2.5 kg rammer test

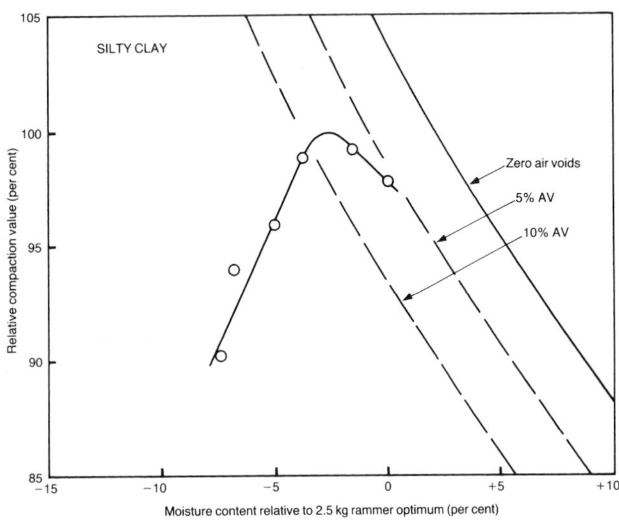

Figure 13.2 Relation between dry density and moisture content for 32 passes of the 6 t track-laying tractor on 150 mm layers of silty clay, expressed in terms of relative compaction with the 2.5 kg rammer test

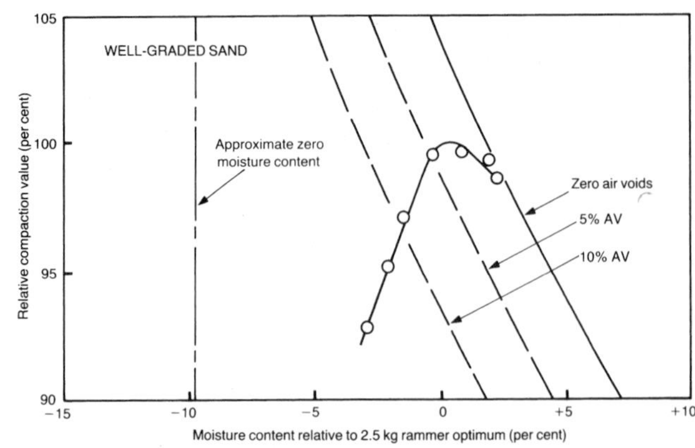

Figure 13.3 Relation between dry density and moisture content for 32 passes of the 6 t track-laying tractor on 150 mm layers of well-graded sand, expressed in terms of relative compaction with the 2.5 kg rammer test

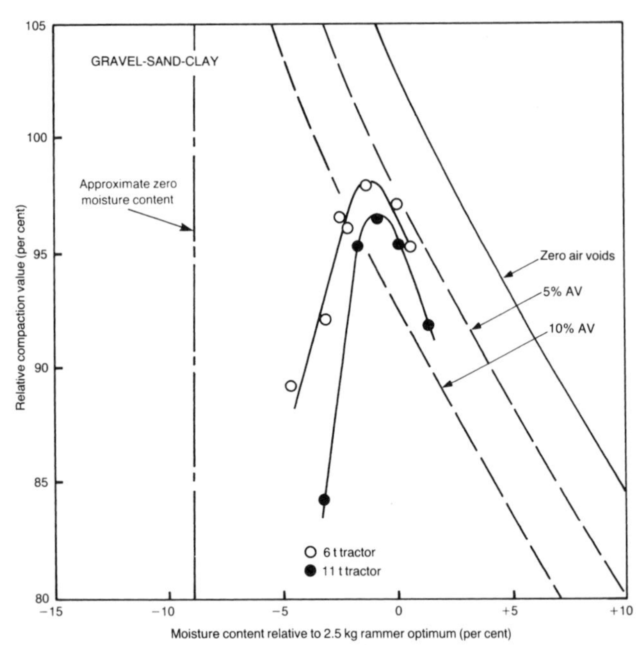

Figure 13.4 Relations between dry density and moisture content for 32 passes of track-laying tractors on 150 mm layers of gravel-sand-clay, expressed in terms of relative compaction with the 2.5 kg rammer test

mean contact pressures (approximately 50 kN/m^2) beneath the tracks of the two machines (Table 13.1). The maximum relative compaction values achieved varied from 97 per cent to 100 per cent, the highest values being produced on silty clay and well-graded sand. Optimum moisture contents for the track-laying tractors ranged from 2 per cent below the optimum of the 2.5 kg rammer compaction test to about 1 per cent above that optimum. At the time of the investigations Lewis (1954) remarked on the similarity of the relations obtained with the tractors to those produced by the same number of passes of the 12 t pneumatic-tyred roller with a 1.4 t wheel load (see Figures 5.1, 5.2, 5.4 and 5.5).

Relations between state of compaction produced and number of passes

13.6 Relations between relative compaction value in 150 mm layers and number of passes of the track-laying tractors are given in Figures 13.5 to 13.7. Both machines were tested on the heavy clay (Figure 13.5); additional tests were carried out with the 6 t tractor on well-graded sand

(Figure 13.6) and with the 11 t machine on gravel-sand-clay (Figure 13.7). Regression analyses have been made of the results given in Figures 13.1 to 13.7 to determine the relations shown in Figure 13.8. This latter Figure relates the number of passes required to achieve 95 per cent relative compaction and 10 per cent air voids to the moisture content (relative to the 2.5 kg rammer optimum). In view of the similarity of the contact pressures of the two machines their results were combined in the analyses. It was found that the results for the heavy clay and silty clay could be taken together as also could those for the well-graded sand and gravel-sand-clay; these are labelled clay soils and well-graded granular soils respectively in Figure 13.8. The regression equations and correlation coefficients for these relations are given in Table 13.2. The various Figures provided only five sets of data for each analysis and the results should, therefore, be treated with some reserve; nevertheless, good correlations have been obtained for the 10 per cent air voids criterion (Level (b) in the Table).

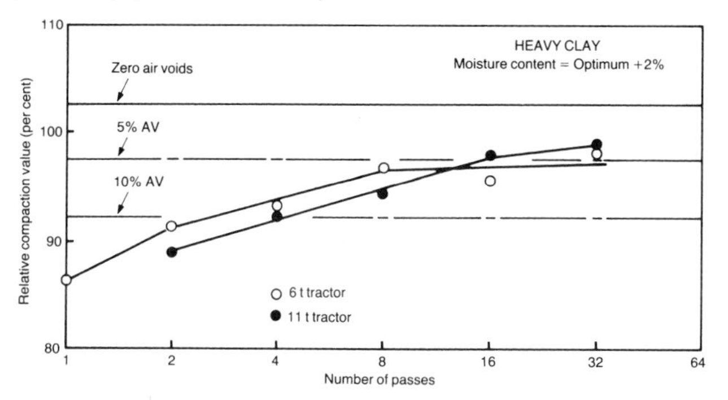

Figure 13.5 Relations between relative compaction and number of passes of track-laying tractors on 150 mm layers of heavy clay. Values are related to the maximum dry density and optimum moisture content obtained in the 2.5 kg rammer compaction test (Figure 3.13)

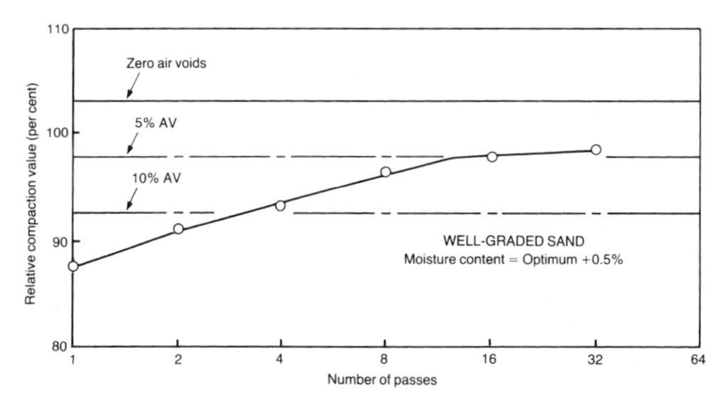

Figure 13.6 Relation between relative compaction and number of passes of the 6 t track-laying tractor on 150 mm layers of well-graded sand. Values are related to the maximum dry density and optimum moisture content obtained in the 2.5 kg rammer compaction test (Figure 3.13)

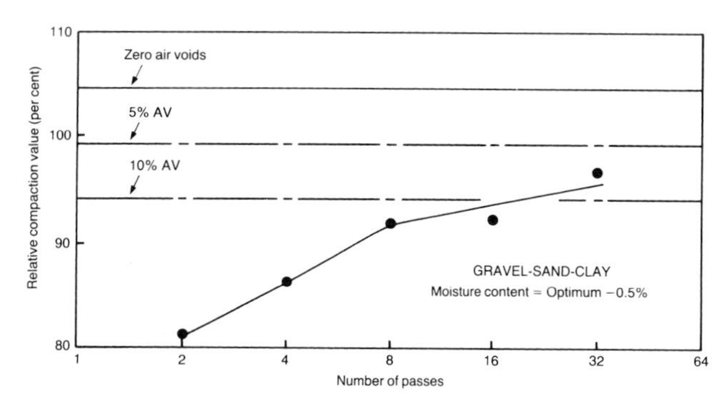

Figure 13.7 Relation between relative compaction and number of passes of the 11 t track-laying tractor on 150 mm layers of gravel-sand-clay. Values are related to the maximum dry density and optimum moisture content obtained in the 2.5 kg rammer compaction test (Figure 3.13)

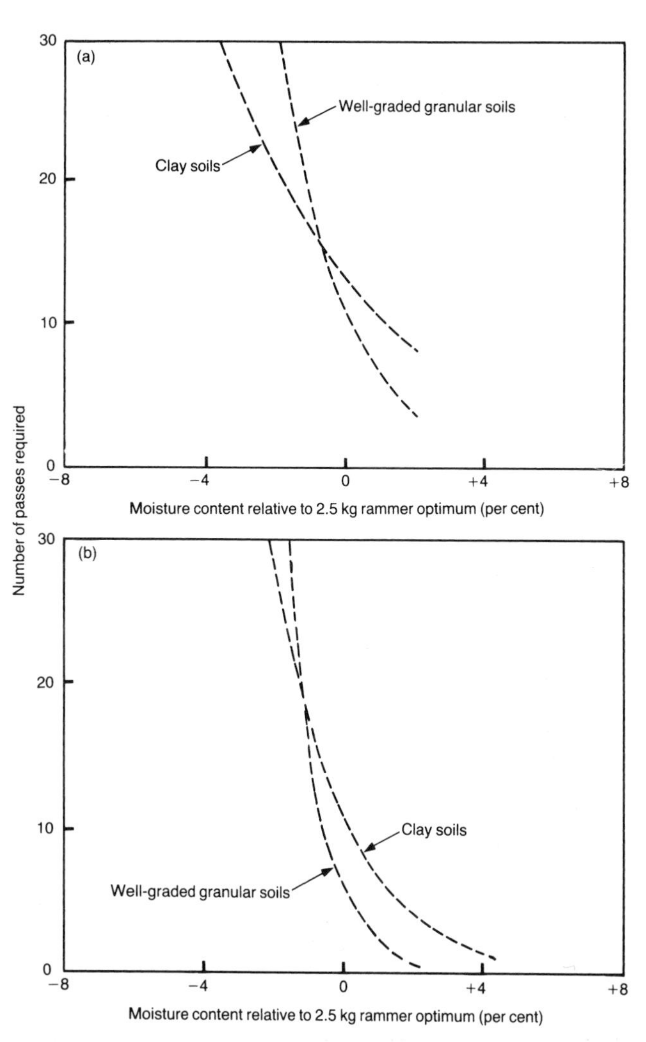

Figure 13.8 Relations between number of passes of track-laying tractors and moisture content required to achieve (a) 95 per cent of the maximum dry density obtained in the 2.5 kg rammer compaction test and (b) 10 per cent air voids. These relations were derived by regression analyses of the results given in Figures 13.1 to 13.7 and apply to 150 mm compacted layers

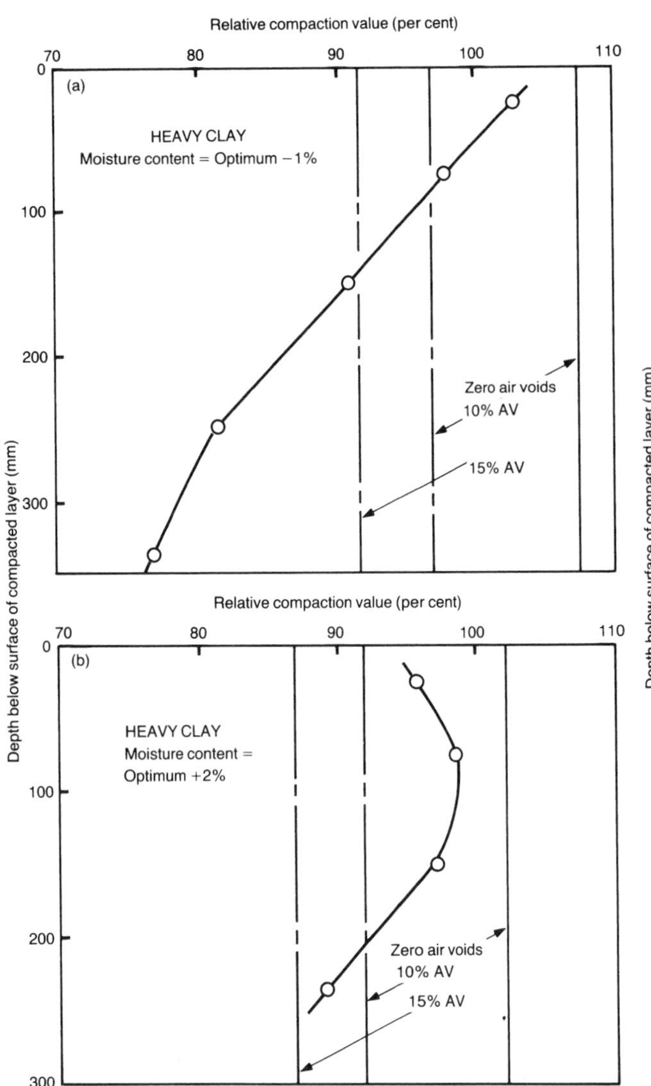

Figure 13.9 Relations between relative compaction and depth in the compacted layer of heavy clay obtained with 32 passes of the 6 t track-laying tractor. All values are related to the maximum dry density and optimum moisture content obtained in the 2.5 kg rammer compaction test (Figure 3.13)

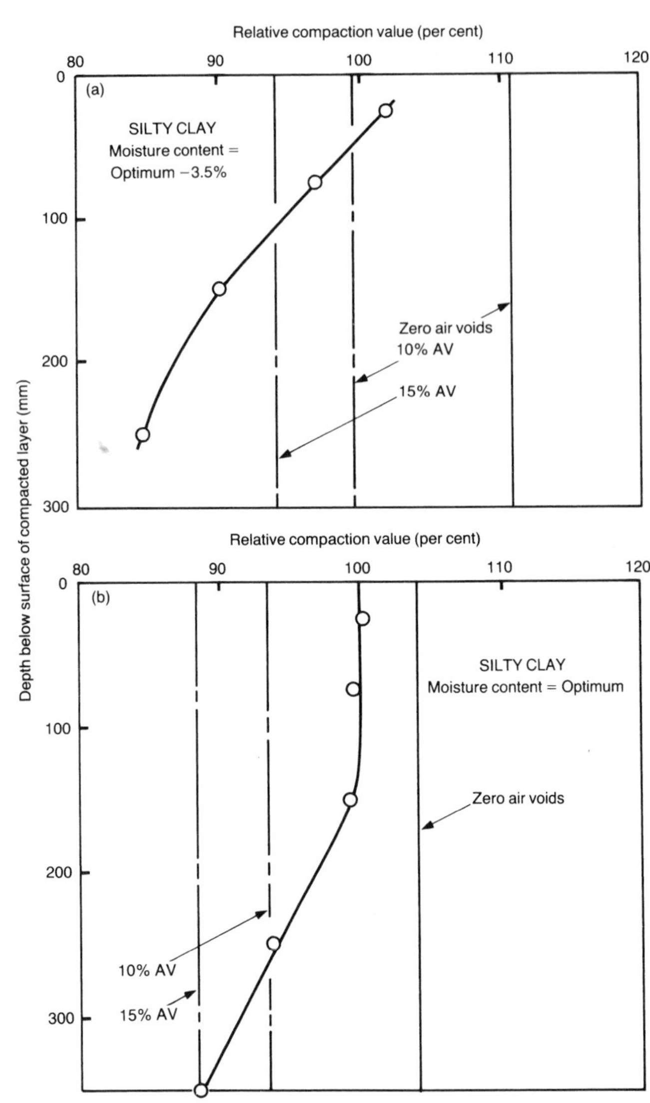

Figure 13.10 Relations between relative compaction and depth in the compacted layer of silty clay obtained with 32 passes of the 6 t track-laying tractor. All values are related to the maximum dry density and optimum moisture content obtained in the 2.5 kg rammer compaction test (Figure 3.13)

Table 13.2 Equations and correlation coefficients for the relations given in Figure 13.8

Soil	Level*	Regression equation†	No of data points	Correlation coefficient
Heavy clay & silty clay	(a)	P=1.12–.101 w	5	0.87
	(b)	P=1.04–.210 w	5	0.99
Well-graded sand & gravel-sand-clay	(a)	P=1.04–.226 w	5	0.86
	(b)	P=.794–.421 w	5	0.94

*Level (a)=95 per cent of the maximum dry density obtained in the 2.5 kg rammer compaction test
(b)=10 per cent air voids
†P=Log_{10} (Number of passes required)
w=Moisture content of soil, per cent, minus the optimum moisture content, per cent, obtained in the 2.5 kg rammer compaction test

13.7 The relations in Figure 13.8 indicate that the compaction of 150 mm layers was only achieved in a practical number of passes of track-laying tractors when the moisture content was fairly high. Thus, at values of moisture content above the optimum of the 2.5 kg rammer test, a maximum of 13 passes of the tractor tracks were required to produce the specified levels of compaction.

Relations between state of compaction and depth within the compacted layer

13.8 Investigations of the variation of state of compaction with depth in the compacted layer were carried out only with the 6 t track-laying

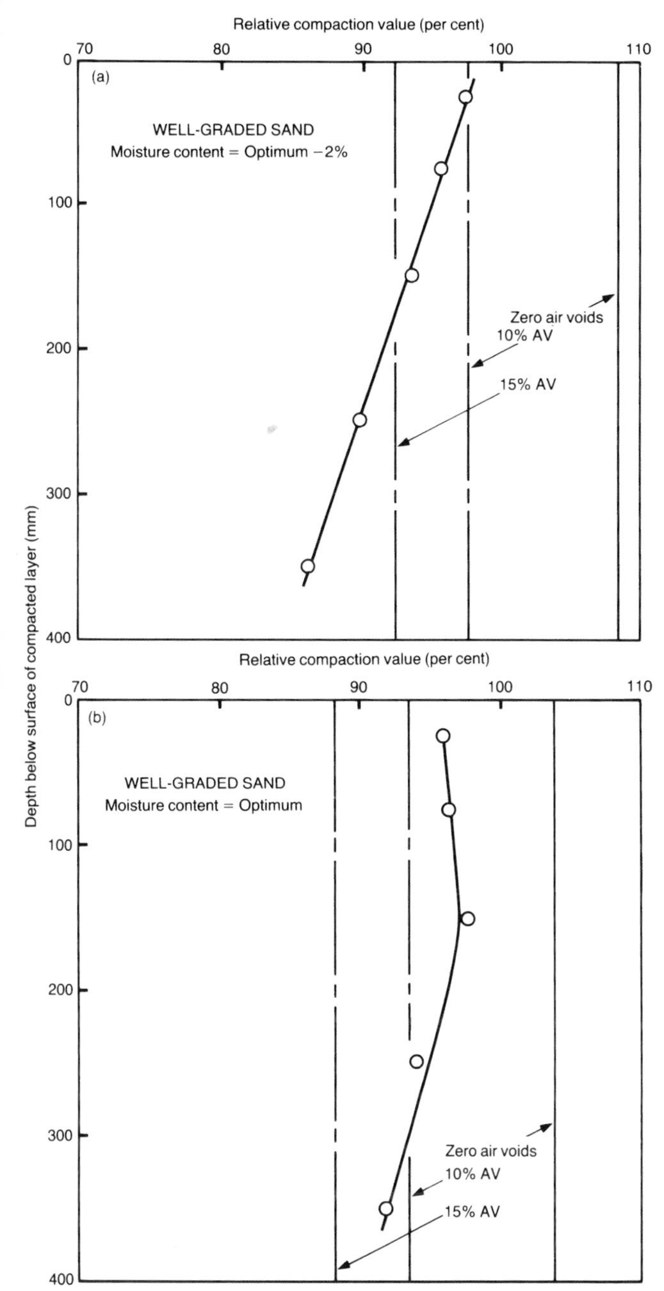

Figure 13.11 Relations between relative compaction and depth in the compacted layer of well-graded sand obtained with 32 passes of the 6 t track-laying tractor. All values are related to the maximum dry density and optimum moisture content obtained in the 2.5 kg rammer compaction test (Figure 3.13)

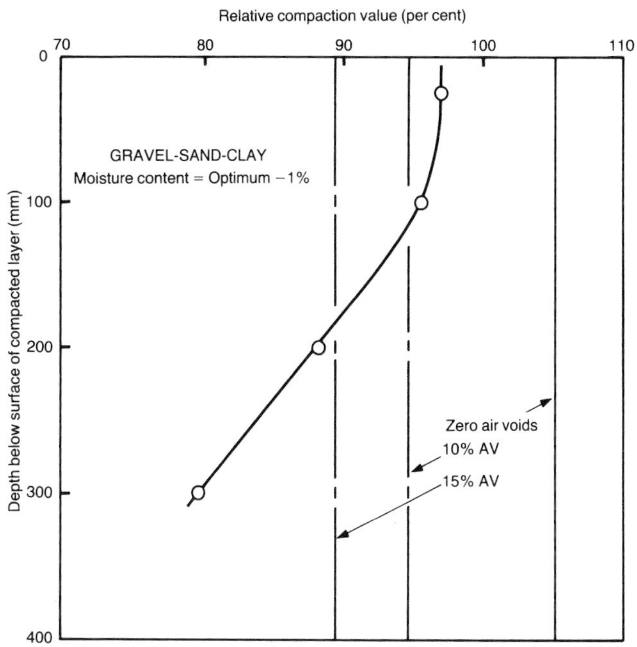

Figure 13.12 Relation between relative compaction and depth in the compacted layer of gravel-sand-clay obtained with 32 passes of the 6 t track-laying tractor. All values are related to the maximum dry density and optimum moisture content obtained in the 2.5 kg rammer compaction test (Figure 3.13)

tractor. The results are shown in Figures 13.9 to 13.12. The heavy clay, silty clay and well-graded sand were each tested at two different values of moisture content, but the gravel-sand-clay was tested at one value only. Layers up to 400 mm final compacted thickness were traversed by 32 passes of the tracks of the 6 t tractor to produce the relations shown in the Figures.

13.9 The results given in Figures 13.9 to 13.12 are summarised in Figure 13.13. The depths of layer capable of compaction to achieve 90 per cent relative compaction at the bottom are shown plotted against moisture content in Figure 13.13(a) and the depths for the achievement of 15 per cent air voids at the bottom of the layer are also plotted against moisture content in Figure 13.13(b). At most, two values of moisture content were available for each soil and, using a logarithmic scale for the depth, the points are shown linked by a straight line. This type of relation has been shown to provide good correlations in the analyses of data in previous Chapters. The results for each soil type are plotted separately in Figure 13.13, but it can be seen that those for silty clay and well-graded sand were fairly similar, and the greatest thickness of layer at a given 'relative' moisture content was achieved with these soils. The smallest thickness at a given 'relative' moisture content was obtained with heavy clay. At moisture contents equal to the optimum of the 2.5 kg rammer compaction test, the depths of layer capable of being compacted to the assumed criteria ranged from about 170 mm with heavy clay to almost 400 mm with well-graded sand. These results, as mentioned earlier, were obtained with 32 passes of the tracks of the 6 t tractor; the results in Figure 13.8, indicating the number of passes required to satisfactorily compact 150 mm thick layers, would be of more practical use for the occasions

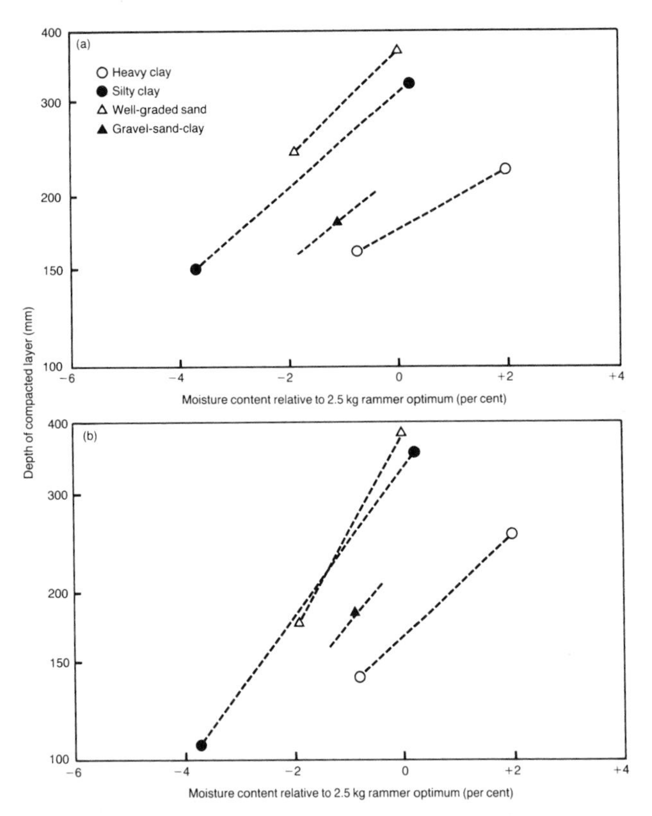

Figure 13.13 Relations between depth of layer and moisture content for 32 passes of the 6 t track-laying tractor to achieve (a) 90 per cent relative compaction and (b) 15 per cent air voids at the bottom of the layer

when track-laying tractors have to be used for the compaction of soil.

Measurement of pressures produced in soil by the 6 t track-laying tractor

13.10 Measurements were made of the pressures produced in heavy clay, silty clay and well-graded sand by the passage of the tracks of the 6 t tractor. Piezo-electric gauges, similar in type to those described in Paragraph 11.23, were installed at various depths in the soil. Typical results for the uppermost gauges, taken from Whiffin (1953), are shown in Figure 13.14. The moisture contents used are shown in the Figure related to the optimum moisture content of the 2.5 kg rammer compaction test.

13.11 An interesting feature of the results was the non-uniform pressure distribution beneath the tracks, with the maximum pressures occurring beneath the front of the tracks. This would have been caused by the effect of the heavy bulldozer blade fitted to the machine during the tests (Plate 13.1). In the photograph the machine is stable as the bulldozer blade is shown resting on the ground; when lifted for travel, however, the blade caused the centre-of-gravity of the machine to be transferred forward to such an extent that only a small proportion of the weight was actually carried on the rear of the tracks. Such a pressure distribution is unusual, and with most track-laying tractors a series of peaks in the pressure distribution occurs in association with the track rollers, as shown in Figure 13.15.

13.12 By measuring the areas under the curves shown in Figure 13.14, the peak pressures at the surface of the soil under the front of the tracks have been estimated as being from three to four times the average pressure under the full length of the track for the clay soils, and over ten times the average pressure for the well-graded sand. The mean contact pressure, given in Table 13.1, was about 50 kN/m², so for clay soils the peak surface pressures were 150 to 200 kN/m². With well-graded sand, the ratio of peak to average pressures determined from the area under the

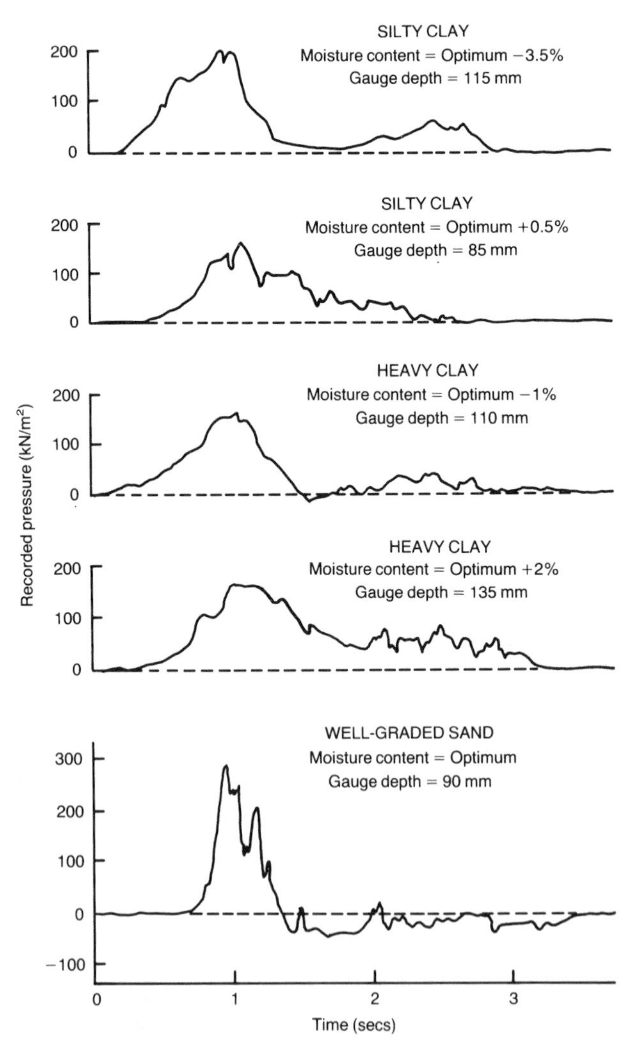

Figure 13.14 Typical pressure records obtained beneath the tracks of the 6 t track-laying tractor

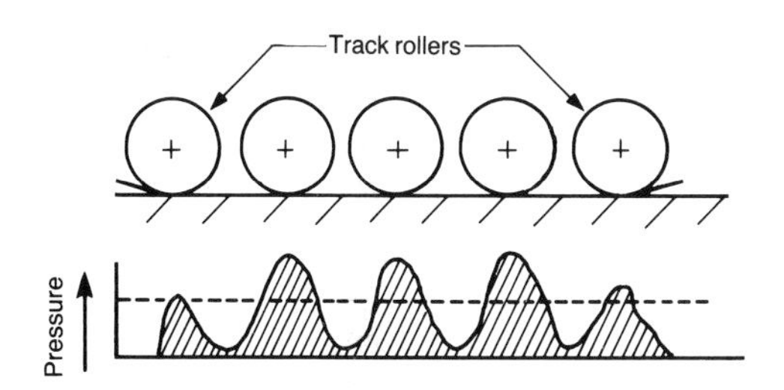

Figure 13.15 Usual form of pressure distribution beneath a track-laying tractor

curve, estimated to be about 10, yields a peak surface pressure of about 500 kN/m^2.

13.13 An assessment can also be made by comparing the contact pressures of pneumatic-tyred rollers, described in Chapter 5, with those of track-laying tractors. The most useful comparisons have been obtained by determining the contact pressure necessary in a pneumatic-tyred roller to produce the same required number of passes as those shown to be required from track-laying tractors. The results for the pneumatic-tyred rollers are summarised in Table 5.3, whilst those for the track-laying tractors are given in Table 13.2.

13.14 For heavy clay, this comparison indicates that the same number of passes were required from the track-laying tractors as from pneumatic-tyred rollers with contact pressures in the range of 230 to 300 kN/m^2. For the well-graded granular soils (well-graded sand and gravel-sand-clay) the contact pressures for pneumatic-tyred rollers producing an equivalent number of passes to the track-laying tractors are in the range of 70 to 140 kN/m^2.

13.15 The conclusion to be reached from these findings is that the performance of the track-laying tractors were related to the peak rather than the average contact pressure under the tracks. Comparison with contact pressures of pneumatic-tyred rollers of similar performance to the tractors appear to confirm the values of pressure measured in clay soils (Figure 13.14). The pressures recorded in the well-graded sand appear very doubtful, however; the states of compaction achieved were what might be expected from peak pressures of only 1½ to 3 times the mean contact pressure.

Feasibility of using track-laying tractors as compactors

13.16 The widths between the outer edges of the tracks of the tractors were 2.3 m and 1.7 m for the 11 t and 6 t machines respectively. Thus the coverage (area of soil compacted in one pass as a percentage of the overall area traversed by the tractor) was 40 to 45 per cent. Mention has already been made in Paragraph 13.5 of the similarity of the results obtained with the track-laying tractors and those produced by the 12 t pneumatic-tyred roller with a wheel load of 1.4 t. The contact pressures of pneumatic-tyred rollers necessary to produce the same results as track-laying tractors, described in Paragraph 13.14, also confirm this similarity. With a coverage of 40 to 45 per cent, a tractor would have to traverse the soil almost three times for each pass of the pneumatic-tyred roller to compact the same area of soil. It was concluded at the time of the investigations (Lewis, 1954) that when towing a pneumatic-tyred roller, the track-laying tractor would not provide a significant addition to the compaction produced by the roller.

13.17 In view of the relatively high cost of operating track-laying tractors, their use solely for the compaction of soil can seldom be justified. They could be of considerable value, however, for compacting soil in areas not easily accessible or trafficable by conventional compaction equipment, eg side slopes of embankments or areas of deposition of soils too wet for use in engineering works.

References

Lewis, W A (1954). Further studies in the compaction of soil and the performance of compaction plant. *Road Research Technical Paper* No 33. HM Stationery Office, London.

West, L (1951). Investigation of the compacting effect on soils of a medium tracklaying tractor. *D.S.I.R., Road Research Laboratory Research Note* No RN/1587/LW (Unpublished).

Whiffin, A C (1953). The pressures generated in soil by compaction equipment. *Proc. Symposium on Dynamic Testing of Soils.* American Society for Testing and Materials, Philadelphia.

CHAPTER 14 SOIL COMPACTION INVESTIGATIONS WITH MISCELLANEOUS TYPES OF EQUIPMENT

Introduction

14.1 Investigations were carried out from time to time on compaction equipment which did not easily fall into any of the categories dealt with in Chapters 4 to 13. These usually were experimental machines and do not compare with the types of equipment accepted for compaction purposes at the present time. It is considered worthwhile briefly describing the machines and the results obtained to provide a permanent record of experiences with these novel methods of compaction. The machines included in this Chapter are an 840 kg agricultural roller, a 120 kg vibrating-plate compactor (not included in Chapter 9 because of the small dimensions of its base plate), a 1.3 t tamping compactor and a pneumatic impact compactor.

Investigations with the 840 kg agricultural roller

14.2 The roller is illustrated in Plate 14.1 and details of relevance to its performance are given in Table 14.1. The roll was ridged, and was made up of segments (individual wheels), butted closely together on a common shaft. The value of mass per unit width given in Table 14.1 is similar to that of the 220 kg pedestrian-operated vibrating roller which was investigated, without vibration, as a smooth-wheeled roller as described in Chapter 4 (see Table 4.1).

14.3 The agricultural roller was investigated in 1950 at the request of a Road Research Board committee to determine whether it might be suitable for compacting soils for civil engineering purposes (West, 1951). It was tested in the original compaction facility comprising a circular test track and was towed by the electrically operated towing lorry (see Chapter 3).

Table 14.1 Details of the 840 kg agricultural roller

Total mass	(kg)	843
Diameter of roll across ridges	(mm)	610
Diameter of roll across troughs	(mm)	560
Width of roll	(m)	2.36
Mass per unit width	(kg/m)	360

Plate 14.1 The 840 kg agricultural roller investigated in 1950

14.4 The relation between state of compaction produced and moisture content was determined for 150 mm layers of sandy clay No 1, using 32 passes of the machine (Figure 14.1). Well-graded sand was used in the determination of the effect of the number of passes on 150 mm layers (Figure 14.2) and the variation of the state of compaction with depth in the layer after 32 passes of the roller (Figure 14.3).

14.5 Generally poor states of compaction were obtained on the sandy clay soil (Figure 14.1) even at moisture contents as high as 4 per cent above the optimum of the 2.5 kg rammer compaction test. With well-graded sand the increase in relative compaction value with increases in number of passes was fairly small

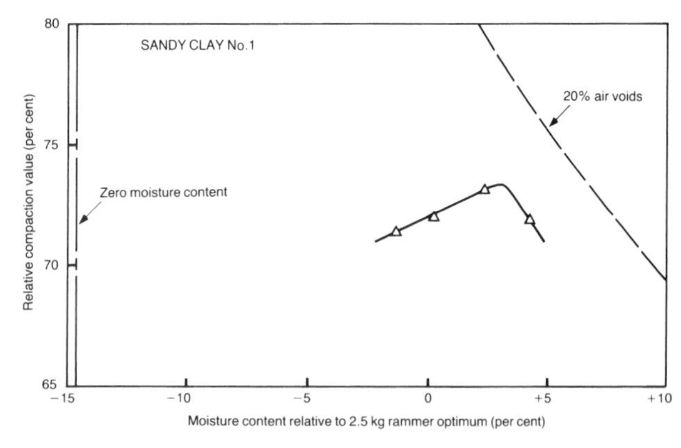

Figure 14.1 Relation between dry density and moisture content for 32 passes of the 840 kg agricultural roller on 150 mm layers of sandy clay No 1, expressed in terms of relative compaction with the 2.5 kg rammer test

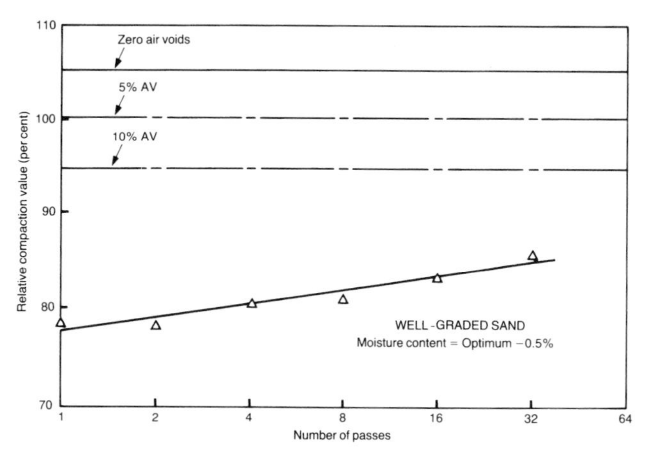

Figure 14.2 Relation between relative compaction and number of passes of the 840 kg agricultural roller on 150 mm layers of well-graded sand. Values are related to the maximum dry density and optimum moisture content obtained in the 2.5 kg rammer compaction test (Figure 3.13)

and poor states of compaction were obtained even after 32 passes. Figure 14.3 indicates that the machine compacted only the upper 50 mm of the layer to the criteria, adopted in previous Chapters, of 90 per cent relative compaction or 15 per cent air voids at the bottom of the layer. The conclusion reached by West (1951) was that the agricultural roller was unsuitable for the compaction of soil for engineering works.

14.6 Interesting comparisons can be made between the results obtained with the agricultural roller and those obtained with the smooth-wheeled rollers given in Chapter 4, particularly the 220 kg pedestrian-operated roller which had an identical roll loading per unit width. Thus, in Figure 4.4, very similar relative compaction values, in the range of 70 to 75 per cent, were obtained on sandy clay No 1 with the 220 kg pedestrian roller as were obtained with the agricultural roller and shown in Figure 14.1. Table 4.3 indicates that on well-graded sand extremely large numbers of passes would be required for a smooth-wheeled roller with a mass per unit width equal to that of the agricultural roller in order to achieve either 95 per cent relative compaction or 10 per cent air voids at the moisture content at which the soil was tested (Figure 14.2). In Figure 4.5 the relation between relative compaction value and moisture content for 32 passes of the 220 kg pedestrian roller gives a relative compaction value of about 86 per cent at the moisture content equivalent to 0.5 per cent below the optimum of the 2.5 kg rammer test; this compares well with the value shown in Figure 14.2 for 32 passes of the agricultural roller. The variation in state of compaction with depth through the compacted layer of well-graded sand for the agricultural roller in Figure 14.3 is of a similar general form to that obtained using the 350 kg pedestrian roller (roll loading 490 kg/m) on the same soil at approximately the same moisture content, shown in Figure 4.21. Figure 4.24 indicates that with well-graded sand a thickness of layer of about 50 mm could be adequately compacted by the roll loading of the agricultural roller (360 kg/m). Thus it can be concluded that the results obtained with the agricultural roller can be predicted from the results obtained with smooth-wheeled rollers, with mass per unit width of roll as the prediction parameter (see Chapter 4).

Investigations with the 120 kg vibrating-plate compactor

14.7 The 120 kg vibrating-plate compactor is illustrated in Plate 14.2 and details of the machine are given in Table 14.2. From the photograph it appears that the machine consisted of a single-cylinder engine with eccentrics attached to the base plate and driven by extensions at both ends of the crank shaft. A form of flexible mounting for the engine can be discerned in the Plate, so the extensions of the crank shaft would have incorporated flexible couplings. This machine was not included in Chapter 9 with other vibrating-plate compactors because of its extremely small contact area and

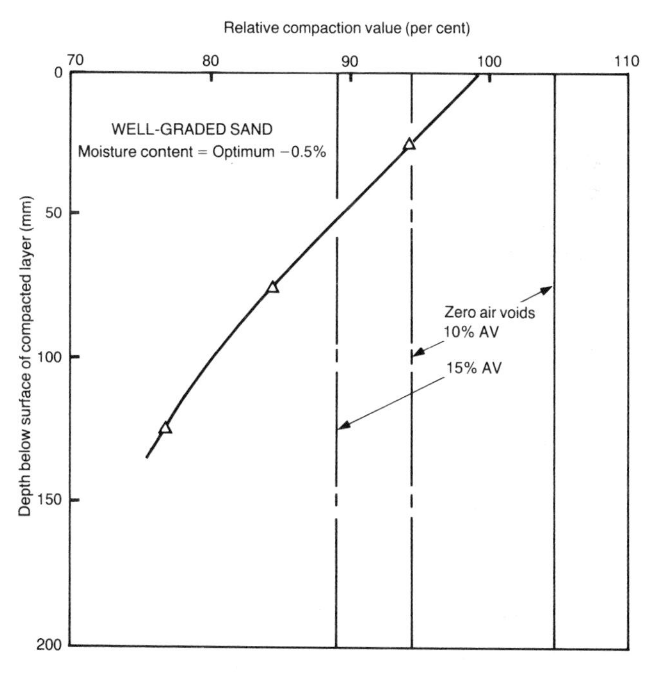

Figure 14.3 Relation between relative compaction and depth in the compacted layer of well-graded sand obtained with 32 passes of the 840 kg agricultural roller. All values are related to the maximum dry density and optimum moisture content obtained in the 2.5 kg rammer compaction test (Figure 3.13)

very low frequency of vibration. No formal report can be traced which describes the investigations with this machine and the results presented here have been taken from original hand-written records.

Table 14.2 Details of the 120 kg vibrating-plate compactor

Total mass	(kg)	120
Width of base plate	(mm)	635
Contact area of base plate	(m^2)	0.129
Mass per unit area	(kg/m^2)	930
Frequency of vibration	(Hz)	30
Normal speed of travel	(km/h)	0.5

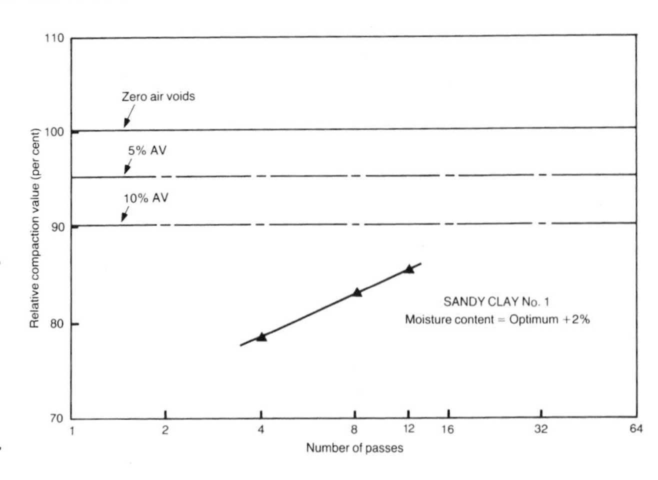

Figure 14.4 Relation between relative compaction and number of passes of the 120 kg vibrating-plate compactor on 150 mm layers of sandy clay No 1. Values are related to the maximum dry density and optimum moisture content obtained in the 2.5 kg rammer compaction test (Figure 3.13)

14.8 Relations between state of compaction produced and number of passes were determined for the 120 kg vibrating-plate compactor on 150 mm layers of sandy clay No 1 (Figure 14.4) and of well-graded sand (Figure 14.5). With both soils the results were rather poor; with the sandy clay this might be expected for this size of vibrating compactor on a cohesive material, and with the well-graded sand the moisture content, at 2 per cent below the optimum of the 2.5 kg rammer test, would have been rather low for adequate compaction by this type of machine.

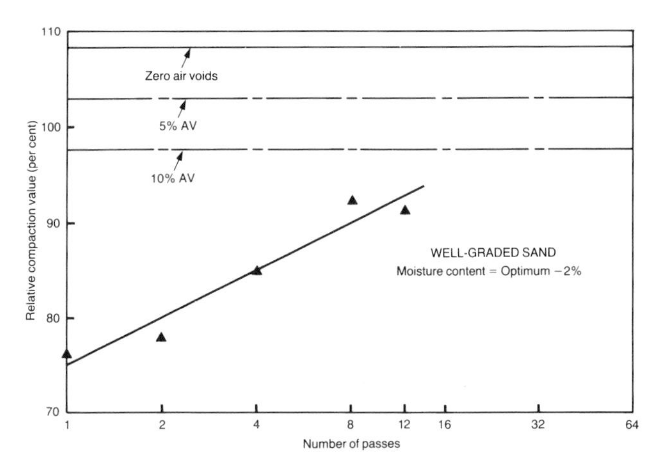

Figure 14.5 Relation between relative compaction and number of passes of the 120 kg vibrating-plate compactor on 150 mm layers of well-graded sand. Values are related to the maximum dry density and optimum moisture content obtained in the 2.5 kg rammer compaction test (Figure 3.13)

14.9 Interesting comparisons can be made with results obtained with the range of vibrating-plate compactors described in Chapter 9. In Figure 9.4(a) the relation shown for 6 passes of the 240 kg compactor on well-graded sand indicates that a relative compaction value of about 90 per cent was obtained in the 150 mm layer at a moisture content 2 per cent below the 2.5 kg rammer optimum; this is similar to the level of compaction produced by 8 passes of the 120 kg machine, as shown in Figure 14.5. The mass per unit area of base plate was 1310 kg/m^2 for the 240 kg compactor (Table 9.1) compared with 930 kg/m^2 with the 120 kg machine. Table 9.2 gives regression equations to determine the required number of passes of vibrating-plate compactors; using the mass per unit area of the 120 kg machine and the test conditions for the relations shown in Figures 14.4 and 14.5, the required numbers of passes have been determined from the regression equations and are compared in Table 14.3 with the number of passes indicated by the test results. (For this calculation the regression equation for sandy

Plate 14.2 The 120 kg vibrating-plate compactor investigated in 1951

clay No 2 was assumed to apply to sandy clay No.1.) In all cases the 120 kg machine would have actually required more passes than the number that might be expected from the normal range of vibrating-plate compactors for its particular mass per unit area of base plate, although the differences were fairly small with the well-graded sand. Thus the 120 kg machine, with its small plate area and relatively low frequency of vibration, in general had a poorer performance than the normally configured vibrating-plate compactors with the same static pressure under the base plate.

Table 14.3 Comparison of the required numbers of passes of the 120 kg vibrating-plate compactor calculated from the equations given in Table 9.2 with those indicated by the results in Figures 14.4 and 14.5

Soil	Moisture content	Level*	Number of passes	
			Table 9.2	Figure 14.4 & 14.5
Sandy clay No 1	Optimum+2%	(a)	17	52
		(b)	13	25
Well-graded sand	Optimum−2%	(a)	10	16
		(b)	18	23

*Level (a)=95 per cent of the maximum dry density obtained in the 2.5 kg rammer compaction test
(b)=10 per cent air voids

Investigations with the 1.3 t tamping compactor

14.10 The tamping compactor was an experimental machine which used an unusual mechanism to provide a tamping action to the surface of a layer of soil. The machine is illustrated in Plate 14.3 and details are given in Table 14.4. It comprised three base plates side-by-side, with the central one considerably wider than the outer ones. The power unit was mounted on the rear of the central base plate. A form of crank shaft, driven by a series of chain drives, passed through bearings near the front of the base plates. The rotation of the crank shaft lifted, moved forward and dropped the central and two outer base plates in turn to produce a tamping action combined with forward movement. The rear portions of the three sections of base plate trailed or 'skidded' over the surface of the soil, and the underside of the base plate had chevron-shaped protrusions to assist traction. The machine travelled in the forward direction only. As with the machine described previously, no formal report of the investigations can be traced and the results presented here for the 1.3 t tamping compactor

Plate 14.3 The 1.3 t tamping compactor investigated in 1952

have been extracted from original hand-written records.

14.11 Relations between state of compaction obtained and moisture content were determined for 6 passes of the machine on 150 mm layers of silty clay (Figure 14.6) and 100 mm layers of well-graded sand (Figure 14.7). Low air contents were achieved with the silty clay at moisture contents in excess of the optimum of the 2.5 kg rammer compaction test and a maximum relative compaction value of about 99 per cent might have been achieved at about 1 per cent above the 2.5 kg rammer optimum. With the well-graded sand a maximum relative compaction value of 103 per cent was achieved at an optimum moisture content about 2 per cent below the optimum of the 2.5 kg rammer compaction test.

14.12 Relations between relative compaction and number of passes were obtained on silty clay, well-graded sand (at two different values of moisture content) and gravel-sand-clay (Figures 14.8 to 14.10). At the values of moisture content and thicknesses of layer used, 95 per cent relative compaction was achieved in 2 to 5 passes of the machine, and 10 per cent air voids was achieved in 1 to 6 passes.

Table 14.4 Details of the 1.3 t tamping compactor

Total mass	(kg)	1380
Total width of base plates	(mm)	648
Contact area (estimated)	(m^2)	0.42
Mass per unit area	(kg/m^2)	3140
Tamping amplitude (approximate)	(mm)	25
Tamping frequency	(Hz)	3.2
Normal speed of travel	(km/h)	0.6

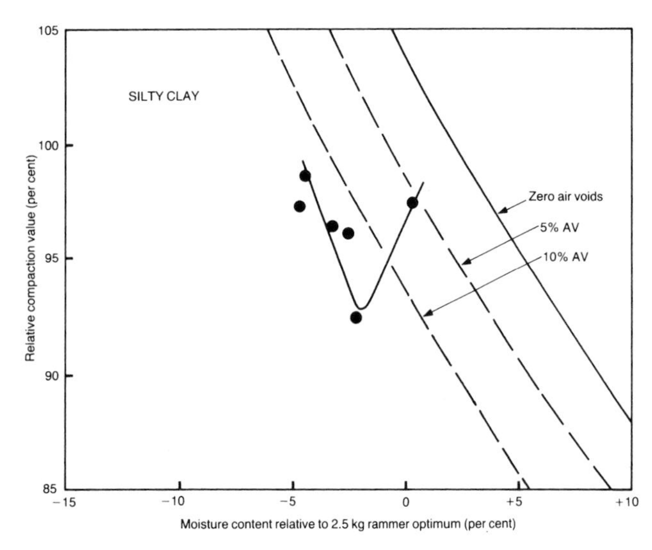

Figure 14.6 Relation between dry density and moisture content for 6 passes of the 1.3 t tamping compactor on 150 mm layers of silty clay, expressed in terms of relative compaction with the 2.5 kg rammer test

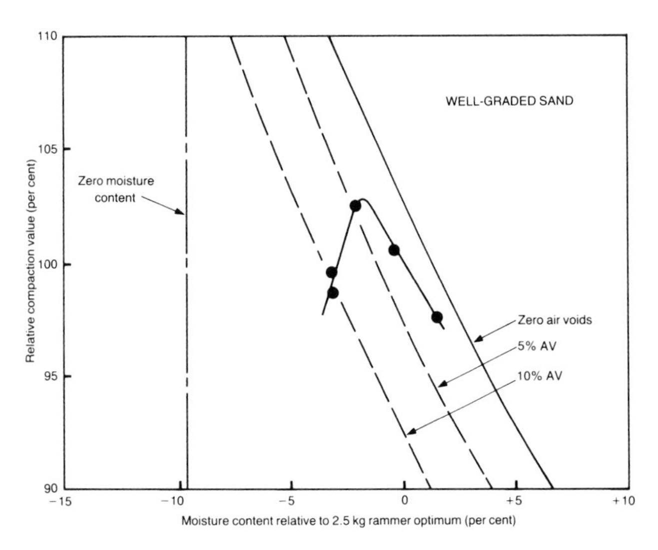

Figure 14.7 Relation between dry density and moisture content for 6 passes of the 1.3 t tamping compactor on 100 mm layers of well-graded sand, expressed in terms of relative compaction with the 2.5 kg rammer test

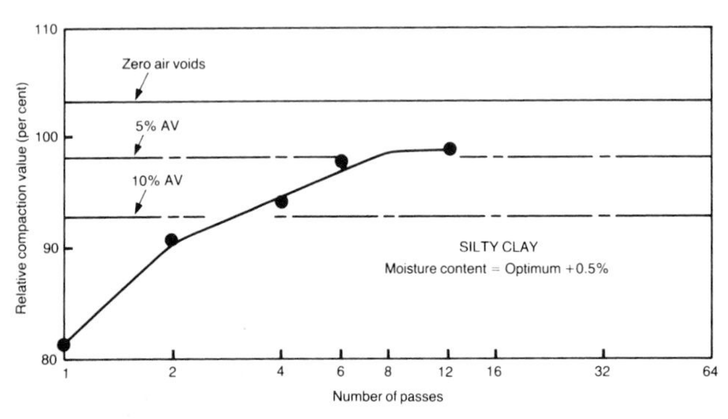

Figure 14.8 Relation between relative compaction and number of passes of the 1.3 t tamping compactor on 150 mm layers of silty clay. Values are related to the maximum dry density and optimum moisture content obtained in the 2.5 kg rammer compaction test (Figure 3.13)

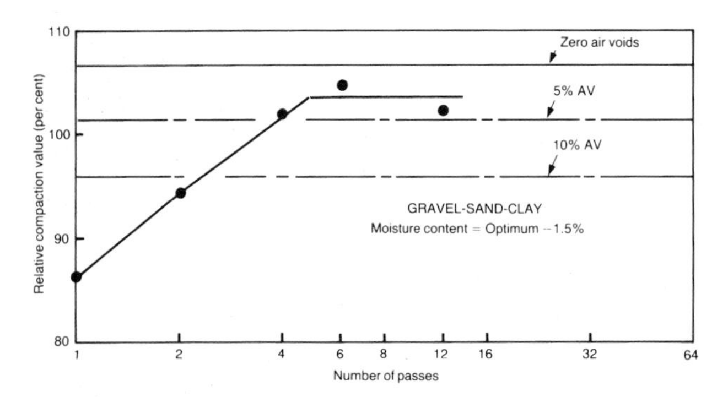

Figure 14.10 Relation between relative compaction and number of passes of the 1.3 t tamping compactor on 150 mm layers of gravel-sand-clay. Values are related to the maximum dry density and optimum moisture content obtained in the 2.5 kg rammer compaction test (Figure 3.13)

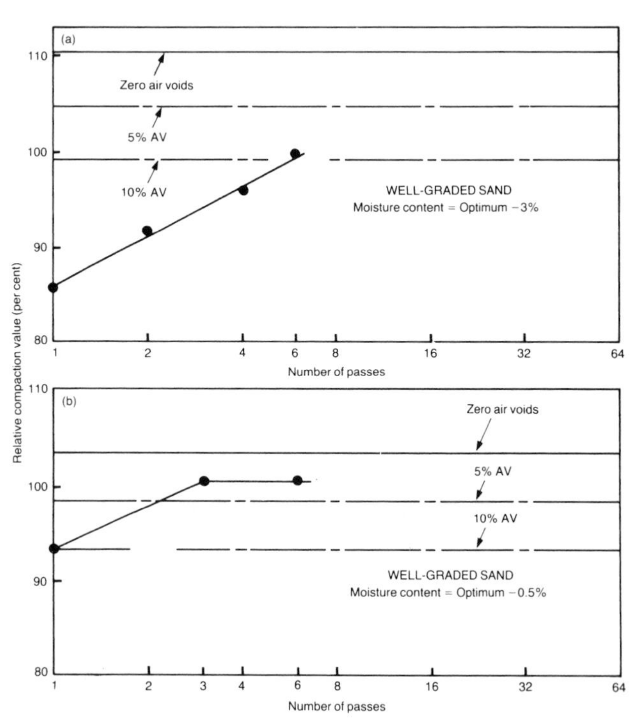

Figure 14.9 Relations between relative compaction and number of passes of the 1.3 t tamping compactor on 100 mm layers of well-graded sand at two different values of moisture content. Values are related to the maximum dry density and optimum moisture content obtained in the 2.5 kg rammer compaction test (Figure 3.13)

14.13 The relation between state of compaction produced and depth in the compacted layer was determined on well-graded sand using 6 passes of the 1.3 t tamping compactor (Figure 14.11). At the moisture content used, 0.5 per cent below

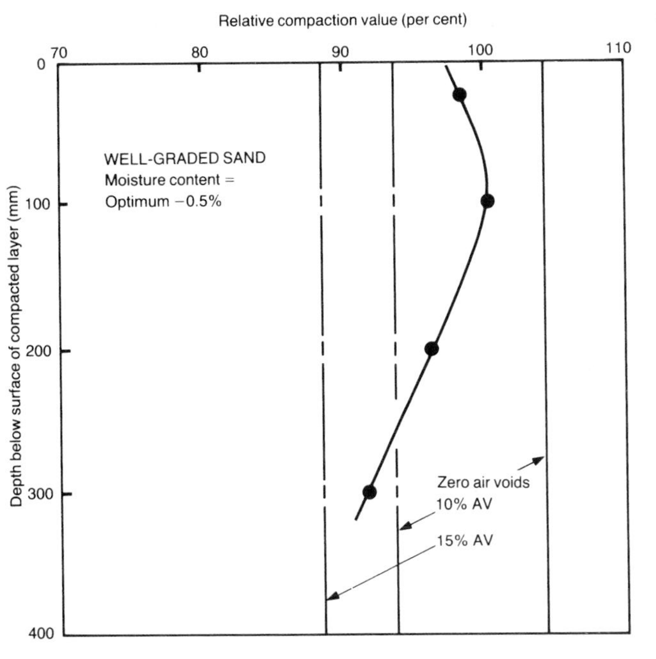

Figure 14.11 Relation between relative compaction and depth in the compacted layer of well-graded sand obtained with 6 passes of the 1.3 t tamping compactor. All values are related to the maximum dry density and optimum moisture content obtained in the 2.5 kg rammer compaction test (Figure 3.13)

the optimum of the 2.5 kg rammer compaction test, satisfactory states of compaction appear to have been achieved at depths of 300 mm in the layer.

14.14 The results discussed in Paragraphs 14.11 to 14.13 all appear to be fairly satisfactory and it is worthwhile comparing the performance of the tamping compactor with the more modern vibro-tamper, described in Chapter 10. A study of Table 10.1 together with Table 14.4 shows that the modern vibro-tamper has a higher frequency (by a factor of about 3), twice the amplitude, but a total mass of less than one-tenth of the 1.3 t tamping compactor. The relation shown in Figure 14.7, obtained with 6 passes of the tamping compactor on 100 mm layers of well-graded sand, compares well with those for 10 passes of vibro-tampers tested under the same conditions (see Figure 10.3). The effects of variations in the number of passes shown in Figures 14.9 and 14.10 are also comparable with those shown in Figures 10.7(a) and (c) and 10.8(e) for the 59 kg vibro-tamper. The similarity of all relations determined under the same conditions confirm that the 1.3 t tamping compactor had a very similar performance to the 59 kg vibro-tamper. Although the latter machine compacted only a 280 mm width of soil in one pass (Table 10.1) compared with the 648 mm width of the 1.3 t tamping compactor, it must be concluded that the 1.3 t tamping compactor, considering its total mass and its lack of portability, was far less efficient than the modern vibro-tamper.

Investigations with the pneumatic impact compactor

14.15 The pneumatic impact compactor is illustrated in Plate 14.4 and details of the machine are given in Table 14.5. It consisted of a double-acting pile-driving hammer standing on a base plate fabricated in the form of a skid. When the hammer was operated, blows were delivered to an anvil block welded to the base plate. The hammer was operated by compressed air supplied by a compressor mounted on the back of a truck, from the rear of which the impact compactor was suspended. Although the energy of each blow and the specific energy calculated from that figure are both given in Table 14.5, the actual energy transmitted to the soil through the anvil block and base plate would have been considerably less.

14.16 The method of operating the compactor was to lower the pile-driving hammer on to the base plate as it rested on the loose soil and to operate the machine for a set period of time. The

Plate 14.4 The pneumatic impact compactor investigated in 1954

Table 14.5 Details of the pneumatic impact compactor

Total mass	(kg)	306
Width of base plate	(mm)	610
Contact area	(m^2)	0.310
Mass per unit area of base plate	(kg/m^2)	987
Energy per stroke of ram (nominal)	(J)	475
Energy per blow per unit area (specific energy)	(kJ/m^2)	1.53
Frequency	(Hz)	6

hammer was then raised, the base plate moved forward a distance equal to its contact length and the procedure repeated. In this way a 610 mm wide strip of compacted soil was produced on which measurements of the state of compaction were made.

14.17 Relations between state of compaction produced and moisture content were determined on 150 mm layers of heavy clay, sandy clay No 2, well-graded sand and gravel-sand-clay (Figures 14.12 to 14.15), using 20 seconds of compaction at each base plate location. With the heavy clay (Figure 14.12) extremely poor states of compaction were produced, but low air contents were attained on the other three soils. With the sandy clay (Figure 14.13) a maximum relative compaction value of about 97 per cent was obtained at an optimum moisture content about 3 per cent above the optimum of the 2.5 kg rammer test and with the granular soils (Figures 14.14 and 14.15) maximum relative

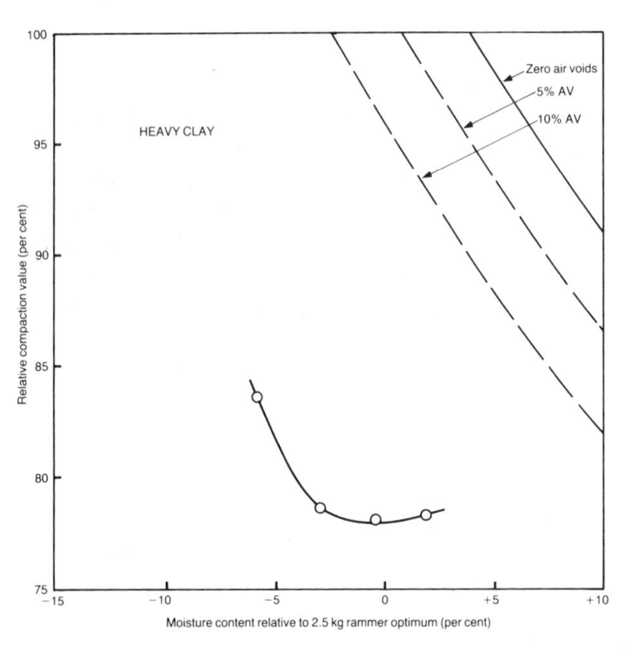

Figure 14.12 Relation between dry density and moisture content for 20 s of compaction with the pneumatic impact compactor on 150 mm layers of heavy clay, expressed in terms of relative compaction with the 2.5 kg rammer test

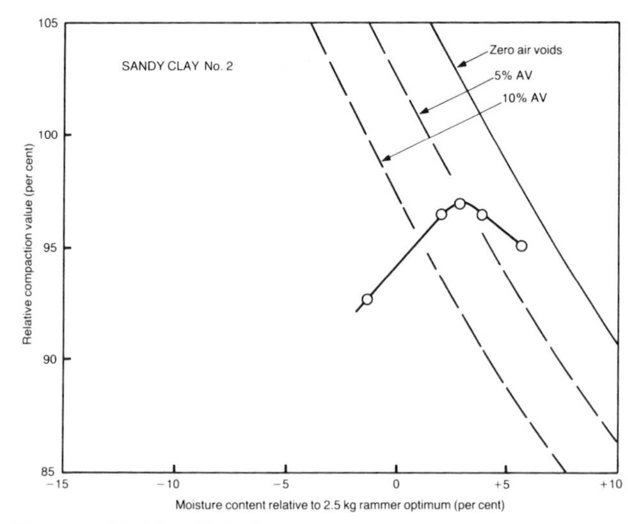

Figure 14.13 Relation between dry density and moisture content for 20 s of compaction with the pneumatic impact compactor on 150 mm layers of sandy clay No 2, expressed in terms of relative compaction with the 2.5 kg rammer test

compaction values of 104 to 107 per cent were reached at moisture contents within +1 to −2 per cent of the optimum of the 2.5 kg rammer test. The machine behaved, therefore, rather as a vibratory compactor in that its performance was considerably better on the granular soils than on the cohesive soils. The relation between state of compaction produced and duration of compaction, determined on gravel-sand-clay (Figure 14.16), indicates that doubling the duration of compaction beyond the value used in determining the effects of variations in moisture content would have had a fairly small effect on the results shown in Figures 14.12 to 14.15.

14.18 The results produced by the pneumatic impact compactor were compared in the original report (Lewis and Parsons, 1954) with those

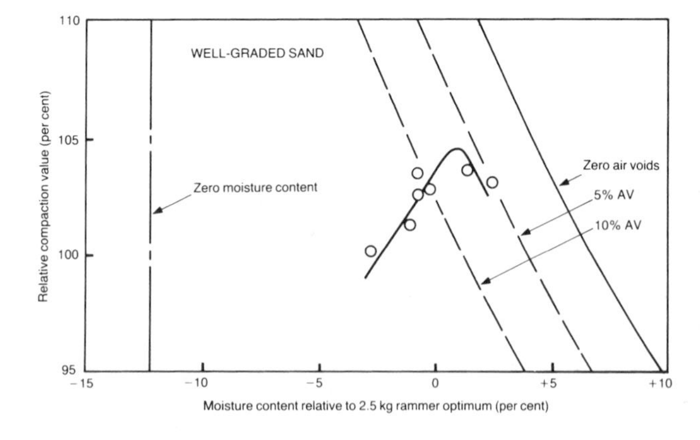

Figure 14.14 Relation between dry density and moisture content for 20 s of compaction with the pneumatic impact compactor on 150 mm layers of well-graded sand, expressed in terms of relative compaction with the 2.5 kg rammer test

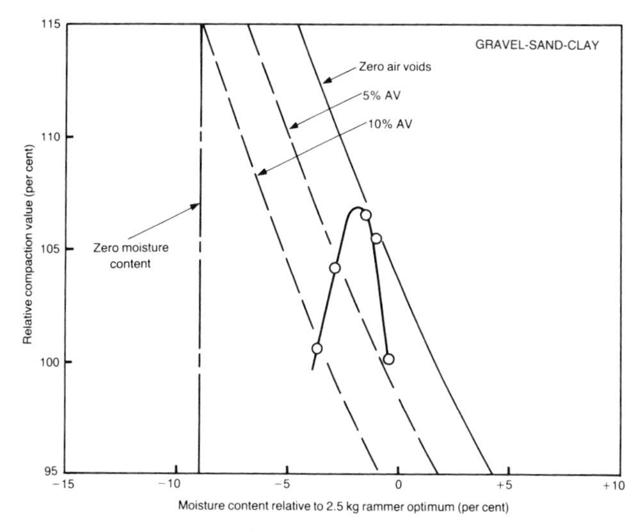

Figure 14.15 Relation between dry density and moisture content for 20 s of compaction with the pneumatic impact compactor on 150 mm layers of gravel-sand-clay, expressed in terms of relative compaction with the 2.5 kg rammer test

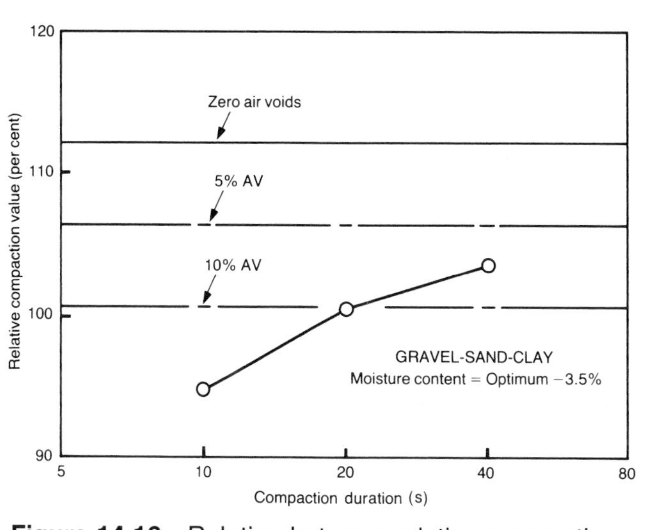

Figure 14.16 Relation between relative compaction and duration of compaction with the pneumatic impact compactor on 150 mm layers of gravel-sand-clay. Values are related to the maximum dry density and optimum moisture content obtained in the 2.5 kg rammer compaction test (Figure 3.13)

obtained with the multi dropping-weight compactor (Chapter 12). For clay soils the dropping-weight compactor produced higher states of compaction at any given value of moisture content (compare Figures 14.12 and 14.13 with Figures 12.1 and 12.2), reflecting the relative values of specific energy of the two machines (5.1 kJ/m^2 for the dropping-weight compactor and 1.5 kJ/m^2 for the pneumatic impact compactor). However, with the well-graded sand and gravel-sand-clay the relative performances of the two machines were reversed, with the pneumatic impact compactor producing the higher maximum relative compaction values (compare Figures 14.14 and 14.15 with Figures 12.3 and 12.4). This again suggests that the pneumatic impact compactor was behaving rather as a vibratory compactor. In view of this, it is of interest to determine the size of vibrating-plate compactor which produced a comparable performance.

14.19 Investigations with vibrating-plate compactors have been described in Chapter 9 and relations similar to those obtained with the pneumatic impact compactor on well-graded sand and gravel-sand-clay (Figures 14.14 and 14.15) were obtained with the 710 kg and 670 kg vibrating-plate compactors (Figures 9.4(a) and (c) and 9.5(a) and (c)). These machines had masses per unit area of 1910 and 1570 kg/m^2 respectively, values somewhat greater than the 987 kg/m^2 exerted by the pneumatic impact compactor. With sandy clay No 2, however, both the 710 and 670 kg vibrating-plate compactors achieved maximum relative compaction values of 105 per cent (Figure 9.3) compared with the 97 per cent of the pneumatic impact compactor (Figure 14.13); the 670 kg vibrating-plate compactor also produced highly satisfactory results on the heavy clay (Figure 9.1) compared with poor results, shown in Figure 14.12, from the impact compactor. The performance of the pneumatic impact compactor cannot be matched, therefore, to any one of the vibrating-plate compactors over the full range of soil types tested.

14.20 In the report at the time of the investigation (Lewis and Parsons, 1954) it was considered that the energy of each blow provided by the pneumatic impact compactor had to be greatly increased if satisfactory states of compaction were to be obtained with cohesive soils. It was also concluded that with granular soils it would probably have been beneficial to increase the frequency of blow to about 20 Hz as well as to increase the energy of the blow.

References

Lewis, W A and Parsons, A W (1954). An investigation of the performance of an experimental impact soil compactor. *D.S.I.R., Road Research Laboratory, Research Note* No RN/2301/WAL.AWP (Unpublished)

West, L (1951). An investigation of the performance of an agricultural roller for compacting soil. *D.S.I.R., Road Research Laboratory, Research Note* No RN/1536/LW (Unpublished)

CHAPTER 15 SELECTION OF PLANT FOR THE COMPACTION OF SOIL

15.1 The guiding principle in the selection of the type of plant for compaction work is that the equipment should be capable of producing efficiently and economically the desired state of compaction at the selected, or pertaining, moisture content of the soil (Lewis, 1954). Apart from the ability of the compaction plant to produce the necessary state of compaction, availability and cost of operating the plant will be the main considerations in the selection of the equipment. In small-scale works the cost of transporting the compaction equipment between sites will be important and higher costs of the compaction operation may be acceptable in achieving minimum overall costs for the work. Practical considerations to be taken into account include:–

(1) The dimensions of the compactor in relation to the width of the strip of soil to be compacted, eg in trench work,

and (2) The ability of the compactor to travel efficiently on the soil, especially on the first pass over the loose layer; rollers towed by crawler tractors may have to be used on particularly difficult soils such as uniformly graded sands.

15.2 To aid in the selection of compaction plant estimates have been made of the thickness of layer capable of being compacted and the number of passes required, together with the resultant output of compacted soil, for the various types of compaction equipment listed in Table 15.1. Four general soil types have been taken; these are:–

(a) Heavy cohesive soils (similar to heavy clay and silty clay),

(b) Light cohesive soils (equivalent to sandy clay Nos 1 and 2)

(c) Well-graded granular soils (equivalent to well-graded sand and gravel-sand-clay), and

(d) Uniformly graded sands.

A number of different moisture contents (relative to the optimum of the 2.5 kg rammer compaction test) has been considered for each type of soil. Two different states of compaction have been assumed to be required, one equivalent to an average of 95 per cent relative compaction in the layer, or 90 per cent relative compaction at the bottom of the layer; the other is equivalent to an average of 10 per cent air voids in the layer, or 15 per cent air voids at the bottom of the layer. (The relative compaction levels are those associated with the maximum dry density obtained in the 2.5 kg rammer compaction test). Wherever possible, the general findings given in the appropriate Chapter (see Chapters 4 to 12) have been used in making the estimates, but

Table 15.1 Types of plant used in Tables 15.2 to 15.16 and the details relevant to the determination of the output of the plant and the costs of compaction

(1) Type of plant	(2) Category	(3) Cost per hour (factorised)*	(4) Width compacted (m)	(5) Compacting speed (km/h)	(6) Plant No. (Tables 15.2–15.17)
	Mass per unit width of roll:–				
2.5t 3-wheel smooth-wheeled roller	2000 kg/m	64	1.2	2.5	1
8t tandem smooth-wheeled roller	4000 kg/m	100	1.32	2.5	2
11t 3-wheel smooth-wheeled roller	6000 kg/m	100	1.88	2.5	3
	Contact pressure:–				
12t self-propelled pneumatic-tyred roller	250 kN/m^2	129	1.99	3.5	4
24t self-propelled pneumatic-tyred roller	500 kN/m^2	172	2.32	3.5	5
40t towed pneumatic-tyred roller (Towing unit 250 kW crawler tractor)	750 kN/m^2	431	2.30	3.5	6

Table 15.1 continued Types of plant used in Tables 15.2 to 15.16 and the details relevant to the determination of the output of the plant and the costs of compaction

(1) Type of plant	(2) Category	(3) Cost per hour (factorised)*	(4) Width compacted (m)	(5) Compacting speed (km/h)	(6) Plant No. (Tables 15.2–15.17)
	Mass per unit effective width of roll:–				
7t towed grid roller (Towing unit 48 kW crawler tractor)	7500 kg/m	134	1.59	4.0	7
Three 5t towed club-foot sheepsfoot rollers (side-by-side) (Towing unit 78 kW crawler tractor)	20000 kg/m	203	6.83	4.0	8
17t self-propelled tamping roller (tandem rolls)	30000 kg/m	388	1.93	6.0	9
Three 4.5 t towed taper-foot sheepsfoot rollers (side-by-side) (Towing unit 78 kW crawler tractor)	45000 kg/m	199	6.83	4.0	10
	Mass per unit width of vibrating roll:–				
600 kg pedestrian-operated double vibrating smooth-wheeled roller	400 kg/m	51	0.60	2.0	11
1.7t self-propelled double vibrating smooth-wheeled roller	1000 kg/m	72	0.85	2.5	12
11t self-propelled single-roll vibrating smooth-wheeled roller	3000 kg/m	140	2.13	2.5	13
12t towed vibrating smooth-wheeled roller (Towing unit 60 kW crawler tractor)	5500 kg/m	203	2.08	2.5	14
	Mass per unit effective width of vibrating roll:–				
6t towed vibrating sheepsfoot roller; coverage 0.15 (Towing unit 60 kW crawler tractor)	20000 kg/m	182	1.83	3.5	15
4.3t towed vibrating sheepsfoot roller; coverage 0.07 (Towing unit 60 kW crawler tractor)	45000 kg/m	173	1.45	3.5	16
	Mass per unit area of base plate:–				
80 kg vibrating-plate compactor	450 kg/m^2	42	0.50	1.0	17
150 kg vibrating-plate compactor	800 kg/m^2	45	0.45	1.0	18
700 kg vibrating-plate compactor	1600 kg/m^2	54	0.68	1.0	19
1.5t vibrating-plate compactor	2500 kg/m^2	61	0.91	1.0	20
	Mass per unit area of base plate:–				
50 kg vibro-tamper	700 kg/m^2	42	0.28	0.5	21
80 kg vibro-tamper	850 kg/m^2	42	0.28	0.5	22
100 kg vibro-tamper	1200 kg/m^2	43	0.40	0.5	23
	Specific energy:–				
600 kg power rammer	4 kJ/m^2	61	0.426 m^2 Area compacted per blow	3000 blows/h	24
100 kg power rammer	7 kJ/m^2	47	0.0457 m^2 Area compacted per blow	3600 blows/h	25
Mobile dropping-weight compactor; 600 kg weight: Height of drop:–	Specific energy:–		Area compacted per blow:–	Blows/h:–	
0.1 m	5 kJ/m^2	123	0.093 m^2	3600	26
0.2 m	12 kJ/m^2	123	0.093 m^2	3300	27
0.9 m	50 kJ/m^2	123	0.093 m^2	2700	28
2.2 m	120 kJ/m^2	123	0.093 m^2	1500	29

*Taking the operating cost of the 11 t smooth-wheeled roller as 100

where general information was not available some judgement was used in assessing the likely figures based on results obtained with individual machines. Where compaction plant were incapable of producing the required state of compaction in the minimum compacted layer thickness (100 mm) in 16 passes it has been noted as 'Not suitable'. The assumption has been made in the calculations of output that the plant is operated for 85 per cent of available time (approximately 50 minutes in each hour) (see Paragraph 19.52) to allow for time lost in turning and manoeuvring the plant and other small stoppages. No allowance has been made for delays caused by bad weather or for the additional costs of transporting the plant to and from the site. The results of the study are presented in Tables 15.2 to 15.16.

15.3 The types of plant selected for use in the calculations and listed in Table 15.1 are representative of the range of machines used in the investigations at the Laboratory, but wherever possible machines currently available have been taken as the models. To avoid the problems of variations in costs with time the costs per hour of operating the plant have been factorised, ie they are expressed as a percentage of the operating cost per hour of the 11 t smooth-wheeled roller. Elements included in the costs are the costs of owning and maintaining the equipment, the cost of consumables such as fuel, and the cost of the operator. These have been based on current hire rates where they are available (Civil Engineering, 1988), but where items of plant were not listed in published hire rates, estimates were made from the published prices of new equipment and the installed engine power (Plant Assessment, 1988).

Plate 15.2 An 8 t tandem smooth-wheeled roller of the type used in the calculations of outputs and costs of compaction (Photograph courtesy of Aveling Barford (Machines) Ltd)

15.4 *Smooth-wheeled rollers.* The three machines given in Table 15.1 are all current models and are shown in Plates 15.1 to 15.3.

15.5 *Pneumatic-tyred rollers.* The range of tyre contact pressures is that covered in the investigations described in Chapter 5. Two self-propelled machines, suitably ballasted, provide the two lower values of contact pressure and a modern towed machine provides the higher

Plate 15.1 A 2.5 t smooth-wheeled roller of the type used in the calculations of outputs and costs of compaction (Photograph courtesy of BSP International Foundations Ltd)

Plate 15.3 An 11 t smooth-wheeled roller of the type used in the calculations of outputs and costs of compaction (Photograph courtesy of Aveling Barford (Machines) Ltd)

Table 15.2(a) Calculated outputs of plant and factorised costs of compaction to achieve 95 per cent relative compaction on heavy cohesive soils at a moisture content 10 per cent below the optimum of the 2.5 kg rammer compaction test

Plant No. (From Table 15.1)	Depth of compacted layer (mm)	Number of machine passes required	Output of compacted soil per hour* (m^3)	Factorised cost per m^3 of compacted soil
1	Not suitable			
2	150	6	70	1.4
3	150	6	100	1.0
4	Not suitable			
5	150	16	65	2.6
6	150	8	128	3.4
7	Not suitable			
8	100	16	145	1.4
9	150	6	246	1.6
10	150	12	290	0.7
11	Not suitable			
12	Not suitable			
13	150	12	57	2.5
14	150	6	111	1.8
15	Insufficient information			
16	Insufficient information			
17	Not suitable			
18	Not suitable			
19	Not suitable			
20	Insufficient information			
21	Not suitable			
22	Not suitable			
23	100	16	1.1	39
24	Not suitable			
25	150	6 blows	3.5	13
26	Not suitable			
27	150	12 blows	3.3	37
28	150	4 blows	8.0	15
29	200	4 blows	5.9	21

*Assuming a 50-minute hour

Table 15.2(b) Calculated outputs of plant and factorised costs of compaction to achieve 10 per cent air voids on heavy cohesive soils at a moisture content 10 per cent below the optimum of the 2.5 kg rammer compaction test

Plant No. (From Table 15.1)	Depth of compacted layer (mm)	Number of machine passes required	Output of compacted soil per hour* (m^3)	Factorised cost per m^3 of compacted soil
1	Not suitable			
2	Not suitable			
3	Not suitable			
4	Not suitable			
5	Not suitable			
6	Not suitable			
7	Not suitable			
8	Not suitable			
9	Insufficient information			
10	Not suitable			
11	Not suitable			
12	Not suitable			
13	Not suitable			
14	Not suitable			
15	Not suitable			
16	Not suitable			
17	Not suitable			
18	Not suitable			
19	Not suitable			
20	Not suitable			
21	Not suitable			
22	Not suitable			
23	Not suitable			
24	Not suitable			
25	Not suitable			
26	Not suitable			
27	100	10 blows	1.6	77
28	150	6 blows	5.3	23
29	150	4 blows	4.4	28

*Assuming a 50-minute hour

value. The smaller of the self-propelled machines is illustrated in Plate 15.4.

15.6 *Sheepsfoot, grid and tamping rollers.* The machines given are similar to those used in the original investigations (Chapter 6) and the photographs in that Chapter should be referred

Text continues on page 234

Plate 15.4 A self-propelled pneumatic-tyred roller of the type used (when ballasted to a total mass of 12 t) in the calculations of outputs and costs of compaction (Photograph courtesy of Bomag (Great Britain) Limited)

Table 15.3(a) Calculated outputs of plant and factorised costs of compaction to achieve 95 per cent relative compaction on heavy cohesive soils at a moisture content 5 per cent below the optimum of the 2.5 kg rammer compaction test

Plant No. (From Table 15.1)	Depth of compacted layer (mm)	Number of machine passes required	Output of compacted soil per hour* (m^3)	Factorised cost per m^3 of compacted soil
1	Not suitable			
2	150	4	105	1.0
3	150	4	150	0.7
4	100	16	37	3.5
5	150	8	129	1.3
6	150	4	257	1.7
7	100	16	34	3.9
8	150	12	290	0.7
9	150	3	492	0.8
10	150	6	580	0.3
11	Not suitable			
12	Not suitable			
13	150	8	85	1.6
14	150	4	166	1.2
15	Insufficient information			
16	100	12	36	4.8
17	Not suitable			
18	Not suitable			
19	Insufficient information			
20	Insufficient information			
21	100	16	0.7	60
22	100	16	0.7	60
23	150	12	2.1	21
24	150	16 blows	10	6.0
25	150	6 blows	3.5	13
26	150	16 blows	2.7	46
27	150	6 blows	6.5	19
28	200	4 blows	11	12
29	250	4 blows	7.4	17

*Assuming a 50-minute hour

Table 15.3(b) Calculated outputs of plant and factorised costs of compaction to achieve 10 per cent air voids on heavy cohesive soils at a moisture content 5 per cent below the optimum of the 2.5 kg rammer compaction test

Plant No. (From Table 15.1)	Depth of compacted layer (mm)	Number of machine passes required	Output of compacted soil per hour* (m^3)	Factorised cost per m^3 of compacted soil
1	Not suitable			
2	100	8	35	2.9
3	150	12	50	2.0
4	Not suitable			
5	Not suitable			
6	150	12	86	5.0
7	Not suitable			
8	Not suitable			
9	Insufficient information			
10	Not suitable			
11	Not suitable			
12	Not suitable			
13	Not suitable			
14	150	8	83	2.4
15	Not suitable			
16	Not suitable			
17	Not suitable			
18	Not suitable			
19	Not suitable			
20	Not suitable			
21	Not suitable			
22	Not suitable			
23	Not suitable			
24	Not suitable			
25	150	12	1.7	28
26	100	16	1.8	68
27	150	12	3.3	37
28	150	4	8.0	15
29	200	4	5.9	21

*Assuming a 50-minute hour

Table 15.4(a) Calculated outputs of plant and factorised costs of compaction to achieve 95 per cent relative compaction on heavy cohesive soils at a moisture content equal to the optimum of the 2.5 kg rammer compaction test

Plant No. (From Table 15.1)	Depth of compacted layer (mm)	Number of machine passes required	Output of compacted soil per hour* (m^3)	Factorised cost per m^3 of compacted soil
1	100	16	16	4.0
2	150	3	140	0.7
3	150	4	150	0.7
4	150	16	55	2.3
5	150	4	259	0.7
6	300	4	513	0.8
7	150	12	68	2.0
8	150	6	581	0.3
9	150	2	738	0.5
10	Probably not suitable because of over-stressing			
11	Not suitable			
12	Not suitable			
13	150	4	170	0.8
14	200	4	221	0.9
15	150	6	136	1.3
16	Probably not suitable because of over-stressing			
17	Not suitable			
18	Not suitable			
19	150	6	14	3.9
20	Insufficient information			
21	150	12	1.5	28
22	150	8	2.2	19
23	150	4	6.4	6.7
24	150	12 blows	14	4.5
25	150	4 blows	5.2	9.0
26	150	8 blows	5.3	23
27	150	4 blows	9.8	13
28	300	4 blows	16	7.7
29	400	4 blows	12	10

*Assuming a 50-minute hour

Table 15.4(b) Calculated outputs of plant and factorised costs of compaction to achieve 10 per cent air voids on heavy cohesive soils at a moisture content equal to the optimum of the 2.5 kg rammer compaction test

Plant No. (From Table 15.1)	Depth of compacted layer (mm)	Number of machine passes required	Output of compacted soil per hour* (m^3)	Factorised cost per m^3 of compacted soil
1	Not suitable			
2	150	4	105	1.0
3	150	4	150	0.7
4	100	16	37	3.5
5	150	4	259	0.7
6	300	4	513	0.8
7	150	12	68	2.0
8	Insufficient information			
9	150	4	369	1.1
10	Probably not suitable because of over-stressing			
11	Not suitable			
12	Not suitable			
13	150	4	170	0.8
14	200	4	221	0.9
15	150	6	136	1.3
16	Probably not suitable because of over-stressing			
17	Not suitable			
18	Not suitable			
19	150	6	14	3.9
20	Insufficient information			
21	150	16	1.1	38
22	150	12	1.5	28
23	150	8	3.2	13
24	150	12 blows	14	4.5
25	150	6 blows	3.5	13
26	150	8 blows	5.3	23
27	150	4 blows	9.8	13
28	350	4 blows	19	6.6
29	450	4 blows	13	9.2

*Assuming a 50-minute hour

Table 15.5(a) Calculated outputs of plant and factorised costs of compaction to achieve 95 per cent relative compaction on heavy cohesive soils at a moisture content 5 per cent above the optimum of the 2.5 kg rammer compaction test

Plant No. (From Table 15.1)	Depth of compacted layer (mm)	Number of machine passes required	Output of compacted soil per hour* (m^3)	Factorised cost per m^3 of compacted soil
1	150	16	24	2.7
2	150	3	140	0.7
3	200	4	200	0.5
4	150	6	148	0.9
5	Probably not suitable because of over-stressing			
6	Probably not suitable because of over-stressing			
7	150	6	135	1.0
8	Probably not suitable because of over-stressing			
9	150	2	738	0.5
10	Probably not suitable because of over-stressing			
11	Not suitable			
12	150	6	45	1.6
13	150	4	170	0.8
14	250	4	276	0.7
15	Probably not suitable because of over-stressing			
16	Probably not suitable because of over-stressing			
17	Not suitable			
18	Not suitable			
19	Insufficient information			
20	Insufficient information			
21	150	4	4.5	9.3
22	150	4	4.5	9.3
23	300	4	13	3.4
24	150	8 blows	20	3.0
25	Probably not suitable because of over-stressing			
26	150	6 blows	7.1	17
27	250	4 blows	16	7.5
28	400	4 blows	21	5.8
29	500	4 blows	15	8.3

*Assuming a 50-minute hour

Table 15.5(b) Calculated outputs of plant and factorised costs of compaction to achieve 10 per cent air voids on heavy cohesive soils at a moisture content 5 per cent above the optimum of the 2.5 kg rammer compaction test

Plant No. (From Table 15.1)	Depth of compacted layer (mm)	Number of machine passes required	Output of compacted soil per hour* (m^3)	Factorised cost per m^3 of compacted soi
1	150	8	48	1.3
2	200	2	281	0.4
3	250	4	250	0.4
4	150	4	222	0.6
5	Probably not suitable because of over-stressing			
6	Probably not suitable because of over-stressing			
7	Insufficient information			
8	Probably not suitable because of over-stressing			
9	Insufficient information			
10	Probably not suitable because of over-stressing			
11	Not suitable			
12	150	3	90	0.8
13	250	4	283	0.5
14	400	4	442	0.5
15	Probably not suitable because of over-stressing			
16	Probably not suitable because of over-stressing			
17	Not suitable			
18	Not suitable			
19	Insufficient information			
20	Insufficient information			
21	150	4	4.5	9.3
22	150	4	4.5	9.3
23	300	4	13	3.4
24	150	6 blows	27	2.2
25	Probably not suitable because of over-stressing			
26	150	4 blows	11	12
27	450	4 blows	29	4.2
28	500	4 blows	27	4.6
29	600	4 blows	18	6.9

*Assuming a 50-minute hour

Table 15.6(a) Calculated outputs of plant and factorised costs of compaction to achieve 95 per cent relative compaction on light cohesive soils at a moisture content 6 per cent below the optimum of the 2.5 kg rammer compaction test

Plant No. (From Table 15.1)	Depth of compacted layer (mm)	Number of machine passes required	Output of compacted soil per hour* (m^3)	Factorised cost per m^3 of compacted soil
1	Not suitable			
2	Not suitable			
3	Not suitable			
4	Not suitable			
5	Not suitable			
6	150	6	171	2.5
7	Not suitable			
8	100	16	145	1.4
9	100	8	123	3.2
10	Not suitable			
11	Not suitable			
12	Not suitable			
13	150	8	85	1.6
14	150	4	166	1.2
15	Not suitable			
16	Not suitable			
17	Not suitable			
18	Not suitable			
19	150	16	5.4	10
20	150	4	29	2.1
21	100	16	0.7	60
22	100	16	0.7	60
23	150	12	2.1	21
24	Not suitable			
25	Not suitable			
26	150	16 blows	2.7	46
27	150	8 blows	4.9	25
28	200	4 blows	11	12
29	350	4 blows	10	12

*Assuming a 50-minute hour

Table 15.6(b) Calculated outputs of plant and factorised costs of compaction to achieve 10 per cent air voids on light cohesive soils at a moisture content 6 per cent below the optimum of the 2.5 kg rammer compaction test

Plant No. (From Table 15.1)	Depth of compacted layer (mm)	Number of machine passes required	Output of compacted soil per hour* (m^3)	Factorised cost per m^3 of compacted soil
1	Not suitable			
2	Not suitable			
3	Not suitable			
4	Not suitable			
5	Not suitable			
6	Not suitable			
7	Not suitable			
8	Not suitable			
9	Not suitable			
10	Not suitable			
11	Not suitable			
12	Not suitable			
13	Not suitable			
14	150	12	55	3.7
15	Not suitable			
16	Not suitable			
17	Not suitable			
18	Not suitable			
19	100	16	3.6	15
20	150	8	15	4.2
21	Not suitable			
22	100	16	0.7	60
23	150	16	1.6	27
24	Not suitable			
25	Not suitable			
26	Not suitable			
27	150	12 blows	3.3	37
28	150	4 blows	8.0	15
29	200	4 blows	5.9	21

*Assuming a 50-minute hour

Table 15.7(a) Calculated outputs of plant and factorised costs of compaction to achieve 95 per cent relative compaction on light cohesive soils at a moisture content 3 per cent below the optimum of the 2.5 kg rammer compaction test

Plant No. (From Table 15.1)	Depth of compacted layer (mm)	Number of machine passes required	Output of compacted soil per hour* (m^3)	Factorised cost per m^3 of compacted soil
1	Not suitable			
2	100	8	35	2.9
3	150	16	37	2.7
4	100	16	37	3.5
5	150	6	173	1.0
6	300	4	513	0.8
7	Not suitable			
8	100	16	145	1.4
9	150	6	246	1.6
10	150	12	290	0.7
11	Not suitable			
12	Not suitable			
13	150	6	113	1.2
14	200	4	221	0.9
15	Not suitable			
16	150	16	40	4.3
17	Not suitable			
18	Not suitable			
19	150	12	7.2	7.5
20	250	4	48	1.3
21	100	16	0.7	60
22	150	16	1.1	38
23	150	6	4.3	10
24	Not suitable			
25	150	12 blows	1.7	28
26	150	12 blows	3.6	34
27	150	6 blows	6.5	19
28	200	4 blows	11	12
29	350	4 blows	10	12

*Assuming a 50-minute hour

Table 15.7(b) Calculated outputs of plant and factorised costs of compaction to achieve 10 per cent air voids on light cohesive soils at a moisture content 3 per cent below the optimum of the 2.5 kg rammer compaction test

Plant No. (From Table 15.1)	Depth of compacted layer (mm)	Number of machine passes required	Output of compacted soil per hour* (m^3)	Factorised cost per m^3 of compacted soil
1	Not suitable			
2	Not suitable			
3	100	16	25	4.0
4	Not suitable			
5	100	16	43	4.0
6	150	4	257	1.7
7	Not suitable			
8	Not suitable			
9	Not suitable			
10	Not suitable			
11	Not suitable			
12	Not suitable			
13	150	12	57	2.5
14	150	4	166	1.2
15	Not suitable			
16	Not suitable			
17	Not suitable			
18	Not suitable			
19	150	12	7.2	7.5
20	150	4	29	2.1
21	100	16	0.7	60
22	150	12	1.5	28
23	150	6	4.3	10
24	Not suitable			
25	150	16 blows	1.3	36
26	150	16 blows	2.7	46
27	150	6 blows	6.5	19
28	150	4 blows	8.0	15
29	250	4 blows	7.4	17

*Assuming a 50-minute hour

Table 15.8(a) Calculated outputs of plant and factorised costs of compaction to achieve 95 per cent relative compaction on light cohesive soils at a moisture content equal to the optimum of the 2.5 kg rammer compaction test

Plant No. (From Table 15.1)	Depth of compacted layer (mm)	Number of machine passes required	Output of compacted soil per hour* (m^3)	Factorised cost per m^3 of compacted soil
1	100	16	16	4.0
2	150	6	70	1.4
3	150	6	100	1.0
4	150	12	74	1.7
5	250	4	431	0.4
6	400	4	684	0.6
7	150	12	68	2.0
8	150	12	290	0.7
9	150	3	492	0.8
10	Probably not suitable because of over-stressing			
11	Not suitable			
12	Not suitable			
13	150	4	170	0.8
14	200	4	221	0.9
15	150	4	204	0.9
16	Probably not suitable because of over-stressing			
17	Not suitable			
18	Not suitable			
19	150	6	15	3.7
20	350	4	68	0.9
21	150	12	1.5	28
22	150	6	3.0	14
23	200	4	8.5	5.1
24	150	16 blows	10	6.0
25	150	8 blows	2.6	18
26	150	8 blows	5.3	23
27	150	4 blows	9.8	13
28	400	4 blows	21	5.8
29	500	4 blows	15	8.3

*Assuming a 50-minute hour

Table 15.8(b) Calculated outputs of plant and factorised costs of compaction to achieve 10 per cent air voids on light cohesive soils at a moisture content equal to the optimum of the 2.5 kg rammer compaction test

Plant No. (From Table 15.1)	Depth of compacted layer (mm)	Number of machine passes required	Output of compacted soil per hour* (m^3)	Factorised cost per m^3 of compacted soil
1	100	16	16	4.0
2	150	4	105	1.0
3	150	6	100	1.0
4	100	16	37	3.5
5	250	4	431	0.4
6	450	4	770	0.6
7	150	12	68	2.0
8	100	16	145	1.4
9	150	4	369	1.1
10	Probably not suitable because of over-stressing			
11	Not suitable			
12	Not suitable			
13	150	4	170	0.8
14	200	4	221	0.9
15	150	6	136	1.3
16	Probably not suitable because of over-stressing			
17	Not suitable			
18	Not suitable			
19	150	6	15	3.7
20	250	4	48	1.3
21	150	8	2.2	19
22	150	4	4.5	9.3
23	250	4	11	4.1
24	150	12 blows	14	4.5
25	150	8 blows	2.6	18
26	150	8 blows	5.3	23
27	150	4 blows	9.8	13
28	450	4 blows	24	5.1
29	500	4 blows	15	8.3

*Assuming a 50-minute hour

Table 15.9(a) Calculated outputs of plant and factorised costs of compaction to achieve 95 per cent relative compaction on light cohesive soils at a moisture content 3 per cent above the optimum of the 2.5 kg rammer compaction test

Plant No. (From Table 15.1)	Depth of compacted layer (mm)	Number of machine passes required	Output of compacted soil per hour* (m^3)	Factorised cost per m^3 of compacted soil
1	150	12	32	2.0
2	150	2	210	0.5
3	150	4	150	0.7
4	150	4	222	0.6
5	350	4	604	0.3
6	500	4	855	0.5
7	150	6	135	1.0
8	Probably not suitable because of over-stressing			
9	150	2	738	0.5
10	Probably not suitable because of over-stressing			
11	Not suitable			
12	150	8	34	2.1
13	150	4	170	0.8
14	250	4	276	0.7
15	Probably not suitable because of over-stressing			
16	Probably not suitable because of over-stressing			
17	Not suitable			
18	Not suitable			
19	200	4	29	1.9
20	450	4	87	0.7
21	150	6	3.0	14
22	150	4	4.5	9.3
23	250	4	11	4.1
24	150	12 blows	14	4.5
25	150	6 blows	3.5	13
26	150	6 blows	7.1	17
27	250	4 blows	16	7.5
28	500	4 blows	27	4.6
29	600	4 blows	18	6.9

*Assuming a 50-minute hour

Table 15.9(b) Calculated outputs of plant and factorised costs of compaction to achieve 10 per cent air voids on light cohesive soils at a moisture content 3 per cent above the optimum of the 2.5 kg rammer compaction test

Plant No. (From Table 15.1)	Depth of compacted layer (mm)	Number of machine passes required	Output of compacted soil per hour* (m^3)	Factorised cost per m^3 of compacted soil
1	150	6	64	1.0
2	200	2	281	0.4
3	200	4	200	0.5
4	200	4	295	0.4
5	450	4	776	0.2
6	600	4	1026	0.4
7	150	4	203	0.7
8	Probably not suitable because of over-stressing			
9	200	2	984	0.4
10	Probably not suitable because of over-stressing			
11	100	8	13	3.9
12	150	6	45	1.6
13	200	4	226	0.6
14	300	4	332	0.6
15	Probably not suitable because of over-stressing			
16	Probably not suitable because of over-stressing			
17	Not suitable			
18	150	16	3.6	13
19	200	4	29	1.9
20	500	4	97	0.6
21	200	4	6.0	7.0
22	250	4	7.4	5.7
23	400	4	17	2.5
24	150	6 blows	27	2.2
25	150	4 blows	5.2	9.0
26	150	6 blows	7.1	17
27	300	4 blows	20	6.3
28	500	4 blows	27	4.6
29	600	4 blows	18	6.9

*Assuming a 50-minute hour

Table 15.10(a) Calculated outputs of plant and factorised costs of compaction to achieve 95 per cent relative compaction on well-graded granular soils at a moisture content 5 per cent below the optimum of the 2.5 kg rammer compaction test

Plant No. (From Table 15.1)	Depth of compacted layer (mm)	Number of machine passes required	Output of compacted soil per hour* (m^3)	Factorised cost per m^3 of compacted soil
1	Not suitable			
2	100	8	35	2.9
3	150	12	50	2.0
4	Not suitable			
5	100	16	43	4.0
6	100	16	43	10
7	Not suitable			
8	Not suitable			
9	Not suitable			
10	Not suitable			
11	Not suitable			
12	Not suitable			
13	150	16	42	3.3
14	150	8	83	2.4
15	Not suitable			
16	Not suitable			
17	Not suitable			
18	Not suitable			
19	150	16	5.4	10
20	150	8	15	4.2
21	100	16	0.7	60
22	100	16	0.7	60
23	150	16	1.6	27
24	100	16 blows	6.8	9.0
25	150	12 blows	1.7	28
26	Insufficient information			
27	Insufficient information			
28	Insufficient information			
29	Not suitable			

*Assuming a 50-minute hour

Table 15.10(b) Calculated outputs of plant and factorised costs of compaction to achieve 10 per cent air voids on well-graded granular soils at a moisture content 5 per cent below the optimum of the 2.5 kg rammer compaction test

Plant No. (From Table 15.1)	Depth of compacted layer (mm)	Number of machine passes required	Output of compacted soil per hour* (m^3)	Factorised cost per m^3 of compacted soil
1	Not suitable			
2	Not suitable			
3	Not suitable			
4	Not suitable			
5	Not suitable			
6	100	16	43	10
7	Not suitable			
8	Not suitable			
9	Not suitable			
10	Not suitable			
11	Not suitable			
12	Not suitable			
13	Not suitable			
14	100	16	28	7.3
15	Not suitable			
16	Not suitable			
17	Not suitable			
18	Not suitable			
19	Not suitable			
20	Not suitable			
21	Not suitable			
22	Not suitable			
23	Not suitable			
24	Not suitable			
25	Not suitable			
26	Insufficient information			
27	Insufficient information			
28	Insufficient information			
29	Not suitable			

*Assuming a 50-minute hour

Table 15.11(a) Calculated outputs of plant and factorised costs of compaction to achieve 95 per cent relative compaction on well-graded granular soils at a moisture content 2.5 per cent below the optimum of the 2.5 kg rammer compaction test

Plant No. (From Table 15.1)	Depth of compacted layer (mm)	Number of machine passes required	Output of compacted soil per hour* (m^3)	Factorised cost per m^3 of compacted soil
1	150	16	24	2.7
2	150	3	140	0.7
3	150	4	150	0.7
4	100	16	37	3.5
5	150	8	129	1.3
6	150	4	257	1.7
7	150	12	68	2.0
8	Not suitable			
9	100	8	123	3.2
10	Probably not suitable because of over-stressing			
11	Not suitable			
12	150	6	45	1.6
13	150	4	170	0.8
14	200	4	221	0.9
15	150	12	68	2.7
16	Probably not suitable because of over-stressing			
17	Not suitable			
18	150	16	3.6	13
19	150	6	15	3.7
20	150	4	29	2.1
21	150	16	1.1	38
22	150	12	1.5	28
23	150	4	6.4	6.7
24	150	12 blows	14	4.5
25	150	6 blows	3.5	13
26	Insufficient information			
27	Insufficient information			
28	Insufficient information			
29	150	4 blows	4.4	28

*Assuming a 50-minute hour

Table 15.11(b) Calculated outputs of plant and factorised costs of compaction to achieve 10 per cent air voids on well-graded granular soils at a moisture content 2.5 per cent below the optimum of the 2.5 kg rammer compaction test

Plant No. (From Table 15.1)	Depth of compacted layer (mm)	Number of machine passes required	Output of compacted soil per hour* (m^3)	Factorised cost per m^3 of compacted soil
1	Not suitable			
2	150	6	70	1.4
3	150	6	100	1.0
4	100	16	37	3.5
5	100	16	43	4.0
6	150	12	86	5.0
7	Insufficient information			
8	Not suitable			
9	Insufficient information			
10	Probably not suitable because of over-stressing			
11	Not suitable			
12	150	8	34	21
13	150	6	113	1.2
14	150	4	166	1.2
15	Not suitable			
16	Not suitable			
17	Not suitable			
18	Not suitable			
19	150	12	7.2	7.5
20	150	6	19	3.2
21	100	16	0.7	60
22	150	16	1.1	38
23	150	8	3.2	13
24	150	16 blows	10	6.0
25	150	12 blows	1.7	28
26	Insufficient information			
27	Insufficient information			
28	Insufficient information			
29	150	4 blows	4.4	28

*Assuming a 50-minute hour

Table 15.12(a) Calculated outputs of plant and factorised costs of compaction to achieve 95 per cent relative compaction on well-graded granular soils at a moisture content equal to the optimum of the 2.5 kg rammer compaction test

Plant No. (From Table 15.1)	Depth of compacted layer (mm)	Number of machine passes required	Output of compacted soil per hour* (m^3)	Factorised cost per m^3 of compacted soil
1	150	6	64	1.0
2	200	2	281	0.4
3	250	4	250	0.4
4	150	4	222	0.6
5	300	4	518	0.3
6	400	4	684	0.6
7	200	4	270	0.5
8	Probably not suitable because of over-stressing			
9	150	3	492	0.8
10	Probably not suitable because of over-stressing			
11	150	3	51	1.0
12	150	2	135	0.5
13	200	4	226	0.6
14	250	4	276	0.7
15	150	4	204	0.9
16	Probably not suitable because of over-stressing			
17	150	16	4.0	11
18	150	8	7.2	6.3
19	250	4	36	1.5
20	600	4	116	0.5
21	150	4	4.5	9.3
22	150	4	4.5	9.3
23	200	4	8.5	5.1
24	150	6 blows	27	2.2
25	200	4 blows	7.0	6.7
26	Insufficient information			
27	Insufficient information			
28	Insufficient information			
29	700	4 blows	21	5.9

*Assuming a 50-minute hour

Table 15.12(b) Calculated outputs of plant and factorised costs of compaction to achieve 10 per cent air voids on well-graded granular soils at a moisture content equal to the optimum of the 2.5 kg rammer compaction test

Plant No. (From Table 15.1)	Depth of compacted layer (mm)	Number of machine passes required	Output of compacted soil per hour* (m^3)	Factorised cost per m^3 of compacted soil
1	150	6	64	1.0
2	200	2	281	0.4
3	250	4	250	0.4
4	200	4	295	0.4
5	200	4	345	0.5
6	250	4	428	1.0
7	Insufficient information			
8	Probably not suitable because of over-stressing			
9	Insufficient information			
10	Probably not suitable because of over-stressing			
11	150	3	51	1.0
12	150	2	135	0.5
13	200	4	226	0.6
14	300	4	332	0.6
15	150	4	204	0.9
16	Probably not suitable because of over-stressing			
17	150	12	5.3	7.9
18	150	4	14	3.1
19	300	4	43	1.2
20	700	4	135	0.5
21	150	4	4.5	9.3
22	200	4	6.0	7.0
23	200	4	8.5	5.1
24	150	4 blows	41	1.5
25	200	4 blows	7.0	6.7
26	Insufficient information			
27	Insufficient information			
28	Insufficient information			
29	700	4 blows	21	5.9

*Assuming a 50-minute hour

Table 15.13(a) Calculated outputs of plant and factorised costs of compaction to achieve 95 per cent relative compaction on well-graded granular soils at a moisture content 2.5 per cent above the optimum of the 2.5 kg rammer compaction test

Plant No. (From Table 15.1)	Depth of compacted layer (mm)	Number of machine passes required	Output of compacted soil per hour* (m^3)	Factorised cost per m^3 of compacted soil
1	200	4	128	0.5
2	Probably not suitable because of over-stressing			
3	Probably not suitable because of over-stressing			
4	350	4	517	0.2
5	Probably not suitable because of over-stressing			
6	Probably not suitable because of over-stressing			
7	300	4	405	0.3
8	Probably not suitable because of over-stressing			
9	Probably not suitable because of over-stressing			
10	Probably not suitable because of over-stressing			
11	150	2	77	0.7
12	150	2	135	0.5
13	200	4	226	0.6
14	250	4	276	0.7
15	Probably not suitable because of over-stressing			
16	Probably not suitable because of over-stressing			
17	150	8	8.0	5.3
18	200	4	19	2.4
19	800	4	116	0.5
20	1000	4	193	0.3
21	150	4	4.5	9.3
22	350	4	10	4.0
23	500	4	21	2.0
24	300	4 blows	82	0.7
25	400	4 blows	14	3.4
26	Insufficient information			
27	Insufficient information			
28	Insufficient information			
29	1000	4 blows	30	4.2

*Assuming a 50-minute hour

Table 15.13(b) Calculated outputs of plant and factorised costs of compaction to achieve 10 per cent air voids on well-graded granular soils at a moisture content 2.5 per cent above the optimum of the 2.5 kg rammer compaction test

Plant No. (From Table 15.1)	Depth of compacted layer (mm)	Number of machine passes required	Output of compacted soil per hour* (m^3)	Factorised cost per m^3 of compacted soil
1	250	4	159	0.4
2	Probably not suitable because of over-stressing			
3	Probably not suitable because of over-stressing			
4	350	4	517	0.2
5	Probably not suitable because of over-stressing			
6	Probably not suitable because of over-stressing			
7	Insufficient information			
8	Probably not suitable because of over-stressing			
9	Probably not suitable because of over-stressing			
10	Probably not suitable because of over-stressing			
11	150	2	77	0.7
12	150	2	135	0.5
13	250	4	283	0.5
14	350	4	387	0.5
15	Probably not suitable because of over-stressing			
16	Probably not suitable because of over-stressing			
17	150	4	16	2.6
18	200	4	19	2.4
19	500	4	72	0.7
20	1000	4	193	0.3
21	300	4	8.9	4.7
22	400	4	12	3.5
23	500	4	21	2.0
24	700	4 blows	190	0.3
25	600	4 blows	21	2.2
26	Insufficient information			
27	Insufficient information			
28	Insufficient information			
29	1000	4 blows	30	4.2

*Assuming a 50-minute hour

Table 15.14 Calculated outputs of plant and factorised costs of compaction to achieve 95 per cent relative compaction on uniformly graded fine sand at a moisture content 8 per cent below the optimum of the 2.5 kg rammer compaction test

Plant No. (From Table 15.1)	Depth of compacted layer (mm)	Number of machine passes required	Output of compacted soil per hour* (m^3)	Factorised cost per m^3 of compacted soil
1	Insufficient information			
2	Insufficient information			
3	Insufficient information			
4	Insufficient information			
5	Insufficient information			
6	Insufficient information			
7	Insufficient information			
8	Insufficient information			
9	200	2	984	0.4
10	Insufficient information			
11	150	2	77	0.7
12	200	2	181	0.4
13	250	4	283	0.5
14	350	4	387	0.5
15	Insufficient information			
16	150	8	81	2.1
17	Not suitable			
18	Not suitable			
19	150	12	7.2	7.5
20	150	6	19	3.2
21	150	4	4.5	9.3
22	150	4	4.5	9.3
23	250	4	11	3.9
24	Insufficient information			
25	Insufficient information			
26	Insufficient information			
27	Insufficient information			
28	200	4 blows	11	12
29	Insufficient information			

*Assuming a 50-minute hour

Table 15.15 Calculated outputs of plant and factorised costs of compaction to achieve 95 per cent relative compaction on uniformly graded fine sand at a moisture content 4 per cent below the optimum of the 2.5 kg rammer compaction test

Plant No. (From Table 15.1)	Depth of compacted layer (mm)	Number of machine passes required	Output of compacted soil per hour* (m^3)	Factorised cost per m^3 of compacted soil
1	Insufficient information			
2	Insufficient information			
3	Insufficient information			
4	Insufficient information			
5	Insufficient information			
6	Insufficient information			
7	150	4	203	0.7
8	Insufficient information			
9	Insufficient information			
10	Insufficient information			
11	150	2	77	0.7
12	200	2	181	0.4
13	300	4	339	0.4
14	350	4	387	0.5
15	Insufficient information			
16	150	8	81	2.1
17	100	16	2.7	16
18	150	8	7.2	6.3
19	150	4	22	2.5
20	200	4	39	1.6
21	Insufficient information			
22	150	4	4.5	9.3
23	Insufficient information			
24	Insufficient information			
25	Insufficient information			
26	Insufficient information			
27	Insufficient information			
28	Insufficient information			
29	Insufficient information			

*Assuming a 50-minute hour

Table 15.16 Calculated outputs of plant and factorised costs of compaction to achieve 95 per cent relative compaction on uniformly graded fine sand at a moisture content equal to the optimum of the 2.5 kg rammer compaction test

Plant No. (From Table 15.1)	Depth of compacted layer (mm)	Number of machine passes required	Output of compacted soil per hour* (m^3)	Factorised cost per m^3 of compacted soil
1	Insufficient information			
2	Insufficient information			
3	Insufficient information			
4	Insufficient information			
5	Insufficient information			
6	Insufficient information			
7	Insufficient information			
8	Insufficient information			
9	Insufficient information			
10	Insufficient information			
11	150	2	77	0.7
12	200	2	181	0.4
13	300	4	339	0.4
14	350	4	387	0.5
15	Insufficient information			
16	150	8	81	2.1
17	150	6	11	4.0
18	150	4	14	3.1
19	350	4	51	1.1
20	1000	4	193	0.3
21	Insufficient information			
22	Insufficient information			
23	Insufficient information			
24	Insufficient information			
25	Insufficient information			
26	Insufficient information			
27	Insufficient information			
28	Insufficient information			
29	Insufficient information			

*Assuming a 50-minute hour

Text continued from page 221

to. It has been assumed that the club-foot and taper-foot sheepsfoot rollers would be towed in a combination of three units, a commonly used ploy to increase productivity in tropical countries. The costs of these units were largely estimated from costs pertaining at the time of the investigations.

15.7 *Vibrating smooth-wheeled rollers.* The four machines listed are all modern vibrating rollers. The 12 t towed vibrating roller is illustrated in Plate 7.16; the three other machines are shown in Plates 15.5 to 15.7.

Plate 15.5 A 600 kg pedestrian-operated double vibrating smooth-wheeled roller of the type used in the calculations of outputs and costs of compaction (Photograph courtesy of Bomag (Great Britain) Limited)

15.8 *Vibrating sheepsfoot rollers.* The two towed machines used in the investigations described in Chapter 8 and shown in Plates 8.2 and 8.3 have been taken as examples of this type of equipment.

Plate 15.7 An 11 t self-propelled single-roll vibrating smooth-wheeled roller of the type used in the calculations of outputs and costs of compaction (Photograph courtesy of Bomag (Great Britain) Limited)

Plate 15.6 A 1.7 t self-propelled double vibrating smooth-wheeled roller of the type used in the calculations of outputs and costs of compaction (Photograph courtesy of Bomag (Great Britain) Limited)

15.9 *Vibrating-plate compactors.* Modern machines have been taken as examples of the first three vibrating-plate compactors listed in Table 15.1. The 80 kg machine is similar to that used in the investigations described in Chapter 9 and shown in Plate 9.2, and examples of 150 kg and 700 kg machines are shown in Plates 15.8 and 15.9. Examples of a modern 1.5 t machine are not readily available and estimates of hypothetical operating costs have been made.

15.10 *Vibro-tampers.* Machines generally similar to those used in the investigations described in Chapter 10 have been taken.

Plate 15.8 A 150 kg vibrating-plate compactor of the type used in the calculations of outputs and costs of compaction (Photograph courtesy of Dynapac (UK) Limited)

Plate 15.9 A 700 kg vibrating-plate compactor of the type used in the calculations of outputs and costs of compaction (Photograph courtesy of Blackwood Hodge UK Limited)

15.11 *Power rammers.* Machines of similar characteristics to those used in the original investigations have had to be taken; modern machines of this type are not readily available. Estimates have been made of the likely operating costs.

15.12 *Dropping-weight compactors.* The range of specific energies given in Table 15.1 is that studied during the numerous investigations described in Chapter 12. However, the mobile dropping-weight compactor shown in Plate 12.5 has been assumed to provide each of the specific energies by variation of the height of drop of the rammer. Note that the frequency of blows increases with diminishing height of drop as described in Paragraph 12.5.

Discussion of Tables 15.2 to 15.16

15.13 The suitability of each item of compaction equipment, its potential output, and the costs of compaction, depend strongly on the soil type, its moisture content relative to the optimum of the 2.5 kg rammer compaction test, and the specified level of compaction. Thus a variety of items of plant appear as most economical depending on the circumstances. At moisture contents below the optimum of the 2.5 kg rammer test a state of compaction equivalent to 10 per cent air voids is much more difficult to reach than the level equivalent to 95 per cent relative compaction (dry density equal to 95 per cent of the maximum dry density obtained in the 2.5 kg rammer compaction test). Thus, only dropping-weight compactors with the higher values of specific energy would be capable of achieving 10 per cent air voids in heavy cohesive soils at a moisture content 10 per cent below the optimum of the 2.5 kg rammer compaction test (see Table 15.2(b)). Except in this particular instance, the most economical methods of compaction, ie those methods having the lowest factorised costs per m^3 of compacted soil, depending on conditions, are smooth-wheeled rollers (Items 2 and 3), pneumatic-tyred rollers (Items 4 to 6), the tamping roller (Item 9), the sheepsfoot rollers (Items 8 and 10), or vibrating smooth-wheeled rollers (Items 12 to 14). Where sheepsfoot rollers are shown to be the most economical, it is solely because of the high output achieved by assuming the use of three towed units with one tractor, as indicated in Table 15.1; this results in a total compacted width per pass of about 6.8 m.

15.14 The higher costs of compaction are associated with the smaller volumes of soil compacted per hour by the smaller items of plant. Such plant are intended for use in confined spaces such as in trenches and around bridge abutments. In these situations the larger, more productive equipment cannot be used. Where no data exist to provide figures on the performance of the equipment the term 'Insufficient information' is used in the Tables.

15.15 Tables 15.2 to 15.16, in conjunction with Table 15.1, provide information on the relative performances of the various items of compaction equipment. All values are based on results obtained during the investigations under ideal conditions at the Laboratory, and take into account the general trends in results where necessary. It is considered that the Tables can be used as a general guide in the selection of compaction equipment for various conditions under which compaction may have to be carried out.

15.16 The Tables also clearly illustrate the predominant influence of the moisture content of the soil on the output of the compaction plant and hence on the costs of compaction. As an example, take the performance of the 11 t self-propelled single-roll vibrating smooth-wheeled roller (Item 13 in the Tables) on light cohesive soils, using the criterion of 95 per cent relative compaction. The performance for these conditions is given in Tables 15.6 (a), 15.7(a), 15.8(a) and 15.9(a) for moisture contents ranging from 6 per cent below to 3 per cent above the optimum of the 2.5 kg rammer compaction test. The calculated outputs are given as follows:–

85 m^3/h at optimum – 6 per cent,

113 m^3/h at optimum – 3 per cent,

170 m^3/h at optimum,

and 170 m^3/h at optimum + 3 per cent.

Thus, in this instance, the cost of the compaction process would be halved if the soil could be used at optimum rather than at 6 per cent below the 2.5 kg rammer optimum. The relative economics of compacting the soil in a dry condition or of adding water to facilitate compaction could be worth considering in such instances. Where the available compaction equipment is found to be incapable of achieving the specified level of compaction, ie they are labelled as 'not suitable' in the Tables, the addition of water will often give a condition where the specification can be achieved. An example of this is given by Item 1, a 2.5 t smooth-wheeled roller, in the compaction of light cohesive soils to 95 per cent relative compaction.

References

Civil Engineering (1988). Measured rates. *Civil Engineering,* September, 1988, 56–63.

Lewis, W A (1954). Further studies in the compaction of soil and the performance of compaction plant. *Road Research Technical Paper* No 33, HM Stationery Office, London.

Plant Assessment (1988). Plant and equipment guide. Plant Assessment (London) Ltd, Bournemouth.

CHAPTER 16 COMPACTION INVESTIGATIONS WITH MATERIALS OTHER THAN SOIL

Introduction

16.1 During the course of the various investigations into compaction and the performance of compaction plant, materials other than the 'standard' soils (ie the soils used in the investigations described in Chapters 4 to 14) were used from time to time. The principal materials in this category were two types of wet-mix macadam which were installed in special test pits in the building used in the middle period of the investigations, from 1954 to 1966. These materials and the methods used in preparing them for test have already been mentioned in Paragraphs 3.7 to 3.12. The findings of the studies with the wet-mix macadams are described in Section A of this Chapter.

16.2 A further material which was the subject of a special compaction investigation prior to its use as embankment fill in a motorway contract was spent domestic refuse. The results are given in Section B of this Chapter.

16.3 Pulverised fuel ash (PFA) is commonly used as a lightweight fill material in earthworks. PFA has been included in some very limited tests at the Laboratory which are described in Section C.

SECTION A COMPACTION OF WET-MIX MACADAM

16.4 In the 1950s and 1960s pre-mixed water-bound macadam or 'wet-mix' was very widely used for the construction of road bases. Problems have been experienced with this material, however, and its use is no longer permitted on principal roads in the United Kingdom. It is, therefore, used only to a very limited extent at the present time. Wet-mix consists of graded crushed aggregate to a fairly close specification mixed with a small amount of water to help minimise the segregation of the material during transportation and laying and to aid the compaction process.

16.5 It is extremely important that a high state of compaction is achieved with wet-mix bases if the material is to perform satisfactorily under traffic. Poor initial compaction of a wet-mix base can result in the development of serious deformations in the road surface during use and may increase the likelihood of damage to the road by ingress of water and by frost action.

16.6 To provide information on the compaction of wet-mix materials, investigations were carried out at the Laboratory in 1957 to 1959 to determine the performance of a range of compaction plant on two of the most common types of wet-mix materials. The materials used were a limestone wet-mix, nominally to the grading given in the Specification for Road and Bridge Works at the time (Ministry of Transport and Civil Aviation, 1957), and a slag wet-mix sold widely by a commercial organisation. The results for the wet-mix macadams given in this Chapter are largely taken from Lewis and Parsons (1961).

Compaction plant used

16.7 The items of compaction plant used with the wet-mix macadams were, with one exception, all used in tests on soils and have been described in the Chapters relevant to the particular types of plant. Details of the various machines are given in Tables 16.1 to 16.4. The machines were as follows:–

(1) 2.7 t smooth-wheeled roller; shown in Plate 4.1. When used for tests on base materials it

Plate 16.1 The 13 t self-propelled pneumatic-tyred roller used in the investigations with wet-mix macadams

was ballasted to a slightly different condition to that used with soils (see Tables 4.1 and 16.1).

(2) 7.4 t smooth-wheeled roller; shown in Plate 4.2. This machine also was ballasted to a different condition to that used with soils (see Tables 4.1 and 16.1).

(3) 13 t pneumatic-tyred roller (Plate 16.1). This machine was rather unusual in that it was a conversion of a three-wheeled smooth-wheeled roller, as can be seen from the photograph. Details are given in Table 16.2.

(4) 350 kg pedestrian-operated vibrating roller; shown in Plate 7.2, with details in Table 16.3.

(5) 990 kg tandem vibrating roller; shown in Plate 7.3, with details in Table 16.3.

(6) 670 kg vibrating-plate compactor; shown in Plate 9.11, with details in Table 16.4.

(7) 710 kg vibrating-plate compactor; shown in Plate 9.12, with details in Table 16.4.

Wet-mix macadams used

16.8 The materials used in the investigations were a wet-mix limestone macadam and a wet-mix slag macadam. The particle-size distributions and particle densities of these materials have already been given in Figure 3.4.

Table 16.1 Details of the smooth-wheeled rollers as used in the investigations on wet-mix macadams

Machine	2.7 t	7.4 t
Total mass (kg)	2690	7410
Front roll		
Width (m)	0.61	1.07
Diameter (m)	0.86	1.07
Mass on roll (kg)	850	3280
Mass per unit width (kg/m)	1400	3070
Rear rolls		
Width (m)	0.38×2	0.46×2
Diameter (m)	0.91	1.37
Mass on rolls (kg)	1840	4130
Mass per unit width (kg/m)	2420	4510

Table 16.2 Details of the 13 t pneumatic-tyred roller used in the investigations on wet-mix macadams

Total mass (kg)	12 670
Number of wheels	8 (4 front, 4 rear)
Diameter of wheels including tyres (m)	1.02
Rolling width (between outer edges of outer wheels) (m)	2.46
Wheel load (front) (kg)	1370
Wheel load (rear) (kg)	1800
Tyre inflation pressure (kN/m^2)	690
Tyre contact pressure (front) (kN/m^2)	550
Tyre contact pressure (rear) (kN/m^2)	610

Table 16.3 Details of the vibrating rollers as used in the investigations on wet-mix macadams

Machine	350 kg Pedestrian-operated	990 kg Tandem
Total mass (kg)	350	990
Front roll		
Width (m)	0.71	0.81
Diameter (m)	0.57	0.66
Mass on roll (kg)	350	310
Mass per unit width (kg/m)	490	380
Rear roll		
Width (m)	–	0.81
Diameter (m)	–	0.66
Mass on roll (kg)	–	680
Mass per unit width (kg/m)	–	840
Vibrating mechanism		
Location	Roll	Rear roll
Normal frequency (Hz)	75	67
Nominal centrifugal force (kN)	12	16
Normal speed of rolling in tests (km/h)	1.0–1.2	1.0–1.5

16.9 The results of laboratory compaction tests carried out on the two materials are given in Figure 16.1. The 2.5 kg and 4.5 kg rammer tests were performed using a similar mould to that used in the vibrating hammer test, ie 150 mm diameter and about 125 mm deep, so that the complete grading of material could be used. The number of blows used in the rammer tests was

Table 16.4 Details of the vibrating-plate compactors as used in the investigations on wet-mix macadams

Machine	670 kg	710 kg
Total mass (kg)	673	711
Width of base-plate (m)	0.62	0.61
Contact area of base-plate (m^2)	0.429	0.372
Mass per unit area (kg/m^2)	1570	1910
Vibrating mechanism		
Number of eccentric shafts	2	2
Frequency (Hz)	20	25
Nominal centrifugal force (kN)	41	NA
Centrifugal force per unit area (kN/m^2)	96	–
Normal speed of travel (km/h)	0.9–1.0	0.5–0.8

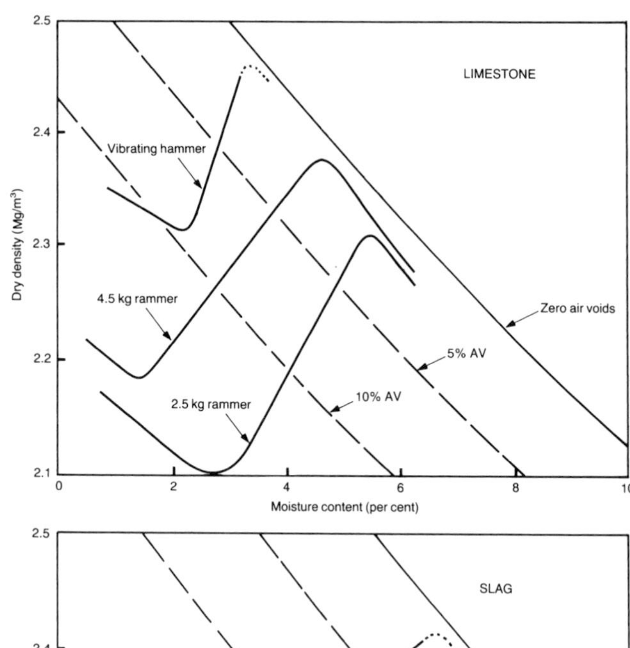

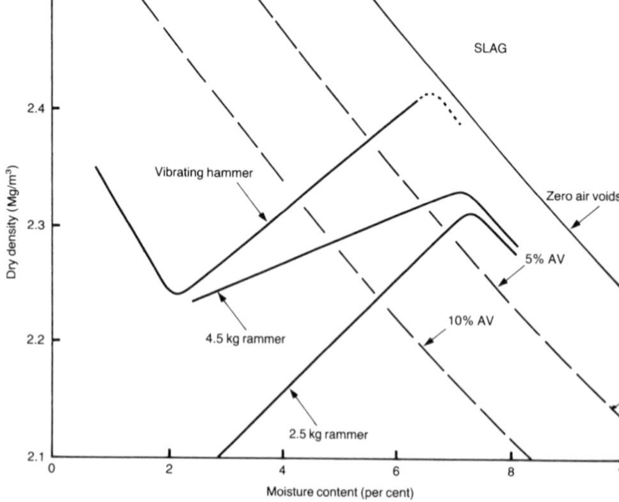

Figure 16.1 Relations between dry density and moisture content obtained with various laboratory compaction tests on wet-mix macadams

increased to 55 per layer to allow for the increase in diameter of the mould from the standard in use at that time (British Standards Institution, 1948).

Factors studied in the investigations with wet-mix macadams

16.10 The factors studied were:–

(i) The relations between dry density and moisture content for the two wet-mix granular base materials when fully compacted by the compaction plant, ie compacted to refusal.

(ii) The relations between dry density and number of passes of the compaction plant for the two materials.

16.11 The tests were mainly carried out using a compacted layer of 150 mm thickness, this being the maximum thickness permitted in the specifications which applied at that time to the compaction of granular base materials. A number of additional tests, however, were also made using 200 mm thick compacted layers to explore the feasibility of relaxing the specification on layer thickness.

16.12 For the full-scale tests the wet-mix macadams were each contained in a pit about 10.5 m long, 3.5 m wide and 0.5 m deep (see Figure 3.3). The floor of each pit was left as the natural foundation soil (sandy clay) to provide conditions as similar as possible to those found in the field. Beneath the upper 150 mm to 200 mm thick test layer there was a further 230 mm of compacted granular material overlying the sandy clay floor of the pits and this lower layer was not disturbed during the investigations. The test conditions were considered, therefore, to be ideal for the compaction of the test layers (Lewis and Parsons, 1961), and the results of the investigations were regarded as being the most favourable that it would be possible to obtain in practice.

Relations between dry density and moisture content for the two base materials when fully compacted by the various items of plant

16.13 The procedures used in preparing the loose layer of wet-mix macadam and the method used in determining the state of compaction produced have been described in Paragraph 3.11. The detailed procedures used when testing the various types of compaction plant were as follows:–

Smooth-wheeled rollers. With both the 2.7 t and 7.4 t smooth-wheeled rollers 32 passes were provided to ensure that each wet-mix macadam was compacted to refusal. Each machine travelled alternately in the forward and reverse directions over the test layer, remaining in the same wheel tracks throughout, at speeds of 1.3 km/h and 1.6 km/h for the 2.7 t and 7.4 t rollers respectively. The determinations of dry density were made only in the rear-wheel tracks.

13 t pneumatic-tyred roller. As with the smooth-wheeled rollers, 32 passes were provided to ensure compaction to refusal. The roller was driven in the forward and reverse directions using a slightly different path for each pass so that a fairly flat surface was obtained. The speed of rolling was about 2.4 km/h and dry density determinations were confined to the central compacted strip.

Vibrating rollers. Three machine widths of the two test materials were compacted by 32 passes, the passes being alternately forward and reverse on each of the three strips. Compaction speeds with the two machines are given in Table 16.3. The dry density determinations were confined to the central compacted strip.

Vibrating-plate compactors. The vibrating-plate compactors had fairly low compacting speeds (Table 16.4) and 16 passes of the 670 kg machine and 6 passes of the 710 kg machine were considered sufficient to obtain compaction to refusal. Three machine widths of the materials were compacted, the passes being alternately forward and backward with the 670 kg machine and forward only with the non-reversible 710 kg machine on each of the three strips. Dry density determinations were made only in the central compacted strip.

16.14 The compaction test procedures outlined above for the various items of plant were carried out on the two wet-mix macadams at various moisture contents and the relations between dry density and moisture content obtained are given in Figures 16.2 to 16.5.

16.15 The most striking feature of the results is the great extent to which the moisture content affected the state of compaction obtained with all the machines tested. This effect was particularly marked with the limestone wet-mix, where a

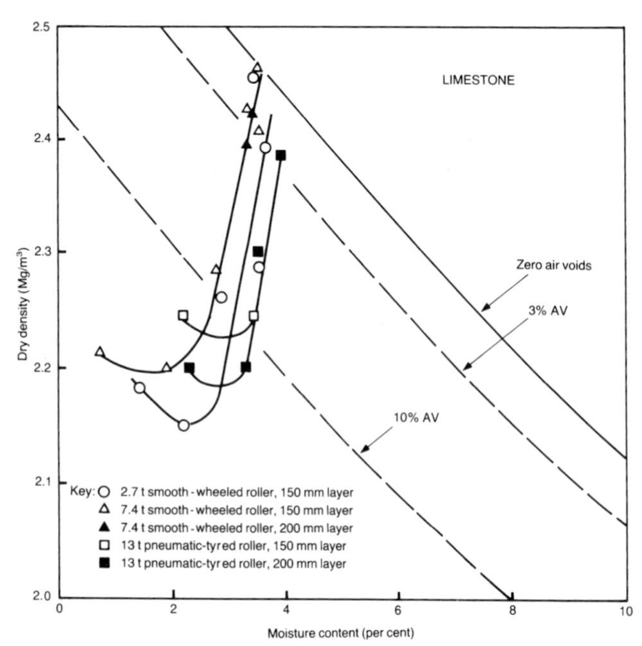

Figure 16.2 Relations between dry density and moisture content obtained after 32 passes of the smooth-wheeled and pneumatic-tyred rollers on the limestone wet-mix macadam

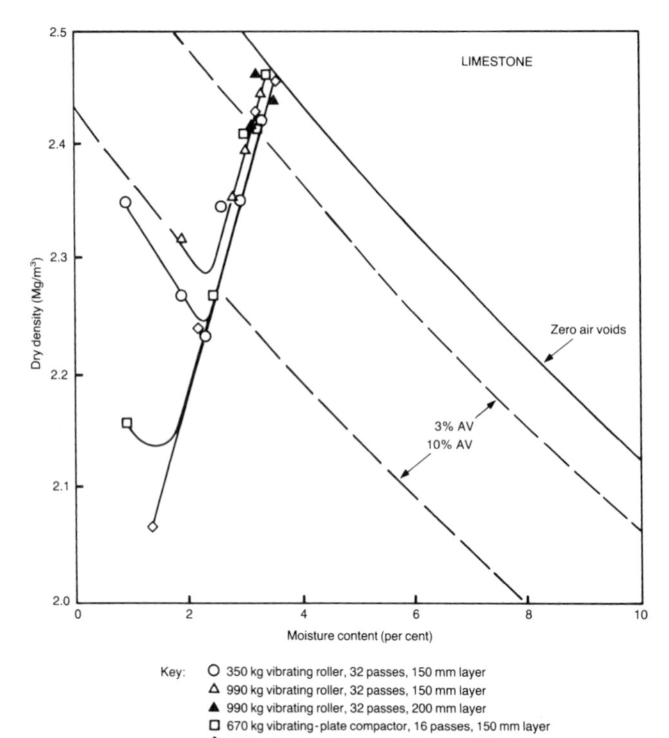

Figure 16.3 Relations between dry density and moisture content obtained with the vibrating rollers and vibrating-plate compactors on the limestone wet-mix macadam

change in moisture content of only 0.5 per cent resulted in a change in the dry density of about 0.1 Mg/m³.

16.16 The results also show that the highest dry densities were obtained as the saturated

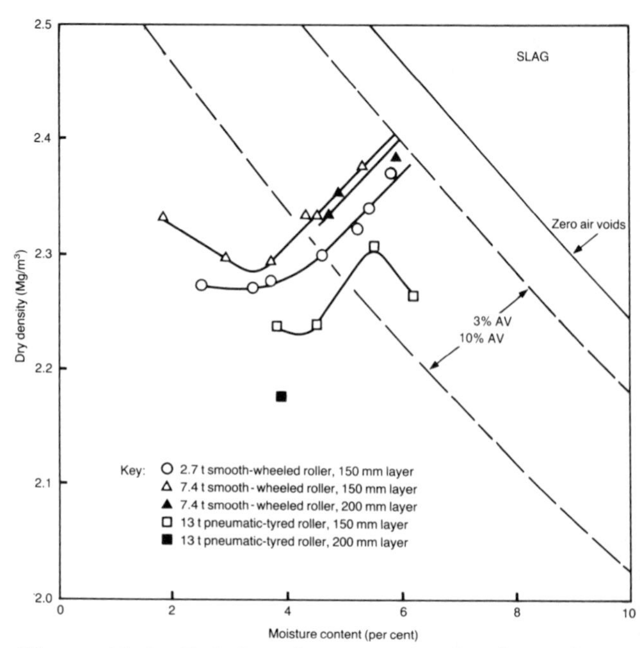

Figure 16.4 Relations between dry density and moisture content obtained after 32 passes of the smooth-wheeled and pneumatic-tyred rollers on the slag wet-mix macadam

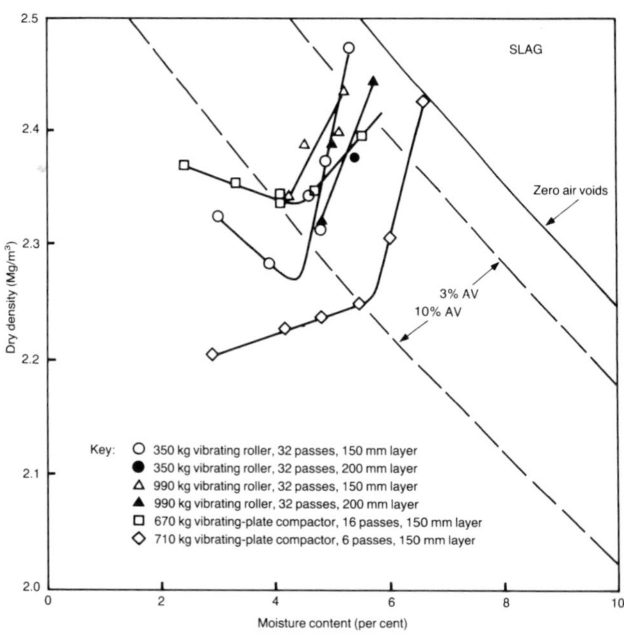

Figure 16.5 Relations between dry density and moisture content obtained with the vibrating rollers and vibrating-plate compactors on the slag wet-mix macadam

condition was approached. Under these conditions, however, pumping of wet fines to the surface was experienced with all the machines tested, and despite the high levels of dry density the compacted materials were extremely spongy and lacked stability. This sponginess was undoubtedly associated with the development of positive pore-water pressures, clearly an undesirable condition. It was concluded, therefore, that in practice wet-mix macadams should be compacted at a moisture content at which a high state of compaction could be obtained but at which the development of the spongy state was avoided. With the limestone wet-mix macadam the moisture content needed to be controlled to within ±0.5 per cent and with the slag to within ±1 per cent to achieve this result. However, taking account of the uncertain weather often experienced in the United Kingdom, it was considered that it would probably be very difficult to control the moisture content of wet-mix macadams to within these relatively close limits.

16.17 To determine the long-term effect on the stability of wet-mix macadams after initial compaction to a spongy condition, the limestone was left for several days after compaction at a high moisture content. The excess water appeared to drain away leaving a comparatively stable surface but, when the material was rolled for a second time to simulate the effect of traffic, the sponginess quickly returned. Thus a fairly prolonged period of drying out was considered to be necessary if a stable base was ultimately to be achieved from an initially spongy condition.

16.18 To help in the comparison of the performances of the various items of plant, Table 16.5 gives the dry densities and corresponding moisture contents for compaction to a state equivalent to an air content of 3 per cent, likely to be the limiting condition at which a high state of compaction could be obtained without the development of serious positive pore-water pressures and instability. Table 16.5 and Figures 16.2 to 16.5 show that, apart from the 2.7 t smooth-wheeled roller and the 13 t pneumatic-tyred roller, the performances of all the machines were fairly similar. The highest dry densities were obtained with the largest vibrating roller tested, the 990 kg machine, but the differences between the various vibrating compactors together with the 7.4 t smooth-wheeled roller were not very significant. The poorest states of compaction were obtained with the 13 t self-propelled, pneumatic-tyred roller and this, together with the rather uneven surface produced, suggests that this type of roller is not particularly suitable for the compaction of wet-mix macadams.

16.19 The results for a 200 mm thick compacted layer, compared with those for a 150 mm thick compacted layer, showed a small drop in the average state of compaction produced in the thicker layer. Although only a limited amount of testing was carried out using 200 mm thick layers, the results suggested that, at least for the 7.4 t smooth-wheeled roller and the 990 kg vibrating roller, a 200 mm thick layer could be compacted to an acceptable state of compaction. It was considered at the time, however, that the use of thicker layers might make it more difficult to achieve close tolerances in the surface levels.

16.20 It is of interest to compare the results obtained with the various items of plant with the results of laboratory compaction tests given in Figure 16.1. In general, at any given moisture content, the various compactors produced significantly higher states of compaction than either of the two laboratory rammer tests, and the optimum moisture contents of the 4.5 kg and 2.5 kg rammer tests were significantly higher than those at which the two wet-mix macadams could be compacted satisfactorily in the field.

Table 16.5 Highest dry densities and corresponding moisture contents consistent with the avoidance of positive pore water pressures in the wet-mix macadams for various items of compaction plant

Compaction plant	Limestone wet-mix macadam				Slag wet-mix macadam			
	150 mm compacted layer		200 mm compacted layer		150 mm compacted layer		200 mm compacted layer	
	Dry density (Mg/m³)	Moisture content (%)	Dry density (Mg/m³)	Moisture content (%)	Dry density (Mg/m³)	Moisture content (%)	Dry density (Mg/m³)	Moisture content (%)
2.7 t smooth-wheeled roller	2.39	3.6	–	–	2.38	6.2	–	–
7.4 t smooth-wheeled roller	2.40	3.3	2.40	3.3	2.40	5.8	2.39	5.9
13 t pneumatic-tyred roller	2.37	3.8	2.37	3.8	High state of compaction not achieved			
350 kg vibrating roller	2.40	3.2	–	–	2.44	5.2	2.41*	5.6*
990 kg vibrating roller	2.41	3.1	2.41	3.1	2.44	5.2	2.42	5.5
670 kg vibrating-plate compactor	2.41	3.1	–	–	2.41	5.7	–	–
710 kg vibrating-plate compactor	2.40	3.2	–	–	2.37	6.3	–	–

*Values estimated from test at one moisture content

The vibrating hammer compaction test results, however, compared favourably with the results obtained with compaction to refusal by the full-scale plant, and its optimum moisture content could be used to provide guidance as to the best moisture content for full-scale compaction.

16.21 With the exception of the pneumatic-tyred roller, the various items of compaction plant produced some crushing of the larger sizes of aggregate although the effect on the particle-size distribution was negligible. The crushing effect was more noticeable on the slag, but appeared to occur only at the surface where it might be considered to be beneficial in producing a tight-knit surface.

Relations between dry density and number of passes of the compaction plant

16.22 In the determination of the effect of the number of passes of the various types of compaction plant on the state of compaction produced, loose layers of suitable thickness of the two wet-mix macadams were prepared as described in Paragraph 3.11. The moisture content at which each test was carried out was selected so that a high state of compaction could be obtained without sponginess developing. This sponginess occurred as the saturated condition was approached (see Paragraph 16.16). The wet-mix macadams were compacted with different numbers of passes of each machine, the methods of operation being similar to those outlined in Paragraph 16.13. Determinations of the dry density of the compacted layer were made at each stage and the results obtained are given in Figures 16.6 to 16.9.

16.23 In the relations shown in Figures 16.6 to 16.9 small unavoidable variations in moisture content between the tests with the various machines caused some overlapping of the results. The mean value of moisture content for each curve, given in the key to each Figure, should be taken into account when comparing the performances of the compaction plant. It should also be noted that changes in moisture content, especially of the limestone wet-mix macadam, might have affected to some extent the shape of the relations between dry density and number of passes.

16.24 The results of the limited number of tests on 200 mm thick compacted layers show that increasing the thickness of the layer from 150 mm did not have a significant effect on the relations between dry density and number of passes.

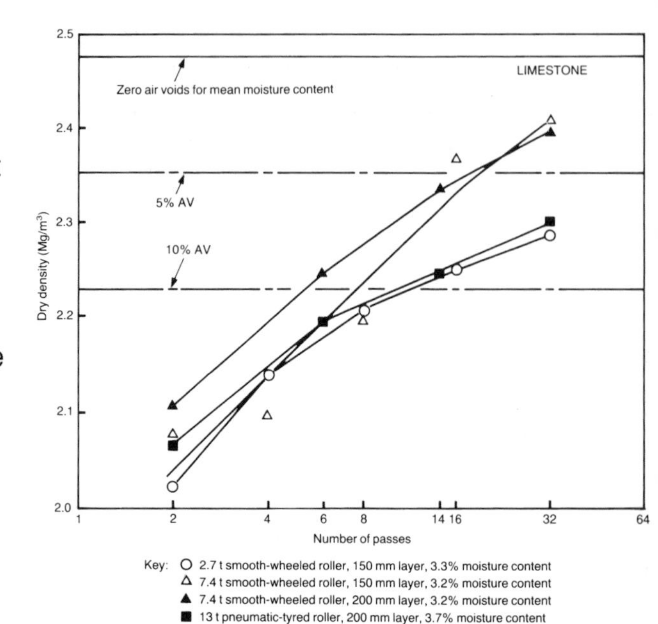

Figure 16.6 Relations between dry density and number of passes of the smooth-wheeled and pneumatic-tyred rollers when compacting the limestone wet-mix macadam

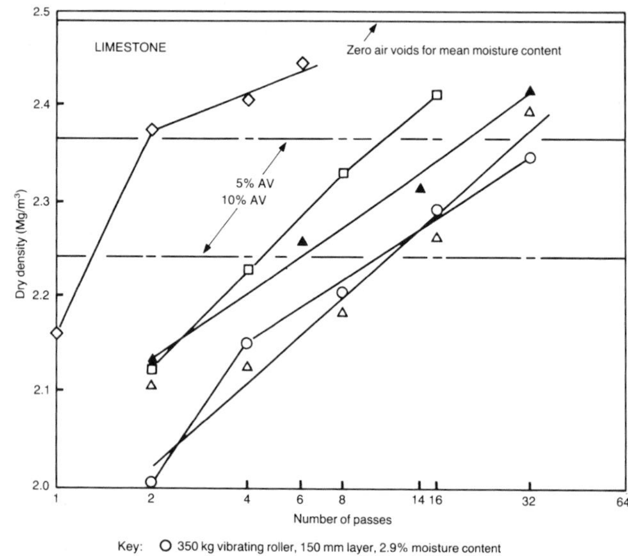

Figure 16.7 Relations between dry density and number of passes of the vibrating rollers and vibrating-plate compactors when compacting the limestone wet-mix macadam

16.25 To summarise the results of the tests estimates have been made of the number of passes required to achieve a specified level of compaction in a 150 mm thick compacted layer at a particular moisture content value, ie taking account of the small variations in moisture content between the various tests. The results

Table 16.6 Approximate possible outputs of the various items of plant in compacting the two wet-mix macadams and the comparative costs per m² for a 150mm compacted thickness

Type of plant	Cost per hour (factor-ised)*	Width compacted by plant (m)	Speed of rolling (km/h)	Limestone			Slag		
				Number of passes required	Area compacted per hour (m²)	Factorised cost per m² of compacted material	Number of passes required	Area compacted per hour (m²)	Factorised cost per m² of compacted material
2.7 t smooth-wheeled roller	64	1.30	1.3	Unsuitable			Unsuitable		
7.4 t smooth-wheeled roller	100	1.78	1.6†	12	200	0.50	24	100	1.00
13 t pneumatic-tyred roller	129	2.46	1.6	Unsuitable			Unsuitable		
350 kg vibrating roller	51	0.71	1.1	12	55	0.93	8	85	0.60
990 kg vibrating roller	66	0.81	1.2	12	70	0.94	6	140	0.47
670 kg vibrating-plate compactor	54	0.61	1.0	6	85	0.64	4	130	0.42
710 kg vibrating-plate compactor	54	0.61	0.7	4	90	0.60	8	45	1.20

*Taking the operating cost of the 7.4 t smooth-wheeled roller as 100

†This was the speed used in the tests; higher speeds could no doubt be employed with some corresponding increase in outputs

are given in Table 16.6. The moisture contents taken for this exercise were 3.4 per cent for the limestone (equal to the optimum of the vibrating hammer compaction test, see Figure 16.1) and 5.5 per cent for the slag (1 per cent below the optimum of the vibrating hammer compaction test, see Figure 16.1). The level of compaction required to be achieved was assumed to be equivalent to 5 per cent air voids in both cases; with limestone this was equivalent to 95 per cent and with slag to about 98.5 per cent of the respective maximum dry densities obtained in the vibrating hammer compaction test.

16.26 Table 16.6 indicates that, given the criteria described above, the 2.7 t smooth-wheeled roller and the 13 t pneumatic-tyred roller were unsuitable to compact 150 mm layers of either of the wet-mix macadams, but all the other machines were successful to varying degrees,

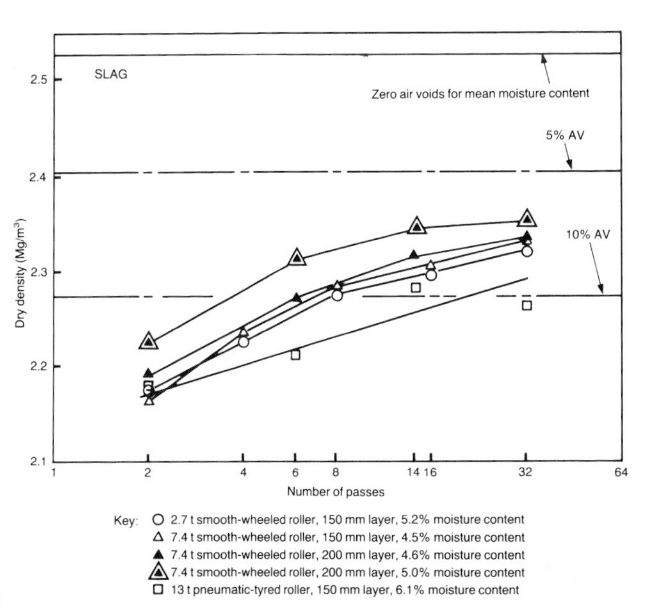

Figure 16.8 Relations between dry density and number of passes of the smooth-wheeled and pneumatic-tyred rollers when compacting the slag wet-mix macadam

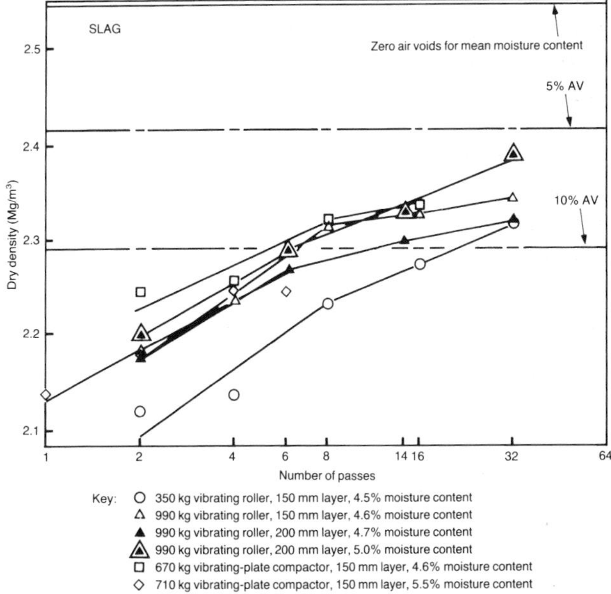

Figure 16.9 Relations between dry density and number of passes of the vibrating rollers and vibrating-plate compactors when compacting the slag wet-mix macadam

as shown by the number of passes required.

Output of compaction plant

16.27 Table 16.6 also gives estimates of the outputs of the machines based on the numbers of passes necessary to comply with the criteria described in Paragraph 16.25. The comparative costs of compacting a unit area of wet-mix macadam have also been included; these are based on factorised operating costs, ie the cost per hour of operating the 7.4 t smooth-wheeled roller has been taken as 100, and the costs per hour of all other items of plant are expressed as percentages of that cost (see Paragraph 15.3 for further details).

16.28 The assumption has also been made in the calculations that the plant is operated for 85 per cent of available time to allow for time lost in turning and manoeuvring the plant and other small stoppages. No allowance has been made for delays caused by bad weather or for the comparative costs of transporting the plant to and from the site.

16.29 As mentioned in Paragraph 16.26, the 2.7 t smooth-wheeled roller and the 13 t pneumatic-tyred roller are shown to be unsuitable for compacting the wet-mix macadams to the set criteria. The highest outputs and the most economical cost per m^2 of compacted material are shown to have been achieved by the 7.4 t smooth-wheeled roller on the limestone and by the 990 kg vibrating roller and 670 kg vibrating-plate compactor on the slag. However, considerations other than outputs and costs, such as the size and type of work, the availability of a particular machine and the ease of operation, could also determine the machine likely to be most suitable for a particular job. It must also be noted that currently much larger vibrating rollers are available than those used in these investigations and such machines would certainly achieve much higher outputs of compacted material.

16.30 In 1961, consideration was being given to the use of spent domestic refuse as a fill material for the construction of the embankments for a proposed section of motorway. The material was available at a dump on the outskirts of London and it was considered that, if shown to be suitable, its use would produce a considerable saving in costs over alternative fill materials.

16.31 With the lack of experience at that time of the use of such material in road embankments, investigations were required to determine whether it would form a stable embankment. To provide information on this aspect, compaction tests were carried out on the material using a heavy smooth-wheeled roller and a tandem vibrating roller. The following information has been taken from Parsons and Krawczyk (1961).

The spent domestic refuse

16.32 A representative quantity of the material from the dump was placed in one of the pits used for the wet-mix macadams and described in Paragraph 16.12. The refuse was compacted in two layers by a number of passes of a 7.4 t smooth-wheeled roller to produce 300 mm of compacted material in the bottom of the pit. A loose layer, 225 mm thick, on which the tests were made, was placed on top of the compacted lower layers which were then left undisturbed throughout the investigation.

16.33 The material placed in the various layers of the test pit varied in moisture content from about 23 to 30 per cent, as sampled at the time of delivery. Prior to the main investigation samples were obtained from the dump and determinations made of their particle-size distributions, using the dry sieving method (British Standards Institution, 1990a), moisture contents and losses on ignition (British Standards Institution, 1990b) (Figure 16.10). The material was shown to have a wide range of

Figure 16.10 Particle-size distributions, moisture contents and losses of weight on ignition of samples obtained at the refuse pit

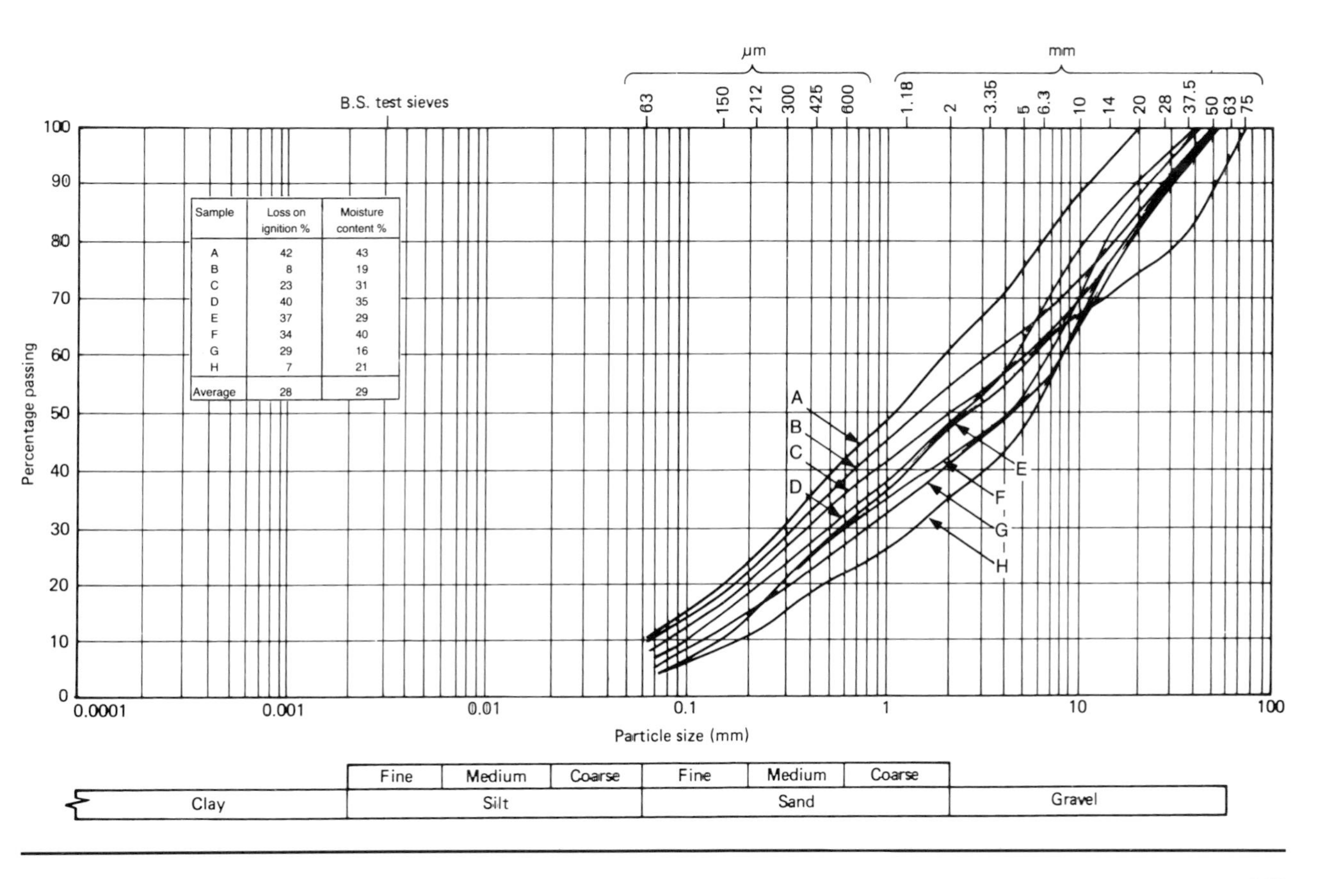

Sample	Loss on ignition %	Moisture content %
A	42	43
B	8	19
C	23	31
D	40	35
E	37	29
F	34	40
G	29	16
H	7	21
Average	28	29

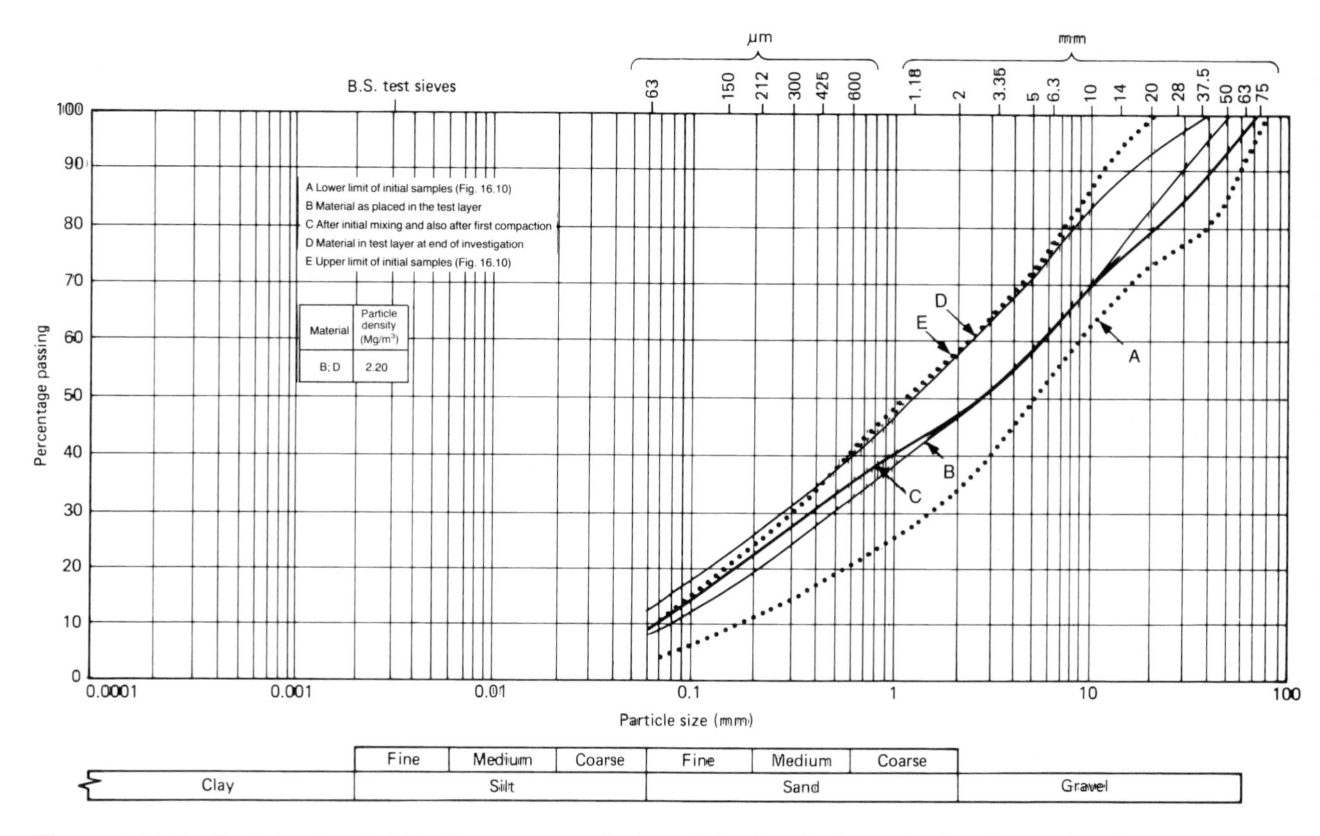

Figure 16.11 Particle-size distributions and result of particle density tests for the domestic refuse as tested

gradings, with an average corresponding to a sandy gravel soil. The moisture content varied between 16 and 43 per cent; this wide variation was probably caused by variations in organic content, evidenced by the losses of weight on ignition, and in grading (Figure 16.10).

16.34 Further particle-size distributions and testing for particle density were carried out on the material placed in the test pit and also at various stages in the test programme (Figure 16.11). These confirmed that the material used in the full-scale compaction tests was representative of the material available within the dump.

16.35 The larger particles in the domestic refuse were principally glass and unburnt coal, and the finer material, with the appearance of coal dust, was non-cohesive. The material was generally black in colour. Figure 16.11 shows that by the end of the investigations, after about 100 passes of the rollers and frequent mixings, some change in particle-size distribution had occurred, but it was considered that no appreciable change in grading was likely in the construction of earthworks where the material would be simply spread and compacted.

16.36 Laboratory compaction tests, using the 4.5 kg and 2.5 kg rammer methods, were carried out on the material, using a mould 150 mm in diameter and 125 mm deep so that material larger than 20 mm could be included. The domestic refuse for test was passed through a 37.5 mm test sieve and particles of glass and china retained on the sieve were broken down to sizes smaller than 37.5 mm and returned to the test material. The number of blows of the rammer was increased to 55 per layer to allow for the increase in diameter of the mould from that of the standard 100 mm diameter mould.

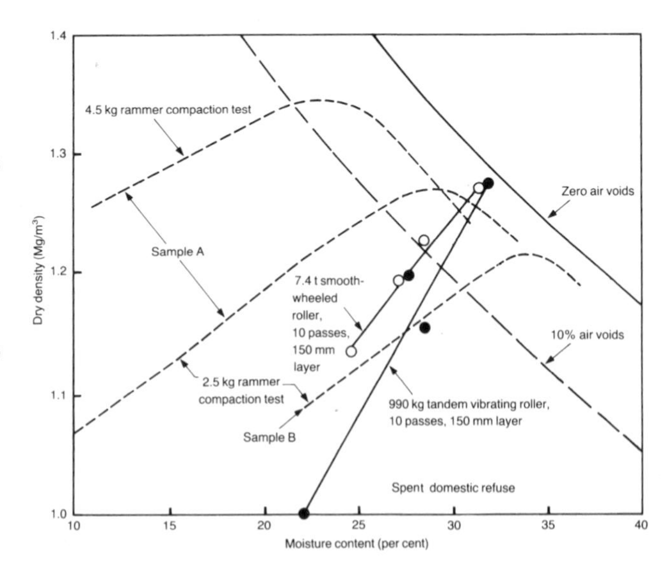

Figure 16.12 Relations between dry density and moisture content obtained with compaction plant and in laboratory compaction tests on domestic refuse

The results of compaction tests on two samples using the 2.5 kg rammer method and on one sample using the 4.5 kg rammer method are shown in Figure 16.12.

Compaction plant employed in the investigations

16.37 The compaction plant employed in the investigations were the 7.4 t smooth-wheeled roller and the 990 kg tandem vibrating roller referred to in Paragraph 16.7; details are given in Tables 16.1 and 16.3.

Relations between dry density and moisture content of the domestic refuse when compacted by full-scale plant

16.38 The relations between dry density and moisture content of 150 mm compacted layers of the domestic refuse were determined for 10 passes of the 7.4 t smooth-wheeled roller and 10 passes of the 990 kg tandem vibrating roller. The material was prepared for test with a procedure similar to that used for soils (see Chapter 3), using a rotary cultivator to mix it to a uniform condition at each of the selected moisture contents. The machine was used at a low speed to produce a gentle stirring action to reduce to a minimum the breakdown of the larger glass particles. The smooth-wheeled roller and vibrating roller were operated on the prepared layer of domestic refuse as described in Paragraph 16.13. The dry density of the compacted material at each stage was determined as the average of six tests using the sand-replacement (large pouring cylinder) method for medium and coarse-grained soils (British Standards Institution, 1990c). The results are given in Figure 16.12.

16.39 Both rollers produced fairly similar relations between dry density and moisture content (Figure 16.12). In the absence of information on the long-term settlement of embankments constructed of the material, it was considered that the same standards of compaction should be required with the spent domestic refuse as with well-graded soils at that time, namely a maximum air void content of 10 per cent. States of compaction equivalent to 10 per cent or less of air voids were achieved with the 10 passes of the compaction plant at moisture contents in excess of 29 per cent. On the basis of the results of the laboratory compaction tests also included in Figure 16.12, this level of compaction at a moisture content of 29 per cent was equivalent to 95 to 100 per cent relative compaction (relative to the maximum dry density obtained in the 2.5 kg rammer compaction test).

16.40 At moisture contents in excess of 32 per cent the domestic refuse became unstable and spongy on compaction, although no difficulty was experienced in operating the compaction plant, even at a moisture content of 34.5 per cent. In this condition it was found impossible to accurately determine the dry density by the sand-replacement method because of the lack of stability of the excavated test hole. However, sponginess only occurred when the saturated condition (zero air voids) was approached, and a determination of moisture content would have been sufficient to obtain a reasonably accurate value of dry density.

16.41 In the laboratory compaction tests, the 2.5 kg rammer method produced maximum dry densities of the same order as those produced by the compaction plant. In all cases the values of dry density produced were extremely low; two factors contributed to this, firstly the high moisture content of the material and secondly its low particle density of 2.20 Mg/m^3.

Relation between the in-situ California bearing ratio and moisture content of the domestic refuse

16.42 The in-situ California bearing ratio (CBR) (British Standards Institution, 1990c) of the domestic refuse was determined at each stage of the determination of the relations between dry density and moisture content. The results are given in Figure 16.13. One single relation between in-situ CBR and moisture content has been drawn through the results produced by the two rollers. The maximum in-situ CBR of 24.5 per cent, as indicated by the curve, occurred at a moisture content of 26 per cent. At a moisture content of 29 per cent, at which 95 per cent relative compaction and 10 per cent air voids were achieved, an in-situ CBR of about 14 per cent was obtained. On increasing the moisture content above 29 per cent, despite further increases in dry density initially (see Figure 16.12), a rapid decrease in in-situ CBR occurred, until at a moisture content of 34 per cent it became almost zero, with the material in an extremely spongy condition.

16.43 No surcharge weights were used in the tests, and it was considered probable at the time

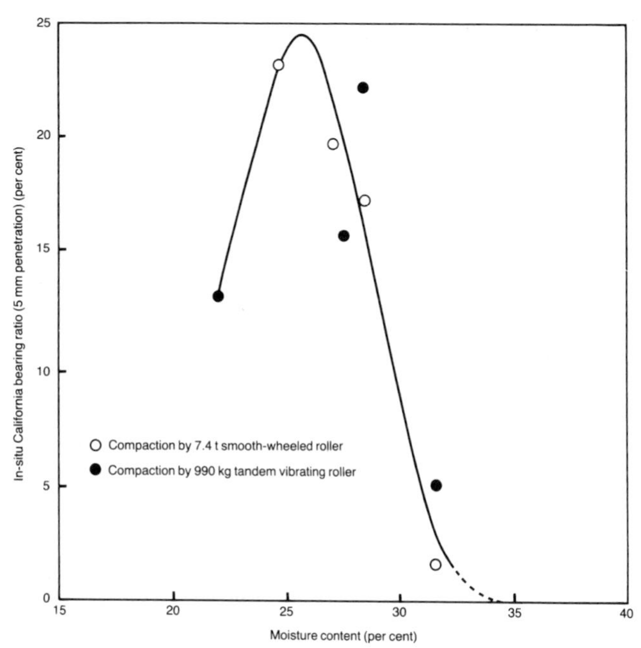

Figure 16.13 Relation between in-situ California bearing ratio and moisture content of domestic refuse when compacted by 10 passes of the 7.4 t smooth-wheeled roller and the 990 kg tandem vibrating roller

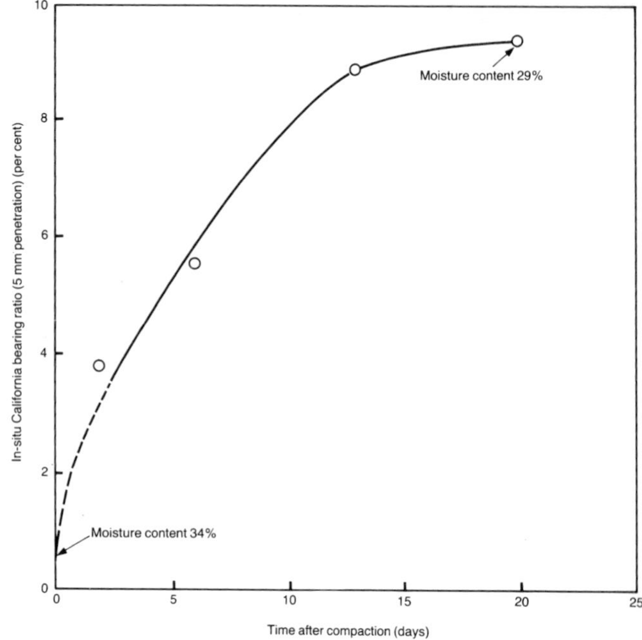

Figure 16.14 Relation between in-situ California bearing ratio and time after compaction of the domestic refuse in a wet condition by the 7.4 t smooth-wheeled roller

that, as the material appeared to possess substantial internal friction, higher values of in-situ CBR would be developed beneath a pavement. It was also noted that laboratory tests would have given values of CBR far higher than those determined by the in-situ method.

16.44 After the final set of measurements was made, at a moisture content of about 34 per cent, when the CBR was so low as to be immeasurable, the compacted material was covered with a tarpaulin to reduce evaporation. In-situ CBR tests were carried out at intervals during the following three weeks to determine any increase in stability of the material and the results are given in Figure 16.14. The in-situ CBR increased from near zero to a value of about 9 per cent after 20 days. The shape of the curve indicates that any further increase would have been negligible. During those 20 days the moisture content of the domestic refuse decreased from 34 per cent to 29 per cent, and most of that loss was thought to be caused by drainage into the lower layers of the test pit. It was considered by Parsons and Krawczyk (1961) that if the domestic refuse was inadvertently compacted in a wet condition so that it was extremely spongy, and provided it was not trafficked by construction vehicles, the strength of the material would increase as a result of the draining away of the excess moisture.

16.45 The general conclusion reached from the investigations was that the spent domestic refuse would be suitable as general fill in mass earthworks, but because of a slight frost susceptibility, it could not be regarded as suitable for use within 450 mm of the surface of the road. Subsequent to these investigations, motorway embankments were constructed using the spent domestic refuse (Anon, 1964) and as far as is known those embankments have all performed satisfactorily.

SECTION C COMPACTION OF PULVERISED FUEL ASH

16.46 Investigations into the compaction of pulverised fuel ash were carried out in two completely independent series. The first was in 1951–52 when the 2.7 t smooth-wheeled roller (Plate 4.1) was used in research into the general problem of whether ash from power stations could be employed in the construction and maintenance of roads; the second series of tests was carried out with vibrating rollers in 1968 to 1971.

The pulverised fuel ash used

16.47 In modern coal-fired power stations, the coal is pulverised before being fed into the combustion chamber. The resultant ash is in the form of small spheres, some of which are hollow. A high proportion of this, extracted from the flue gases, is known as pulverised fuel ash (PFA). The remaining ash, consisting of larger coagulations of particles, is collected from the bottom of the combustion chamber and is known as furnace bottom ash. The PFA used in the investigations was 'conditioned ash'. This is PFA from power station hoppers to which water has been added to bring it to a state suitable for compaction.

16.48 The particle-size distributions of the PFA used are given in Figure 16.15. The PFA used in the second series of investigations (with vibrating rollers) was the finer and the more uniformly graded of the two materials used. Particle density was fairly low at about 2.4 Mg/m^3 for the first series of tests. The particle density of the PFA used with vibrating rollers was not measured, but values for some PFAs have been recorded at as low as 2.0 Mg/m^3 (Fox, 1984).

Tests with the 2.7 t smooth-wheeled roller

16.49 Investigations were made in 1951–52 to determine the relation between dry density and moisture content for 32 passes of the 2.7 t

Figure 16.15 Particle-size distributions of pulverised fuel ash used in compaction investigations

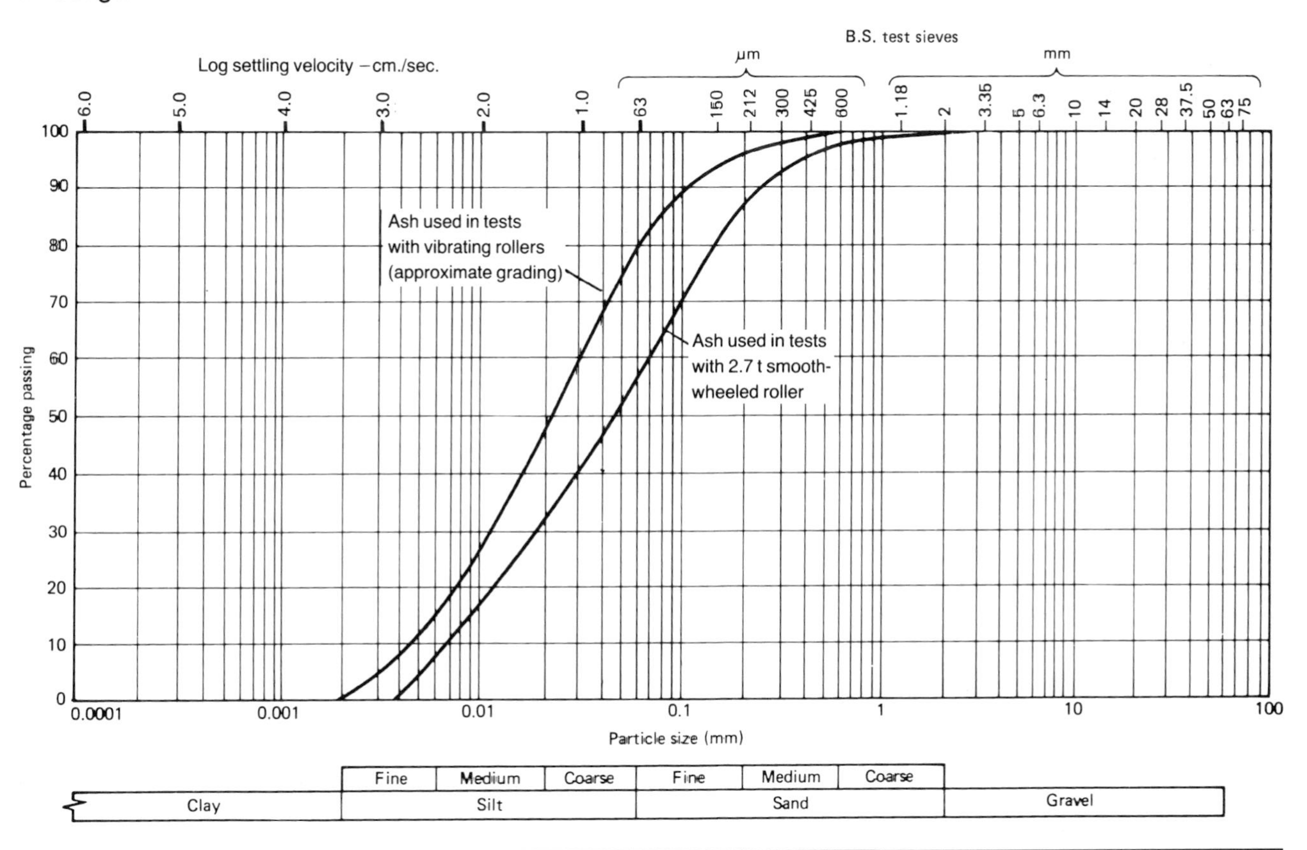

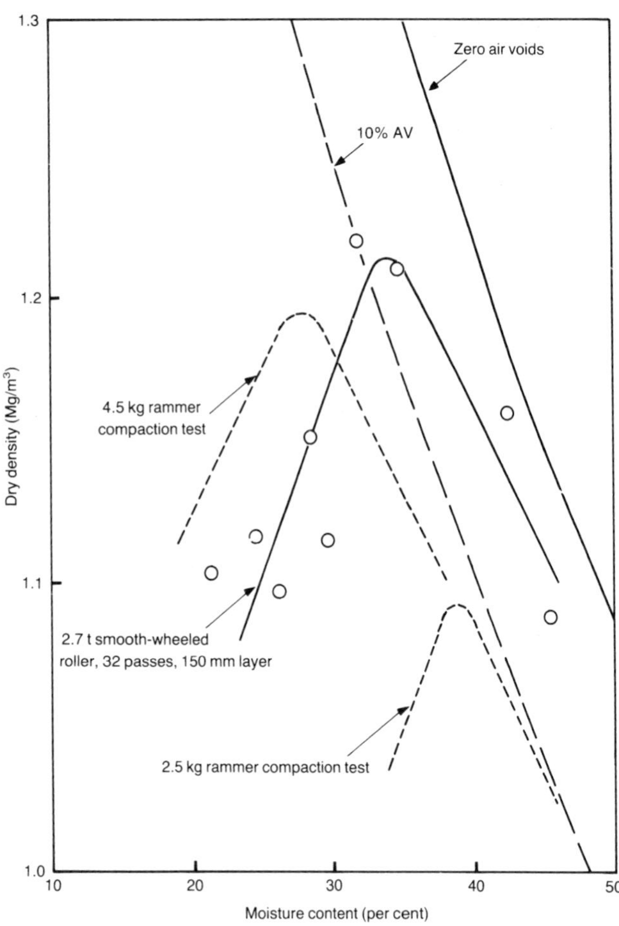

Figure 16.16 Relations between dry density and moisture content obtained with the 2.7 t smooth-wheeled roller and in laboratory compaction tests on pulverised fuel ash

smooth-wheeled roller (Plate 4.1 and Table 16.1) on a 150 mm compacted layer. The results are shown in Figure 16.16. Compaction tests, using the 2.5 kg and 4.5 kg rammer methods, were also carried out and the results are included in the same Figure. These results indicate that the smooth-wheeled roller produced a maximum dry density in excess of the maximum achieved in the 4.5 kg rammer compaction test, and about 110 per cent of the maximum dry density achieved in the 2.5 kg rammer test. The low air contents obtained with the roller at moisture contents in excess of 34 per cent are indicative of the relatively wide range of particle sizes in this particular PFA (Figure 16.15); with more uniform gradings that are common with such materials higher air contents would have occurred, even when well compacted at moisture contents above the optimum for the compaction plant.

16.50 During the determination of the relation between dry density and moisture content with the 2.7 t smooth-wheeled roller measurements of in-situ California bearing ratio (CBR) were also made and the results are given in Figure 16.17. These show that a maximum CBR of 33 per cent was achieved at a moisture content of about 27 per cent; at higher moisture contents the CBR decreased rapidly. However, many conditioned PFAs are known to have pozzolanic properties (Fox, 1984) and it is likely that, in practice, increases in CBR could occur with time following completion of compaction at high moisture contents.

16.51 In the investigations it was found that the workable range of moisture contents within which the compaction could be carried out satisfactorily was 25 to 35 per cent. At lower moisture contents the material was too light to handle whilst at moisture contents higher than 35 per cent excess pore water pressures were set up in the PFA by the roller and the material became very spongy. As the moisture content increased above this value the ash became progressively weaker and eventually the roller become 'bogged down' when the moisture content was about 45 per cent. This change in the stability of the material was reflected in the CBR values which fell to a negligible level at moisture contents higher than 40 per cent (Figure 16.17).

Tests with vibrating rollers

16.52 In the period from 1968 to 1971, a limited number of tests were carried out with three of

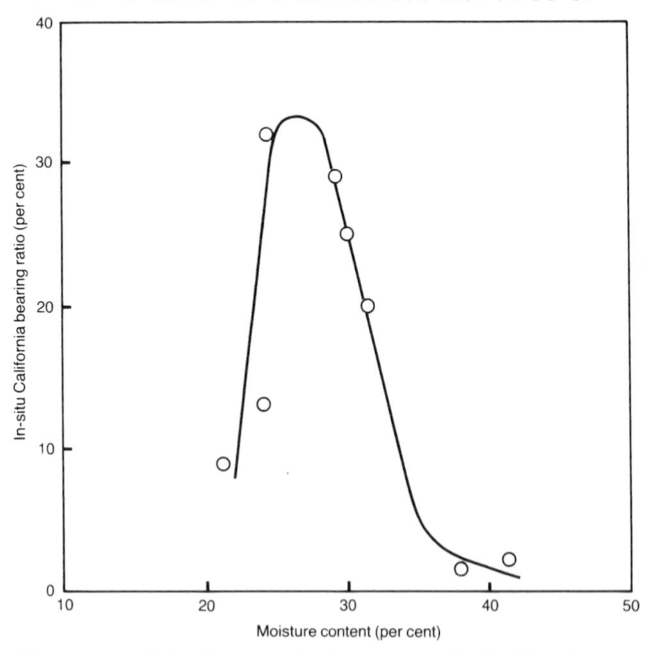

Figure 16.17 Relation between in-situ California bearing ratio and moisture content of pulverised fuel ash when compacted by 32 passes of the 2.7 t smooth-wheeled roller

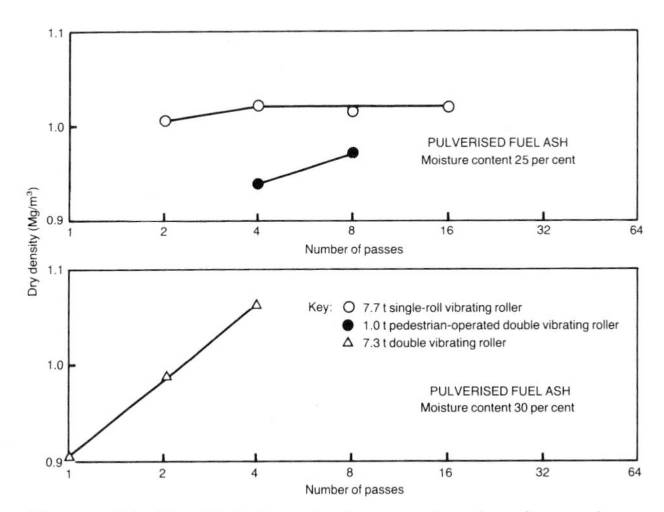

Figure 16.18 Relations between dry density and number of passes of vibrating rollers on 230 mm layers of pulverised fuel ash

the vibrating rollers investigated with the standard test soils (Chapter 7). The vibrating rollers used were:–

(i) The 7.7 t single-roll vibrating roller shown in Plate 7.14;

(ii) The 1.0 t pedestrian-operated double vibrating roller shown in Plate 7.4;

(iii) The 7.3 t double vibrating roller shown in Plate 7.13.

Details of the three machines are given in Table 7.1.

16.53 Relations between dry density and number of passes were determined for each machine and the results are given in Figure 16.18. A compacted layer thickness of 230 mm was used for all the rollers, but the tests differed in some practical details.

16.54 The 7.7 t machine failed to operate on the loose layer of PFA, but pre-compaction by a tracklaying tractor enabled operation of the roller to proceed. The dry density determinations were made in the top 230 mm of compacted PFA; this layer is likely to have contained some over-stressed surface material, resulting in relatively low dry densities and the lack of any further increase in the state of compaction produced after 4 passes of the vibrating roller (Figure 16.18).

16.55 The 1.0 t pedestrian-operated vibrating roller travelled successfully on loose PFA provided the thickness of the loose layer did not exceed 350 mm; such a loose thickness resulted in a compacted layer of 230 mm. As with the 7.7 t machine, the dry density of the complete layer was determined, but it is unlikely that any significant over-stressing of the surface occurred with this size of machine. Relatively low dry densities were produced although the trends in the results (Figure 16.18) indicate that an increased number of passes would have produced higher values; it is also likely that the layer thickness was too great for this class of machine and about 150 mm would have been more appropriate.

16.56 The 7.3 t double vibrating roller operated successfully on the loose PFA. A total thickness of 300 mm was compacted and the top 70 mm of loose, over-stressed material was removed before the dry density of the lower 230 mm was determined. The trends of the results indicate that relatively high values of dry density would have been obtained if more than 4 passes (the maximum used in the tests) had been applied.

16.57 The results of laboratory compaction tests on the PFA used in the investigations with vibrating rollers are given in Figure 16.19. Results for 2.5 kg and 4.5 kg rammer tests and the vibrating hammer test are included. Most commonly, the minimum state of compaction required in specifications for the use of conditioned PFA is 95 per cent of the maximum dry density obtained in the 2.5 kg rammer compaction test (Department of Transport (1986) and Fox (1984)). This would entail the achievement of a dry density of 1.17 Mg/m^3 in the material used in the vibrating roller tests. This value of dry density was not reached in any of the test results shown in Figure 16.18, although it is likely that the 7.3 t double vibrating roller would have achieved this value in about 12 passes. If the over-stressed surface material had not been included in the determinations of dry density with the 7.7 t machine the values produced would have been considerably higher, although the need to provide some pre-compaction to enable the roller to operate on the PFA must be taken into account when considering the use of this type of machine. As mentioned in Paragraph 16.55, it is possible that the use of a reduced compacted layer thickness would have considerably enhanced the results produced by the 1.0 t pedestrian-operated machine. The 7.7 t and 1.0 t machines were tested with the PFA at a moisture content that was 5 per cent below the optimum of the 2.5 kg rammer compaction test (see Figure 16.19) and an increase in moisture content to around the optimum of that test (ie to the same level as that used with the 7.3 t machine) would also have improved their apparent performances.

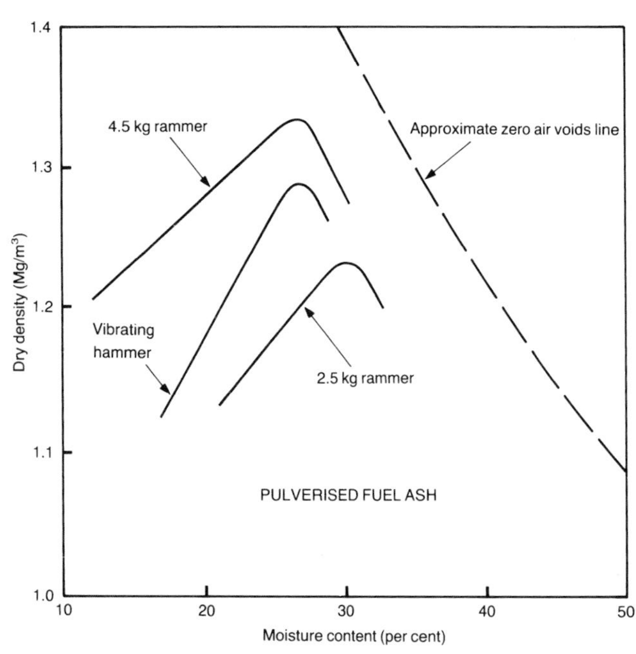

Figure 16.19 Relations between dry density and moisture content obtained in laboratory compaction tests on the pulverised fuel ash used in investigations with vibrating rollers (see Figure 16.18)

Conclusions on the compaction of PFA

16.58 The investigations carried out with the 2.7 t smooth-wheeled roller and the three vibrating rollers provide little information on possible outputs of compacted PFA for the various machines or on the thicknesses of layer to use. Problems of trafficability of the self-propelled compactors, and of over-stressing of the surface of the compacted layers have, however, been highlighted. Comprehensive information on the performance of compaction plant on pulverised fuel ash is available from information published by the Central Electricity Generating Board.

References

Anon (1964). M4 motorway from Chiswick flyover to Langley-Slough by-pass. The Surveyor and Municipal Engineer (10) 19, 20, 58–9.

British Standards Institution (1948). Methods of test for soil classification and compaction. BS 1377 : 1948. BSI, London.

British Standards Institution (1990a). British Standard methods of test for soils for civil engineering purposes : BS 1377 : Part 2 Classification tests. BSI, London.

British Standards Institution (1990b). British Standard methods of test for soils for civil engineering purposes : BS 1377 : Part 3 Chemical tests. BSI, London.

British Standards Institution (1990c). British Standard methods of test for soils for civil engineering purposes : BS 1377 : Part 9 In situ tests. BSI, London.

Department of Transport (1986). Specification for highway works. 6th edition. HM Stationery Office, London.

Fox, N H (1984). Pulverised fuel ash as structural fill. *Proc. 2nd Int. Conf. on Ash Technology and Marketing.* Central Electricity Generating Board, London, 495–9.

Lewis, W A and Parsons, A W (1961). The performance of compaction plant in the compaction of two types of granular base material. *Road Research Technical Paper* No 53. HM Stationery Office, London.

Ministry of Transport and Civil Aviation (1957). Specification for road and bridge works. 2nd edition. HM Stationery Office, London.

Parsons, A W and Krawczyk, J (1961). An investigation of the suitability of spent domestic refuse for embankment construction. *D.S.I.R., Road Research Laboratory, Research Note* No RN/4063/ AWP.JK (Unpublished).

CHAPTER 17 LABORATORY COMPACTION TESTS

17.1 The compaction of soil in the laboratory using standardised procedures is generally regarded as an essential ingredient in the design of earthworks and also is often used in the subsequent control of the full-scale compaction process. Usually a laboratory compaction test comprises the determination of a relation between the dry density produced and the moisture content of the soil using a constant compactive effort. The main purposes to which the results of laboratory compaction tests are put are:–

1. The classification of soils;
2. Guidance on potential levels of compaction to be achieved in full- scale work;
3. Guidance on the most appropriate moisture content at which to compact a soil or granular material;
4. The provision of a standard against which to assess the dry density achieved in-situ.

17.2 Examples of the application of such tests for classifying soils are provided in Paragraphs 3.17 to 3.20, where variations in the characteristics of the Laboratory's standard test soils over the years have been illustrated by the trends in maximum dry densities and optimum moisture contents obtained in laboratory compaction tests.

17.3 The concept of the use of laboratory compaction tests to set standards for full-scale compaction has been described in Paragraphs 2.23 to 2.28.

17.4 The commonly used laboratory compaction tests are described in this Chapter, and the work carried out at the Laboratory in the development of vibratory compaction tests is included. The Chapter ends with a general discussion on the problem of removal of coarse particles from soil used in laboratory compaction tests and the effect on the results.

Compaction test using a 2.5 kg rammer

17.5 This compaction test is described in Part 4 of British Standard BS 1377 (British Standards Institution, 1990), and is specified, with slight variations from the British Standard, as ASTM Test D 698 (American Society of Testing and Materials, Annual publication) and as AASHTO Test T 99 (American Association of State Highway and Transportation Officials, 1986). In many European countries it is referred to as the normal or basic Proctor test (Ministere de l'Equipement, 1976), a reference to R R Proctor, who first pioneered this type of test in 1933 in connection with the construction of earthdams (Proctor, 1933).

17.6 The requirements of the British Standard version of this test are that soil is compacted in either a 1000 cm^3 mould, internal diameter 105 mm, or a California bearing ratio (CBR) mould, internal diameter 152 mm and depth 127 mm, depending on the maximum particle size of the soil. Three approximately equal layers are compacted, using a 2.5 kg rammer falling through a height of 300 mm. The rammer is 50 mm in diameter and 27 blows (1000 cm^3 mould) or 62 blows (CBR mould) of the rammer are distributed uniformly over each of the layers of soil. Surplus compacted soil is struck off flush with the top of the mould and the bulk density of the compacted soil calculated from the mass of soil and the volume of the mould. The moisture content of the soil in the mould is determined and a dry density calculated. A series of tests at various values of moisture content enables a relation between dry density and moisture content to be plotted. (Figure 2.11 contains results for the various test soils used in full-scale investigations.) The maximum dry density and optimum moisture content are recorded from the plotted curve (see Figure 2.2).

17.7 The British Standard method restricts the maximum particle size that may be used in the 1000 cm^3 mould to 20 mm and in the CBR mould to 37.5 mm. The procedure further stipulates that material is too coarse to be tested if more than 30 per cent is retained on the 20 mm sieve or more than 10 per cent is retained on the 37.5 mm sieve. Where more than 5 per cent of the soil exceeds the maximum allowable particle size and is removed from the test sample, procedures for substitution of the material or correction of the results may be applied (see Paragraphs 17.33 to 17.35).

Compaction test using a 4.5 kg rammer

17.8 This compaction test is described in Part 4 of British Standard 1377 (British Standards Institution, 1990) and is specified, with slight variations from the British Standard, as ASTM Test D 1557 (American Society of Testing and Materials, Annual publication) and as AASHTO Test T 180 (American Association of State Highway and Transportation Officials, 1986). In many European countries it is referred to as the Modified Proctor test. The principle is similar to that of the 2.5 kg rammer method, but a much larger compactive effort is employed.

17.9 In the British Standard procedure the soil is compacted in a 1000 cm^3 mould, internal diameter 105 mm, or a California bearing ratio (CBR) mould, internal diameter 152 mm and depth 127 mm, depending on the maximum particle size. Five approximately equal layers are compacted, using a 4.5 kg rammer falling through a height of 450 mm. The rammer is 50 mm diameter, and 27 blows (1000 cm^3 mould) or 62 blows (CBR mould) of the rammer are distributed uniformly over each of the layers of soil. The bulk density and dry density are determined as for the 2.5 kg rammer test (see Paragraph 17.6) and the relation between dry density and moisture content plotted; the maximum dry density and optimum moisture content are determined from the plotted curve.

17.10 The comments relating to restrictions on the maximum particle size of the soil, made for the 2.5 kg rammer method (see Paragraph 17.7), also apply in the case of the 4.5 kg rammer compaction test.

Compaction test using a vibrating hammer

17.11 The laboratory compaction tests involving the use of impact compaction by the 2.5 kg and 4.5 kg rammers (Paragraphs 17.5 to 17.10) have been shown to produce results with granular base materials which are considerably lower than can be achieved with full-scale compaction plant (see Paragraph 16.20), and the optimum moisture contents of the tests are significantly higher than those at which base materials should be compacted in the field (Lewis and Parsons, 1961). For this reason investigations into a vibratory compaction method were carried out at the Laboratory during the period from 1959 to 1964. To ensure that the new test would be generally accepted in the United Kingdom, it was considered that a method should be developed which employed apparatus in widespread use in site laboratories (Parsons, 1964). It was decided, therefore, to investigate the use of vibrating hammers, often employed at that time in the compaction of test cubes of concrete, together with moulds as used in the California bearing ratio (CBR) test. Other types of mould were tested, however, including those used in the manufacture of 150 mm cubes and, for cohesive soils, moulds of 100 mm diameter as used to prepare specimens for frost susceptibility tests.

17.12 In addition to the use of different sizes and shapes of mould as mentioned above, studies were made of the effects on the states of compaction obtained of variations in the type of vibrating hammer, in the size of the tamping foot, in the static load applied to the vibrating hammer, in the time of operation of the hammer, and in the voltage supply to the hammer. Numerous test materials were used in the studies, but the following illustrations of the test results are restricted to one material, the limestone wet-mix macadam. The particle-size distribution of the limestone is given in Figure 3.4. The information has been taken from Parsons (1964).

Type of vibrating hammer and size of tamping foot

17.13 Relations between dry density of limestone wet-mix macadam and the nominal power consumption of the vibrating hammers used are given in Figure 17.1. The 152 mm diameter CBR mould and two sizes of tamping foot (146 mm diameter and 76 mm diameter) were used; the tamping feet with one of the vibrating hammers are shown in Plate 17.1. The material was compacted in the mould in three equal layers, with 60 s of vibration per layer, to provide a total compacted thickness between 127 and 133 mm. The bulk density of the compacted specimen was determined from its mass, the diameter of the mould and the measured depth. Plate 17.2 shows the measurement of the depth of a compacted specimen using a datum bar across the top of the mould collar and a depth gauge.

17.14 The conclusions reached from the results were that the use of the small tamping foot caused some overstressing of the surface of

Plate 17.1 The various tamping feet used in the study of a vibrating hammer compaction test. Sizes of tamping feet were, from left to right, 146 mm square, 146 mm diameter and 76 mm diameter

each compacted layer, especially with finer-grained soils, and a tamping foot covering almost the entire surface area of the compacted layer was adopted to avoid this effect. The relations obtained, such as those shown in Figure 17.1, also indicated that only small rates of change in dry density with variation in nominal power consumption occurred at the higher values of power consumption, and a rated power consumption in the range of 600 – 750 W was adopted. It was also concluded, however, that a suitability test was needed to check that the vibrating hammer, although having the correct

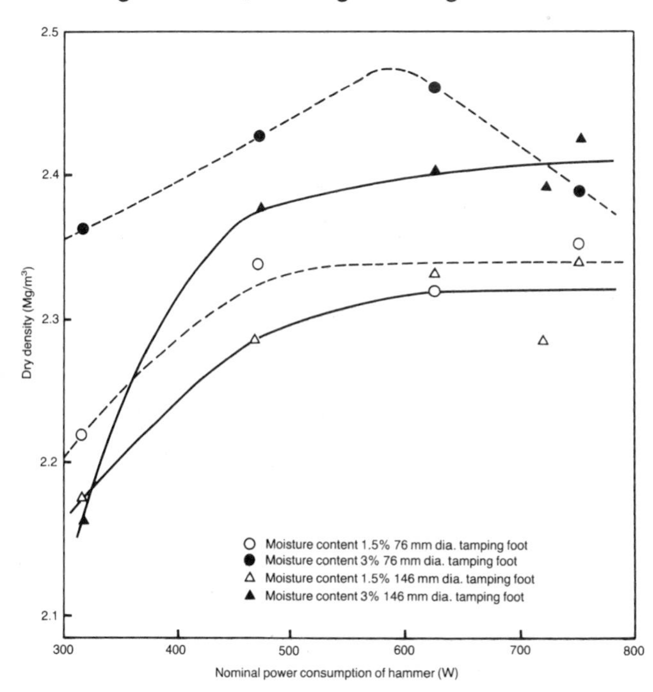

Figure 17.1 Relations between dry density of limestone wet-mix macadam and nominal power consumption of the vibrating hammer

Plate 17.2 Method of measuring the height of compacted material in the mould in the vibrating hammer compaction test

rating, was also working efficiently (see Paragraph 17.21).

Static load applied to the vibrating hammer

17.15 Sets of weights were attached to the body of the vibrating hammer to simulate various levels of downward pressure applied by the operator; the vibrating hammer is shown in use with the attachments in Plate 17.3. Relations between dry density and static load are given in Figure 17.2 for the limestone wet-mix macadam at two different values of moisture content. Dry density increased with increase in static weight up to the practical maximum applied, but it was concluded that provided the static load, including the weight of the vibrating hammer, was controlled between 300 and 400 N, little variation in test results would occur.

Time of compaction by the vibrating hammer

17.16 Compaction time per layer was varied from 15 to 120 s; results for limestone are shown in Figure 17.3. It was noted that increases in dry

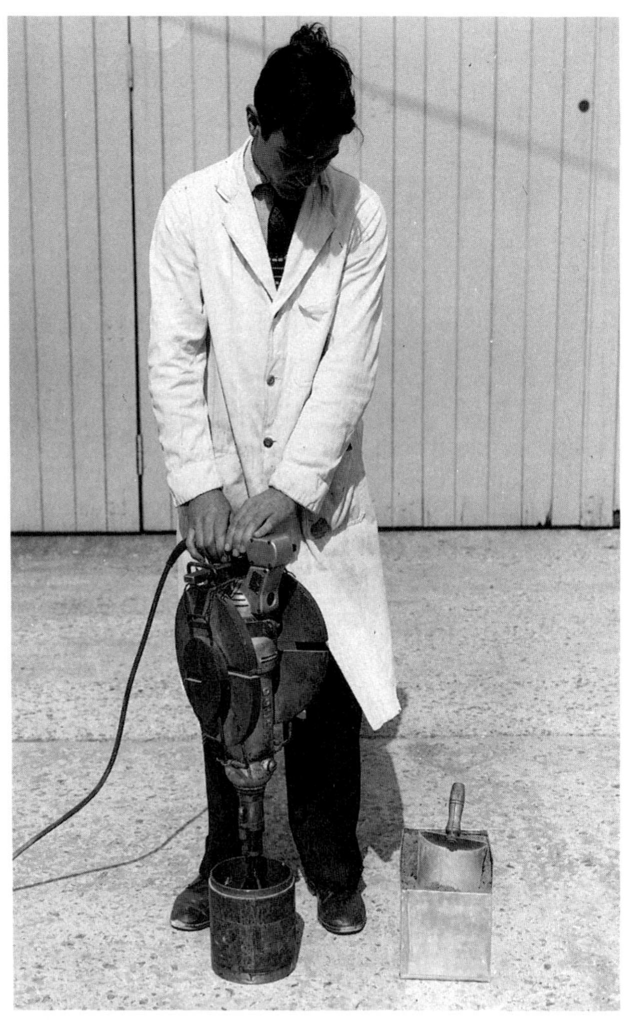

Plate 17.3 Vibrating hammer with weights attached as used in studies of the effect of variations in static load on the results of the vibrating hammer compaction test

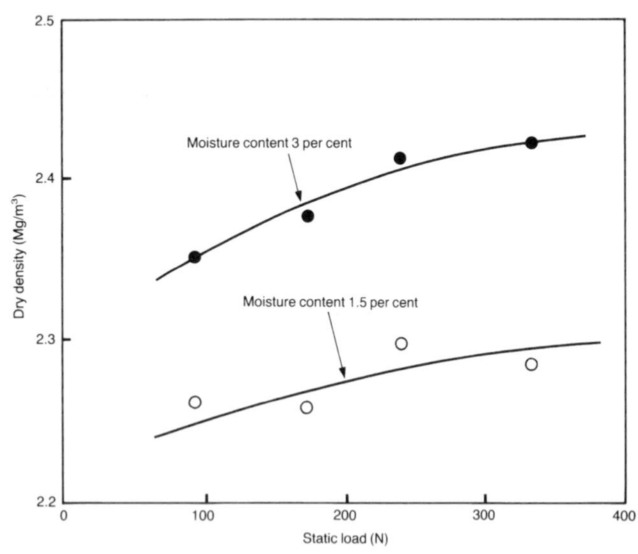

Figure 17.2 Relations between dry density of limestone wet-mix macadam and static load applied to vibrating hammer

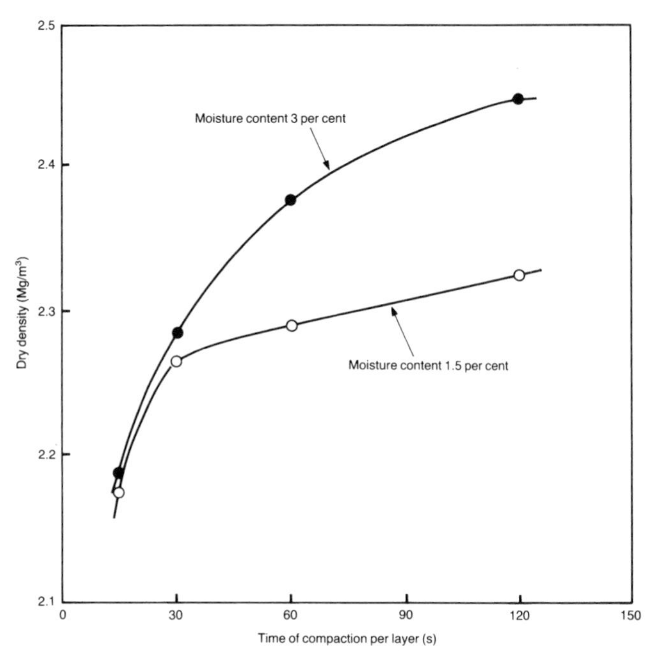

Figure 17.3 Relations between dry density of limestone wet-mix macadam and time of compaction by a vibrating hammer

density were not large for some soils for times in excess of 60 s, although the limestone illustrated (Figure 17.3) was an exception to this. Taking into account the need for a reasonably quick test and the necessity of avoiding operator fatigue while applying the static load to the vibrating hammer, it was decided to adopt a vibration time of 60 s per layer. This time would have to be strictly controlled to minimise potential variations in results with materials such as limestone wet-mix.

Mains voltage applied to the vibrating hammer

17.17 A voltage control unit was inserted in the supply lead to the vibrating hammer and the power consumption was measured as the machine was used. Relations between dry density and moisture content of limestone are given in Figure 17.4 for two levels of supply voltage. It should be noted that the power consumption of the vibrating hammer used was below the range adopted for the standard test (600–750 W). Had a more powerful vibrating hammer been used, the reduction in power consumption, equal to about 15 to 20 per cent for a reduction in voltage from 250 to 230 volts, would probably have caused a smaller difference than that shown in Figure 17.4. The results given in Figure 17.1 are likely to be illustrative of the effect of variations in power consumption resulting from changes in mains voltage.

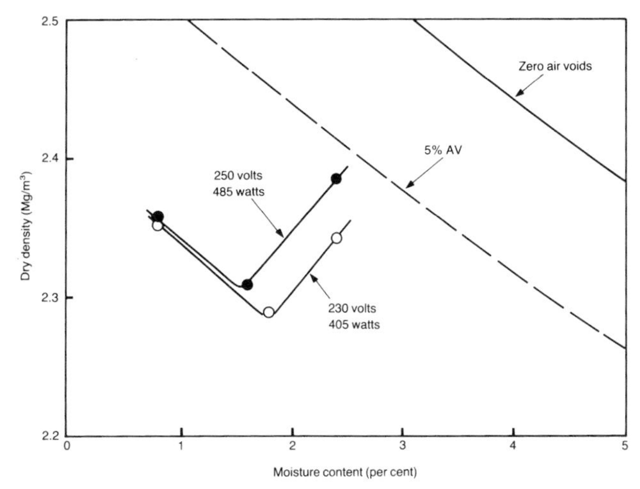

Figure 17.4 Relations between dry density and moisture content of limestone wet-mix macadam obtained with a vibrating hammer when the supply voltage was varied from 230 to 250 volts. The power consumptions shown were measured during the tests

Size and shape of mould

17.18 Comparisons have been made in Figure 17.5 of relations between dry density and moisture content of limestone obtained using the CBR mould with the procedure adopted as a result of the studies described above, and using a 150 mm cube mould with times of vibration of 60 s and 120 s per each of three layers. In this instance the procedure used with the CBR mould produced the highest test results; with other materials, however, the differences were not as significant as those illustrated. Because of

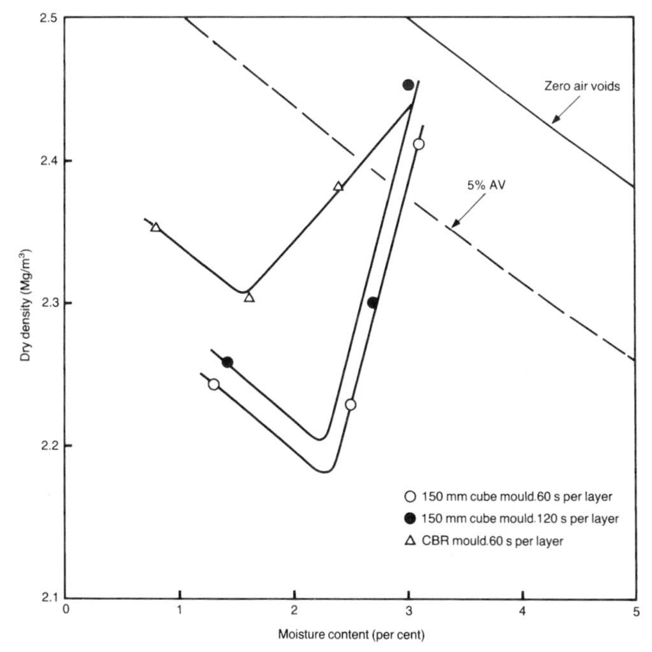

Figure 17.5 Relations between dry density and moisture content of limestone wet-mix macadam compacted by a vibrating hammer in two types of mould

the relative lightness of the CBR mould and the smaller quantity of soil required for testing with it, it was considered that this type of mould should be adopted for use in a vibrating hammer compaction test. It was recommended that the CBR mould used should be of steel and of the type with a screw-on base and collar.

The definitive vibrating hammer test

17.19 The studies described in Paragraphs 17.13 to 17.18 culminated in the British Standard vibrating hammer compaction test as specified in Part 4 of British Standard BS 1377 (British Standards Institution, 1990). The test is shown being performed in Plate 17.4. The test requires a vibrating hammer with a power consumption of 600 to 750 W. Soil is compacted in a cylindrical CBR mould, ie a mould of 152 mm diameter, in three approximately equal layers, with 60 s of vibration per layer. During compaction, a firm downward pressure has to be applied to the vibrating hammer so that the total downward

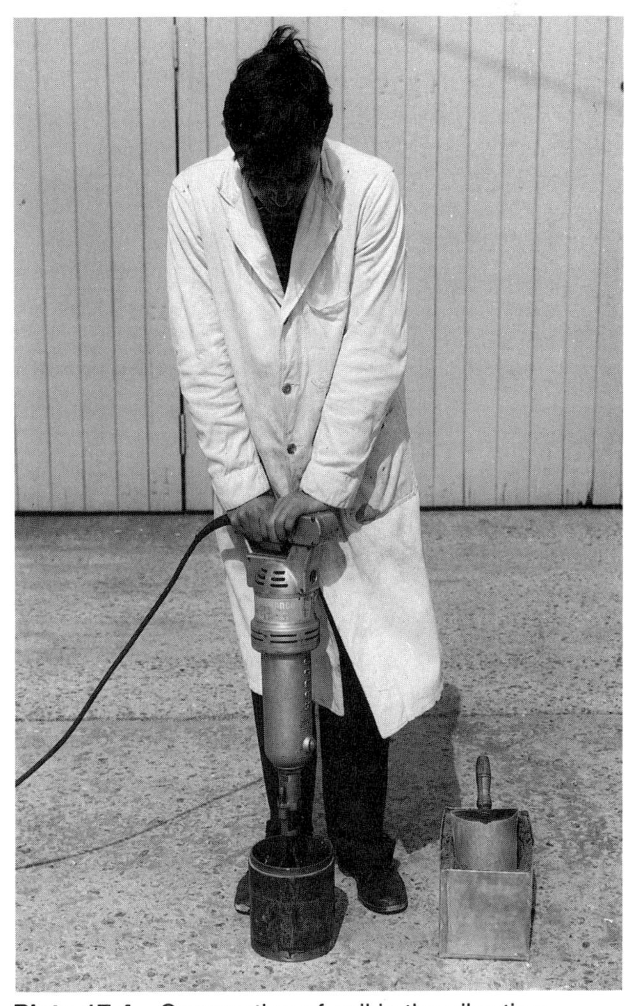

Plate 17.4 Compaction of soil in the vibrating hammer compaction test

force, including that resulting from the mass of the hammer and tamping foot, is between 300 N and 400 N. The tamping foot attached to the vibrating hammer has a circular base which almost completely covers the area of the mould, producing a flat surface to the compacted material. The total depth of the specimen after compaction should lie between 127 mm and 133 mm for the compaction test to be acceptable. Because of the size of the mould used the maximum particle size that may be incorporated in the sample is 37.5 mm. The bulk density is determined from the mass of the soil, the measured depth of the specimen and the known circular area of the mould. The test is carried out over a range of moisture contents to produce a relation between dry density and moisture content and hence the maximum dry density and optimum moisture content may be determined.

17.20 Comparisons have been made in Figures 17.6 and 17.7 of the maximum relative compaction values (ie the maximum dry densities expressed as percentages of the maximum dry density obtained in the 2.5 kg rammer compaction test) and optimum moisture contents (relative to the optimum of the 2.5 kg rammer compaction test) resulting from various compaction methods on various materials. Thus the results produced by the vibrating hammer compaction test are compared with those produced by the 2.5 kg and 4.5 kg rammer methods and by various items of full-scale plant; the results for the plant have been taken from relevant earlier Chapters. These comparisons indicate that with the more granular materials the vibrating hammer test is more efficient than either of the laboratory rammer tests, although as the soils become finer and more cohesive, the 4.5 kg rammer test produces the highest maximum relative compaction values at the lowest optimum moisture contents. However, for the complete range of soils and base materials illustrated, of the three laboratory tests, the vibrating hammer test provides the rather better guide to the performance of the full-scale compaction plant.

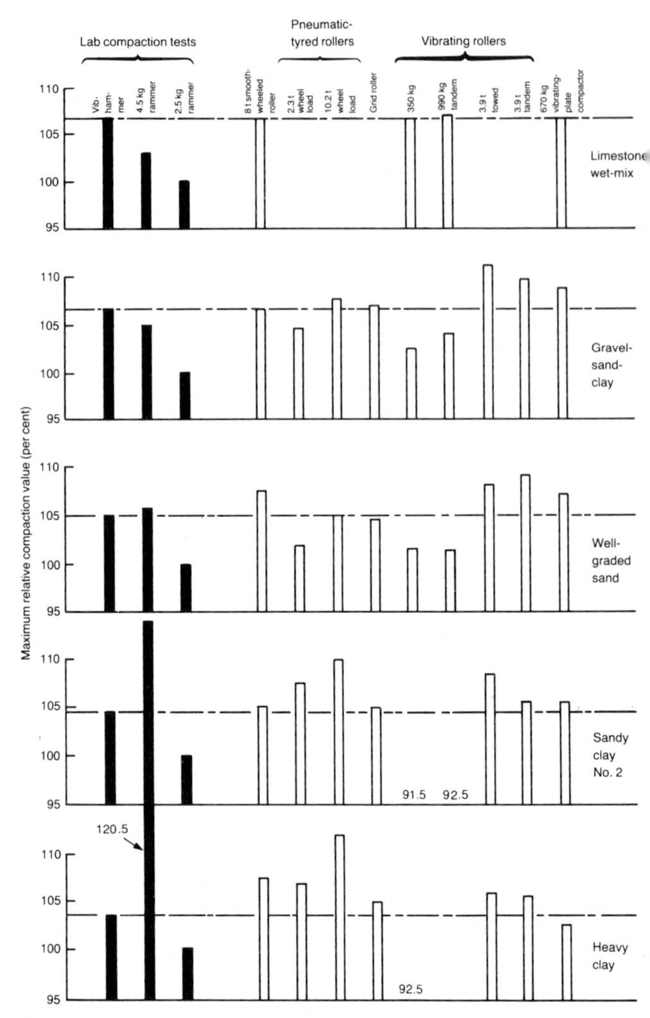

Figure 17.6 Maximum relative compaction values achieved in laboratory compaction tests and by compaction to refusal by compaction plant

Calibration test for the vibrating hammer

17.21 To overcome possible errors resulting from deteriorations in performance of vibrating hammers with age, a calibration test was developed and has been included in the test procedure in Part 4 of British Standard BS 1377 (British Standards Institution, 1990). In the calibration test the vibrating hammer to be tested is used with the standard procedure to compact a silica sand of specified grading from the Leighton Buzzard district, prepared at a moisture content of 2.5 ± 0.5 per cent. Three tests are carried out and the dry density in each test is determined to the nearest 0.002 Mg/m^3; the range of values should not exceed 0.01 Mg/m^3. The vibrating hammer is considered suitable for use in the compaction test if the mean dry density of the sand exceeds 1.74 Mg/m^3.

Swedish vibratory apparatus

17.22 A laboratory vibratory compaction test of a similar character to that of the British Standard vibrating hammer test has been described by Forssblad (1967). The apparatus consisted of a 3-phase electric motor integral with an eccentric unit, a tamping foot to which the motor was

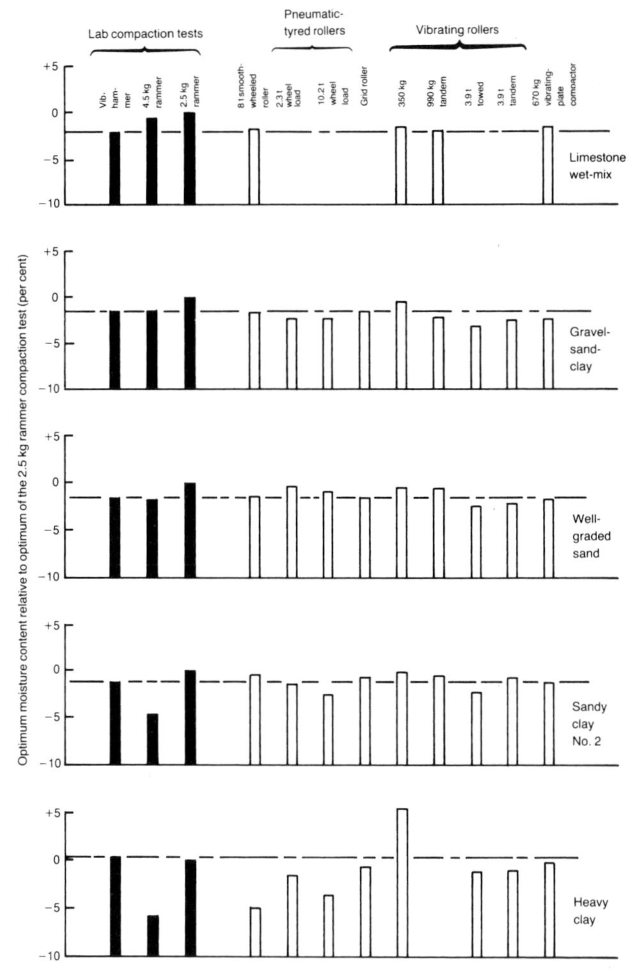

Figure 17.7 Optimum moisture contents, relative to the optimum of the 2.5 kg rammer compaction test, obtained in laboratory compaction tests and by compaction to refusal by compaction plant

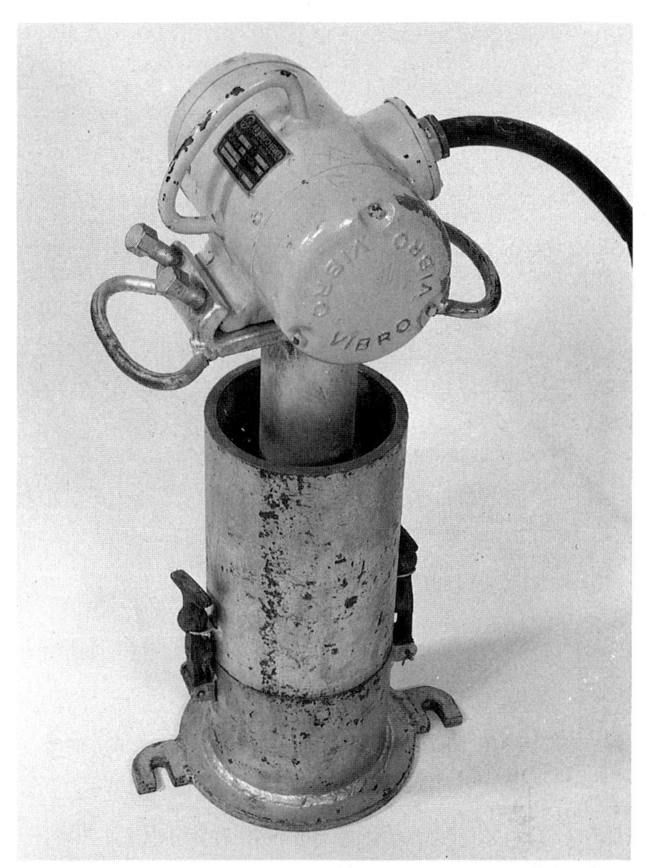

Plate 17.5 The Swedish vibratory apparatus for the laboratory compaction of soils and granular materials

clamped, a mould and a collar. The apparatus is shown in use in Plate 17.5 and dismantled, to show the various components, in Plate 17.6. The tamping foot had a circular compacting face 150 mm diameter and was also fitted with a circular guide plate which operated within the collar of the mould. The collar was attached to the mould by a pair of rubber lugs and, in normal use, the mould was rigidly clamped to a concrete block with a mass of at least 400 kg. The recommended procedure was to compact material in the mould in two layers, using two minutes of vibration per layer, with the vibrating unit free-standing on the surface of the layer. The depth of the compacted specimen was determined by means of a depth gauge (see Plate 17.2 for a similar method used with the British Standard vibrating hammer test) and the bulk density determined from the mass of material, the cross-sectional area of the mould and the measured depth. Details of the apparatus are given in Table 17.1.

17.23 Comparative tests, using the Swedish apparatus and the newly developed vibrating hammer test, were carried out at the Laboratory and have been reported by Odubanjo (1968). Four materials were used in the study, namely heavy clay, uniformly graded fine sand, well-graded sand and limestone wet-mix macadam. The particle-size distributions and other information on these materials are given in Figure 3.4. The procedure used with the Swedish apparatus was as given above; the quantity of material to be used was not laid down, however, in the recommended procedure, but as the diameters of the moulds were identical, the same total mass of material was used as that in the vibrating hammer test. Relations between dry density and moisture

Table 17.1 Details of the Swedish vibratory compaction apparatus

Power consumption	250 W
Nominal frequency of vibration	50 Hz
Nominal centrifugal force	2.45 kN
Diameter of tamping foot	150 mm
Diameter of mould	152 mm
Depth of mould	138 mm
Depth of collar	243 mm
Mass of tamper and motor	35.4 kg
Mass per unit area of tamping foot	2000 kg/m^2

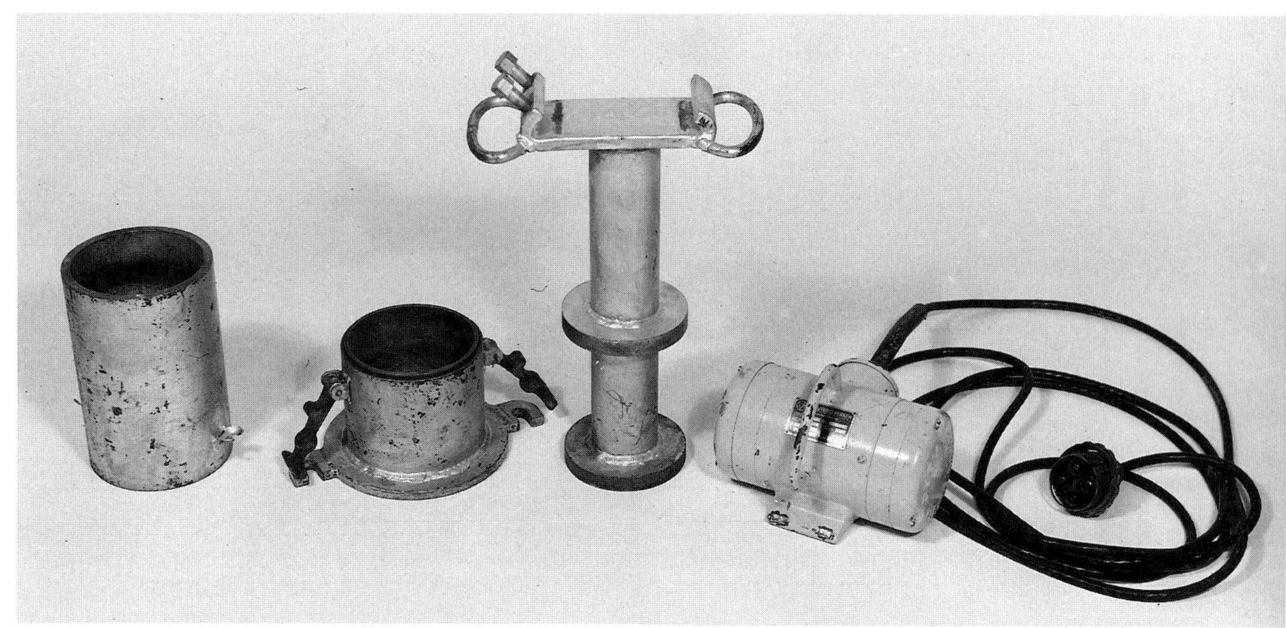

Plate 17.6 The dismantled Swedish vibratory apparatus showing its various components

content were determined for both methods and are shown in Figures 17.8 to 17.11.

17.24 The results obtained with both the Swedish vibratory compaction apparatus and the British Standard vibrating hammer compaction test were very similar for heavy clay over most of the range of moisture contents tested (Figure 17.8) and for well-graded sand (Figure 17.10). With uniformly graded fine sand (Figure 17.9) and limestone wet-mix macadam (Figure 17.11), however, the BS vibrating hammer compaction test produced higher values of dry density than did the Swedish apparatus over the range of moisture contents tested.

17.25 To obtain a more exact comparison between the performance of the Swedish apparatus and that of the BS vibrating hammer compaction test, tests were carried out using limestone wet-mix macadam in which the compaction procedures were the same for the two methods, ie three layers of material were compacted using one minute of vibration per layer (Figure 17.11). However, the amended procedure caused a reduction in the values of dry density obtained at any given moisture content with the Swedish apparatus, and thus further increased the divergence from the results obtained with the BS vibrating hammer compaction test. The results indicate that the

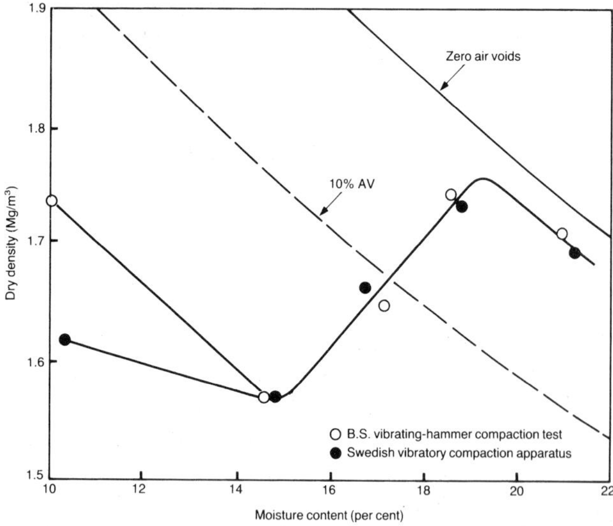

Figure 17.8 Relations between dry density and moisture content of heavy clay using two vibratory compaction methods

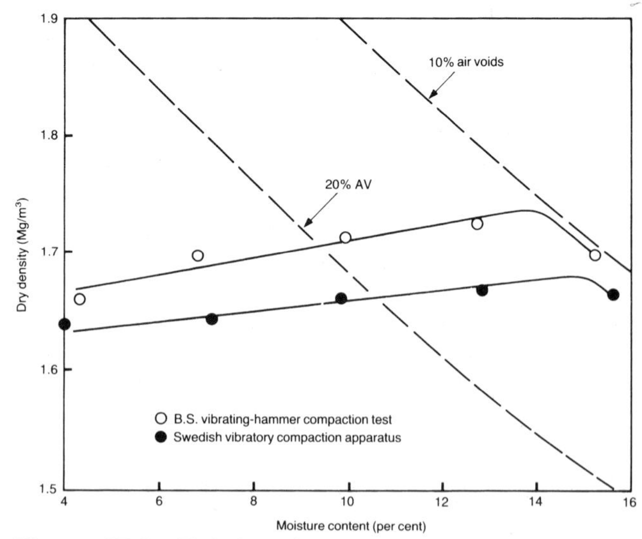

Figure 17.9 Relations between dry density and moisture content of uniformly graded fine sand using two vibratory compaction methods

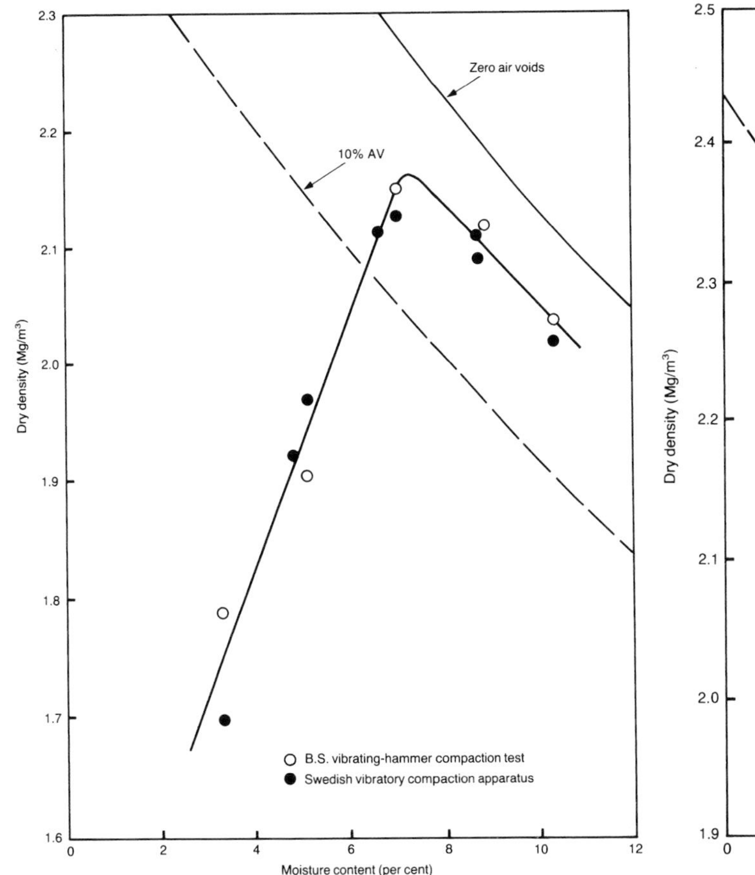

Figure 17.10 Relations between dry density and moisture content of well-graded sand using two vibratory compaction methods

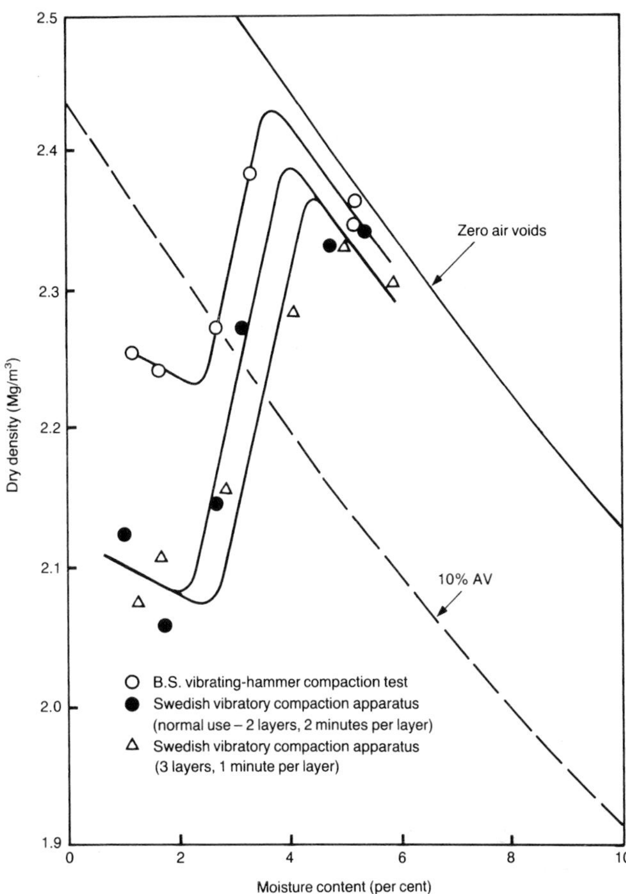

Figure 17.11 Relations between dry density and moisture content of limestone wet-mix macadam using two vibratory compaction methods. Results for non-normal use (three layers, one minute per layer) of the Swedish apparatus are also shown

compactive energy provided by the Swedish compaction apparatus was less than that provided in the vibrating hammer test. During compaction, the tamping foot of the Swedish apparatus was observed to bounce erratically, and it was suggested at the time (Odubanjo, 1968) that the introduction of some form of retaining pressure to prevent the apparatus bouncing, as is present in the vibrating hammer test, could have enhanced its performance.

17.26 It was concluded that the differences in maximum dry density with some materials were such that the two test methods could not be regarded as alternatives for specifying compaction standards. The Swedish apparatus, using a standard vibrating unit, could be more strictly specified than the vibrating hammer test, however, and it was suggested that it could well be worth while considering apparatus of the Swedish type as standard.

Compactability test for graded aggregate

17.27 As part of an investigation into the mechanical properties of graded aggregates, British Standard vibrating hammer compaction tests were carried out on a wide range of materials by Pike (1972). He considered that studies could be simplified if the results were expressed on a volumetric basis. Thus the unit used was the proportion of volume occupied by solids (V_s).

$$V_s = \frac{100.\rho_d}{\rho_s} \text{ (per cent)}$$

where ρ_d=dry density (Mg/m^3)
and ρ_s=particle density (Mg/m^3)

17.28 Examples of results expressed in this form, obtained on numerous aggregates over a range of moisture contents, are shown in Figure 17.12. In these results the moisture content has also been expressed on a volumetric basis, as the proportion of volume occupied by residual free water, ie the proportion of volume occupied by water at the end of the test and excluding water contained within the aggregate.

$$V_{wf} = \frac{W_f \cdot \rho_d}{\rho_w}$$

$$W_f = W_R - W_A$$

where V_{wf} = proportion of volume occupied by free residual moisture (per cent);
W_f = free residual moisture content (per cent);
ρ_d = dry density (Mg/m³);
ρ_w = density of water (1 in Mg/m³);
W_R = residual moisture content (total)(per cent);
W_A = water absorption value (per cent).

Aggregates at moisture contents lower than their water absorption values yield negative values of volume occupied by free water (Figure 17.12).

17.29 It is interesting to note that the relations between the proportions of volume occupied by solids and residual free water did not reach values of moisture content high enough to produce the normal shape of such relations as experienced with soils (see Figure 2.2), so no clearly defined maximum dry density (or maximum proportion of volume occupied by solids) was shown. These results are similar to those obtained in vibrating hammer compaction tests and with full-scale plant when working with wet-mix macadams (Figure 17.5, together with Figures 16.1 to 16.5). Pike (1972) concluded that research was needed to improve the apparatus and methods used in the BS vibrating hammer compaction test.

17.30 Subsequent to this work Pike and Acott (1975) carried out development work to produce an aggregate compactability test which is now specified in British Standard BS 5835 (British Standards Institution, 1980). The apparatus in its final form is shown diagrammatically in Figure 17.13 and the apparatus set up for a test in a noise reduction cabinet is shown in Plate 17.7.

17.31 The main difference between the apparatus for the aggregate compactability test and the vibrating hammer test lies in the compaction mould assembly. The mould has

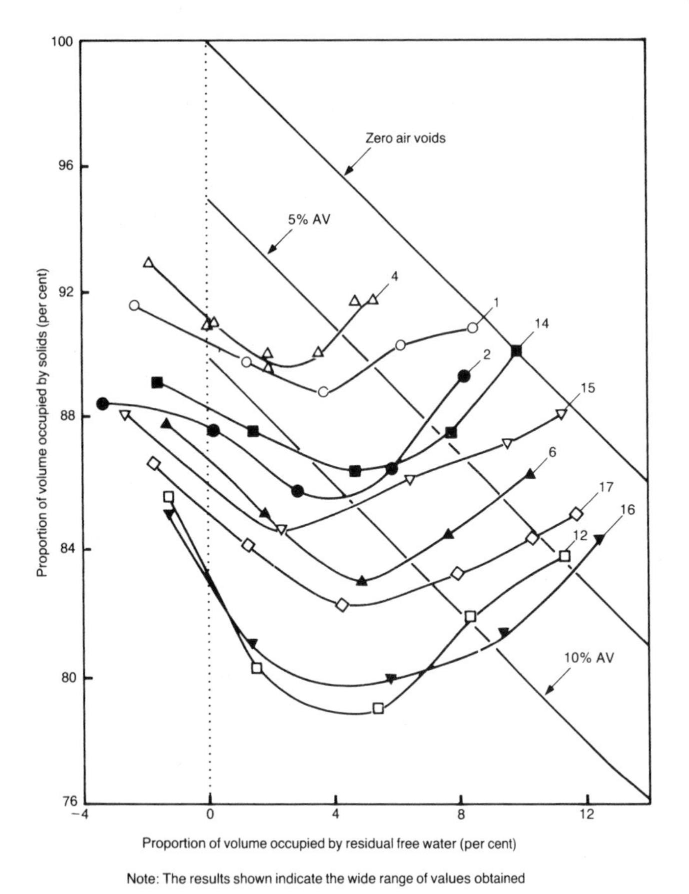

Figure 17.12 Results of vibrating-hammer compaction tests from Pike (1972). The numbers alongside the curves refer to types of aggregate and their sources

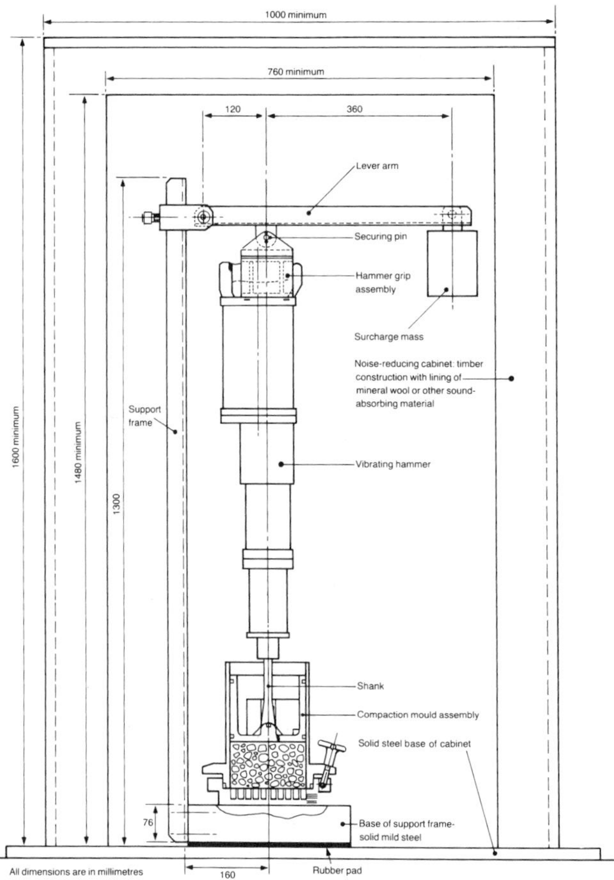

Figure 17.13 Apparatus for the aggregate compactability test

Plate 17.7 Aggregate compactability apparatus set up for test in a noise reduction cabinet

been modified (Figure 17.13) to prevent the loss of fines during compaction by enclosing the test specimen within the walls of the mould, an anvil and, at the base of the mould, a composite filter. The apparatus and procedure developed for the aggregate compactability test were designed to produce similar, but possibly more reliable, results to those obtained in the vibrating hammer compaction test. A 900 W vibrating hammer is specified for the test; the vibrating hammer and mould are supported in a frame and the necessary downward pressure of 350 N to 370 N is provided by a lever arm and surcharge mass (Figure 17.13). The whole apparatus is normally contained in a noise reduction cabinet (Plate 17.7). Material to be tested is compacted in a single layer with three minutes of vibration. The bulk density is determined from the mass of material used, the depth of the compacted specimen and the cross-sectional area of the mould. The mass of dry material used in the test should be between 2.4 kg and 2.6 kg. (This compares with a mass of dry granular material of between 5 kg and 6 kg required to comply with the procedure for the vibrating hammer compaction test.) An additional requirement of the aggregate compactability test (British Standards Institution, 1980) is that five replicate tests should be carried out for each test condition. Relations between dry density and moisture content or the equivalent volumetric relations (see Figure 17.12) are normally determined.

Laboratory compaction test using a vibrating table

17.32 The determination of a maximum index density of cohesionless free-draining soils, as specified in ASTM test procedure D 4253 (American Society of Testing and Materials, Annual publication), includes the use of a vibrating table which may be excited either by electromagnetic means or by eccentric or cam-driven mechanisms. A sample of soil, either in the oven-dry or saturated condition, is vibrated in a mould, with a surcharge of 14 kPa, for a specified period. The bulk density is determined from the measurement of the depth of the compacted specimen and the mass of soil used. The mass of soil and size of mould depend on the maximum particle size of the soil in the sample. The required characteristics of the vibrating table are laid down in the above-mentioned ASTM standard (D 4253).

Limitations on particle size and their effects on results

17.33 Because of restrictions imposed by the sizes of mould used in laboratory compaction tests, the coarser fraction of certain materials has to be removed. It is generally regarded that the removal of small amounts of stone (up to 5 per cent) which exceeds the maximum allowable particle size will affect the density obtained in the test only by amounts comparable with the experimental error involved in the test. The exclusion of larger proportions of stone may have a major effect on the density obtained compared with that obtainable from the soil as a whole. In the British Standard laboratory compaction tests (British Standards Institution, 1990), notes to the test procedures advise that where 5 to 10 per cent of coarse gravel or cobbles (retained on the 37.5 mm sieve) is removed, it may be replaced by an equal amount of 20 mm to 37.5 mm particles of similar characteristics. Alternatively, the coarser material may be removed and a correction applied to the maximum dry density based on the displacement of the soil matrix by stones of known particle density.

17.34 Results of early work at the Laboratory, where various amounts of 20 mm – 2 mm graded aggregate were added to a cohesive soil, are given in Figure 17.14 (Road Research Laboratory, 1952); the results of the optional correction procedure given for the British Standard tests have been added. An alternative method of correction for the removal of coarse particles is given by the American Association of State Highway and Transportation Officials (1986) as AASHTO T 224 (Correction for coarse particles in the soil compaction test). This involves an empirical equation as follows:–

$$D=(1-P_c)D_f+0.9\ P_cD_c$$

where D =adjusted maximum dry density of the total material (Mg/m³);
P_c=proportion by weight of coarse particles removed;
D_f=maximum dry density of the fine material as determined in the test (Mg/m³);
and D_c=the particle density of the removed coarse material (Mg/ m³).

The result of using this correction method is also included in Figure 17.14.

17.35 It is clear from Figure 17.14 that the principle involved in these correction procedures only holds up to a certain proportion of coarse aggregate. With increasing proportions beyond about 40 per cent of aggregate the dry density of the mixture reduces rapidly as intergranular contact increases. The limit of 10 per cent advised for corrections to the results of laboratory compaction tests in Part 4 of BS 1377 appears, on this basis, to be fairly conservative. However, both methods of correction illustrated tend to overcorrect the dry density and provide higher values than those resulting from the practical work, even with small proportions of coarse aggregate.

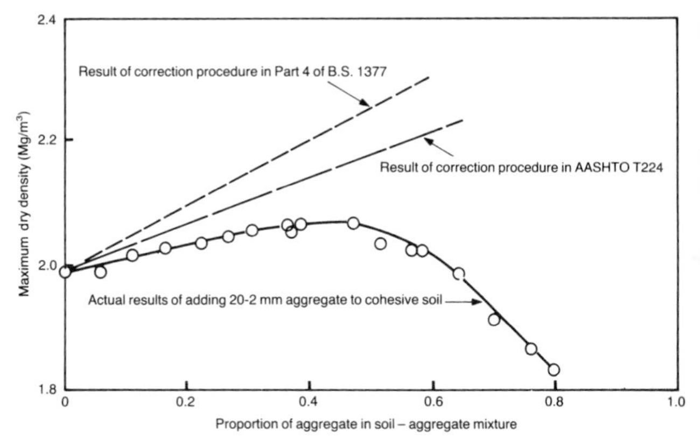

Figure 17.14 Effect on results of 2.5 kg rammer compaction test of adding various proportions of 20 mm to 2 mm graded aggregate

References

American Association of State Highway and Transportation Officials (1986). Standard specifications for transportation materials and methods of sampling and testing. Part II, Methods of sampling and testing. AASHTO, Washington.

American Society of Testing and Materials (Annual). Annual book of ASTM standards. ASTM, Philadelphia.

British Standards Institution (1980). Recommendations for testing of aggregates : BS 5835 : Part 1 Compactibility test for graded aggregates. BSI, London.

British Standards Institution (1990). British Standard methods of test for soils for civil engineering purposes : BS 1377 : Part 4 Compaction related tests. BSI, London.

Forssblad, L (1967). New method for laboratory soil compaction by vibration. *Symposium on Compaction of Earthwork and Granular Bases, Highway Research Record* No 177. Highway Research Board, Washington, 219–24.

Lewis, W A and Parsons, A W (1961). The performance of compaction plant in the compaction of two types of granular base material. *Road Research Technical Paper* No 53. HM Stationery Office, London.

Ministere de l'Equipement (1976). Recommendations for the construction of road earthworks (in French). Setra, Laboratoire Central des Ponts et Chaussees, Paris.

Odubanjo, T O (1968). A study of a laboratory compaction test using a Swedish vibratory apparatus. *Ministry of Transport, RRL Laboratory Report* LR 129. Road Research Laboratory, Crowthorne.

Parsons, A W (1964). An investigation of a laboratory vibratory compaction test for soils and base materials. *D.S.I.R., Road Research Laboratory Note* No LN/612/AWP (Unpublished).

Pike, D C (1972). Compactability of graded aggregates. 1. Standard laboratory tests. *Department of the Environment, TRRL Laboratory Report* LR 447. Transport and Road Research Laboratory, Crowthorne.

Pike, D C and Acott, S M (1975). A vibrating hammer test for compactability of aggregates. *Department of the Environment, TRRL Supplementary Report* 140UC. Transport and Road Research Laboratory, Crowthorne.

Proctor, R R (1933). The design and construction of rolled earth dams. Engineering News Record, 111 (9) 245–8, (10) 216–9, (12) 348–51, (13) 372–6.

Road Research Laboratory (1952). Soil mechanics for road engineers. HM Stationery Office, London, Chapter 9.

CHAPTER 18 MEASUREMENT OF THE IN-SITU STATE OF COMPACTION

Introduction

18.1 The measurement of the state of compaction of soil is a frequent, often essential, requirement during the placement of fill. Allowance has to be made for the real variability in the state of compaction arising from variations in soil type, moisture content, compactive effort applied and other factors (see Chapter 2). This is usually accomplished by carrying out a number of determinations over an area of compacted soil.

18.2 The usual method of measuring compaction in the field is to determine the in-situ dry density of the soil. It is important that the measurements are carried out in such a way that the quality of compaction is determined as accurately as possible; it is imperative, therefore, that the complete depth of layer being compacted is included in the sample or that the least compacted part of the layer is sampled. Dry density normally decreases towards the bottom of the layer, where the compaction stresses are lowest, and the density in the lower regions of the compacted layer may be critical to the satisfactory performance of the compacted fill. Normal procedures are to attempt to measure the average density through the complete depth of the compacted layer. However, where particularly deep layers are being compacted (say, in excess of 300 mm thickness) it is usually advisable to determine the variation of density throughout the layer (by carrying out the measurement of density at various levels) or to concentrate on the density of the lower 100 to 150 mm of the layer.

18.3 Six basic methods of determining the in-situ dry density are described in this Chapter and investigations carried out at the Laboratory into various aspects of the test methods are discussed. The methods are as follows:–

1. The sand-replacement or sand-cone method;
2. The water-replacement method;
3. The rubber balloon method;
4. The core-cutter or drive-cylinder method;
5. The immersion in water and water displacement methods;
6. Nuclear methods.

Sand-replacement or sand-cone method

18.4 *Procedure.* The method is specified in Part 9 of British Standard BS 1377 (British Standards Institution, 1990b), as ASTM Test D 1556 (American Society of Testing and Materials, Annual publication) and as AASHTO Test T 191 (American Association of State Highway and Transportation Officials, 1986). The method comprises the excavation of a cylindrical hole and the careful collection and weighing of all the removed material. The dry weight of the soil is determined after measuring the moisture content of the whole or a portion of the removed material. The volume of the hole is measured by first determining the weight of a dry, uniformly graded sand that is required to exactly fill the hole. The density assumed by the sand in filling the hole is determined separately by filling a container of known volume and of approximately the same dimensions as the excavated hole; thus the weight of sand can be converted to a volume. The dry density of the compacted soil is determined from:–

$$\frac{\text{Mass of dry soil from hole}}{\text{Volume of hole}}$$

18.5 For accuracy it is essential that the surface of the layer in which the hole is excavated is flat and that the soil surrounding the excavated hole remains undisturbed. A special sand-pouring cylinder is normally used (Figure 18.1) which incorporates a cone at the base to ensure that the excavated hole is completely filled by the sand. The calibration procedure to determine the density of the sand includes a separate procedure to determine the amount of sand filling the cone. Plates 3.4 and 3.9 show sand-replacement tests being carried out in the Laboratory's pilot-scale facilities.

18.6 The volume of hole required depends on the particle-size distribution of the soil. For the

British Standard tests, with fine and medium grained soils the excavated hole is 100 mm diameter, whereas with coarse-grained soils a 200 mm diameter hole, with equivalent sand-pouring cylinder, is used.

Sand-replacement method – methods of determining the volume of density holes excavated in coarse granular materials

18.7 An investigation described by West and Parsons (1953), which led to the British Standard procedures for measuring the in-situ dry density of coarse-grained soils, included a study of four different methods of placing the sand in an excavated density hole of 200 mm diameter. Two facsimiles of density holes were used and their volumes were accurately determined by filling with water. One was formed by casting an aluminium block round a plaster-of-Paris core which was itself the 'cast' of a hole dug in a coarse granular base; the other was a tin can with creased and dented sides (Plate 18.1). The methods of filling were:–

1. A sand-pouring cylinder of similar pattern to that shown in Figure 18.1, but scaled-up to match the 200 mm diameter of the density hole (Plate 18.2);

2. A hand scoop; surplus material was struck-off using a straight edge;

3. A tin can with a hole in the lid; surplus material was again struck off using a straight edge;

4. A pouring cone (Plate 18.3); the cone was used as shown in Plate 18.3, with the sand dropping from a constant height, and also mounted on a tripod so that the orifice was 200 mm above the top of the hole. Again surplus material was struck off using a straight edge.

Each method was separately calibrated using the normal cylindrical calibrating container of approximately similar dimensions to the facsimile density holes (200 mm diameter by 200 mm deep).

18.8 The materials used for filling the hole ranged from a fine sand (600 μm to 300 μm) to a coarse, well-rounded gravel (5 mm to 3.35 mm). For some of the tests measurements were made

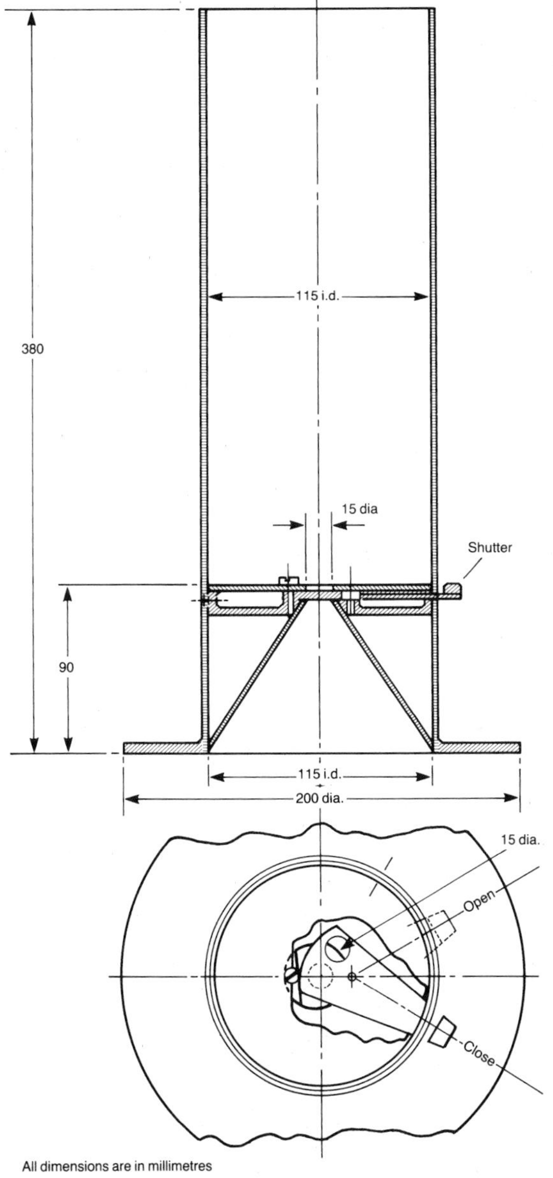

Figure 18.1 Sand-pouring cylinder for use with 100 mm diameter holes in the sand-replacement test

Plate 18.1 Creased and dented tin used to simulate an excavated density hole in granular material

Plate 18.2 Large pouring cylinder for filling excavated density holes of 200 mm diameter

Plate 18.3 The pouring cone in use to determine the volume of an excavated density hole

by several operators and in other tests differently shaped displacers (Plate 18.4) were used to partially fill the hole in an examination of the possibility of reducing the quantity of sand to be poured.

18.9 The results of the study indicated that the sand-pouring cylinder, the method involving the hand scoop, and the method involving a tin can with a hole in the lid all gave volumes within 1 per cent of the true volume, but the least scatter in the results was obtained with the sand-pouring cylinder. The least accurate results were obtained using the pouring cone, which was also the slowest of the methods tested. The displacers reduced the accuracy of the results obtained. This study also found that the particle size of the filling sand, within the range used, had no appreciable effect on the accuracy of the methods, although the scatter of the results increased with increase in particle size.

18.10 As a result of this study the hand-scoop method was added as a subsidiary method (Test 15(C)) in the British Standard test procedures for the determination of in-situ dry density (British Standards Institution, 1975), but the method has subsequently been omitted from the most recent edition of BS 1377 (British Standards Institution, 1990b).

18.11 A later investigation into the effect of particle size of the sand also used a facsimile of an actual test hole excavated in a cement-bound granular base (Parsons, 1966a). A wax impression was taken of a 200 mm diameter by 200 mm deep test hole in the base material (Plate 18.5) and a plaster cast formed round the wax (Plate 18.6). The volume of the facsimile hole was accurately measured by weighing the amount of water required to fill it, after ensuring that the plaster was fully saturated to avoid

Plate 18.4 Displacers used in 200 mm diameter density holes as a possible method of conserving the filling sand

errors caused by absorption. Four different sands were used, and the bulk density of each was determined by calibration using a 200 mm sand-pouring cylinder and a calibrating container 200 mm diameter by 200 mm deep. The sand was then used with the sand-pouring cylinder to determine the volume of the standard plaster hole, five repeat tests being made. The results are given in Table 18.1.

Table 18.1 Effect of the particle size of the sand used in the sand-replacement test on the accuracy in determining the volume of a facsimile of a test hole; volume of test hole = 7565cm³

Type of sand	Calibration bulk density (Mg/m³)	Measured volume of test hole (cm³)	Error in determining volume of test hole (per cent)
Chertsey sand passing 600 μm sieve	1.455	7509	−0.7
Chertsey sand passing 2 mm and retained on 1.18 mm sieves	1.325	7403	−2.1
Leighton Buzzard sand passing 600 μm and retained on 300 μm sieves	1.478	7499	−0.9
Leighton Buzzard sand passing 2 mm and retained on 1.18 mm sieves	1.510	7435	−1.7

Plate 18.5 Solidified paraffin wax after removal from an excavated density hole in cement-bound granular base

18.12 The errors in the determination of the volume of the standard plaster hole (Table 18.1) were smaller with the finer than with the coarser sands. Errors less than 1 per cent were registered with the sands finer than the 600 μm sieve, but the errors increased to around 2 per cent with the sands passing 2 mm and retained on 1.18 mm sieves. It was concluded (Parsons, 1966a) that where rough-sided density holes are excavated, as in coarse granular materials, greater errors are likely to occur when the coarser sand is used. To reduce errors to a minimum, sand finer than 600 μm was recommended to be used so that the irregularities in the sides of the hole could be filled more effectively. It should be borne in mind, however, that in coarse material with a deficiency of fines, where significant voids may occur even in the fully compacted state, fine sand passing 600 μm may flow into fissures in the sides of the density hole and cause opposite, ie positive, errors in the determination of the volume of the hole; in such instances the use of the coarser sand might be more appropriate.

Sand-replacement method – effects of variations in dryness of the sand used for volume determination

18.13 The sand used to determine the volume of excavated holes in the sand-replacement test has to be oven-dried before use (British Standards Institution, 1990b). The effects of long periods of storage after oven-drying on the moisture contents of various sands and the effects of moisture content variations on the calibration bulk density were studied by Parsons (1966b).

18.14 Relations between the calibration bulk density and moisture content of the three sands

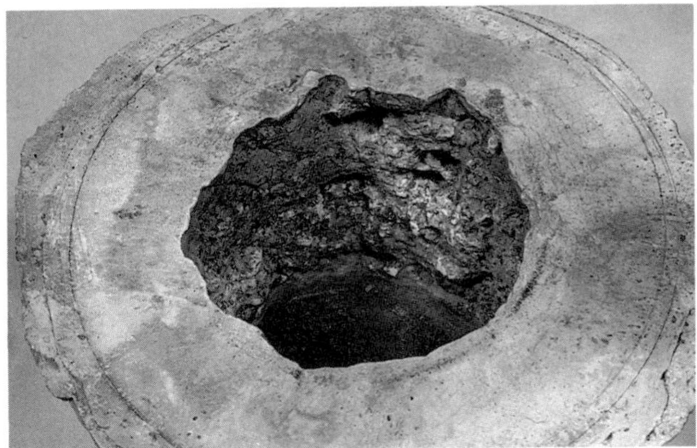

Plate 18.6 Plaster facsimile of a density hole formed from the wax cast shown in Plate 18.5

studied are shown in Figures 18.2 to 18.4. These show that in both the 200 mm and 100 mm sand-replacement tests the calibration bulk density of each sand was markedly affected by small changes in moisture content. For the total range of moisture content within which each sand would pour freely from the sand-pouring cylinder, the variation in calibration bulk density that occurred was about 0.15 Mg/m^3.

18.15 The effects of storing sand, originally oven-dry, in sacks and in tins with close-fitting lids in an unheated shed are shown in Figures 18.5 and 18.6. Clearly the sands stored in sacks were more affected by the humidity of the atmosphere than the sands stored in tightly closed tins. The moisture contents of the sands in the sacks increased rapidly at first and then tended to become reasonably stable and in equilibrium with the ambient humidity. The Chertsey sand (Figure 18.5) reached equilibrium in about seven days and the Leighton Buzzard sand in only about three days.

18.16 A controlled laboratory test was also carried out on the sands by subjecting them to various known degrees of humidity in a vacuum

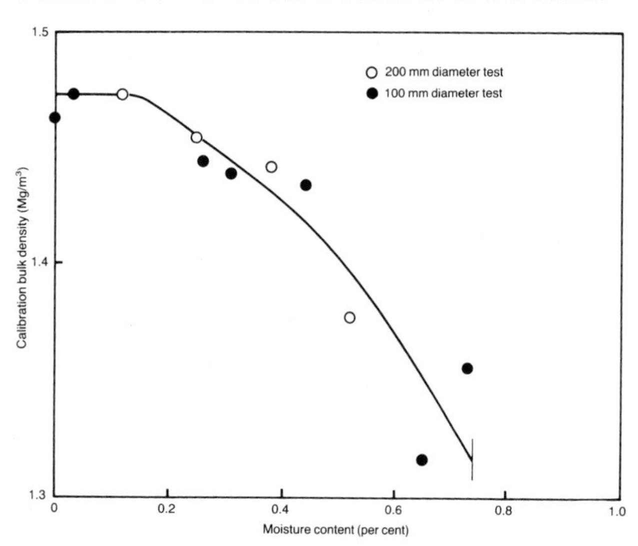

Figure 18.2 Effect of small variations in moisture content on the calibration bulk density of Chertsey sand passing the 600 μm sieve

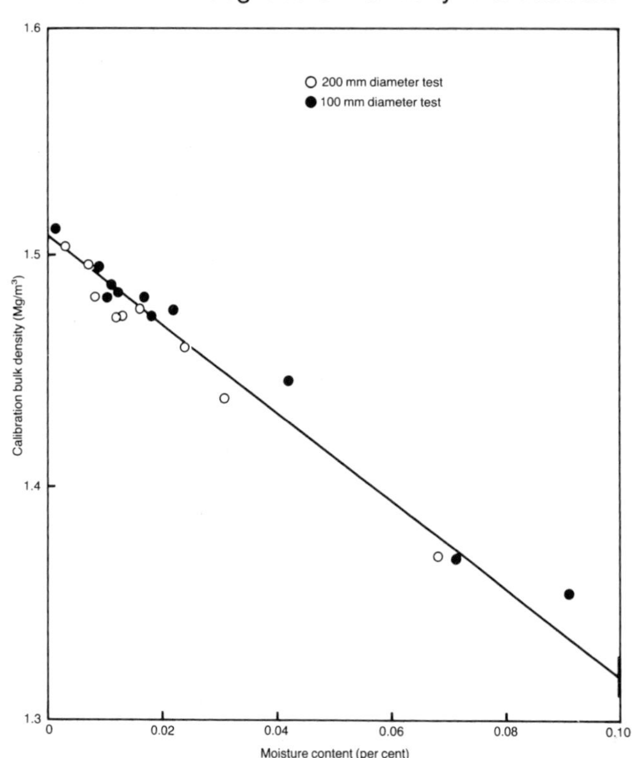

Figure 18.4 Effect of small variations in moisture content on the calibration bulk density of Leighton Buzzard sand passing 600 μm and retained on 300 μm sieves

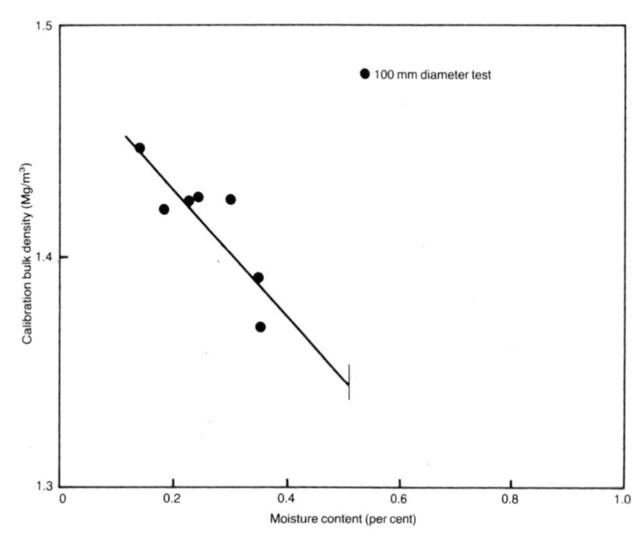

Figure 18.3 Effect of small variations in moisture content on the calibration bulk density of Chertsey sand passing 600 μm and retained on 300 μm sieves

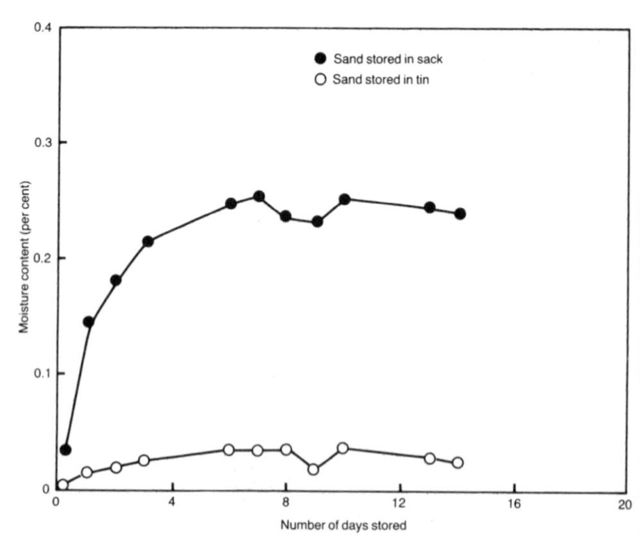

Figure 18.5 Relations between moisture content of Chertsey sand (passing 600 μm) and the number of days stored in an unheated building (March – April)

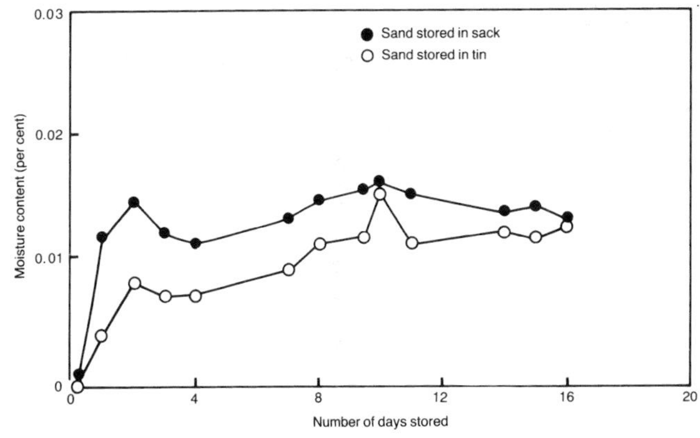

Figure 18.6 Relations between moisture content of Leighton Buzzard sand (passing 600 µm and retained on 300 µm sieves) and the number of days stored in an unheated building (March – April)

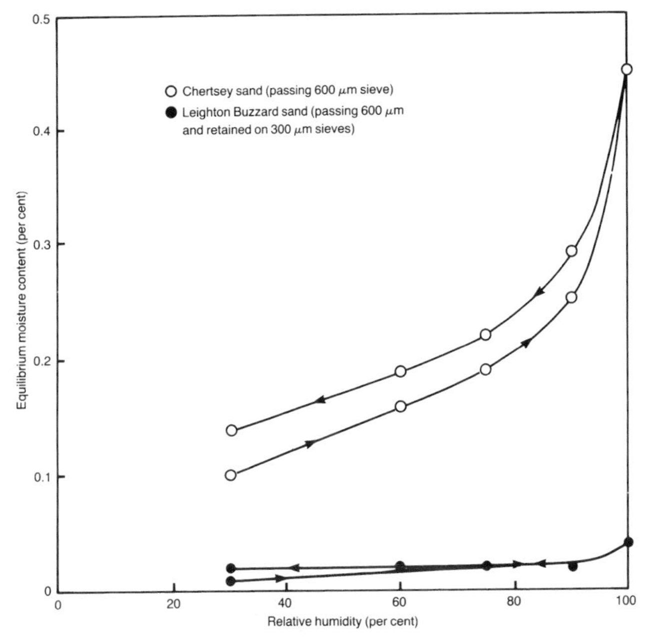

Figure 18.7 Relations between the equilibrium moisture content of sand and the relative humidity for both increasing and decreasing relative humidities

desiccator, using sulphuric acid at various concentrations to produce the required levels of humidity. The results are shown in Figure 18.7. These results generally confirm those given in Figures 18.5 and 18.6.

18.17 The results of this study have been summarised in Table 18.2, which gives an indication of the errors in calibration bulk density which might occur if sands are calibrated immediately after oven-drying and are then stored in sacks, or alternatively exposed to a relative humidity of 90 per cent, for a few days before use. The Table shows that errors up to almost 0.04 Mg/m^3 (ie up to about 2.5 per cent) can occur under these conditions. It is considered important, therefore, to allow the sand to reach an equilibrium moisture content before calibration and, if the relative humidity is fluctuating, frequent calibrations, eg once a day, are necessary (Parsons, 1966b). Such procedures are included in the sand-replacement test methods (British Standards Institution, 1990b).

18.18 The unavoidable procedure of exposing the material to the atmosphere when the moisture content samples were taken was considered to have influenced the relations shown in Figures 18.5 and 18.6 for the sands stored in tins with close-fitting lids (Parsons, 1966b). Storage in tightly closed tins was considered to preserve the material in a non-equilibrium condition in relation to the atmospheric humidity, so that errors could occur as a result of moisture content changes during use. This method of storage, therefore, was not recommended.

Sand-replacement method – unavoidable errors caused by 'slump effect'

18.19 During studies of the states of compaction achieved in earthwork construction on six major road schemes, measurements of the in-situ dry density and moisture content were made using the sand-replacement and oven-drying procedures (British Standards Institution, 1990a, 1990b); this work has been described by Lewis and Parsons (1963). One of the features of the study was the high proportion of tests which yielded erroneous results, indicated by negative

Table 18.2 Effects on the calibration bulk density of possible changes in the moisture content of sands, originally oven-dried, when stored in sacks in an unheated building and when exposed to a relative humidity of 90 per cent

Type of sand	Maximum variation in moisture content (per cent)		Maximum variation in calibration bulk density (Mg/m^3)	
	Stored in sacks*	Relative humidity of 90 per cent	Stored in sacks*	Relative humidity of 90 per cent
Chertsey sand passing 600 µm sieve	0.25	0.25	0.019	0.019
Leighton Buzzard sand passing 600 µm and retained on 300 µm sieves	0.015	0.02	0.029	0.039

*Relative humidity varied from 45 to 75 per cent

values for the calculated air voids, on certain sites. For instance, in a total of 176 individual measurements made on the Oxford Southern and Western By-pass, 28 yielded negative air voids. A histogram of the distribution of values of air content on that site is shown in Figure 18.8; the results on a dry density/moisture content plot are given later in Figure 19.4.

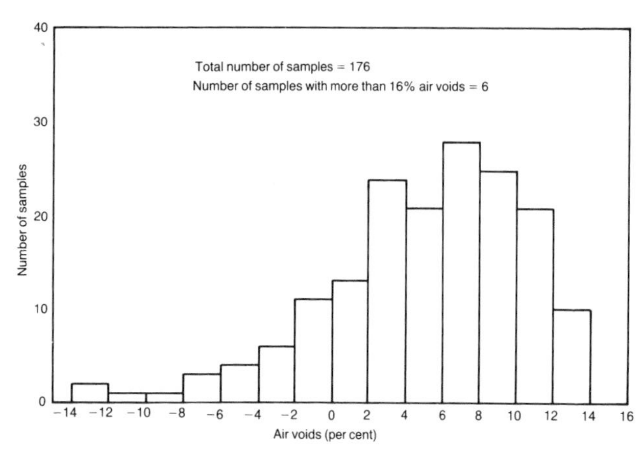

Figure 18.8 Distribution of calculated air-void values from measurements of states of compaction by the sand-replacement method on the Oxford Southern and Western By-pass

18.20 It is generally considered that such negative air voids derive from a serious over-estimation of the dry density by the sand-replacement method, caused by the soil surrounding the sample hole flowing (slumping) inwards during its excavation. The volume measured by the calibrated sand would then be smaller than that occupied by the soil removed from the hole and the calculated bulk density would be too high. The study referred to above (Lewis and Parsons, 1963) showed that the phenomenon was restricted to wet granular materials having little or no cohesion. The possibility exists that, with the same soil types, positive errors in dry density may also have occurred in the results yielding positive air voids. No simple solution to this problem has so far been found and it poses a serious limitation on the use of the sand-replacement test or any other test method which involves the measurement of the volume of a sample hole excavated in granular soils.

Sand-replacement method – additional factors affecting accuracy

18.21 The calibrating container used to determine the bulk density of the sand should be of approximately the same dimensions as the excavated sample hole in the soil. Bulk density of the sand has been found to decrease by about 1 per cent for a 25 mm decrease in the depth of the container (Road Research Laboratory, 1952).

18.22 The height of sand in the sand-pouring cylinder should be kept approximately the same at the beginning of each test, including the calibration to determine the bulk density of the sand. The bulk density of the sand has been found to decrease by about 1 per cent for a 50 mm decrease in the initial level of the sand in the sand-pouring cylinder (Road Research Laboratory, 1952).

Water replacement method

18.23 This method has been recently introduced into Part 9 of British Standard BS 1377 (British Standards Institution, 1990b), but it has been previously described in British Standard 5930 (British Standards Institution, 1981). It is intended to cover the determination in-situ of the density of coarse and very coarse soils where the other methods for determining the field density would be unsuitable because the volume sampled would be too small to be representative. The principle is generally similar to that of the sand-replacement method; a circular density ring is used on the ground surface and the hole is excavated inside the ring. A flexible plastics sheet is used to line the excavated hole and retain water which determines the volume of the hole. As a general guide, the ring diameter should be at least four times the size of the largest particle encountered in the sample. No research has been carried out into this method at the Laboratory, but full details are given in British Standards Institution (1990b).

Rubber balloon method

18.24 This method is specified as ASTM Test D 2167 (American Society of Testing and Materials, Annual publication) and as AASHTO Test T 205 (American Association of State Highway and Transportation Officials, 1986). In principle, this method of determining the density of soil in-situ is very similar to the sand-replacement method (see Paragraphs 18.4 to 18.6). A hole is excavated in the compacted material and the removed soil is weighed and its dry weight determined following a moisture content measurement. The volume of the excavated hole is determined from the volume of water required to inflate a rubber balloon to completely fill the hole. The water is contained in a calibrated vessel, enclosed at its base by the rubber balloon, and a means of pressurising the liquid is normally incorporated in the design of the apparatus. An initial reading is taken by placing the rubber balloon apparatus on the surface of the soil at the location of the intended density hole, and the final reading is taken after excavation of the hole. The difference between the two readings gives the volume of the excavated hole. As with the sand-replacement test, the slumping inwards of the perimeter of the excavated hole can cause errors with very wet granular soils having little or no cohesion (see Paragraphs 18.19 to 18.20).

18.25 An early form of rubber balloon apparatus has been described by Road Research Laboratory (1952) and the apparatus is shown dismantled in Plate 18.7. This apparatus was placed over the excavated density hole and water was forced into the balloon by air pressure so that it filled the hole and the instrument was raised about 20 mm off the ground. The air tap was then closed and the operator placed both feet firmly on the base-plate so that the balloon was forced into any irregularities in the hole. The volume of the hole was found from the difference between initial and final water levels in the glass cylinder. This apparatus was stated to provide a convenient and quick method, but it gave relatively unreliable results (Road Research Laboratory, 1952); the probable inaccuracy with which the balloon fitted the hole was attributed to the lack of control of the air pressure applied.

Plate 18.7 An early type of rubber balloon apparatus, dismantled for maintenance

18.26 An investigation described by Parsons (1956) examined the use of a 'membrane densitometer' developed in France for use with excavated density holes of 100 to 150 mm diameter and up to about 150 mm deep. The volume of water was registered by graduations on the piston rod and an associated vernier scale, and pressure was applied to the water to expand the rubber balloon to closely fit the density hole by means of the piston which had a water-tight seal with the cylinder wall. When in use the apparatus was attached to a reference plate as shown in Plate 18.8. The apparatus was found to under-estimate by 1 per cent the volume of a plaster-cast facsimile of a 100 mm diameter by 150 mm deep density hole in granular soil (Plate 18.6 shows an example of a plaster-cast facsimile of a density hole); the sand-replacement method under-estimated the volume of the same facsimile hole by 0.7 per cent. With smooth-sided holes in sandy clay soil the membrane densitometer yielded volume determinations which were on average about 2 per cent less than those obtained using the sand-replacement method. It was suggested at the time (Parsons, 1956) that with smooth-sided holes significant errors could be introduced when using the membrane densitometer due to air becoming trapped in the bottom of the hole by the rubber membrane forming a seal with the sides of the hole; with soils pervious to air this effect was considered unlikely. The conclusion reached from the investigation by Parsons

Plate 18.8 The membrane densitometer, the rubber balloon apparatus investigated in 1955

(1956) was that the membrane densitometer had no general advantage over the sand-replacement apparatus for the determination of the dry density of compacted soil.

18.27 A further rubber balloon apparatus, shown in Plate 18.9, was investigated by Parsons and Odubanjo (1965). The apparatus was designed for use with 100 mm diameter density holes and the neck of the rubber balloon was attached to a 38 mm diameter orifice in the base-plate. The glass cylinder was graduated and the air space above the water was connected via a vent tube to a rubber bulb which was pumped to pressurise or evacuate the air space depending upon which end of the bulb was connected to the apparatus. Tests were made with facsimiles of density holes, two in heavy clay and one in gravel-sand-clay (hoggin), formed of epoxy resin from a paraffin wax cast of each hole (Plate 18.10). A second series of tests was carried out in density holes excavated in soils compacted in a large box mould. In all cases the sand-replacement test method was also used. The results obtained indicated that, with the balloon apparatus, errors were likely to occur in soils

Plate 18.10 The facsimiles of density holes used with the rubber balloon apparatus illustrated in Plate 18.9

containing gravel because of the inability of the water-filled rubber membrane to fill the crevices in the sides of the excavated density hole. With these stony soils errors up to 2 per cent occurred and the apparatus was hardly suitable, therefore, for measuring the state of compaction of such materials. The equipment could not be used, anyway, with coarse granular soils or base materials which generally require density holes of at least 150 mm diameter. It was concluded (Parsons and Odubanjo, 1965) that the

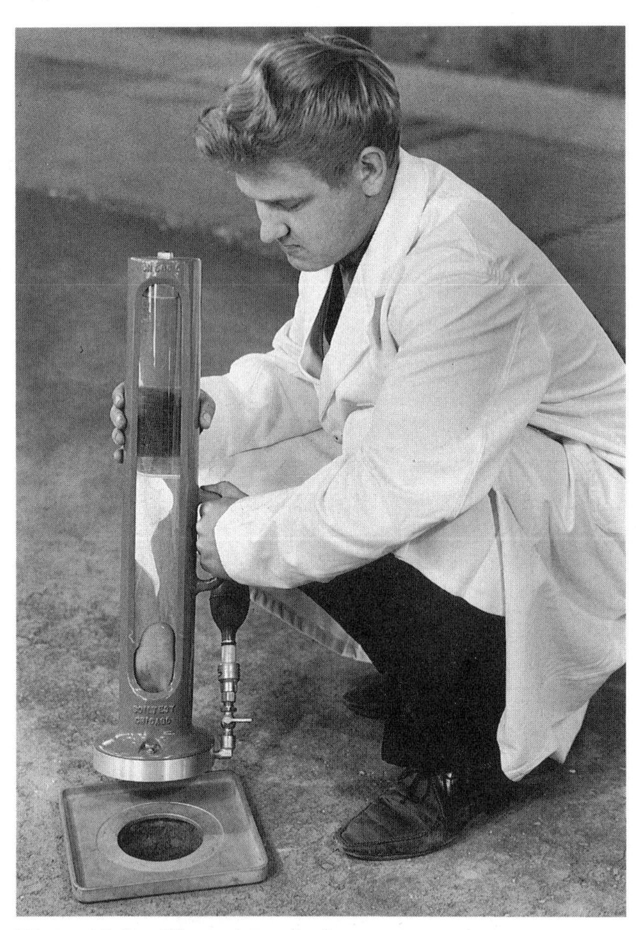

Plate 18.9 The rubber balloon apparatus investigated in 1965

Plate 18.11 Core-cutter tests to measure the in-situ dry density being carried out in the test facility used during early compaction investigations

apparatus was only suitable for use with fine-grained soils, ie soils with little or no gravel fraction. With these types of soil the equipment was capable of measuring the volume of excavated density holes to an accuracy better than 1 per cent. It was considered that the balloon apparatus was more portable and easier to use than the sand-replacement method and these advantages would tend to off-set the slight disadvantage of the slightly more variable results.

Core cutter or drive cylinder method

18.28 This method is specified in Part 9 of BS 1377 (British Standards Institution, 1990b), as ASTM Test D 2937 (American Society of Testing and Materials, Annual publication) and as AASHTO Test T 204 (American Association of State Highway and Transportation Officials, 1986). It consists of driving into the compacted soil a thin-walled open-ended steel cylinder usually about 100 mm diameter and 130 mm long (Plate 18.11). The cylinder containing the soil is dug out of the ground and the soil protruding from its ends is trimmed off. The mass of soil contained in the known volume of the cylinder provides the bulk density, and the dry density is determined following a measurement of moisture content. By its very nature, the method is applicable only to soils that are relatively free of stones.

18.29 The disturbance created in the soil by driving the cylinder can give rise to errors in the measured dry density, especially when the soil is poorly compacted. Comparisons made using core cutters with varying wall thicknesses ranging from 1.5 mm to 6.5 mm produced results which suggested that the compression of the soil caused by the insertion of the core cutter into the ground was outweighed by an expansion of the soil caused by the stresses set up as the cutter was driven into the ground (Road Research Laboratory, 1952). To obtain the most accurate measurement of dry density with the core cutter method it was concluded that a cutter with a wall as thin as practically possible should be used. The requirement in the British Standard test (British Standards Institution, 1990b) is that the wall thickness should be 3 mm.

Immersion in water and water displacement methods

18.30 These methods are specified in Part 2 of BS 1377 (British Standards Institution, 1990a) and as AASHTO Test T 233 (American Association of State Highway and Transportation Officials, 1986). They require the recovery of an undisturbed sample of soil and can only apply, therefore, to cohesive or stabilised soils. The mass of the specimen, after suitable trimming, is determined and its volume calculated, either from the difference in weight recorded when the sample is weighed in air and suspended under water, or from the volume of water displaced when the sample is immersed. From the bulk density so determined the dry density is calculated following the measurement of moisture content. To ensure the accuracy of the measurement of its volume the sample is coated with paraffin wax and allowance made for the weight and volume of the wax used.

Nuclear methods for the determination of bulk density

18.31 The determination of the bulk density of soil by nuclear methods depends upon the scattering and absorption by the soil of gamma radiation emitted from a small source. The intensity of gamma radiation reaching a convenient detector depends principally on the bulk density of the soil, the higher the bulk density the lower the intensity of radiation detected. Usually calibration charts, relating bulk density of the soil to the intensity of detected gamma radiation, are established using blocks of material of known density or by comparison with other accepted methods of determining in-situ density.

18.32 There are four basic configurations employing this principle (Figure 18.9). They are as follows:–

1. Surface direct-transmission;
2. Double-probe direct-transmission;
3. Surface back-scatter;
4. Back-scatter depth probe.

Nuclear determination of bulk density – surface direct-transmission method

18.33 This method has been incorporated in Part 9 of BS 1377 (British Standards Institution, 1990b), as ASTM Test D 2922, Method B (American Society of Testing and Materials, Annual publication) and as ASSHTO Test T 238, Method B (American Association of State Highway and Transportation Officials, 1986).

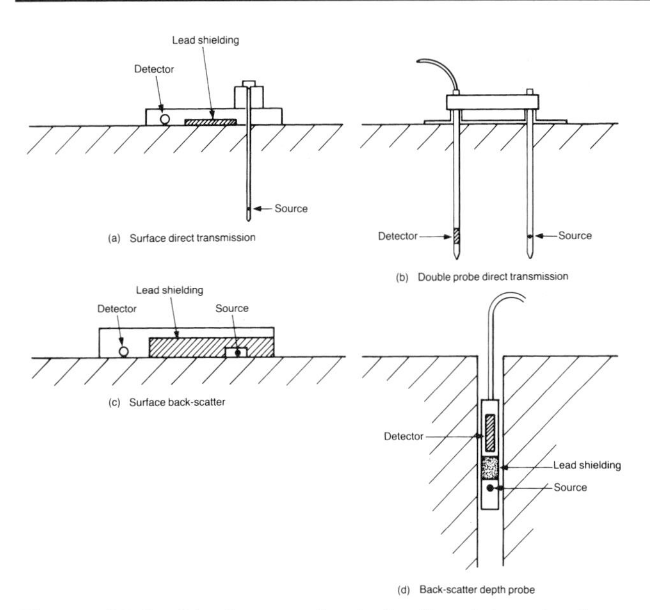

Figure 18.9 Nuclear methods for the determination of bulk density

Plate 18.13 Early nuclear direct-transmission apparatus in use

Either the source or detector (usually the source) is inserted to a known depth in the soil whilst the other element remains on the surface (Figure 18.9(a)). The method can be calibrated to provide a means of measuring the average bulk density of soil between the surface and the depth to which the source (or detector) is inserted.

18.34 An early form of direct-transmission apparatus was developed at the Laboratory and has been described by Lewis (1965). The various components are shown in Plate 18.12 and the apparatus is shown in use in Plate 18.13. The general arrangement of the source and detector when the apparatus was in use is shown in Figure 18.10. The radioactive source (2.5 mC of caesium 137) which produced the gamma radiation was contained near the end of a probe which was inserted to a depth of up to 150 mm into the material being tested. A Geiger Muller tube for detecting the gamma radiation was contained in a surface unit about 175 mm from the probe. The rate of transmission of gamma radiation through the soil was recorded by a separate scaler unit (Plate 18.13) which contained the power supplies and counters, and a timing device which stopped the counters automatically after one minute.

18.35 Investigations with this early form of direct-transmission gamma radiation apparatus included calibrations on blocks of cemented materials and on specimens of various soils made up to uniform densities in a large mould, and against bulk densities determined by the sand-replacement method on compacted areas

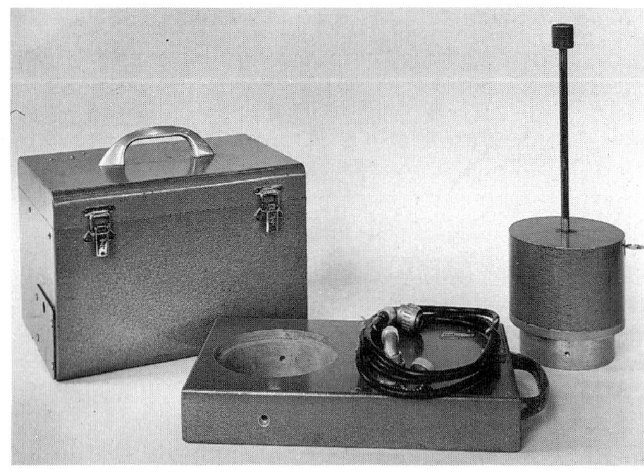

Plate 18.12 Various components of the early form of direct-transmission nuclear apparatus for the determination of bulk density

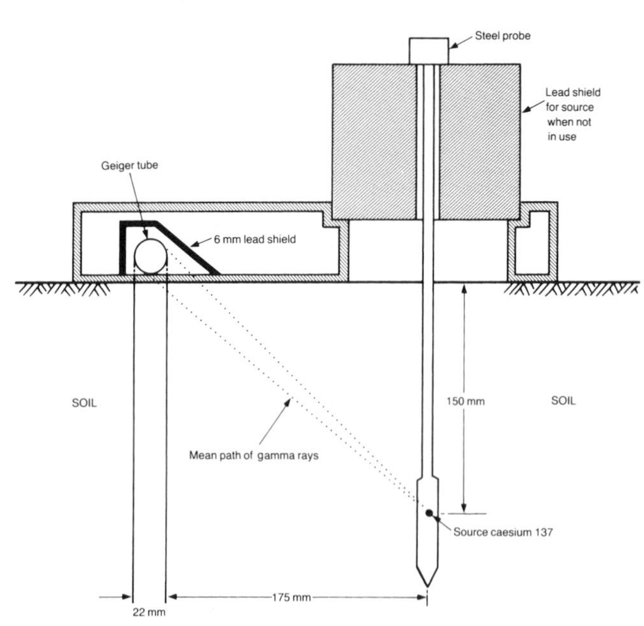

Figure 18.10 General arrangement of source and Geiger tube in early form of nuclear direct-transmission apparatus

of the soils in the test pits. The particle-size distributions of the soils used are given in Figure 3.4. The calibration so obtained is shown in Figure 18.11. The results indicate that the intensity of gamma radiation recorded for a given bulk density was affected by the type of material. At the time of the investigations it was considered that the most likely cause was the effect of the different mineral constituents of the soils, an effect which is particularly serious with the surface back-scatter method (see Paragraphs 18.54 to 18.56).

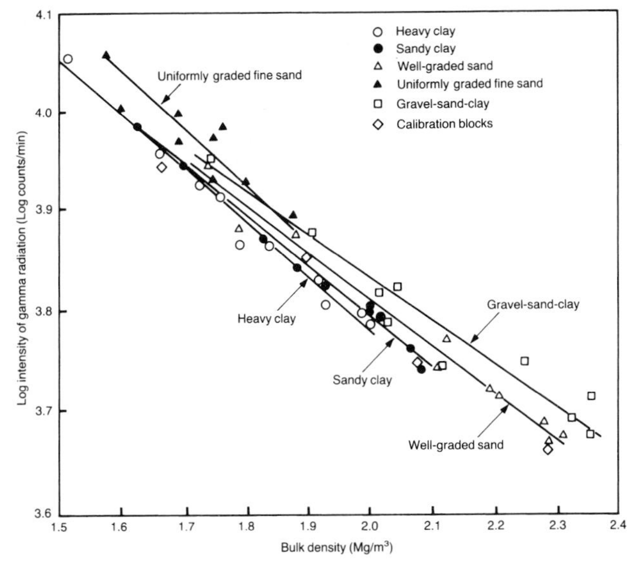

Figure 18.11 Calibration relations between intensity of gamma radiation and bulk density for the early direct-transmission apparatus

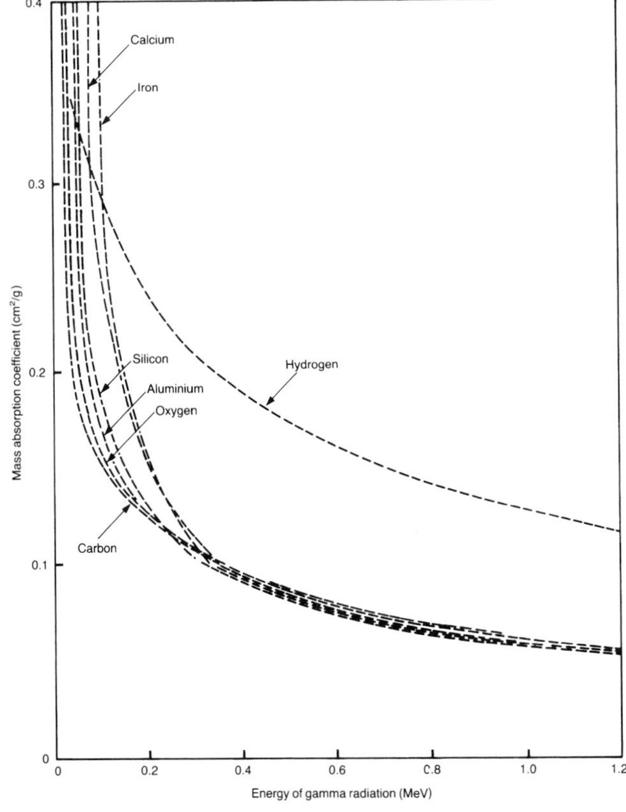

Figure 18.12 Relations between mass absorption coefficient and the energy of gamma radiation for elements commonly found in soil

18.36 Relations between the mass absorption coefficient and the energy of gamma radiation for several elements which commonly occur in soil are shown in Figure 18.12 (Morgan, 1955). Apart from hydrogen, all the elements have very similar mass absorption coefficients at energy levels above 0.3 MeV. However, at lower energy levels, the various elements differ widely in mass absorption coefficient, the higher absorption occurring with the higher atomic weights. For a given density, materials containing a large proportion of elements such as calcium and iron would be expected to absorb more gamma radiation at energy levels below 0.3 MeV than materials containing elements with lower atomic weights such as aluminium or silicon. The direct-transmission apparatus which produced the results shown in Figure 18.11 had a caesium 137 source with a gamma radiation energy level of 0.66 MeV, and it can be expected that a proportion of the radiation would have undergone sufficient scattering to reduce its energy level below 0.3 MeV before reaching the Geiger counter. Thus elements such as iron or calcium would have caused some reduction in radiation count rates, while higher count rates could be expected with materials consisting of silicon and aluminium.

18.37 A quantitative analysis of the main constituents of the soils is given in Table 18.3. The iron and calcium contents were highest for the heavy clay and sandy clay No 2 and lowest for the uniformly graded fine sand, confirming that the intensity of radiation at any given bulk density (Figure 18.11) increased with decrease of the iron and calcium contents of the soils.

Table 18.3 Quantitative analysis of the main constituents of the test soils

Soil	Percentage present in oven-dry material			Ignition loss
	Silica (SiO_2)	Ferric oxide (Fe_2O_3)	Calcium oxide (CaO)	(%)
Heavy clay	58	7.1	1.2	8
Sandy clay No 2	72	6.2	4.0	6
Well-graded sand	86	3.3	1.0	3
Uniformly graded fine sand	91	2.1	0.8	1
Gravel-sand-clay	90	3.5	0.6	2

Note: The remainder would have consisted mainly of alumina (Al_2O_3)

However, the effect of soil type was more marked than might be expected for the direct-transmission technique and further work was recommended to be carried out on the subject (Lewis, 1965).

18.38 Hydrogen has a widely different mass absorption coefficient than the other elements (Figure 18.12), so some effect of moisture content on the calibration can also be expected. However, the weight of hydrogen in a given volume of soil is normally relatively small and it was estimated by Lewis (1965) that the variation in calibration would not be more than about 0.02 Mg/m^3 in bulk density for a variation in moisture content of 10 per cent.

18.39 The additional work recommended by Lewis (1965) into the effect of soil type on the calibration of the early direct-transmission gamma radiation apparatus was carried out by Harland and Urkan (1966). They found that, after allowing for the abnormal absorption of gamma radiation by hydrogen, the effect of soil type was attributable to two causes. The minor one was the photo-electric absorption of low-energy scattered gamma photons, which depended on the chemical composition of the soil, as discussed in Paragraphs 18.36 to 18.37. The major cause was found to be associated with the different density disturbances generated in soils of different particle-size distributions when the probe hole was formed. The normal method of forming the hole was, and remains so with modern equipment, by forcing a spike into the soil with the aid of a hammer. In general, it was found that the more granular materials suffered greater disturbance with associated losses of density in the area immediately surrounding the probe hole, thus causing a higher intensity of radiation to be detected than would have been the case if the soil had remained undisturbed. Thus gravels and sands would have the highest calibration curves as illustrated in Figure 18.11. This effect would be further compounded by the effect of chemical composition in the particular test soils used.

18.40 In modern direct-transmission gamma radiation apparatus a form of discrimination is often used to avoid the counting of low-energy gamma radiation. This can be in the simple form of a lead screen (Gabilly, 1969), and would largely restrict the cause of calibration variations to the disturbance created by the forming of the probe hole. In the extreme, this disturbance can be so large with coarse-grained soils that the use of this technique becomes untenable.

18.41 Other investigations, described by Lewis (1965), with the early direct-transmission apparatus indicated that the apparatus sampled a depth slightly greater than the depth to which the radioactive source was inserted, ie about 165 mm. It was also concluded that, for all practical purposes, the equipment measured the average bulk density over the depth sampled irrespective of the magnitude of density gradients present in the soil.

18.42 In recent investigations into the performance of compaction plant a modern nuclear gauge has been used; this was capable of measuring bulk density using either direct-transmission or back-scatter of gamma radiation, and of measuring moisture content using neutron radiation (see Paragraphs 18.70 to 18.72). The apparatus is shown in Plates 18.14 and 18.15. For all determinations of bulk density this apparatus was used in the direct-transmission mode, and a diagram of its configuration when in use is shown in Figure 18.13. The gamma radiation source, 8 mC of caesium 137, was near the tip of a steel rod which could be lowered into a pre-formed hole in the test material for operation in the direct-transmission mode; the depth of measurement was variable between 50 mm and 300 mm in 25 mm increments. The gamma radiation detectors were located in the gauge housing. To compensate for any effect of local environmental conditions and for decay of the radioactive source, the test procedure included the determination of a standard count, made on a polyethylene reference block, shown in Plate 18.15, with the apparatus in the back-scatter mode, ie with the source in the retracted position. The count ratio

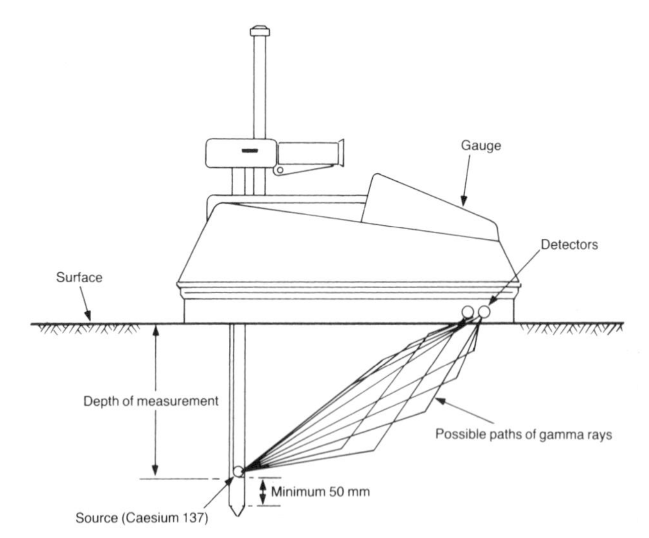

Figure 18.13 The measurement of density by direct transmission with a recent type of nuclear density gauge

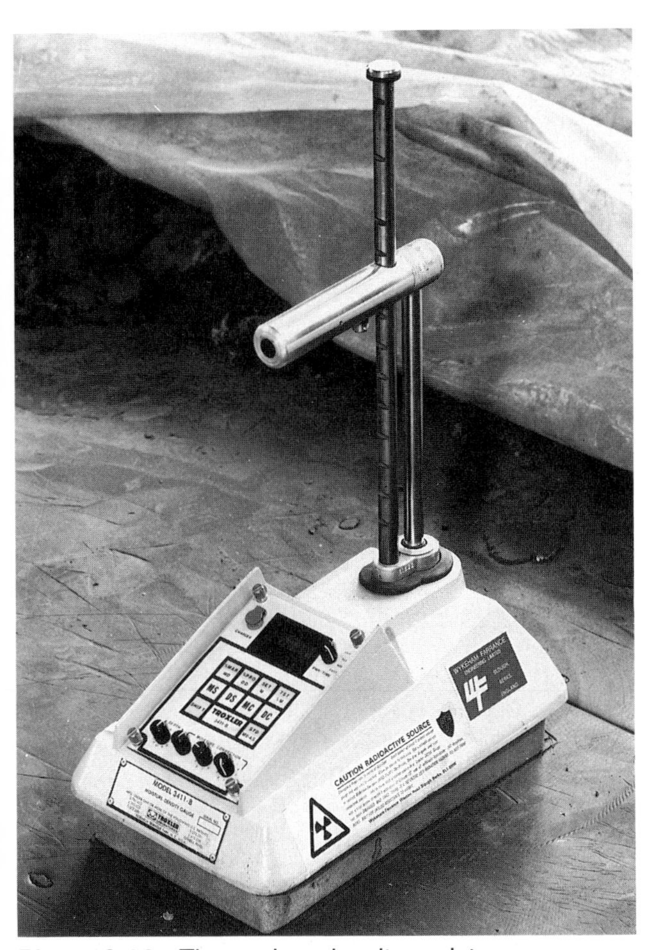

Plate 18.14 The nuclear density-moisture gauge used in recent investigations; the apparatus is shown in use with the direct-transmission mode for density measurement

Plate 18.15 The nuclear density-moisture gauge used in recent investigations with items of ancillary equipment, including the reference block, template, spike and hammer for forming the hole for the source probe and a levelling spatula

(in-situ radiation count rate divided by the standard count rate) was used to determine the bulk density from a calibration pre-determined by the manufacturer. The nuclear gauge was equipped with a micro-processor which was programmed to provide a direct readout of the bulk density.

18.43 The investigations with this later gauge mainly comprised 'calibrations' against conventional sand-replacement tests during the determinations of the performance of compaction plant. The particle-size distributions of the soils used are given in Figure 3.6. As an example, the results for the sandy clay No 2 are given in Figure 18.14, where the bulk densities determined by the nuclear gauge for 100 mm and 200 mm thick layers are plotted against bulk densities determined by the sand-replacement method (British Standards Institution, 1990b) for the same thickness of layer. Each point on the graph represents the mean of four determinations by each method. The results of regression analyses of the results obtained with each of the test soils are given in Table 18.4. The equations show that differences in

Table 18.4 Regression equations relating values of bulk density determined by a direct-transmission nuclear gauge and by the sand- replacement method

Soil	No. of results	Regression equation	Correlation coefficient
Heavy clay	32	y=0.733x+0.525	0.97
Sandy clay No 2	30	y=0.697x+0.602	0.98
Well-graded sand	51	y=0.767x+0.423	0.98
Uniformly graded fine sand	24	y=0.714x+0.480	0.99
Gravel-sand-clay (all results)	55	y=0.810x+0.374	0.97
Gravel-sand-clay (Moisture content < 7.5%)	19	y=0.796x+0.376	0.98
Gravel-sand-clay (Moisture content ⩾ 7.5%)	36	y=0.754x+0.505	0.97

y=bulk density measured by nuclear gauge (Mg/m^3)
x=bulk density measured by sand-replacement method (Mg/m^3)

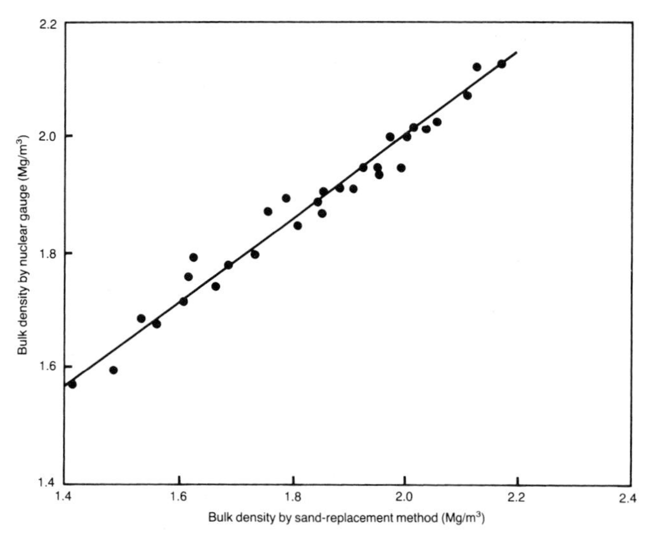

Figure 18.14 Correlation of results of bulk density measurements made by a modern nuclear gauge in direct-transmission mode and by the sand-replacement method in sandy clay No 2

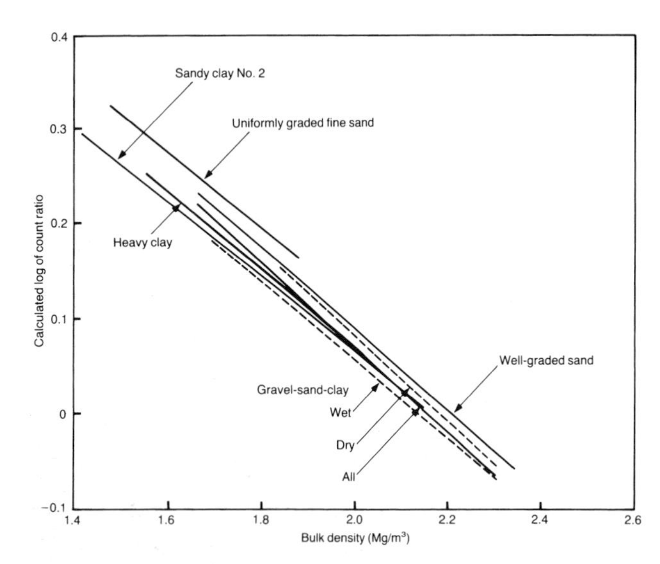

Figure 18.15 Calculated relations between log count ratio and bulk density for the direct-transmission nuclear gauge used in recent investigations for a sample depth of 100 mm

correlation occurred from soil to soil; also, with the gravel-sand-clay, there were differences between the results obtained with the wetter and drier soil conditions.

18.44 To compare the results given in Table 18.4 with those obtained with the early form of direct-transmission nuclear gauge it is necessary to plot radiation intensity against the bulk density determined by the sand-replacement method. Calibration information provided with the instrument permitted the conversion of the bulk density obtained with the nuclear gauge to the count ratio (radiation intensity measured divided by the standard count); the calculated relations between the logarithm of the count ratio and bulk density are shown in Figure 18.15.

18.45 The relative positions of the calibrations for the various soil types are very similar to those shown in Figure 18.11 for the early direct-transmission apparatus. The one possible exception is the calibration for the gravel-sand-clay, which is off-set to the right in Figure 18.11 but does not occupy that position in Figure 18.15. A study of the original data for the results shown in Figure 18.11 has revealed that the maximum moisture content used was 7.4 per cent. The results for gravel-sand-clay from the recent investigations have, therefore, been divided into two sets, one for moisture contents less than 7.5 per cent (dry) and the other for moisture contents equal to and above that value (wet). The regression equations for these two additional bulk density correlations have been included in Table 18.4. The 'calibration line' given for dry gravel-sand-clay in Figure 18.15 is very close to that for well-graded sand, although it is not as high as that given in Figure 18.11. It can be concluded from these comparisons that the changes in calibration with different soil types registered by the early form of direct-transmission apparatus were also found in the modern equipment. As an example, the difference in calibration, in units of bulk density, between the sandy clay No 2 and the uniformly graded fine sand is about 0.07 Mg/m^3 in both cases (Figures 18.11 and 18.15). The combined effect of disturbance in forming the probe hole and of chemical composition of the soil, described in Paragraph 18.39, is an inherent feature, therefore, of the operation of direct-transmission nuclear gauges. Because of this, it is essential that nuclear gauges are calibrated for individual site conditions.

Nuclear determination of bulk density – double-probe direct-transmission method

18.46 With this method both the source and detector are inserted into the compacted soil using parallel probes normally lowered into pre-formed holes made through a template (Figure 18.9(b)). Thus the mean path of the directly transmitted gamma radiation is horizontal and the bulk density at a given depth in the compacted layer can be determined. The comments regarding problems of disturbance due to insertion of the probe and the difficulties with granular soils, described in Paragraphs 18.39 to 18.40, associated with the use of the surface direct-transmission technique, also apply to the double-probe direct-transmission method. No studies have been made at the Laboratory with this type of equipment; however, its usefulness in studying the variation of density with depth through the compacted layer has been demonstrated by Morel et al (1980). The application of this technique to the monitoring of the bulk density of soil for agricultural purposes has been described by Soane et al (1971).

Nuclear determination of bulk density – surface back-scatter method

18.47 In this method both the radioactive source and the detector are placed on the surface of the soil (Figure 18.9(c)). The method has been incorporated in Part 9 of the latest edition of BS 1377 (British Standards Institution, 1990b), and is specified as ASTM Test D 2922, Method A (American Society for Testing and Materials, Annual publication) and as AASHTO Test T 238, Method A (American Association of State Highway and Transportation Officials, 1986). The detected radiation is that which has undergone a scattering process within the soil, the intensity of radiation registered depending on the bulk density of the soil within the depth sampled.

18.48 An early form of back-scatter apparatus was investigated at the Laboratory and the results have been described by Lewis (1965). A diagrammatic arrangement of the surface unit of the apparatus is shown in Figure 18.16. The radioactive source (10 mC of radium-beryllium) produced both gamma and neutron radiation, the latter for the determination of the moisture density of the soil in association with the boron trifluoride tubes (see Paragraphs 18.62 to 18.65). The gamma radiation detector (Geiger tube) was separated from the source by about 200 mm of lead shielding. Some of the gamma radiation entering the material below the source was scattered back and detected by the Geiger tube. The detector was connected by cable to a scaler unit which contained the power supplies and counters, and a timing device which stopped the counters automatically after one minute. The apparatus is shown in operation in Plate 18.16. The position of the Geiger tube could be varied to alter the spacing between the source and detector; it was claimed that this feature enabled the depth sampled to be varied. A standard block was provided, consisting of a wooden box containing paraffin wax and a small block of lead set in the wax at one end of the block. The block was used to obtain standard radiation counts to ensure that the apparatus was functioning correctly. When the apparatus was not in use the standard block acted as a carrying case for the surface unit and provided some shielding from radiation (Plate 18.17).

18.49 Investigations with this early form of back-scatter nuclear equipment comprised the determination of the volume sampled, a study of the effect of density gradients in the test layer and the effect of the type of material on the calibration of the equipment. The particle-size

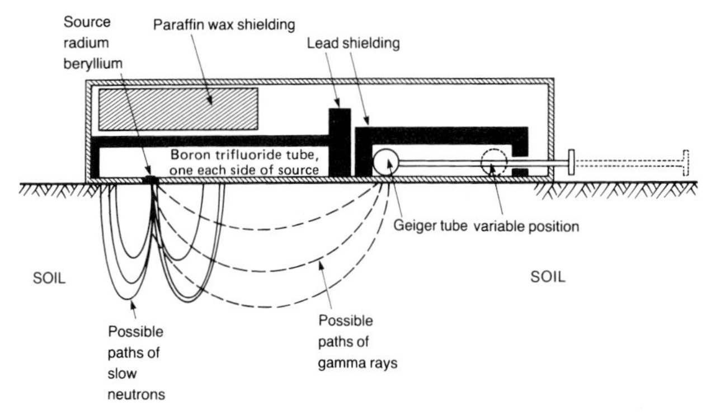

Figure 18.16 Early form of back-scatter gamma and neutron radiation apparatus – diagrammatic arrangement

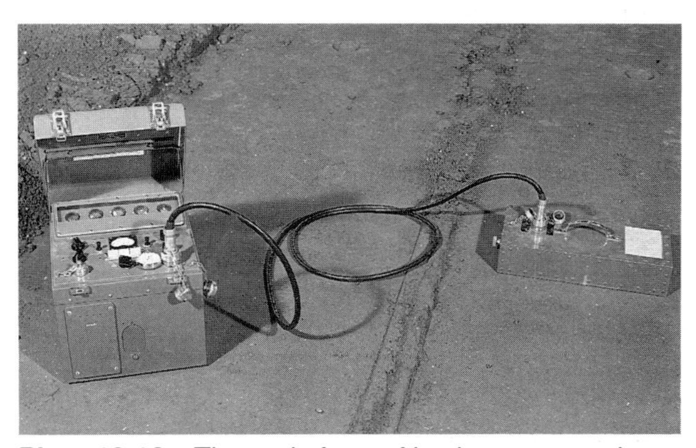

Plate 18.16 The early form of back-scatter nuclear gauge in use on compacted soil

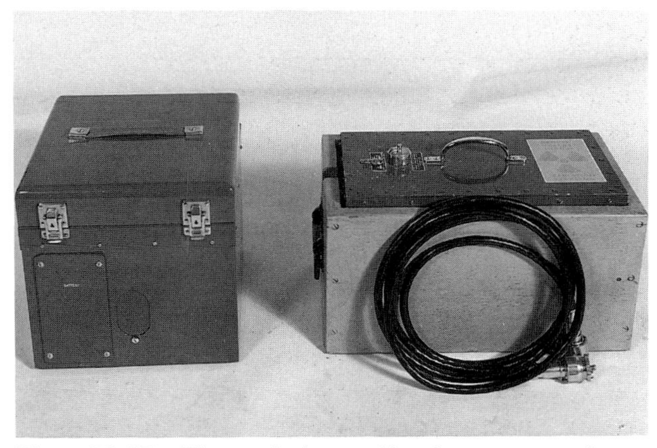

Plate 18.17 The early form of back-scatter nuclear apparatus as transported

distributions of the soils used are given in Figure 3.4. Special calibration blocks were also manufactured from cemented materials; these measured 900 mm square by about 380 mm deep with different materials and mix proportions to give bulk densities covering the range 1.67 to 2.32 Mg/m^3. The bulk densities of the blocks were determined immediately before use by weighing the block and measuring the external dimensions.

18.50 The horizontal area sampled by the equipment was determined on the calibration blocks of cemented materials. The intensities of gamma radiation were measured with the surface unit placed at various distances from the edge of each block. Typical results are given in Figure 18.17 for the minimum distance between the source and detector. The horizontal area sampled was found to be approximately the area of the surface unit (about 400 mm by 200 mm), there being a slight decrease in area as the density of the material increased.

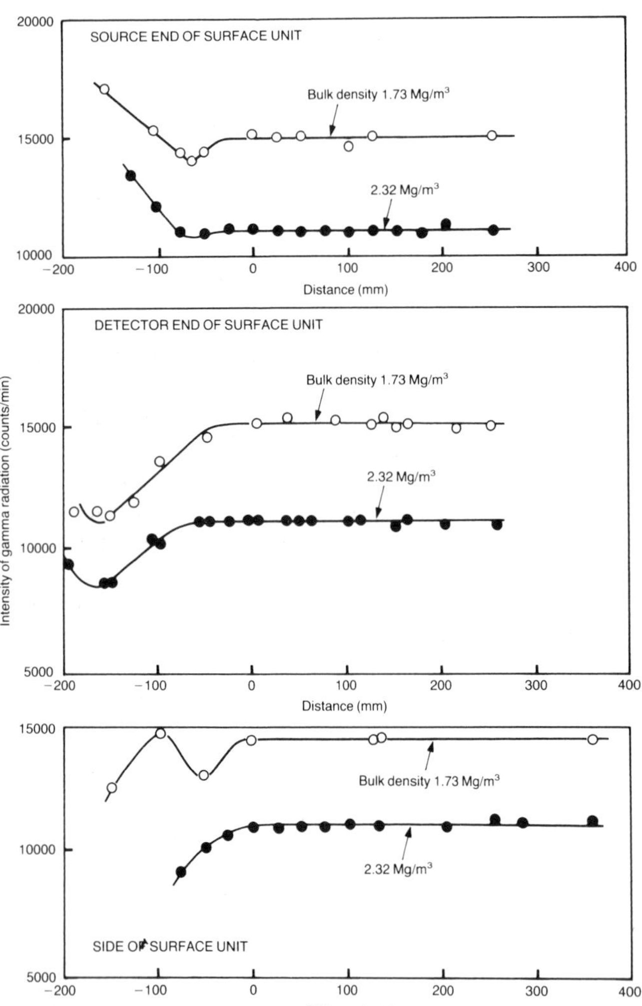

Figure 18.17 Relations between the intensity of gamma radiation measured by the early form of back-scatter apparatus and the distance of the surface unit from the edges of calibration blocks

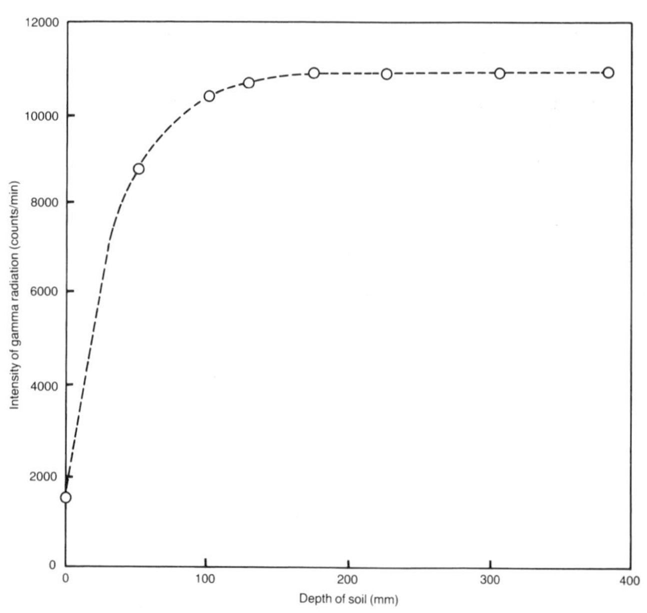

Figure 18.18 Typical example of the relation between the intensity of gamma radiation registered by the early form of back-scatter apparatus and the depth of soil as successive layers of soil were removed from the bottom of the sample

18.51 The depth sampled by the equipment was determined using samples of soil prepared in a wooden mould 460 mm by 380 mm by 380 mm deep. Heavy clay, sandy clay No 2 and well-graded sand were compacted with a vibrating hammer either in thin layers so that no appreciable variations in bulk density occurred throughout the specimen, or in a single thick layer to produce a gradient of bulk density. The surface unit was placed on top of each compacted specimen in such a position that there was no interference from the sides of the mould. After the intensity of radiation had been recorded, a thin layer of soil was removed from the bottom of the mould and the radiation measurement was repeated. This procedure continued until less than about 50 mm of soil remained. A typical result is shown in Figure 18.18.

18.52 The depths sampled by 95, 80 and 50 per cent of the maximum limiting intensity of radiation recorded were obtained directly from relations of the type shown in Figure 18.18; these depths have been plotted against the bulk density of the soil in Figure 18.19. These relations (Figure 18.19) show that 50 per cent of the radiation recorded penetrated only the top 20 to 30 mm of the compacted soil. The results obtained also showed that the spacing between the source and the Geiger tube (see Paragraph

18.48), within the limits imposed by the design of the apparatus, had very little effect on the depth sampled; there was only very slight evidence that the depth penetrated was greatest with the minimum spacing possible with the equipment. It was clear that the results of measurements with the back-scatter equipment were dependent mainly on the state of compaction of the top 25 mm or so of the layer under test. It was concluded at the time (Lewis, 1965) that, since in practice the state of compaction of the top 25 mm of a layer often bears little relation to the average conditions throughout the layer, the back-scatter type of equipment was not likely to provide measurements of sufficient accuracy for the control of earthwork and road-base construction.

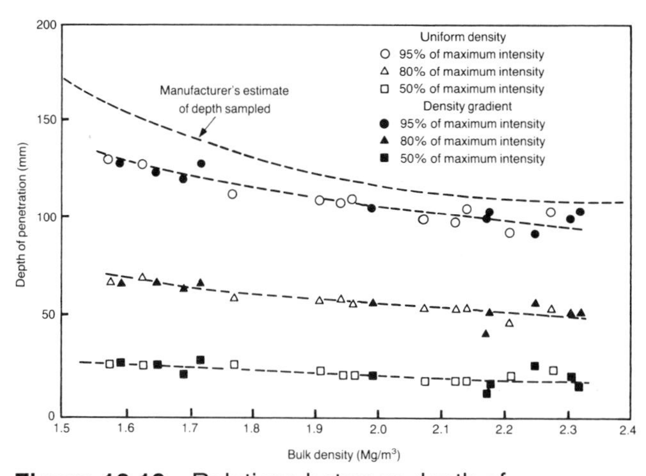

Figure 18.19 Relations between depth of penetration of 95, 80 and 50 per cent of the measured gamma radiation and the bulk density of the soil for the early form of back-scatter apparatus

18.53 The potential errors associated with the presence of density gradients in compacted layers were assessed from the measurements described above in Paragraph 18.51. Calibration curves were established using the uniformly compacted specimens prior to the removal of the bottom layers of soil; the radiation intensities recorded on the samples with density gradients were compared with the uniform density calibrations to determine the likely error. The density gradients were determined during the removal of the soil in thin layers. Figure 18.20 shows the errors in bulk density related to the gradient of bulk density over the depth penetrated by 95 per cent of the recorded gamma radiation (about 100 mm to 130 mm). With the type of compaction plant normally employed in earthwork and road-base construction density gradients of the order of 2 $Mg/m^3/m$ can occur in work of an acceptable standard. Figure 18.20 indicates that errors of the order of 0.03 to 0.04 Mg/m^3 occurred in these circumstances with the back-scatter type of equipment. Lewis (1965) commented that with the usual form of density gradient the errors would be such as to give the impression that a very much better state of compaction was being achieved than was actually the case.

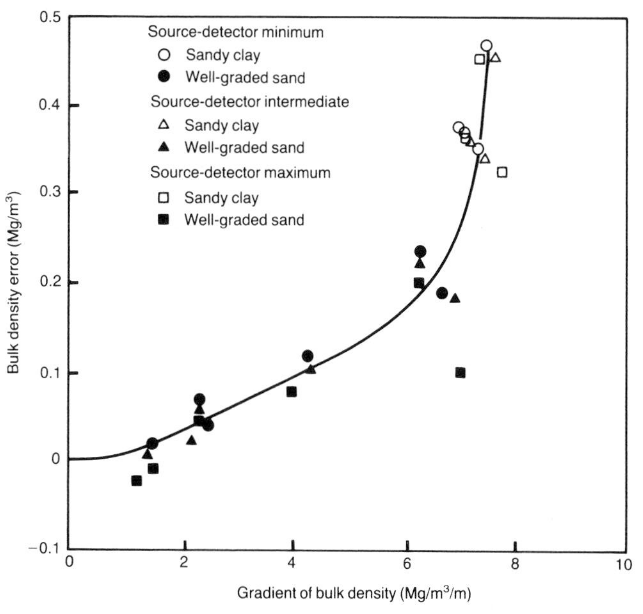

Figure 18.20 Relation between bulk density error and gradient of bulk density for the early form of back-scatter apparatus

18.54 Calibrations of the early form of back-scatter gamma radiation apparatus were carried out on moulded samples of the various soils used in the full-scale compaction work (see Chapter 3), with the addition of a crushed quartzite gravel and a crushed limestone. Compaction of the samples was carried out in such a way as to produce near-uniform density conditions, and the actual densities were determined from the mass and volume of each specimen. The results for the minimum spacing of the source and detector are given in Figure 18.21. The results for the intermediate and maximum spacings were found to follow a very similar pattern. The results indicate that the intensity of gamma radiation recorded for a given bulk density was affected by the type of material. The use of a single calibration line could have introduced errors of the order of $\pm$ 0.1 Mg/m^3. The relations between the mass absorption coefficient and the energy of gamma radiation for several elements which commonly occur in soil are shown in Figure 18.12 and the possible effect on calibrations of nuclear density gauges has been discussed in Paragraphs 18.36 to 18.38. The back-scatter apparatus used to obtain the results shown in Figure 18.21 had a

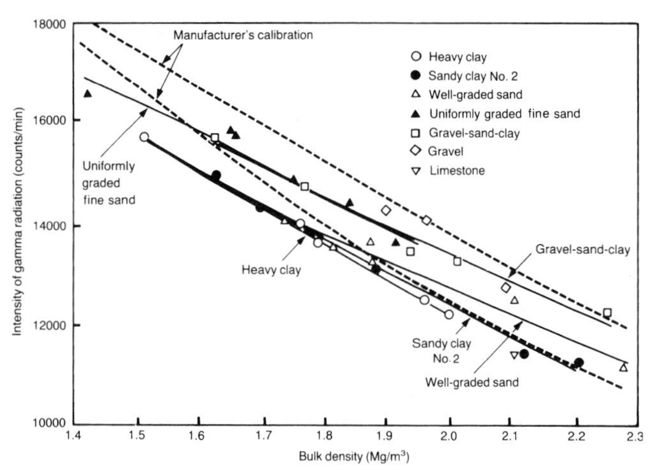

Figure 18.21 Calibration of the early form of back-scatter gamma radiation apparatus with various materials

radium-beryllium source, the gamma radiation from which had a spectrum of energy ranging from 0.18 MeV to 2.198 MeV, although a considerable proportion of the radiation had energies of 0.607 and 0.350 MeV (Morgan, 1955). It would be expected, therefore, that, allowing for the loss of energy which would have occurred at each back-scatter collision and the initial low energy of some of the radiation, a considerable proportion of the gamma radiation passing through the soil would have been below an energy of 0.3 MeV. Thus the presence of elements such as iron and calcium would be expected to give lower counts while higher counts would be expected with materials consisting of silicon or aluminium.

18.55 A quantitative analysis for most of the soils used with the back-scatter apparatus is given in Table 18.3. Similar analyses for the additional materials included in the results shown in Figure 18.21 are given in Table 18.5. The results obtained confirm that the intensity of radiation at any given bulk density increased with decrease of the iron and calcium contents of the materials. The variation in mass absorption coefficients is regarded as the only possible cause of the 'soil type effect' exhibited in Figure 18.21. As discussed in relation to the early results with a direct-transmission gamma radiation apparatus (Paragraph 18.38), the effect of the different mass absorption coefficient of hydrogen was considered to be relatively small (Lewis, 1965).

Table 18.5 Quantitative analysis of the main constituents of additional materials tested with the early form of back-scatter gamma radiation apparatus

Soil	Percentage present in oven-dry material			Ignition loss
	Silica (SiO_2)	Ferric oxide (Fe_2O_3)	Calcium oxide (CaO)	(%)
Crushed quartzite gravel	91	1.6	1.4	1
Crushed limestone	2	0.4	54	43

Note: The remainder of the materials would have consisted mainly of alumina (Al_2O_3)

18.56 Lewis (1965) concluded that the shallow depth sampled by the back-scatter equipment in relation to the thicknesses of compacted layers being employed in earthwork construction, together with the effect of the type of material on the calibration, appeared to introduce serious limitations on the use of this type of equipment for the measurement of the state of compaction of soil on site.

18.57 The modern nuclear gauge, described in Paragraph 18.42, could be used in the back-scatter mode as well as the direct-transmission mode for the determination of bulk density. A diagram of the relative positions of the radioactive source and the radiation detectors in the back-scatter mode is given in Figure 18.22. No work has been carried out using the nuclear gauge in this mode; the principles of operation were little changed from those of the early form of apparatus and the direct-transmission mode was, therefore, regarded as most likely to provide meaningful results. Thus the depth sampled with the modern back-scatter apparatus would still be relatively small, although some improvement over the distribution shown in Figure 18.19 might be expected as a result of collimation of the emitted and detected radiation, a feature which has been incorporated in some modern gauges. Equally, the 'soil type effect' would be expected to be reproduced with

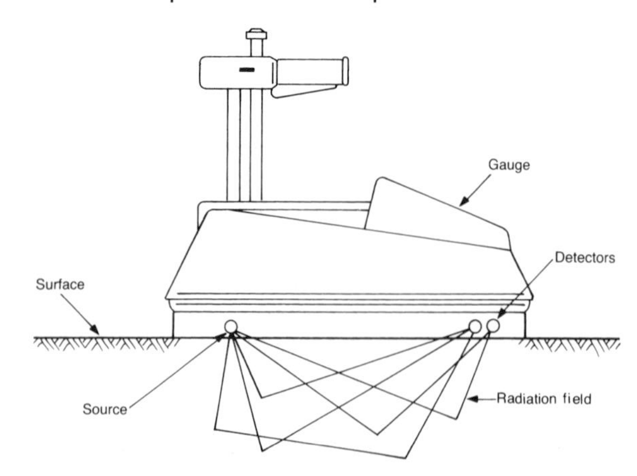

Figure 18.22 The back-scatter method of measuring density and/or moisture content with a recent type of nuclear gauge

modern back-scatter gauges; even though the source may be different than the radium-beryllium of the early gauge, the amount of scattering within the soil would certainly give rise to the low gamma energy levels which suffer differential rates of absorption depending on the constituents of the soil.

18.58 The use of the back-scatter gamma radiation technique for the determination of bulk density is only ideally suited to the control of the state of compaction of relatively thin layers, such as are used in bituminous surfacings and some other pavement layers. Developments of the back-scatter principle for application in these circumstances have given rise to mobile equipment to provide a continuous record of bulk density; two types of such mobile equipment have been described by Du Mesnil-Adelee et al (1980). The first, a 'small mobile shoe', consists of a back-scatter density gauge suspended above the surface of the compacted material on a four-wheel mini-tractor. The depth sampled is stated to be about 70 mm. A continuous record is produced of the variation of bulk density with distance travelled and the apparatus is intended for use primarily on bituminous wearing courses.

18.59 The second mobile apparatus described by Du Mesnil-Adelee et al (1980) is termed a 'mobile gammadensimeter'. It is mounted on a trailer, towed by a truck containing the power supplies and recording equipment, and is equipped with a powerful 750 mC radioactive source of cobalt 60. By using a dual detector system it is stated that the soil type effect is eliminated. The depth sampled is stated to be 150 to 200 mm, probably achieved by virtue of the powerful source and the collimation of the emitted radiation. The equipment is intended for use in determining the bulk density of pavement layers. Thus by sacrificing portability and using heavy, mounted equipment, it is possible to overcome some of the deficiencies of the back-scatter technique. It must be remembered, however, that the accuracy of the mobile equipment relies upon the air gap beneath the source and detector system remaining constant; only layers of material laid and compacted to surface levels with close tolerances can, therefore, be tested in this way. This must restrict the application of mobile methods to subgrade and pavement layers only.

18.60 Mobile back-scatter equipment, providing a continuous record of bulk density along a length of compacted material, illustrates the way in which the back-scatter technique can be effectively utilised. By mounting the equipment on a vehicle the need for a relatively small portable apparatus no longer exists; powerful sources with appropriate shielding, and large collimated detectors can be used so that an adequate depth of influence is achieved. A further sophistication with this type of mobile equipment is the recording of the data on magnetic disc and the automatic analysis of the results by computer.

Nuclear determination of bulk density – back-scatter depth probe

18.61 The principle of this method (Figure 18.9(d)) is identical to that of the surface back-scatter apparatus, except that the gamma radiation source and the detector are contained in an apparatus capable of being lowered down a cased borehole to provide a measurement, at various depths, of the bulk density of the soil surrounding the borehole. Extreme care is needed in forming the borehole and inserting the casing to avoid disturbing the soil having the greatest influence on the resulting measurements. Because of its nature this test method would rarely be applicable to the control of the compaction process.

Nuclear method for the determination of moisture content

18.62 Because the end product required in any measurement of the in-situ quality of compaction is normally dry density, the value of bulk density determined by the gamma radiation method has to be converted to dry density from a knowledge of the moisture content. For maximum convenience most examples of nuclear density apparatus also incorporate a nuclear means of measuring moisture content. The method is based on the principle of attenuation of neutrons by the light-weight hydrogen atoms in the compacted material. The neutrons lose energy upon colliding with the nuclei of hydrogen atoms, whereas collisions with the nuclei of the atoms of other elements result in negligible losses of energy. The intensity of low energy (slow) neutrons scattered back by the soil and detected by an appropriate detector increases, therefore, with increase in the amount of water present. The slow neutron count rate is related to the mass of water per unit volume of compacted soil (moisture density) and calibrations can be made using materials of known moisture content and bulk density.

18.63 The neutron method of measuring

moisture density, with the source of neutrons and the detector both contained in one unit, can be conveniently incorporated with surface direct-transmission, surface back-scatter and back-scatter depth probe methods for determining bulk density. With the results in terms of moisture density, the determination of dry density is as follows:–

$$\rho_d = \rho - m$$

where ρ_d=dry density,
ρ =bulk density determined by the gamma radiation technique,
m =moisture density, mass of water per unit volume of compacted soil, determined by the neutron radiation technique.

18.64 The moisture content in usual terms, w, ie the mass of water expressed as a percentage of the mass of dry soil, is determined from:–

$$w = \frac{m}{\rho_d} \times 100 \text{ (per cent)}$$

18.65 The early type of back-scatter gamma radiation equipment, described in Paragraph 18.48, incorporated the moisture measuring method. The radium-beryllium source emitted neutrons as well as gamma radiation and boron trifluoride tubes contained within the surface unit detected the slow neutrons scattered back from the soil (see Figure 18.16). All investigations of the measurement of moisture density with that apparatus were made at the same time as the studies of the measurement of bulk density, described in Paragraphs 18.49 to 18.56.

18.66 The horizontal area sampled in the determination of moisture density was determined on the special calibration blocks by measuring the intensity of slow neutron radiation at various distances from the edge of each block. Typical results are shown in Figure 18.23. The horizontal area sampled was approximately 400 mm square, there being a slight decrease as the moisture density increased.

18.67 The depth sampled in the determination of moisture density using the early form of back-scatter equipment was studied at the same time and in a similar manner to the study of the depth sampled in the determination of bulk density (see Paragraph 18.51). A typical result is shown in Figure 18.24. Using curves of the type shown in Figure 18.24, relations were deduced between the depth sampled by 95, 80 and 50 per cent of

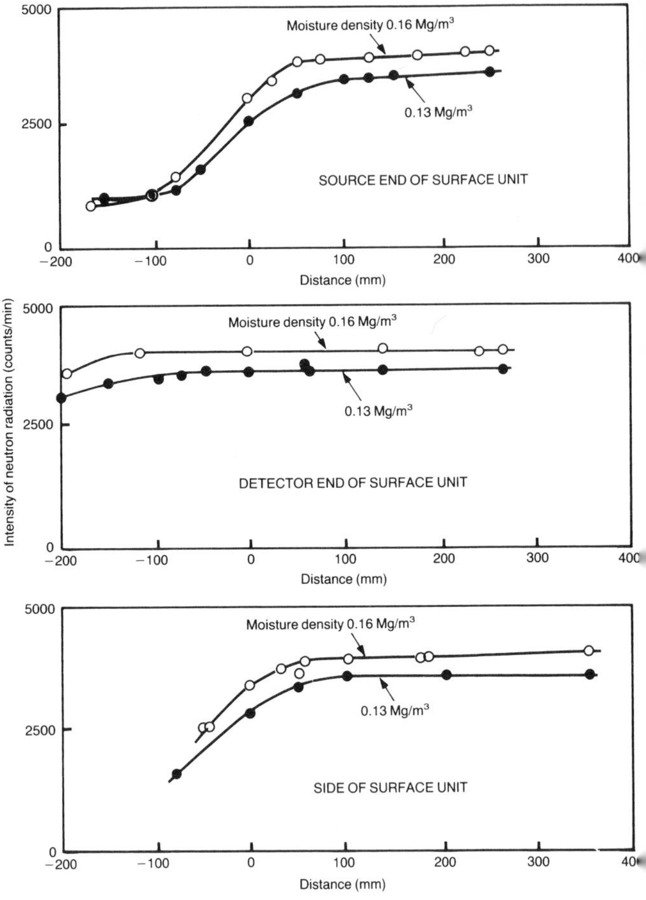

Figure 18.23 Relations between the intensity of neutron radiation measured by the early form of nuclear gauge (back-scatter apparatus) and the distance of the surface unit from the edges of calibration blocks

the maximum limiting intensity of neutron radiation recorded and the moisture density of the soil (Figure 18.25). These relations show that 50 per cent of the radiation recorded penetrated only the top 30 to 50 mm of the soil, a slight increase on the penetration for 50 per cent of the gamma radiation in bulk density measurements (see Paragraph 18.52). The moisture density of

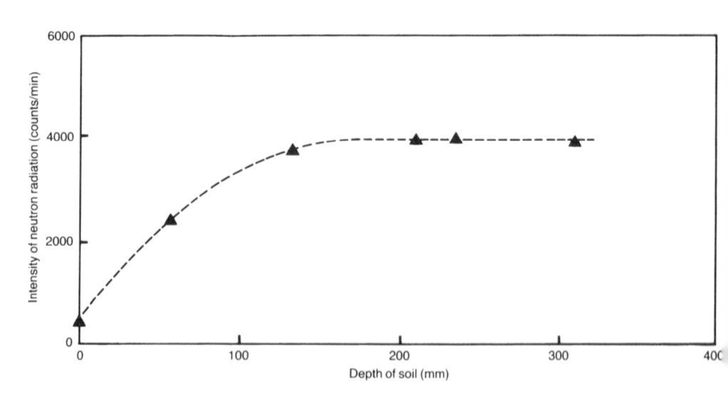

Figure 18.24 Typical example of the relation between the intensity of neutron radiation registered by the early form of nuclear gauge (back-scatter apparatus) and the depth of soil as successive layers of soil were removed from the bottom of the sample

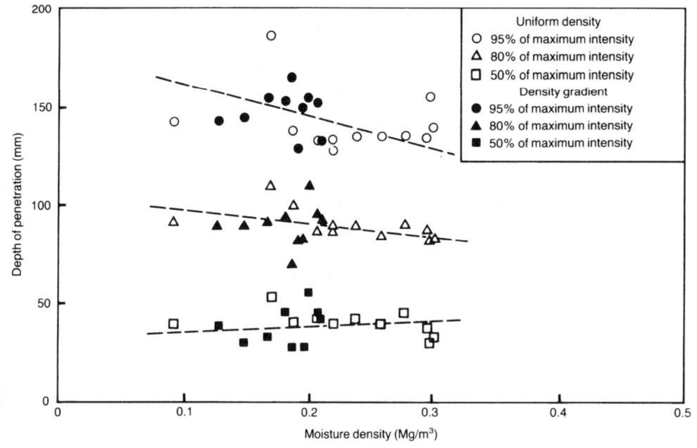

Figure 18.25 Relations between the depth of penetration of 95, 80 and 50 per cent of the measured neutron radiation and the moisture density of the soil for the early form of nuclear gauge (back-scatter apparatus)

the top 25 to 50 mm of the compacted layer had a strong influence, therefore, on the results of measurements made with the apparatus.

18.68 The effect of moisture gradients in the soil, a natural outcome of the presence of density gradients, was also determined with the early form of nuclear gauge. Comparisons were made between the intensities of neutron radiation recorded on samples of uniform bulk density and moisture density with those recorded on samples with gradients of bulk density, and hence of moisture density. The errors in moisture density for various gradients of moisture density have been calculated and are given in Figure 18.26. The conclusion reached at the time (Lewis, 1965) was that the error in moisture

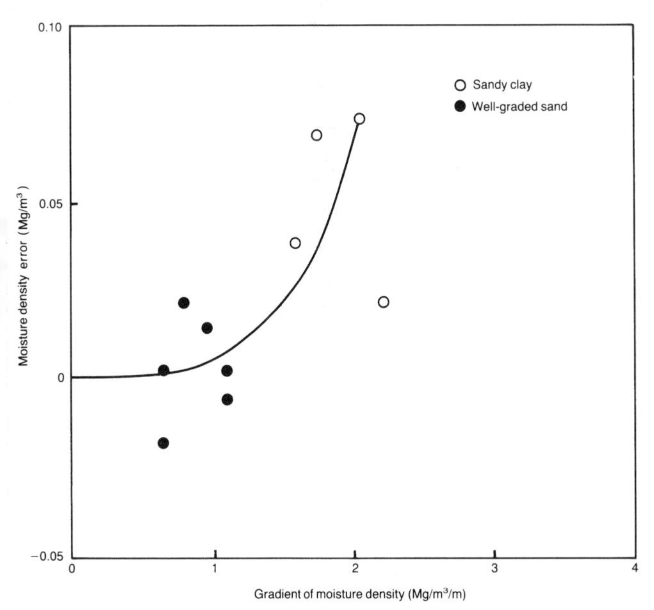

Figure 18.26 Relation between moisture density error and gradient of moisture density for the early form of nuclear gauge (back-scatter apparatus)

density arising from gradients of moisture produced by density gradients would not be very significant in practice as the moisture density gradients would generally be less than 0.6 Mg/m³/m. Where severe surface drying or wetting has occurred, however, it was concluded that serious errors could be introduced.

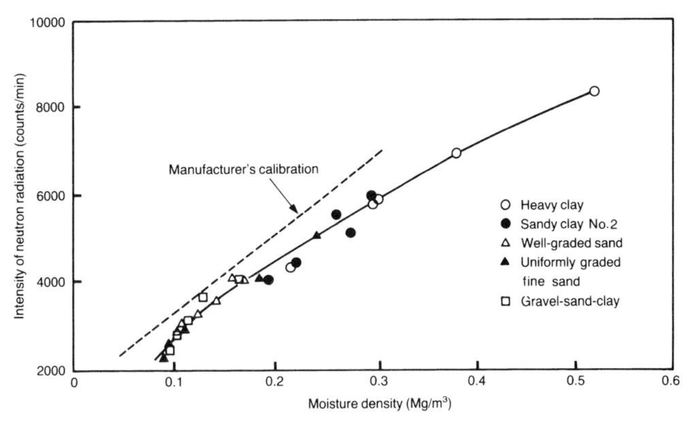

Figure 18.27 Calibration curve for moisture measurement obtained with the early form of nuclear gauge (back-scatter apparatus) with various soils

18.69 The effect of the type of material on the calibration for the measurement of moisture density with the early type of back-scatter equipment was studied simultaneously with the effects on the bulk density calibration. The experimental method is described in Paragraph 18.54. The results for five different soil types are given in Figure 18.27. Although there was some scatter in the intensities of neutron radiation for a given mass of water per unit volume of soil (moisture density), no clear pattern emerged and, in contrast to the calibration for the measurement of bulk density (Figure 18.21), a single calibration line appeared to be justified. However, the results show that the calibration provided by the manufacturer was inaccurate and calibration of the equipment by the user was again shown to be necessary.

18.70 The modern nuclear gauge used in recent investigations of compaction equipment has been described in Paragraph 18.42 and is illustrated in Plates 18.14 and 18.15 and Figures 18.13 and 18.22. It included a method of determining the moisture density of soil using similar principles to those employed in the early form of nuclear gauge and which have been described in Paragraphs 18.62 to 18.65. The source of neutrons in the modern gauge was americium 241-beryllium and a helium-3 tube was used to detect the low energy neutrons scattered back from the soil. As in the determination of bulk density with the modern gauge, the test procedure for moisture density determination included the determination of a standard count made on a polyethylene reference block (see Plate 18.15). The count ratio (in-situ neutron count rate divided by the standard neutron count rate) was used to determine the moisture density from a calibration pre-determined by the manufacturer. The nuclear gauge was equipped with a micro-processor which was programmed to provide a direct read-out of moisture density. The count rates for moisture density measurement were automatically carried out concurrently with those for bulk density, a marked improvement in productivity compared with the early form of nuclear gauge, where the measurements had to be carried out in consecutive time periods. (With the modern gauge, moisture content in mass terms could also be obtained from the micro-processor, but this additionally utilised the results of the bulk density measurement made by the gamma radiation technique, as shown in Paragraphs 18.63 to 18.64; errors in moisture content on a mass basis were compounded, therefore, by errors in the measurement of bulk density.)

18.71 The modern gauge also had the facility to adjust the moisture calibration after comparisons with oven-dry moisture content measurements to compensate for possible soil type effects; however, to establish the calibrations described in Paragraph 18.72, no adjustments were made to the pre-determined manufacturer's calibration.

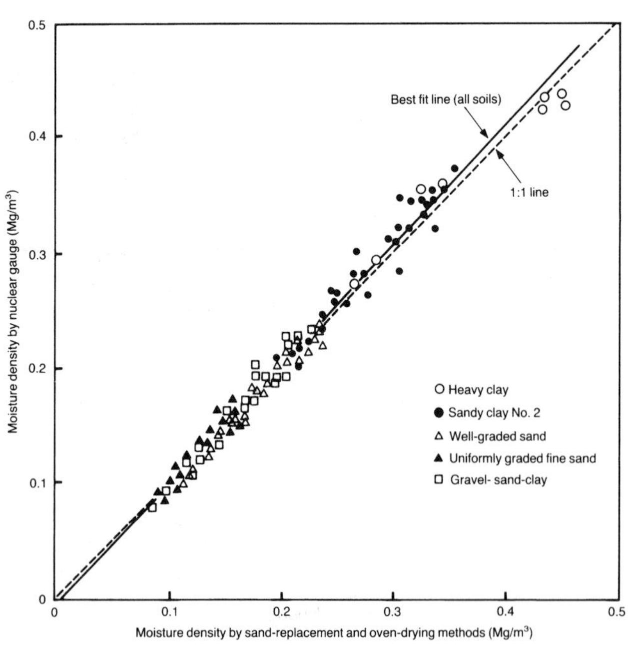

Figure 18.28 Correlation of results of moisture density measurements made by a modern nuclear gauge and by the sand-replacement and oven-drying methods for all test soils

18.72 The investigations with the modern nuclear gauge, already described in Paragraphs 18.42 to 18.43, also included a series of 'calibrations' where the moisture densities registered by the nuclear gauge were compared with moisture densities determined by the conventional sand-replacement tests together with the oven-drying method. (The inter-relation between moisture density, moisture content and dry density is given in Paragraph 18.64.) The results for the five different soils used are shown in Figure 18.28. These results confirm those obtained for the early form of nuclear gauge, shown in Figure 18.27, and the conclusion reached at the time of that investigation that a single calibration line could be used for the determination of moisture density for the complete range of soil types used in the tests. In contrast to the results given in Figure 18.27, however, Figure 18.28 shows that the manufacturer's calibration incorporated in the micro-processor of the modern nuclear gauge was completely reliable.

General discussion on the application of nuclear gauges to measurements of density and moisture content

18.73 The results obtained with the modern nuclear gauge show that the direct-transmission gamma method can be reliable, although calibrations need to be established on each soil type encountered. With the test soils used in the investigations the calibration for moisture density did not need any significant adjustment from that provided with the instrument. Investigations were also made of the use of a nuclear gauge in measuring the state of compaction of soil placed in a narrow trench. It was found that the calibrations for bulk density (direct-transmission method) remained identical to those established for each particular soil type on surface layers. With moisture density measurements, however, because low energy neutrons were scattered back from the walls of the trench, a trench correction factor had to be applied. The procedure for this was detailed in the instructions for the use of the gauge. It was found in tests with the modern, portable nuclear gauge that the testing time and effort required to carry out tests were considerably reduced in comparison with the alternative sand-replacement tests.

18.74 It is concluded that nuclear gauges have now improved to such an extent that results are acceptably accurate and reliable and the productivity of results is high. It is clear that the principles of the method should be fully understood by the operator and precautions regarding the checking of the electronic stability of the apparatus and the establishment of the calibrations must be rigorously applied.

18.75 The use of nuclear equipment requires a license (from the Department of the Environment in England and from the equivalent legislative authority in each of the other countries of the United Kingdom) and there are monitoring procedures for the operators, including medical examinations and the issue of film badges to ensure that the radiation exposure is within acceptable limits. Despite these requirements, the sources are sealed and, for the portable gauges, are relatively small so that exposure levels are extremely low. The licensing and monitoring procedures, however, are likely to inhibit the use of portable nuclear gauges so that they are restricted to the larger construction sites or to use by a limited number of testing consultants.

References

American Association of State Highway and Transportation Officials (1986). Standard specifications for transportation materials and methods of sampling and testing. Part II, Methods of sampling and testing, AASHTO, Washington.

American Society of Testing and Materials (Annual). Annual book of ASTM standards. ASTM, Philadelphia.

British Standards Institution (1975). Methods of test for soils for civil engineering purposes. BS 1377: 1975. BSI, London.

British Standards Institution (1981). Code of practice for site investigations. BS 5930 : 1981. BSI, London.

British Standards Institution (1990a). British Standard methods of test for soils for civil engineering purposes : BS 1377 : Part 2 Classification tests. BSI, London.

British Standards Institution (1990b). British Standard methods of test for soils for civil engineering purposes : BS 1377 : Part 9 In situ tests. BSI, London.

Du Mesnil-Adelee, M, Lebas, M and Paybernard, J (1980). Continuous measurement of density (in French). *International Conference on Compaction*, Editions Anciens ENPC, Paris, Vol II, 523–30.

Gabilly, Y (1969). The R-type gamma-density gauges (in French). *Bulletin de Liaison des Laboratoires Routiers Ponts et Chaussees*, No 36, 53–76.

Harland, D G and Urkan, S R (1966). Effect of soil type on the single probe gamma ray transmission method for measuring in-situ densities. *Ministry of Transport, RRL Report* No 52. Road Research Laboratory, Crowthorne.

Lewis, W A (1965). Nuclear apparatus for density and moisture measurements – study of factors affecting accuracy. *Rds and Rd Constr*, 43, (506), 37–43.

Lewis, W A and Parsons, A W (1963). An analysis of the states of compaction measured during the construction of embankments on six major road schemes. *D.S.I.R., Road Research Laboratory Note* No LN/399/WAL.AWP (Unpublished).

Morel, G, Franceschina, R and Schaeffner, M (1980). Gamma double probe (in French). *International Conference on Compaction*, Editions Anciens ENPC, Paris, Vol II, 583–9.

Morgan, H (Ed) (1955). Handbook of radiology. Year Book Publishers Inc., Chicago, 90 – 6.

Parsons, A W (1956). An investigation of a membrane densitometer for determining the volume of holes for dry density measurements in soil. *D.S.I.R., Road Research Laboratory Research Note* No RN/2715/AWP (Unpublished).

Parsons, A W (1966a). An investigation into the accuracy achieved in measuring the state of compaction of cement-bound granular material. *Rds and Rd. Constr.*, 44 (520), 95 – 8.

Parsons, A W (1966b). An investigation into the effect of variations in the dryness of the sand used in the sand-replacement test. *Rds and Rd. Constr.*, 44 (521), 125 – 7.

Parsons, A W and Odubanjo, T (1965). An investigation of a balloon apparatus for measuring the volumes of excavated density holes. *Road Research Laboratory, Laboratory Note* No LN/821/AWP.TO (Unpublished).

Road Research Laboratory (1952). Soil mechanics for road engineers. HM Stationery Office, London, Chapter 9.

Soane, B D, Campbell, D J and Herkes, S M (1971). Hand-held gamma-ray transmission equipment for the measurement of bulk density of field soils. *J. Agric. Engng. Res.*, 16, 146 – 56.

West, L and Parsons, A W (1953). The determination of the dry density of compacted layers of coarse granular materials. D.S.I.R., *Road Research Laboratory Note* No RN/1950/LW.AWP (Unpublished).

CHAPTER 19 COMPACTION SPECIFICATIONS

19.1 Specifications for compaction can be either in terms of the end product to be achieved or alternatively in terms of the method to be applied. In the former type of specification (end product) a quantifiable property of the compacted soil has to be achieved; in the latter (method specification) the procedure to be used in the compaction process (type of plant, number of passes, thickness of layer) is detailed.

End-product specifications

19.2 The basic measurement of the in-situ state of compaction is in terms of the dry density of the soil (see Paragraphs 2.3 and 18.2). However, because of the effects of soil characteristics, dry density alone is not indicative of the state of compaction (see Paragraphs 2.18 – 2.19) and interpretations of the dry density have to be made for purposes of determining the levels of compaction achieved in relation to the levels that are potentially attainable or that are necessary to attain the required engineering properties. Three basic forms of end-product specification are commonly used; these are:–

1. Relative compaction;
2. Relative density or density index; and
3. Air voids.

End-product specifications – relative compaction

19.3 The dry density achieved in the compaction process can be expressed as a percentage of the maximum dry density obtained by using a specified compactive effort in the laboratory. The most commonly used laboratory compaction test in earthworks specifications is that involving the 2.5 kg rammer (Paragraphs 17.5 to 17.7). However, the 4.5 kg rammer compaction test and various forms of vibratory compaction test (Chapter 17) can also be used as standards in relative compaction specifications. The actual percentage specified can vary depending on the application of the earthworks being constructed and the soil type, but for the 2.5 kg rammer test it is usually in the range of 90 to 100 per cent. Where soil type varies from place to place in a compacted layer it is often necessary to establish the maximum dry density at each location of a measurement of the in-situ dry density. To reduce the effort required in the laboratory compaction tests under these circumstances, rapid methods have been evolved for controlling compaction to this form of specification (Hilf, 1957). A discussion of the application of relative compaction concepts to the judgement of compaction has been given in Paragraph 2.24.

End-product specifications – relative density or density index

19.4 The relative density or density index form of specification is often applied to cohesionless soils, where the effect of moisture content is less than with other soils, and compares the void ratio of the in-situ material with the maximum and minimum void ratios determined in the laboratory (American Society of Testing and Materials, 1973). The minimum dry density associated with the maximum void ratio (e_{max}) is normally determined by pouring dry soil into a mould or shaking soil in a container. Test procedures are given in Part 4 of BS 1377 (British Standards Institution, 1990) and as ASTM Test D 4254 (American Society of Testing and Materials, Annual publication). The maximum dry density associated with the minimum void ratio (e_{min}) can be determined by a form of vibratory compaction, usually with fully saturated soil; test procedures are again given in Part 4 of BS 1377 and in ASTM Tests D 4253 and D 4254 (American Society of Testing and Materials, Annual publication) (see Paragraph 17.32).

19.5 Relative density is calculated as follows:–

$$\text{Relative density} = \frac{e_{max} - e}{e_{max} - e_{min}} \times 100 \text{ (per cent)}$$

where e=the void ratio of the in-situ material

In terms of dry density

$$\text{Relative density} = \frac{\rho_{dmax}\,(\rho_d - \rho_{dmin})}{\rho_d\,(\rho_{dmax} - \rho_{dmin})} \times 100 \text{ (per cent)}$$

where ρ_{dmax}=the maximum index dry density associated with e_{min}
ρ_{dmin} =the minimum index dry density associated with e_{max}
ρ_d =the in-situ dry density

Note : In Part 4 of BS 1377 density index is defined similarly to relative density as given above. In ASTM Tests D 4253 and D 4254 density index is defined separately, and is different from the definition of relative density given above.

End-product specifications – air voids

19.6 The measurement of the air remaining in the compacted soil is a logical expression of the state of compaction, considering that the reduction of air voids is implicit in the compaction process (see Paragraphs 2.1 – 2.2). Air void content is determined from the equation:–

$$\text{Air voids} = 100\left\{1-\rho_d\left(\frac{1}{\rho_s}+\frac{w}{100\rho_w}\right)\right\} \text{ (per cent)}$$

where ρ_d = dry density of the compacted soil
ρ_w = density of water
ρ_s = particle density
and w = moisture content (per cent)

For practical purposes ρ_s may be assumed a constant unless there are wide variations in soil type.

19.7 As indicated by the equation, an increase in moisture content at constant dry density leads to a reduction in air voids and it is necessary to ensure that a low air content is not achieved at the expense of using soil of excessively high moisture content. Commonly specified values for well-graded soils are a maximum of 10 per cent air voids for bulk earthworks and 5 per cent air voids for high quality applications. Normally these values have to be achieved at moisture contents within specified limits or below a specified upper limit. Additional discussion of the application of air voids in the judgement of compaction, with a comparison of the air void and relative compaction methods, has been given in Paragraphs 2.25 to 2.27.

End-product specifications – investigations on construction sites

19.8 For most of the 1950s and 1960s an end-product specification was in use for the compaction of earthworks in road construction in the United Kingdom. The specification for the bulk earthworks required that a state of compaction of 10 per cent air voids or better had to be achieved in at least 9 results out of every 10 consecutive measurements made in the compacted material (Ministry of Transport, 1963). The required air content was decreased to 5 per cent air voids in the upper 600 mm of the embankment.

19.9 To determine the degree to which the specification was being achieved and to identify any problems arising in its application a series of visits was made in 1959 to 1962 to six major road construction sites. During these visits determinations of the state of compaction were made in the compacted layers of the embankments, using the sand-replacement method (see Paragraphs 18.4 to 18.6) ; on the South Mimms By-pass, however, where clay soil predominated, the core cutter method (see Paragraphs 18.28 to 18.29) was also used. The results obtained, taken from Lewis and Parsons (1963) and Parsons (1973), are shown plotted on dry density/moisture content charts in Figures 19.1 to 19.6 and a summary is given in Table 19.1.

19.10 The results of the in-situ dry density and moisture content measurements show a very large range of values at each site (Figures 19.1 to 19.6). These ranges were attributable more to variations in the type of material than to variations in the compactive effort applied. The wide range of soil types encountered is illustrated by the particle-size distributions and the results of laboratory compaction tests given in the Figures for typical samples of the principal soil types. Variations also occurred within any individual soil type, and the results of the classification and laboratory compaction tests should be regarded as giving only a broad indication of the conditions.

19.11 Figures 19.1 to 19.6 contain many results which are clearly in error by virtue of their positions above the zero air voids line. Those results which were significantly in error occurred with granular materials in a wet condition. The errors are considered to be attributable to the 'slump' effect, where the soil surrounding the sample hole formed in the sand-replacement test flowed or slumped inwards during its excavation. The phenomenon has been fully discussed in Paragraphs 18.19 to 18.20. Where many results were in error, as occurred on the Maidenhead

Figure 19.1

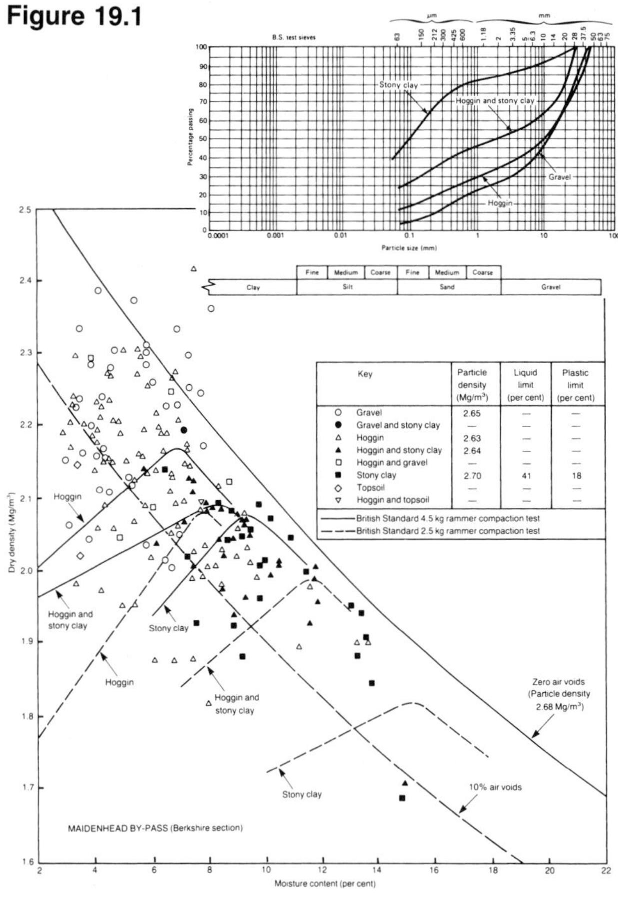

Figure 19.2

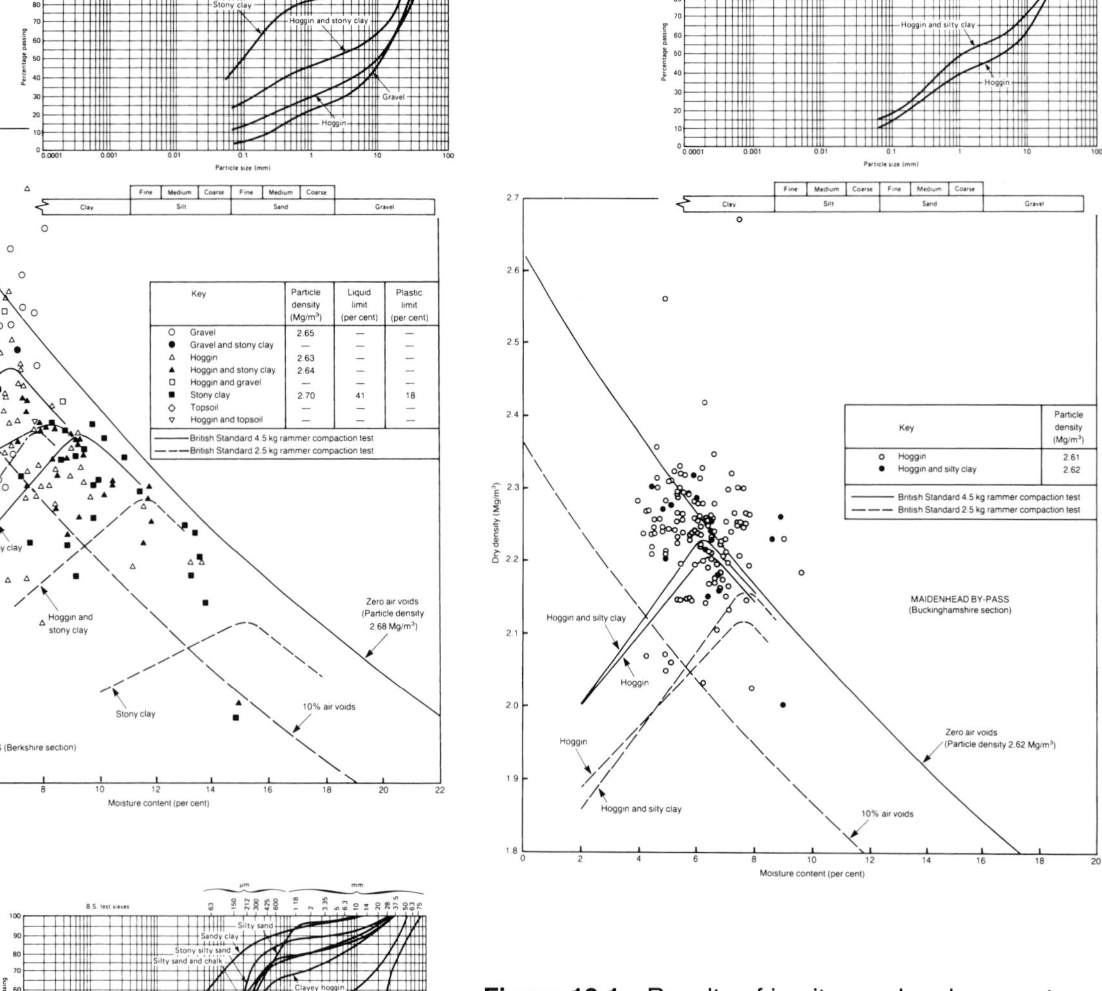

Figure 19.3

Figure 19.1 Results of in-situ sand-replacement tests and laboratory tests on soils used in earthwork construction on Maidenhead By-pass (Berkshire section)

Figure 19.2 Results of in-situ sand-replacement tests and laboratory tests on soils used in earthwork construction on Maidenhead By-pass (Buckinghamshire section)

Figure 19.3 Results of in-situ sand-replacement tests and laboratory tests on soils used in earthwork construction on Slough By-pass

Figure 19.4 Results of in-situ sand-replacement tests and laboratory tests on soils used in earthwork construction on Oxford Southern and Western By-pass

Figure 19.5 Results of in-situ sand-replacement tests and laboratory tests on soils used in earthwork construction on Biggleswade By-pass

Figure 19.6 Results of in-situ sand-replacement and core-cutter tests and laboratory tests on soils used in earthwork construction on South Mimms By-pass

Figure 19.4

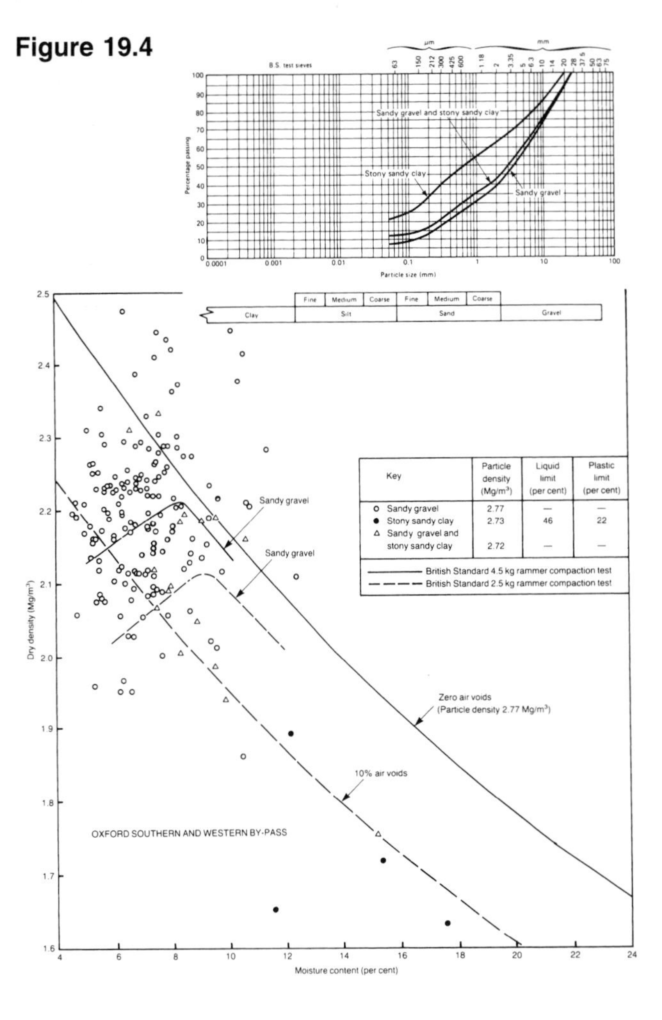

Figure 19.5

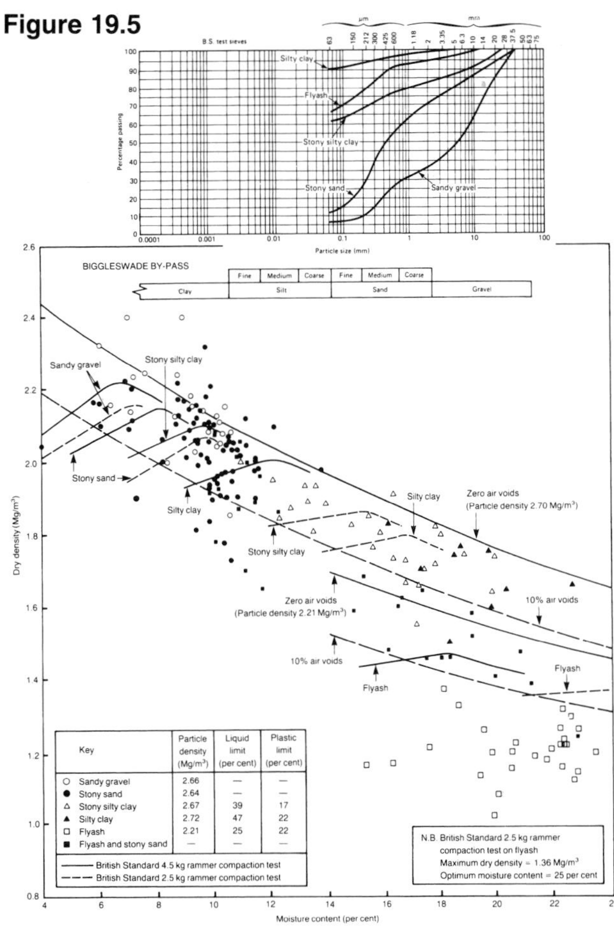

Figure 19.6

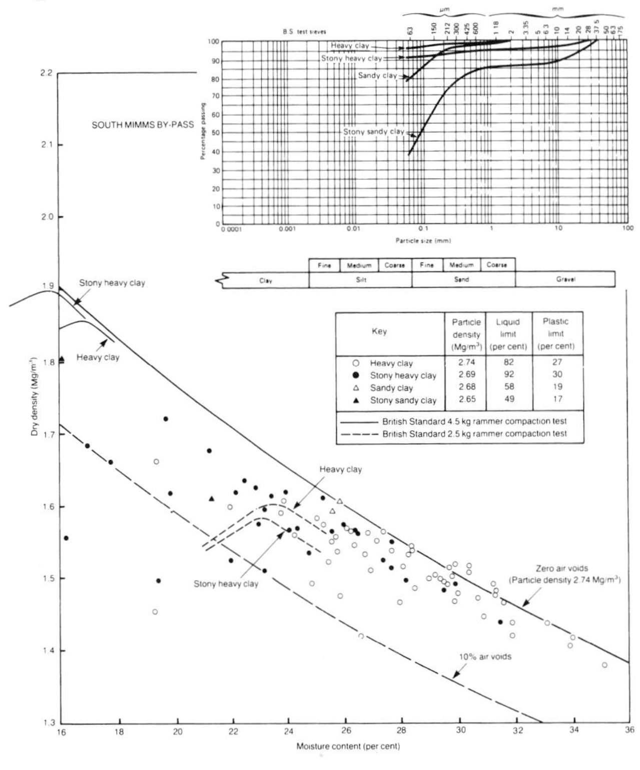

By-pass (Buckinghamshire Section) (Figure 19.2) and on the Oxford Southern and Western By-pass (Figure 19.4), doubt must arise as to the accuracy of the remainder of the measurements made at those sites in similar materials.

19.12 Lewis and Parsons (1963) and Parsons (1973) concluded that comparatively high rates of testing were required for adequate control of compaction of earthworks using an end-product specification. Based on the average standard deviation for each construction site it was estimated that 40 measurements of dry density and moisture content would have been required to determine the average air content of a volume of fill material to an accuracy of ± 1 per cent.

19.13 Table 19.1 gives the results of a re-analysis of the data based on a division of the results into a number of groups according to general soil type. In the re-analysis all results exhibiting negative air voids have been adjusted to an air content of zero. As would be expected, the lower the mean air content the smaller the value of standard deviation, and within each group the standard deviations and means of the

Table 19.1 Means and standard deviations of air content determined during earthwork construction. Results are classified in groups according to general soil type

Soil type	Site No*	No of results	Air content (per cent) Mean	Standard deviation
Clean granular soils, ⩽ 10% passing 63 μm	1	67	5.4	±4.9
	2	51	2.3	±2.9
	4	178	5.5	±4.5
	5	23	1.8	±3.0
Granular soils with fines, > 10% passing 63 μm	1	88	6.4	±4.4
	2	92	1.5	±2.1
	3	24	6.1	±3.7
	5	75	3.8	±4.2
Intermediate and low plasticity clays; with or without stones	1	54	5.5	±3.2
	3	80	6.2	±3.7
	5	26	5.0	±3.6
	6	7	3.1	±2.7
High plasticity clays; with or without stones	5	9	6.1	±5.2
	6	85	2.9	±3.7
Uniformly graded materials	3	21	10.2	±4.9
	5	58	13.9	±8.9

*Site No 1 – Maidenhead By-pass (Berkshire section)
Site No 2 – Maidenhead By-pass (Buckinghamshire Section)
Site No 3 – Slough By-pass
Site No 4 – Oxford Southern and Western By-pass
Site No 5 – Biggleswade By-pass
Site No 6 – South Mimms By-pass

air content are approximately related in the form:–

$$\text{Standard deviation} = b\sqrt{(\text{Mean})}$$

The values of b, given in Table 19.2, vary with the general soil type.

19.14 As mentioned in Paragraph 19.8, the specification required that 10 per cent air voids or better had to be achieved in at least 9 results out of every 10 consecutive measurements. Based on the mean value of b for each general soil type, the mean value required to meet the specified requirements and the mean value necessary so that 10 tests and 20 tests would provide a 0.9 probability of ensuring compliance with the specification are given in Table 19.2. Thus, taking the clean granular soils as an example, the actual mean state of compaction required to achieve the required level of 9 results in 10 at 10 per cent air voids or better is an air content of 4.5 per cent. However, this could only be the target mean value if a large number of tests were carried out. The mean air content required if 10 measurements were made is 2.6 per cent or less to retain a 90 per cent confidence that the value of 4.5 per cent is not exceeded. If 20 test measurements were made, then the mean air content resulting from those measurements should be 3.1 per cent or less.

19.15 An inspection of Table 19.2 reveals that very low mean air contents are required to achieve even a 10 per cent air void criterion in 90 per cent of the test measurements. Of the 14 soil type/site combinations given in Table 19.1 to which the 10 per cent air void criterion applied (excluding uniformly graded materials) only 6 combinations achieved compliance. Only on two sites, Nos 2 and 6, did all soil types conform to the specified compaction requirements. Sites 1, 3 and 4 failed to meet the compaction requirements in any of the types of material encountered; however, this does not exclude the possibility that local areas or individual layers in the embankments may have done so. Despite these failures to achieve specification, however, the author is not aware of any problems created by settlement within the embankment fill in the areas studied over the 25 years or so that the roads have been in service. The results indicate

Table 19.2 Estimates of mean air contents necessary to comply with the requirement that at least 9 results in 10 should have an air content of 10 per cent or less

Soil type	b*	Mean air content necessary to achieve specification requirements (per cent)†	Mean air content (per cent) required to ensure compliance at 90% confidence level using:– 10 tests	20 tests
Clean granular soils, ⩽10% passing 63 μm	2.0	4.5	2.6	3.1
Granular soils with fines, >10 % passing 63 μm	1.7	5.1	3.3	3.8
Intermediate and low plasticity clays; with or without stones	1.5	5.5	3.8	4.3
High plasticity clays; with or without stones	2.1	4.4	2.5	3.0
Uniformly graded materials	10 % air void criterion not applicable			

$^{*}b=\dfrac{\text{Standard deviation of air content}}{\sqrt{(\text{Mean air content})}}$

†9 or more results in 10 to have 10 % air voids or better

that the requirement that 5 per cent air voids be achieved in 90 per cent of the test results in the upper 600 mm of the fill was extremely onerous and likely to be achieved only with considerable difficulty.

19.16 The conclusions reached at the time of the investigations (Lewis and Parsons,1963) were:–

1. A higher rate of testing than was being applied in practice was required for proper control of earthwork compaction to the specifications pertaining at that time (see Paragraph 19.12).

2. Errors in the sand-replacement test in wet non-cohesive materials made the control of compaction in such materials using that method extremely difficult.

3. The requirement that 9 results in 10 should have an air content of 10 per cent or less represented a very high state of compaction and an even higher state of compaction needed to be aimed at in order to ensure compliance with the specification with a reasonable degree of confidence.

19.17 The early investigations with nuclear methods of determining the in-situ state of compaction (see Chapter 18) concluded that such methods were not sufficiently accurate or rapid to provide any immediate likelihood of increasing the rate of testing for compaction control (Parsons, 1973). To increase the rate of testing using conventional techniques would have resulted in interference with the contractor's earthmoving activities and in the cost of control approaching the cost of the compaction process itself. It was suggested that further consideration could usefully be given to the method of specifying the compaction requirements for earthworks. It was considered, in conclusion, that the difficulties encountered in the air voids type of end-product specification would be largely overcome by the introduction of a method specification for the compaction of earthworks.

Method specifications

19.18 In a method specification for compaction the precise procedure to be used is laid down. Thus the type of compactor, mass, relevant dimensions, and any other factors influencing performance, together with the thickness of layer to be compacted and the number of passes of the machine, are all specified.

Department of Transport's method specification for the compaction of earthworks – principles

19.19 The investigations into the performance of compaction plant carried out at the Laboratory over many years have already been described in Chapters 4 to 14. The results of these investigations provided the basis for a new specification which was prepared following the conclusions regarding end-product specifications described above. The principle of the new method specification for earthwork compaction was that the contractor should have as free a choice of compaction plant as possible. Having selected the type of plant to be used, then the maximum depth of layer and the minimum number of passes to be used were given in the specification for that item of plant in combination with the soil type encountered. The compactive effort that was used to formulate the specification was that which would achieve, in well-graded soils, an average state of compaction of 10 per cent air voids at a moisture content at the bottom of the estimated range of natural moisture contents in the United Kingdom for each type of soil; this moisture content was considered to represent a condition in which the soil would be at its most difficult to compact.

19.20 The specification prepared on this basis was published as Clause 609 in the 1969 edition of the Specification for Road and Bridge Works (Ministry of Transport, 1969). To avoid inhibiting the development of new types of compactor or the improvement in efficiency of existing types, the specification permitted the use of equipment not listed, or a change in the method of use of existing equipment, provided the contractor demonstrated beforehand that such equipment or method produced at least a similar state of compaction to that of one of the approved methods already in the specification.

19.21 The Department of Transport's method specification in use at the present time is basically similar to the original first published in 1969, although some detailed refinements and additions have been introduced in the latest version (Department of Transport, 1986). The basic principles in the formulation of the specification are illustrated in Figure 19.7.

19.22 *Design condition (Figure 19.7).* For each of the well-graded test soils used in the full-scale investigations of compaction plant the design condition has been set at a moisture content at the bottom of the estimated range of natural

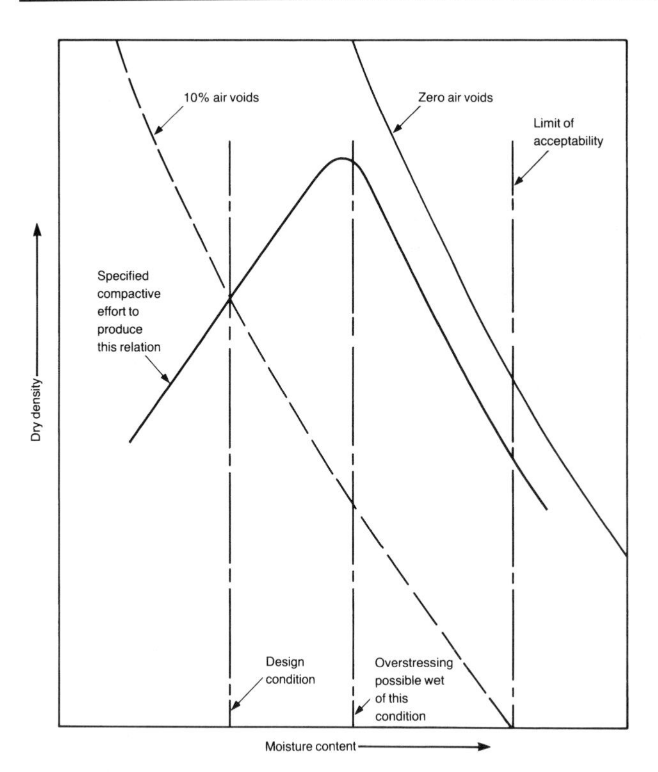

Figure 19.7 Principle of the method specification for the compaction of well-graded soils in earthworks for road construction in the United Kingdom

moisture contents. This range, however, has been estimated for the near-surface condition of in-situ materials and many clay soils are known to have moisture contents dry of this condition when excavated from near the bottom of deep cuttings. This design condition is now recognised as conforming to moisture condition values (MCV) (see Chapter 20) in the range 12.5 to 14.

19.23 *Specified compactive effort (Figure 19.7).* The compactive effort in the specification is selected to produce a relation between dry density and moisture content which passes through the intersection of the design moisture content and the 10 per cent air voids line as shown in the Figure. Thus for soils dry of the design condition an air content higher than 10 per cent will be produced using the specified compactive effort, whilst as the moisture content increases beyond the design condition the air content will decrease, with increasing dry density, until the minimum air content is reached. With increasing moisture content beyond the optimum moisture content/maximum dry density condition for the specified compactive effort, dry density will decrease at an approximately constant air content, as shown.

19.24 *Limit of acceptability (Figure 19.7).* This upper limit would not normally depend upon the compaction requirements but would be based on

Department of Transport specification – refer to paragraphs 19.28 to 19.31 for details.

Table 19.3 Methods 1 to 7 for the compaction of earthwork materials. Taken from Tables 6/4 of Part 2 of Department of Transport (1986) and Department of Transport (1988)

Type of compaction plant	Ref No	Category	Method 1		Method 2		Method 3		Method 4		Method 5		Method 6			Method 7		Ref No
			D	N‡	D	N‡	D	N‡	D	N	D	N	N for D=110 mm	N for D=150 mm	N for D=250 mm	N for D=150 mm	N for D=250 mm	
Smooth wheeled roller (or vibratory roller operating without vibration)		Mass per metre width of roll:–																
	1	over 2100 kg up to 2700 kg	125	8	125	10	125	10*	175	4	unsuitable		unsuitable	unsuitable	unsuitable	unsuitable	unsuitable	1
	2	over 2700 kg up to 5400 kg	125	6	125	8	125	8*	200	4	unsuitable		16	unsuitable	unsuitable	unsuitable	unsuitable	2
	3	over 5400 kg	150	4	150	8	unsuitable		300	4	unsuitable		8	16	unsuitable	12	unsuitable	3
Grid roller		Mass per metre width of roll:–																
	1	over 2700 kg up to 5400 kg	150	10	unsuitable		150	10	250	4	unsuitable		unsuitable	unsuitable	unsuitable	unsuitable	unsuitable	1
	2	over 5400 kg up to 8000 kg	150	8	125	12	unsuitable		325	4	unsuitable		20	unsuitable	unsuitable	16	unsuitable	2
	3	over 8000 kg	150	4	150	12	unsuitable		400	4	unsuitable		12	20	unsuitable	8	unsuitable	3
Tamping roller		Mass per metre width of roll:–																
	1	over 4000 kg	225	4	150	12	250	4	350	4	unsuitable		12	20	unsuitable	4	8	1

Pneumatic-tyred roller		Mass per wheel:–																
	1	over 1000 kg up to 1500 kg	125	6	unsuitable		150	10*	250	4	unsuitable		unsuitable	unsuitable	unsuitable	unsuitable	unsuitable	1
	2	over 1500 kg up to 2000 kg	150	5	unsuitable		unsuitable		300	4	unsuitable		unsuitable	unsuitable	unsuitable	12	unsuitable	2
	3	over 2000 kg up to 2500 kg	175	4	125	12	unsuitable		350	4	unsuitable		unsuitable	unsuitable	unsuitable	6	unsuitable	3
	4	over 2500 kg up to 4000 kg	225	4	125	10	unsuitable		400	4	unsuitable		unsuitable	unsuitable	unsuitable	5	unsuitable	4
	5	over 4000 kg up to 6000 kg	300	4	125	10	unsuitable		unsuitable		unsuitable		12	unsuitable	unsuitable	4	16	5
	6	over 6000 kg up to 8000 kg	350	4	150	8	unsuitable		unsuitable		unsuitable		12	unsuitable	unsuitable	unsuitable	8	6
	7	over 8000 kg up to 12000 kg	400	4	150	8	unsuitable		unsuitable		unsuitable		10	16	unsuitable	unsuitable	4	7
	8	over 12000 kg	450	4	175	6	unsuitable		unsuitable		unsuitable		8	12	unsuitable	unsuitable	4	8
Vibratory roller		Mass per metre width of a vibrating roll:–																
	1	over 270 kg up to 450 kg	unsuitable		75	16	150	16	unsuitable		unsuitable		unsuitable	unsuitable	unsuitable	unsuitable	unsuitable	1
	2	over 450 kg up to 700 kg	unsuitable		75	12	150	12	unsuitable		unsuitable		unsuitable	unsuitable	unsuitable	unsuitable	unsuitable	2
	3	over 700 kg up to 1300 kg	100	12	125	10	150	6	125	10	unsuitable		16	unsuitable	unsuitable	unsuitable	unsuitable	3
	4	over 1300 kg up to 1800 kg	125	8	150	8	200	10*	175	4	unsuitable		6	16	unsuitable	unsuitable	unsuitable	4
	5	over 1800 kg up to 2300 kg	150	4	150	4	225	12*	unsuitable		unsuitable		4	6	12	12	unsuitable	5
	6	over 2300 kg up to 2900 kg	175	4	175	4	250	10*	unsuitable		400	5	3	5	11	10	unsuitable	6
	7	over 2900 kg up to 3600 kg	200	4	200	4	275	8*	unsuitable		500	5	3	5	10	10	unsuitable	7
	8	over 3600 kg up to 4300 kg	225	4	225	4	300	8*	unsuitable		600	5	2	4	8	8	unsuitable	8
	9	over 4300 kg up to 5000 kg	250	4	250	4	300	6*	unsuitable		700	5	2	4	7	8	unsuitable	9
	10	over 5000 kg	275	4	275	4	300	4*	unsuitable		800	5	2	3	6	6	12	10
Vibrating plate compactor		Mass per m² of base plate:–																
	1	over 880 kg up to 1100 kg	unsuitable		unsuitable		75	7	unsuitable		unsuitable		unsuitable	unsuitable	unsuitable	unsuitable	unsuitable	1
	2	over 1100 kg up to 1200 kg	unsuitable		75	10	100	6	75	10	unsuitable		unsuitable	unsuitable	unsuitable	unsuitable	unsuitable	2
	3	over 1200 kg up to 1400 kg	unsuitable		75	6	150	6	150	8	unsuitable		unsuitable	unsuitable	unsuitable	unsuitable	unsuitable	3
	4	over 1400 kg up to 1800 kg	100	6	125	6	150	4	unsuitable		unsuitable		8	unsuitable	unsuitable	10	unsuitable	4
	5	over 1800 kg up to 2100 kg	150	6	150	5	200	4	unsuitable		unsuitable		5	8	unsuitable	8	unsuitable	5
	6	over 2100 kg	200	6	200	5	250	4	unsuitable		unsuitable		3	6	12	6	unsuitable	6
Vibro-tamper		Mass:–																
	1	over 50 kg up to 65 kg	100	3	100	3	150	3	125	3	unsuitable		4	8	unsuitable	unsuitable	unsuitable	1
	2	over 65 kg up to 75 kg	125	3	125	3	200	3	150	3	unsuitable		3	6	12	unsuitable	unsuitable	2
	3	over 75 kg up to 100 kg	150	3	150	3	225	3	175	3	unsuitable		2	4	10	unsuitable	unsuitable	3
	4	over 100 kg	225	3	200	3	225	3	250	3	unsuitable		2	4	10	8	unsuitable	4
Power rammer		Mass:–																
	1	100 kg up to 500 kg	150	4	150	6	unsuitable		200	4	unsuitable		5	8	unsuitable	8	unsuitable	1
	2	over 500 kg	275	8	275	12	unsuitable		400	4	unsuitable		5	8	14	6	10	2
Dropping-weight compactor		Mass of rammer over 500 kg height of drop:–																
	1	over 1 m up to 2 m	600	4	600	8	450	8	unsuitable		unsuitable		unsuitable	unsuitable	unsuitable	unsuitable	unsuitable	1
	2	over 2 m	600	2	600	4	unsuitable		unsuitable		unsuitable		unsuitable	unsuitable	unsuitable	unsuitable	unsuitable	2

D=Maximum depth of compacted layer
N=Minimum number of passes of compactors or blows of dropping-weight compactors
‡=N to be doubled for upper 600 mm of general fill
* = Self-propelled rollers unsuitable; rollers should be towed by track-laying tractors

the minimum engineering standards required in the fill material; the trafficability of earthmoving plant and of compaction plant may be a prime consideration in setting the limit of acceptability in some applications (Symons, 1979; Parsons and Toombs, 1988).

19.25 *Overstressing limit (Figure 19.7).* When the minimum air void condition is reached using the specified compactive effort, water in the soil will take up some of the stress exerted by the compactor and excess pore-water pressures will be generated (see Paragraph 2.6). With increasing moisture content the excess pore-water pressures will be generated at an earlier stage in the compaction process. The symptoms of this condition are the remoulding of the surface of the compacted layer and permanent deformation in the form of rutting under the wheels or rolls of the compactor. Elastic deformation of the surface of the layer is not indicative of excess pore-water pressures in the surface layer, but possibly in one or more of the underlying layers. Overstressing in this way is not serious with most soil types encountered, although a temporary reduction in compactive effort, by reducing the number of passes of the compactor or increasing the thickness of layer, may be considered. With some materials, however, chalk in particular, overstressing leads to further breakdown of the larger particles, giving rise to wet, plastic fines and a severe weakening of the freshly placed fill material (Ingoldby and Parsons, 1977); reduction of the compactive effort in such material is advisable, therefore, when the symptoms of overstressing are exhibited.

19.26 With uniformly graded soils and other materials for which the 10 per cent air voids criterion does not apply, the relation between dry density and moisture content of the specified compactive effort must still pass through the intersection of the design moisture content and the required alternative level of compaction (say a value of dry density or an air content value higher than 10 per cent).

19.27 The principles outlined above for the formulation of the compactive effort to be used in the method specification have been employed in Chapter 15 to produce the numerous Tables intended as an aid to the selection of plant for the compaction of soil. In Chapter 15, however, various design moisture contents, related to the optimum of the 2.5 kg rammer compaction test, were adopted.

Department of Transport method specification for the compaction of earthworks – details

19.28 The latest form of the method specification for the compaction of earthwork materials is given in Clause 612 of Part 2 of the Specification (Department of Transport, 1986). Earthwork materials are classified according to their properties and, if the method specification is to be applied, allocated one of the 'Methods' given in a Table. Table 19.3 (Pages 288–289) shows the relevant details from the specification. The applications of the seven methods are given in Table 19.4.

19.29 The compaction plant are categorised, for purposes of the method specification, in terms of type and mass (mass per metre width of roll in the case of machines with rolls, mass per wheel for pneumatic-tyred rollers, mass per square metre of contact area for vibrating-plate compactors, and total mass for vibro-tampers, power rammers and dropping-weight compactors). This largely follows the conclusions reached from the investigations of the performance of plant when assessing the predominant factor influencing performance (see Chapters 4 to 12).

19.30 Table 19.3 contains two important qualifications as indicated in the definitions given at the foot of the Table. Firstly, with Methods 1, 2 and 3, the number of passes, N, has to be doubled for the compaction of general fill within the top 600 mm of that fill, ie immediately below the capping or, if no capping is to be constructed, immediately below the sub-base. This is a near approximation to the earlier end-product specification in which the upper 600 mm of the embankment fill was compacted to a criterion of 5 per cent air voids instead of the 10 per cent air voids used for the bulk of the fill (see Paragraph 19.8). Relations between relative compaction and numbers of passes in the relevant Figures in Chapters 4 to 12 support this approach in that air voids largely reduce by about 5 per cent for each doubling of the number of passes of the various machines until the near zero air void condition is approached.

19.31 The second qualification in Table 19.3 is that with Method 3, intended for use with uniformly graded granular materials and silts (see Table 19.4), only compactors towed by track-laying tractors should be used where items are marked with an asterisk. This applies to smooth-wheeled rollers, pneumatic-tyred rollers, and heavier types of vibrating roller, where self-propelled machines often experience trafficability difficulties when working on uniformly graded soils.

Table 19.4 Applications of the seven different methods specified for the compaction of earthworks in Department of Transport (1986) (see Table 19.3)

Compaction method No	Class designated in Table 6/1 of Spec.	General description of material	Application
1	2A	Wet cohesive material and some chalk with saturation moisture content more than 20%	General fill
	7C	Selected wet cohesive fill	Fill to reinforced earth
2	1A	Well graded granular material and chalk with saturation moisture content less than 20%	General fill
	2B	Dry cohesive material	General fill
	2C	Stony cohesive material	General fill
	6I	Selected well-graded granular material	Fill to reinforced earth and anchored earth
	7D	Selected stony cohesive material	Fill to reinforced earth
3	1B	Uniformly graded granular material	General fill
	2D	Silty cohesive material	General fill
	6C	Selected uniformly graded granular material	Starter layer
	6D	Selected uniformly graded granular material	Starter layer below pulverised fuel ash
	6H	Selected uniformly graded granular material	Drainage layer to reinforced earth and anchored earth structures
	6J	Selected uniformly graded granular material	Fill to reinforced earth and anchored earth
4	3	Chalk	General fill
5	1C	Coarse granular material	General fill
	6B	Selected coarse granular material	Starter layer
6	6F	Selected granular material	Capping
	9A	Cement-stabilised well-graded granular material	Capping
7	9B	Cement-stabilised silty cohesive material	Capping
	9D	Lime-stabilised cohesive material	Capping

Department of Transport method specification for the compaction of earthworks – additional requirements

19.32 Definitions and notes relating to the Tables in the method specification (Table 19.3) are also included in Clause 612 of the Specification (Department of Transport, 1986). These generally apply to individual types of compaction plant as follows:–

19.33 *Smooth-wheeled rollers.* The mass per metre width of roll is to be determined by dividing the total mass on the roll by the total roll width. Where the machine has more than one axle, as with tandem or three-wheeled self-propelled machines, the relevant mass per metre width is the highest value so determined.

19.34 *Grid rollers.* These are defined as machines with compacting rolls constructed of heavy steel mesh of square pattern (see Plates 6.7 to 6.9).

19.35 *Tamping rollers.* These are defined as machines with rolls from which 'feet' project and where the projected end area of each 'foot' exceeds 0.01 m^2 and the sum of the areas of the feet exceeds 15 per cent of the area of the cylinder swept by the ends of the feet (see Plates 6.5 and 6.6). The value of 15 per cent is equivalent to a coverage of 0.15 as given in Table 6.1. The definition of tamping rollers given above ensures that sheepsfoot rollers (Chapter 6) are excluded, their performance being incompatible with the soil conditions encountered in the United Kingdom (see Paragraph 6.44). The numbers of passes for tamping rollers given in Table 19.3 apply to machines that have two rolls in tandem; if only one tamping roll traverses each point on the surface of the layer on any one pass of a machine, the number of passes should be twice the number given in the Tables.

19.36 *Pneumatic-tyred rollers.* The mass per wheel is calculated as the average for all the wheels, ie the total mass of the roller divided by the number of wheels. The effective width of a pneumatic-tyred roller, ie the width of soil deemed to have been rolled in one pass, is the sum of the widths of the individual wheel tracks together with the sum of the spacings between the wheel tracks, provided that each spacing does not exceed 230 mm. Where the spacings exceed 230 mm the effective width is the sum of the widths of the individual wheel tracks only. The limitations on the widths of wheel spacings are based on the configurations of the pneumatic-tyred rollers tested at the Laboratory (see Chapter 5).

19.37 *Vibratory rollers.* These machines are defined as self-propelled or towed smooth-wheeled rollers having means of applying mechanical vibration to one or more rolls. Vibrating rollers employed for Method 5 (compaction of coarse granular material) must be of the single-roll type, ie towed machines such as those shown in Plates 7.11, 7.15 and 7.16 or self-propelled machines of a similar configuration to that shown in Plate 7.14. This requirement in Method 5 ensures that only machines with the higher amplitudes of vibration are employed, a necessary requirement for good compaction of the thicker layers employed with the coarse granular material.

19.38 The speed of operation of vibrating rollers is restricted such that the lowest gear should be used with self-propelled machines with mechanical transmission, and a maximum speed of travel of 2.5 km/h should be used with towed machines and self-propelled machines with hydrostatic transmission. If higher gears or speeds are used an increased number of passes has to be provided in proportion to the increase in speed of travel. A discussion of the effect of variations in speed of travel on the performance of vibrating rollers is given in Paragraph 7.46.

19.39 With double vibrating rollers, where the mechanical vibration is applied to two rolls in tandem, the number of passes shall be half the number given in Table 19.3. Where, in these circumstances, the mass per metre width of one roll differs from that of the other, the machine is categorised according to the smallest value of mass per metre width. Alternatively, if the two values of mass per metre width are widely disparate, it may be advantageous to treat the machine as if it has only a single vibrating roll with a mass per metre width equal to that of the roll with the highest value and to use the actual number of passes given in the Table.

19.40 Where vibrating rollers are used with the vibrating mechanism inoperative they are to be classified as smooth-wheeled rollers. To class as vibratory rollers they should be used with their vibrating mechanisms operating only at the frequency of vibration recommended by the manufacturers. Where more than one amplitude setting is available and/or a range of frequencies is recommended, the machine should be operated at the maximum amplitude setting and at the maximum frequency recommended for that setting. A discussion on the effects of variations in frequency of vibrating rollers is given in Paragraphs 7.55 to 7.56.

19.41 A further requirement for vibrating rollers in the Department of Transport's method specification for earthwork compaction is that the machines should be equipped or provided with devices indicating the frequency at which the vibrating mechanism is operating and the speed of travel, thus enabling the above requirements to be continuously monitored.

19.42 *Vibrating-plate compactors.* These are defined as machines having base-plates to which are attached sources of vibration, each consisting of one or two eccentrically weighted shafts. The mass per square metre of the base-plate of a vibrating-plate compactor is calculated by dividing the total mass of the machine in its working condition by its area in contact with compacted material. The machines should be operated at the frequency of vibration recommended by the manufacturers and the numbers of passes given in Table 19.3 are based on a speed of travel not exceeding 1 km/h; if a machine has a speed of travel in excess of 1 km/h the number of passes should be increased in proportion to the increase in speed of travel.

19.43 *Vibro-tampers.* These are defined as machines in which an engine-driven reciprocating mechanism acts on a spring system through which oscillations are set-up in a base plate.

19.44 *Power rammers.* Defined as machines which are actuated by explosions in an internal combustion chamber, each explosion being controlled manually by the operator.

19.45 *Dropping-weight compactors.* These are defined as machines in which a dead-weight is dropped from a controlled height using a hoist mechanism; this category includes self-propelled machines with mechanical traversing mechanisms, as shown in Plate 12.5, capable of compacting soil in trenches and close to structures.

19.46 *General.* Where combinations of different types or categories of plant are used in Methods 1 to 5, the depth of layer should be that for the type of plant requiring the least depth of layer and the number of passes should be that for the type of plant requiring the greatest number of passes.

19.47 Where different types of material which require different compaction methods are being placed in such a way that it is not practicable to define the separate areas in which each type occurs, then the Method involving the greater compactive effort should be used. The Specification does not explain how to compare compactive efforts where both the depths of layer and numbers of passes are different; in the author's opinion the highest value of the ratio number of passes/thickness of layer would provide a good indication of the highest compactive effort. Where one of the Methods involved in these circumstances indicates that a given type of plant is unsuitable, then this unsuitability must also extend to the other type or types of material being compacted in the area.

Department of Transport method specification for the compaction of earthworks – control procedures

19.48 In earthwork construction operations involving the placement and compaction of general fill, the counting of the number of passes of a compactor or a team of compactors may be difficult and on occasion impossible. In situations where the number of passes cannot be accurately monitored it is suggested that the potential output of the compaction team should be compared with the rate of placement of the earthwork material. The potential output of the compaction team should equal or exceed the rate of placement of fill for compliance with the specification; the works inspectors should then ensure that a uniform coverage of the area is achieved and that the maximum thickness of compacted layer is not exceeded. A flow chart detailing the control procedure is given in Figure 19.8.

19.49 *Identification of class of soil (Figure 19.8).* Within the Department of Transport specification this is not intended to be an onerous task; although in the 1986 edition (Department of Transport, 1986) many classes of soil appear in the classification Table (Table 6/1 of that specification), only a limited number of different methods apply to them (see Table 19.4). Where results of the appropriate tests show that the soil lies across a boundary between two different methods of compaction, the method which requires the higher compactive effort is assumed to apply (see Paragraph 19.47).

19.50 *Class of compactor (Figure 19.8).* Each machine in use has to be identified and classified to determine the maximum thickness of layer and minimum number of passes to be applied. Its general type (smooth-wheeled roller, vibratory roller, vibrating-plate compactor, etc), its total mass and distribution of mass, for example between rolls, and, in most instances, either its width of roll, number of wheels, or area of plate, must be determined. Direct weighing and measurement are preferable, although the machine specifications may be resorted to. Careful attention needs to be paid to the detailed requirements regarding the correct parameters to use, as described in Paragraphs 19.33 to 19.46.

19.51 *Estimation of potential output (Figure 19.8).* When the compactor is operating normally its speed of travel should be noted (vibrating rollers should be equipped with devices indicating the speed of travel, but for other types of compactor timing over a measured distance is a practical method). If the measured speed exceeds any maximum stipulated, or if a higher gear than that stipulated is used (eg with vibrating rollers), the necessary action should be taken, ie the speed should be reduced or the number of passes should be increased (see Paragraphs 19.38 and 19.42). Note that with many compactors the Specification does not require a speed restriction.

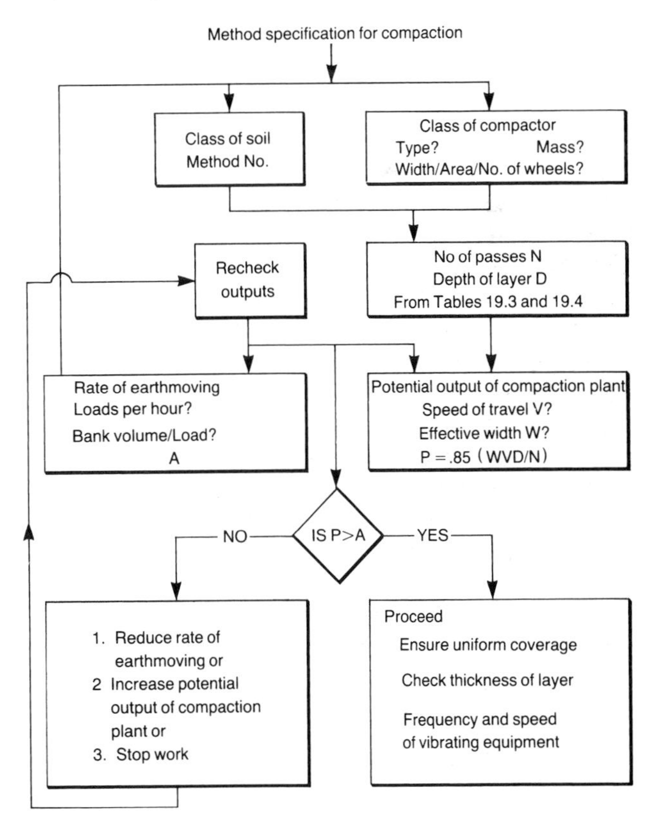

Figure 19.8 Flow chart for the control of the method specification for the compaction of earthworks according to Department of Transport (1986)

19.52 The potential output of each compactor can be determined from the following:–

$$P = \frac{0.85\,W\,V\,D}{N}$$

where P = output of compacted soil (m^3/h)
W = effective width of compactor (m)
V = speed of travel (km/h)
D = maximum depth of compacted layer (mm) required in the specification
and N = minimum number of passes required in the specification

The factor 0.85 is applied to allow for overlap between adjacent compacted strips, turn rounds, minor stoppages, etc.

19.53 *Control of method specification (Figure 19.8).* The rate of input of earthwork material to a given area of fill should be assessed from the number of loads per hour and the approximate capacity (compacted volume) of the earthmoving machines. The rate of input in terms of compacted volume in m^3/h must not exceed the potential output of the compactor or the compaction team, as calculated above.

19.54 The thickness of layer should be checked regularly, using a probe if appropriate in the loose material or by small excavations in the compacted material. The relation between loose and compacted thickness will quickly become apparent and the probing of the loose layer can take precedence. The uniform coverage of the area to be compacted and the continuous operation of the compaction plant should be ensured by general supervision of the fill area.

19.55 With vibrating rollers and vibrating-plate compactors it is also necessary to check the frequency of vibration and ensure that it is as stated in the manufacturer's specifications. The Specification requires vibrating rollers to be fitted with a device automatically indicating the frequency (see Paragraph 19.41), but where such devices are not installed, as with vibrating-plate compactors, the use of a frequency measuring device of the vibrating wand type, held against an appropriate part of the machine by a person walking alongside, is usually effective.

19.56 *Over-compaction.* Over-compaction results when the moisture content of the soil is in excess of the optimum moisture content for the compactive effort being applied (see Paragraph 19.25, over-stressing limit). Symptoms generally comprise remoulding of the surface of the compacted soil or severe permanent deformation upon passage of the compactor. Note that a 'rubber mattress' effect is not necessarily indicative of over-compaction of the topmost layer.

19.57 Before concluding that over-compaction is taking place it is advisable to excavate a few trial holes to the bottom of the layer to ensure that full compaction (near saturation conditions) occurs throughout the total depth of the layer. If the bottom of the layer is not in a saturated condition then over-compaction has not taken place and the compactive effort should be maintained at its original level. If a decision is made to reduce the compactive effort (see Paragraph 19.25) it should be deemed to be a temporary measure and reversions to the specified compactive effort should be made from day-to-day to ensure that soil conditions still require the modified procedure.

19.58 *Under-compaction.* This occurs when the soil is dry of the design condition (Figure 19.7) so that a high air content is produced in the compacted material. As the material is, therefore, well dry of the optimum moisture content for the fairly high compactive effort being applied, the under-compacted layer is strong and does not deform under traffic. The surface of the layer will appear well-compacted and the condition is difficult to identify. Large voids can occur near the bottom of the layer and these can only be detected by digging small trial holes and inspecting the condition of the material in the lower part of the layer. Under-compaction is a potentially serious condition if the material so placed is subjected to the ingress of water, eg if placed near the side-slopes of embankments or near formation level; increases in moisture content will then result in severe weakening of the soil and settlement. In general, material placed at a moisture condition value (MCV) (see Chapter 20) in excess of 14 may be susceptible to under-compaction. Compactive effort may be increased to remedy the situation and usually better results will be achieved by reducing the thickness of the layer rather than by increasing the number of passes.

Method specification for the compaction of earthworks in France

19.59 Earthwork compaction in France is carried out mainly by method specification, with a far more complex system than that specified in the

United Kingdom (Ministere de l'Equipement, 1976). Soil classification is based on soil type, with the addition, for most soil types, of moisture condition; this is in terms of 'wet', 'average' or 'dry', depending on the relation between the actual moisture content and the optimum moisture content of the 2.5 kg rammer compaction test. Weather conditions are also considered in determining the acceptability of the soil and the compactive effort to be applied; the weather conditions in the classification are (a) average or heavy rain, (b) light rain, (c) neither rain nor evaporation, and (d) significant rates of evaporation (average or high temperatures, dry, windy conditions).

19.60 The soil classification, combined with the existing moisture condition and the pertaining weather conditions, determine the compactive effort to be applied; this is in terms of 'intense', 'average' or 'light' compaction in combination with 'thin' or 'average' layer thicknesses. For each combination of soil type, compactive effort and category of compaction plant a maximum thickness of layer, 'e', is specified together with a value for the ratio between volume of soil compacted, 'Q', and surface area, 'S', covered by the machine to compact this volume. The ratio Q/S has been quoted as representing the thickness compacted in one pass of the machine (Chaigne et al, 1980). In other terms:–

$$\frac{Q}{S} = \frac{e}{N}$$

where N is the number of passes required and Q, S and e are as defined above. It can be seen that if less than the maximum thickness specified is used, the required value of N is reduced in proportion.

19.61 The specifications are, therefore, essentially in the form of tables giving the values of 'e' and 'Q/S' for the different classes of soil and compactor that are usually encountered (Ministere de l'Equipement, 1976). Compactors included in the specification for mass earthworks are pneumatic-tyred rollers (classified by load per wheel), vibrating rollers (classified by mass per unit width of roll with sub-classes depending on the type of machine – towed, self-propelled with a single roll, tandem roller with one roll vibrating, double vibrating roller), and non-vibrating tamping rollers (classified by average load per unit width of roll).

19.62 Compliance with the specification used in France is normally monitored on a daily basis with a measurement of the volume compacted (Q) and an estimate of the surface covered (S) from the dimensions of the machine, the speed of travel and the total operating time (Leflaive et al, 1980). In many applications of the French method specification for the compaction of earthworks the compaction machines are equipped with tachographs which automatically record on a circular chart the speed of travel, frequency of vibration (if appropriate) and working times (Machet, 1980). These daily charts allow simple calculations of the surface areas covered (S) and provide an essential element in permanently recording compliance or non-compliance with the specification.

Considerations in the selection of type of compaction specification in the United Kingdom

19.63 End-product specifications require the measurement of the in-situ state of compaction, and must lead to some interference with the earthmoving operations, even on large construction sites. With the use of modern nuclear gauges for measuring the state of compaction, the rate of testing can be sufficiently high to ensure an accurate assessment, given proper calibration, of the level of compaction achieved (see Paragraph 18.74). However, nuclear equipment requires licensing and the monitoring of radiation exposure of the operator (see Paragraph 18.75), so that the use of portable nuclear gauges is likely to be restricted to large construction sites or to use by a limited number of testing consultants.

19.64 In confined areas and on small construction sites the measurement of the in-situ state of compaction is likely to create an even greater degree of interference with the placement and compaction of the fill in successive layers. In these circumstances the method type of specification appears to be more attractive. In small areas the counting of the number of passes of the plant and the control of the thickness of the compacted layer would become easier and, providing some form of moisture control is applied to the incoming material, eg by moisture condition test (Chapter 20), adequate states of compaction should be assured.

19.65 If the site is so small, eg a trench opening in a road, that continuous supervision is not possible, neither the end-product nor method specifications can be rigorously applied in their existing form. However, the method type of specification is likely, in the author's opinion, to have the best chance of success in this area in the future. Method specifications on small sites would be aided considerably if all compactors employed in such operations were fitted with meters indicating the accumulated time for which they have been operating (this would be somewhat similar to the use of tachographs in France, see Paragraph 19.62). A simple comparison of accumulated machine hours and volume of compacted backfill would provide a good indication of whether or not compliance with a method specification has been achieved.

19.66 With the introduction of the moisture condition test (see Chapter 20) for the control of the moisture acceptability of earthwork materials, further refinements to the method specification in the future could introduce variations in compactive effort for different ranges of moisture condition value (MCV). This introduction of variable compactive effort would largely overcome the problems of over-compaction and under-compaction discussed in Paragraphs 19.56 to 19.58. The basis for a specification with variations in compactive effort depending on moisture content, such as is used in France (see Paragraphs 19.59 to 19.60), is contained in the Tables in Chapter 15.

References

American Society for Testing and Materials (1973). Evaluation of relative density and its role in geotechnical projects involving cohesionless soils. *ASTM Special Technical Publication* STP No 523. ASTM, Philadelphia.

American Society of Testing and Materials (Annual). Annual book of ASTM standards. ASTM, Philadelphia.

British Standards Institution (1990). British Standard methods of test for soils for civil engineering purposes : BS 1377 : Part 4 Compaction related tests. BSI, London.

Chaigne, P, Leflaive, E and Schaeffner, M (1980). A new concept for compaction specifications for road embankments (in French). *International Conference on Compaction.* Editions Anciens ENPC, Paris, Vol II, 493–9.

Department of Transport (1986). Specifications for highway works. 6th edition. HM Stationery Office, London.

Department of Transport (1988). Specification for highway works. Appendix L. HM Stationery Office, London.

Hilf, J W (1957). A rapid method of construction control for embankments of cohesive soil. *Conference on Soils for Engineering Purposes, ASTM Special Technical Publication* STP No 232. American Society for Testing and Materials, Philadelphia, 123–42.

Ingoldby, H C and Parsons, A W (1977). The classification of chalk for use as a fill material. *Department of the Environment, Department of Transport, TRRL Laboratory Report* 806. Transport and Road Research Laboratory, Crowthorne.

Leflaive, E, Schaeffner, M, Leny, G and Puig, J (1980). Compaction control of earthworks (in French). *International Conference on Compaction.* Editions Anciens ENPC, Paris, Vol II, 571–6.

Lewis, W A and Parsons, A W (1963). An analysis of the states of compaction measured during the construction of embankments on six major road schemes. *D.S.I.R., Road Research Laboratory Note* No LN/399/WAL.AWP (Unpublished).

Machet, J M (1980). Compactor-mounted control devices (in French). *International Conference on Compaction.* Editions Anciens ENPC, Paris, Vol II, 577–81.

Ministere de l'Equipement (1976). Recommendations for the construction of road earthworks (in French). Setra, Laboratoire Central des Ponts et Chaussees, Paris.

Ministry of Transport (1963). Specification for road and bridge works. 3rd edition. HM Stationery Office, London.

Ministry of Transport (1969). Specification for road and bridge works. 4th edition. HM Stationery Office, London.

Parsons, A W (1973). Research in earthwork specification and construction. *Ground Engineering,* November.

Parsons, A W and Toombs, A F (1988). Pilot-scale studies of the trafficability of soil by earthmoving vehicles. *Department of Transport, TRRL Research Report* 130. Transport and Road Research Laboratory, Crowthorne.

Symons, I F (1979). General report: Performance of clay fills. *Clay Fills.* Institution of Civil Engineers, London, 297–302.

CHAPTER 20 THE MOISTURE CONDITION TEST FOR THE CONTROL OF ACCEPTABILITY OF FILL MATERIAL

Introduction

20.1 Whenever embankments are constructed, whether for highways, dams or for other purposes, certain properties related to shear strength, permeability or compressibility, or any combination of these factors, are assumed for the fill material so that a proper design for the works can be prepared. Two major factors affect the relevant properties; these are, firstly, the condition of the earthwork material itself and, secondly, the quality of its placement in the embankment, principally the state of compaction achieved. Moisture content is a principal factor influencing the condition of the material; its effects, additionally, on the compaction process have been discussed in Paragraphs 2.5 to 2.9 and are illustrated in the various Chapters dealing with the performance of compaction equipment. For the control of the quality of fill material, therefore, limits have to be set within which the engineering requirements for a successful design can be achieved. Having set the limits, an effective method of test is necessary to ensure that they are not exceeded.

20.2 With the relative compaction form of specification (see Paragraph 19.3) the achievement of a specified level of compaction can only occur within a range of moisture contents related to the optimum moisture content of the particular laboratory compaction test used (see Paragraphs 2.23 to 2.27); the relative compaction specification, therefore, has an inbuilt control of quality of the material used. With many forms of compaction specification, however, the separate control of moisture content is usually necessary.

20.3 Methods of moisture control have included measurements of shear strength, either using in-situ methods (shear vane) or tests on remoulded samples (unconfined compression tests or triaxial tests) (Kennard et al, 1979). In many instances the moisture content is measured and compared with other soil properties such as the plastic limit or the optimum moisture content in a laboratory compaction test.

20.4 In road construction in the United Kingdom, prior to the issue of the 6th edition of the Department of Transport's specification (Department of Transport, 1986), it was normal to specify in some way an upper limit for the moisture content of earthwork material. The list of unsuitable materials included "materials having a moisture content greater than the maximum permitted for such materials in the Contract, unless otherwise permitted by the Engineer" (Ministry of Transport, 1969). Notes to the Fourth Edition of the Specification, published in 1969, advised that, for cohesive soils, the upper limit of moisture content should be related to the plastic limit and a value of plastic limit multiplied by 1.2 was suggested as satisfactory. For well-graded granular soils and uniformly graded materials the advice given was to relate the upper limit of moisture content to the optimum moisture content as determined in the 2.5 kg rammer compaction test (British Standards Institution, 1990b), a value of 0.5 to 1.5 per cent above the optimum being recommended.

20.5 An early investigation of the variability of moisture content and plastic limit of cohesive soils in earthwork construction (Parsons, 1965) showed that very large variations in plastic limit and moisture content can occur over very small areas of earthwork construction. The determination of moisture content alone, therefore, would not be sufficient to achieve control of moisture content, even in localised areas, except on sites where the soil is very uniform. For cohesive soils it was concluded that the determination of plastic limit would also have to be made. An example of the results obtained is shown in Figure 20.1. The investigation also showed that, on average, about 12 samples would have to be taken and the moisture content and plastic limit of each sample determined in order to assess, with a reasonable accuracy, the acceptability of the soil.

20.6 The plastic limit test is carried out on material passing the 425 μm sieve (British Standards Institution, 1990a), whereas the moisture content is measured on a representative sample of the soil as a whole. A factor affecting the relation between moisture content and plastic limit on any site, therefore, is the stone content of the soil. The results shown in Figure 20.1 were obtained with a stone-free sandy clay; results given in Figure 20.2 were obtained on a range of plastic soils including a hoggin (gravel-sand-clay) and a stony clay; the Figure shows clearly the gross distortion of the moisture content-plastic limit relation caused by the stone content of some soils. Parsons (1965) concluded that any correction allowing for the effect of the coarse material on the moisture content would be liable to serious error and suggested that the acceptability of materials containing more than 20 per cent of gravel-size particles (larger than 2 mm) should not be controlled by comparison of the moisture content and plastic limit.

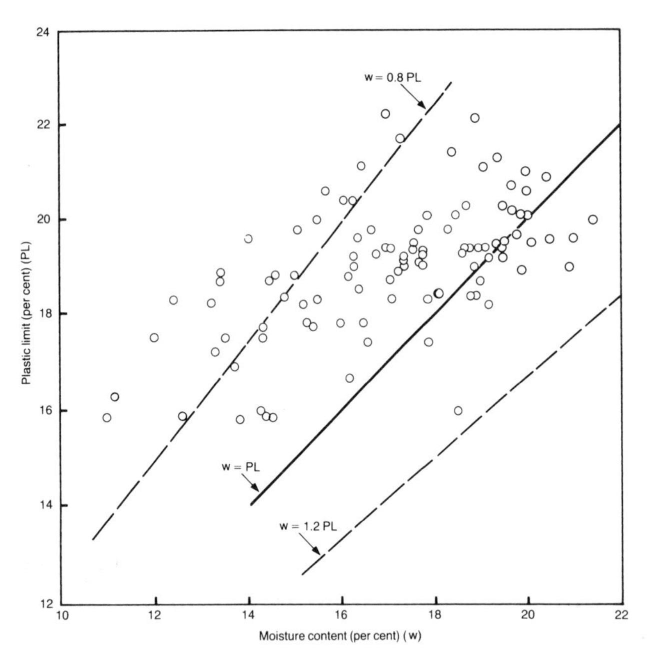

Figure 20.1 Results of moisture content and plastic limit measurements in one area of a motorway construction site. Length of sample area 500 m. Soil type : sandy clay

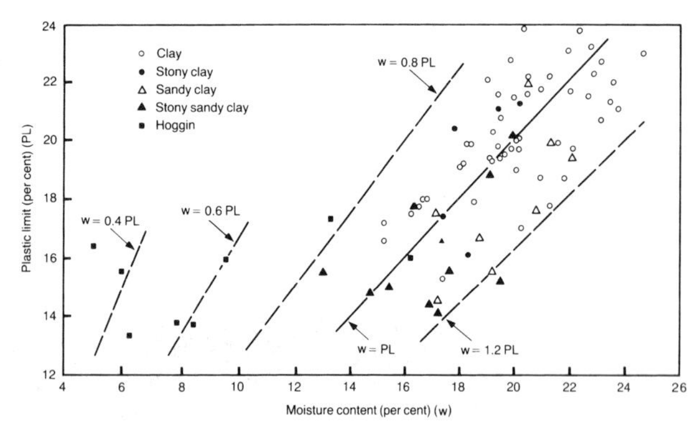

Figure 20.2 Results of moisture content and plastic limit measurements in one area of a motorway construction site with variable cohesive soils. Length of sample area 500 m.

20.7 No similar type of research was carried out for granular soils, with which the moisture content was compared with the optimum of the 2.5 kg rammer compaction test, but the same high degree of variability can be expected to occur. The adequate control of the acceptability of earthwork material was concluded to be a laborious process which included delays, possibly up to 24 hours, while awaiting the determination of the moisture contents associated with the tests. More recent research (Sherwood, 1971) also showed that both plastic limit tests and laboratory compaction tests were prone to serious errors, as exhibited by large variations in test results carried out on soils with uniform properties by various testing laboratories. In the light of these problems, control of the moisture acceptability of earthwork material was frequently considered to be impractical.

20.8 The need for a new type of test was recognised, therefore, for use in the control of soil acceptability. The test had to be fairly rapid and had to provide an immediate result. The ability to test a wide range of material was considered essential. Research was carried out into means of satisfying this need, and as a result the moisture condition test was developed (Parsons, 1976).

Principles of the moisture condition test

20.9 The test is basically a compaction test in which the compactive effort is increased incrementally until a state of full compaction is reached. The compactive effort, in terms of the number of blows of a rammer, necessary to compact a sample of soil determines the moisture condition value (MCV) of that sample.

20.10 The effect of variations in moisture content on the density produced in a sample of soil has been discussed already in Paragraphs 2.5 to 2.9. Idealised relations between bulk density and moisture content are shown in Figure 20.3. The relations for different compactive efforts converge at moisture contents above the highest optimum, ie at moisture contents beyond the optimum of the lowest compactive effort. Thus, at moisture content A′, compactive effort A would just achieve full compaction, and no further increase in density can occur with the application of any further compaction, such as compactive efforts B or C (Figure 20.3). At moisture content B′, however, an increase in density would occur with

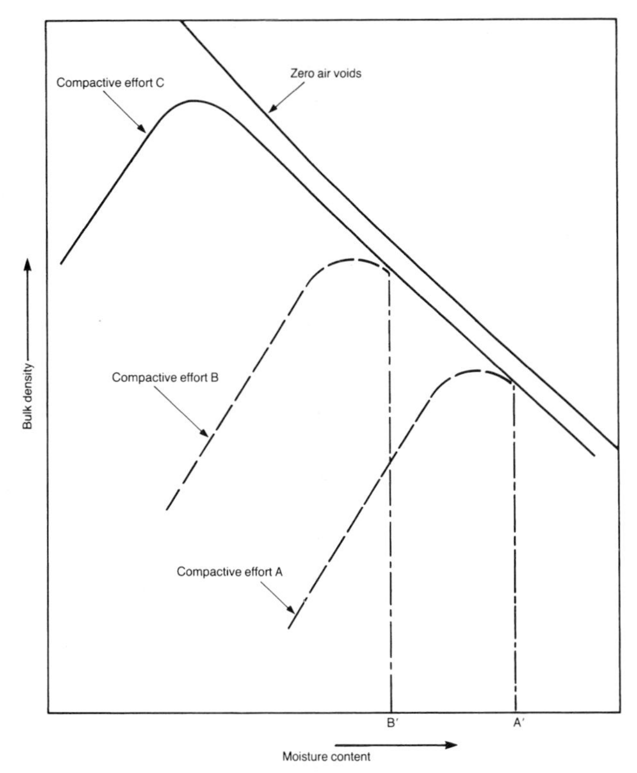

Figure 20.3 Idealised relations between bulk density and moisture content

the application of further compaction over that of compactive effort A, but no further increase in density would occur with any additional compactive effort beyond that of compactive effort B. Thus, at any given value of moisture content for a given sample of soil, there is a unique compactive effort at which the density of the sample ceases to increase. The higher the moisture content, the lower the compactive effort (eg the number of blows of a rammer) beyond which no further increase in density occurs.

Apparatus for the moisture condition test

20.11 The apparatus developed for the test (Plate 20.1) has been designed so that changes in the length of rammer protruding from the mould can be monitored, thus avoiding the need to monitor changes in density. The apparatus has a 100 mm diameter mould with a detachable base, a free-falling rammer with a mass of 7 kg and a diameter of 97 mm, and an automatic release mechanism, adjustable to maintain a constant height of drop on to the surface of the soil specimen. A 1.5 kg sample of soil passing a 20 mm sieve is normally used, with a drop height for the rammer of 250 mm. The length of rammer protruding from the mould is normally measured using a depth gauge readable to 0.1 mm. The adjustment of the automatic release mechanism to maintain a constant height of drop is aided by two markers which have to be aligned; one marker is attached to the rammer and the other to a rod connected to the cross-member containing the automatic release mechanism. Adherence of soil to the rammer and extrusion of soil between the rammer and the inner surface of the mould are avoided by placing a lightweight rigid fibre disc, 99 mm in diameter, on top of the soil in the mould. To eliminate, as far as possible, the effects of using the apparatus on surfaces of varying stiffness, eg on soft soil on an exposed site or on the concrete floor of a laboratory, the base of the apparatus has been made as heavy as possible while maintaining overall portability. Thus, the total weight of the apparatus is about 50 kg, of which at least 31 kg is contributed by the base.

Test procedure to determine the moisture condition value (MCV) of a sample of soil

20.12 The sample is initially passed through a 20 mm sieve, breaking up aggregations of soil

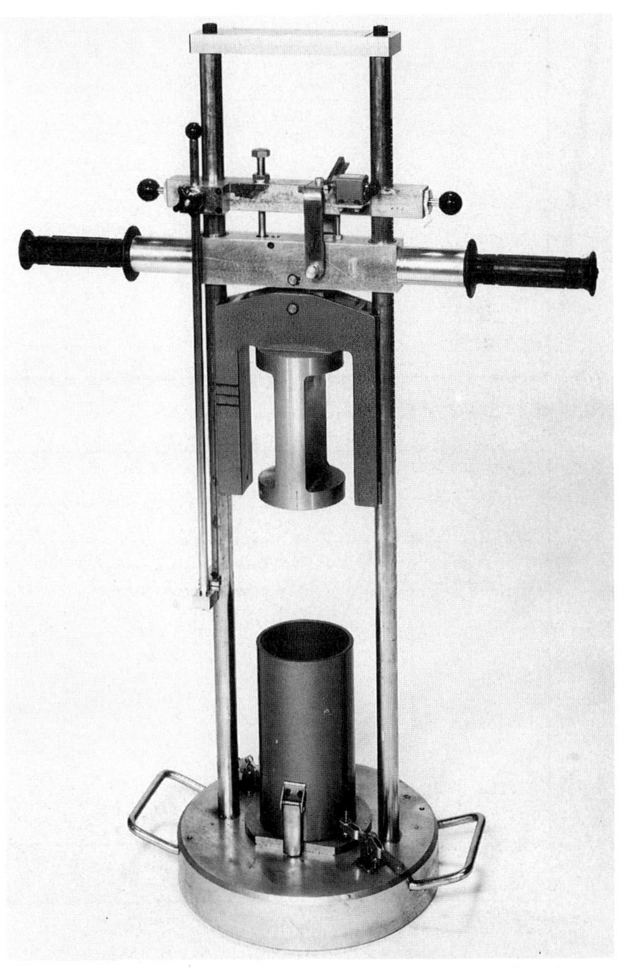

Plate 20.1 A moisture condition apparatus for the determination of the moisture condition value (MCV)

as necessary and removing only individual particles that are retained on the sieve. A 1.5 kg sample of soil which has been passed through the sieve is placed loosely in the mould and the lightweight disc placed on top of the soil. Successive blows of the rammer are applied, maintaining the height of drop at 250 mm by adjusting the position of the automatic release mechanism as necessary. The length of rammer protruding from the mould is measured by means of a depth gauge at various stages of the compaction process and noted on a data form such as that shown in Figure 20.4. The protrusion of the rammer at any given number of blows is compared with the protrusion for four times as many blows and the difference in penetration of the rammer (a measure of the change in density) is determined. The 'change in penetration' is plotted against the lower cumulative number of blows in each case (Figure 20.4), the cumulative number of blows being on a logarithmic scale. The descending part of the relation normally plots as a straight line.

Soil: Heavy clay Moisture content: 31.5 per cent

Total number of blows of rammer 'n'	Protrusion of rammer above mould (mm)	Change in penetration between 'n' and '4n' blows of rammer (mm)
1	106.3	22.5
2	96.4	29.2
3	89.0	31.8
4	83.8	31.7
6	74.7	23.1
8	67.2	15.6
12	57.2	5.6
16	52.1	0.5
24	51.6	0
32	51.6	
48	51.6	
64	51.6	

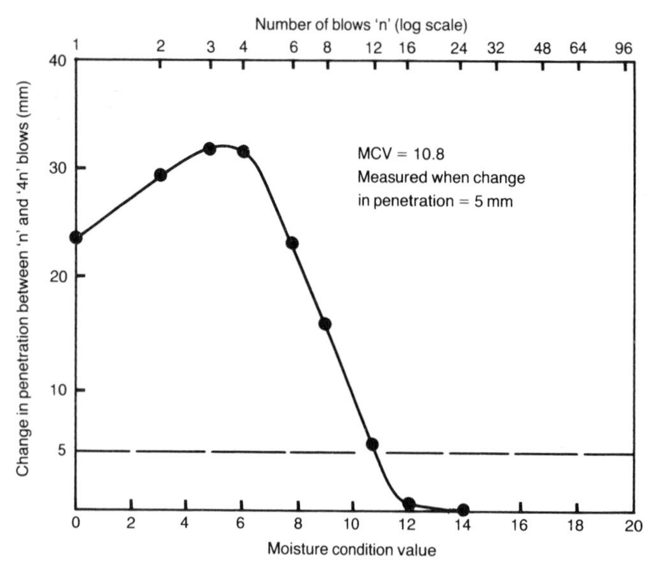

Figure 20.4 Results of a determination of MCV of a heavy clay soil

20.13 Clearly, the state of compaction ceases to increase at the point when the change in penetration just reaches zero. However, some curvature occurs in the relation as the change in penetration approaches zero, and to avoid predicting the point of zero change in penetration a value of 5 mm has been arbitrarily selected as indicating the point beyond which no significant change in density occurs. The MCV is defined as 10 times the logarithm (to the base 10) of the number of blows corresponding to a change in penetration of 5 mm on the plotted relation. The MCV can be read to the nearest 0.1 directly from the bottom axis if charts are prepared such as that shown in Figure 20.4.

Calibration and setting the limits of acceptability

20.14 Relations between MCV and moisture content can be established by determining MCVs at various moisture contents. Figure 20.5 gives an example for a heavy clay soil. To show the construction of the calibration the change in penetration-number of blows relations, plotted for the determination of the individual MCVs, are also included in the Figure. If the soil is in a fully remoulded condition the relation between MCV and moisture content is a straight line over a substantial range of moisture contents. With over-consolidated clays, however, a concave-upwards curve can be produced on a similar plot. It is likely that the curvature of the relation is caused by the different degrees by which the over-consolidated clay structure is destroyed in the preparation of the samples at different moisture contents; in these circumstances the concave-upwards curve may be considered more likely to represent the effects on MCV of moisture content variations during site operations.

20.15 If, for the same bulk sample of soil, the relations between engineering parameters of interest and moisture content are determined, then the MCV can be related to those engineering parameters through the common factor of moisture content. Undrained shear strength and California bearing ratio (CBR) are often used for this purpose, together with the results of laboratory compaction tests, to provide a complete design package, as shown in Figure 20.6 (Greenwood, 1987). If, for instance, a minimum undrained shear strength of 50 kN/m^2 is assumed in the design, the results in Figure 20.6 show that a minimum MCV of 7.5 would be an appropriate limit to set for control during construction. If it is also assumed that the

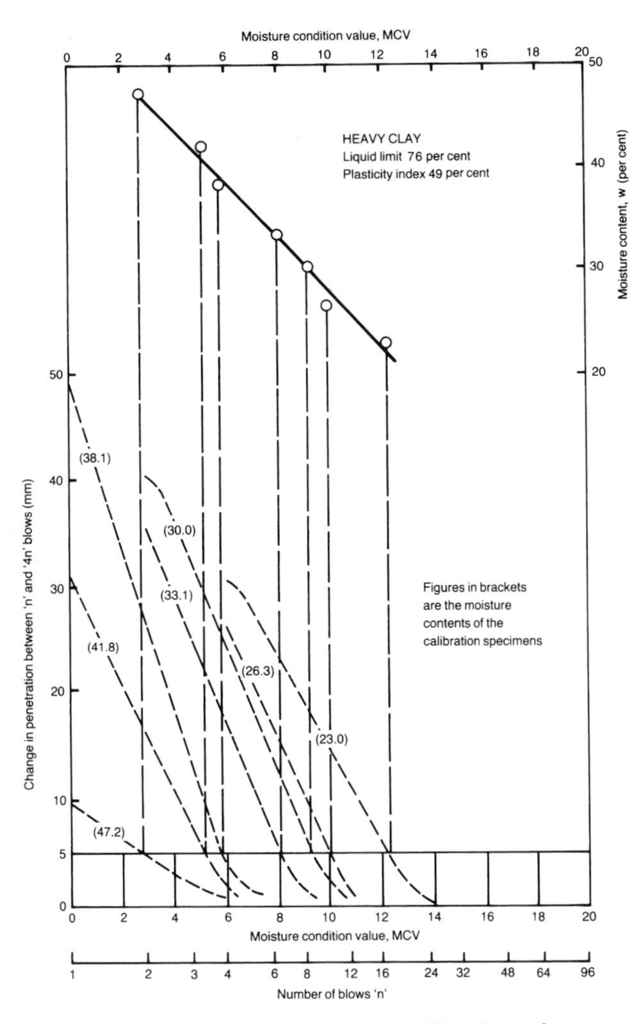

Figure 20.5 Moisture condition calibration of a heavy clay soil

minimum moisture content to ensure adequate compaction is the optimum of the 4.5 kg rammer compaction test, a maximum MCV of about 13 would be indicated by the results in the Figure.

20.16 An additional factor to be taken into account in setting limits of acceptability may be the trafficability of the earthmoving plant. In the construction of fairly shallow embankments, where relatively low values of shear strength may be sufficient to avoid risks of shear failure, the most critical factor may be the ability of construction plant to operate effectively (Symons, 1979). Investigations have been made, both on construction sites and in the pilot-scale facility, to relate the trafficability of various types of earthmoving machine to the MCV of the soil (Parsons and Toombs, 1988).

Studies of the trafficability of soil by earthmoving plant

20.17 Information has been obtained on criteria for acceptance of fill material for situations where

Plate 20.2 The two-axle articulated dump truck

the operation of earthmoving plant is an important consideration (Parsons and Toombs, 1988). Pilot-scale studies were carried out using two types of articulated dump truck, a scraper box towed by a crawler tractor, a two-axle tipper lorry and three types of vehicle representative of three-axle dump trucks with rigid chassis. One of the articulated dump trucks is shown in Plate 20.2. The machines were operated on the sandy clay No 2 soil for most of the tests and a limited study was carried out on heavy clay and well-graded sand. Particle-size distributions and other properties of these soils are given in Figure 3.6.

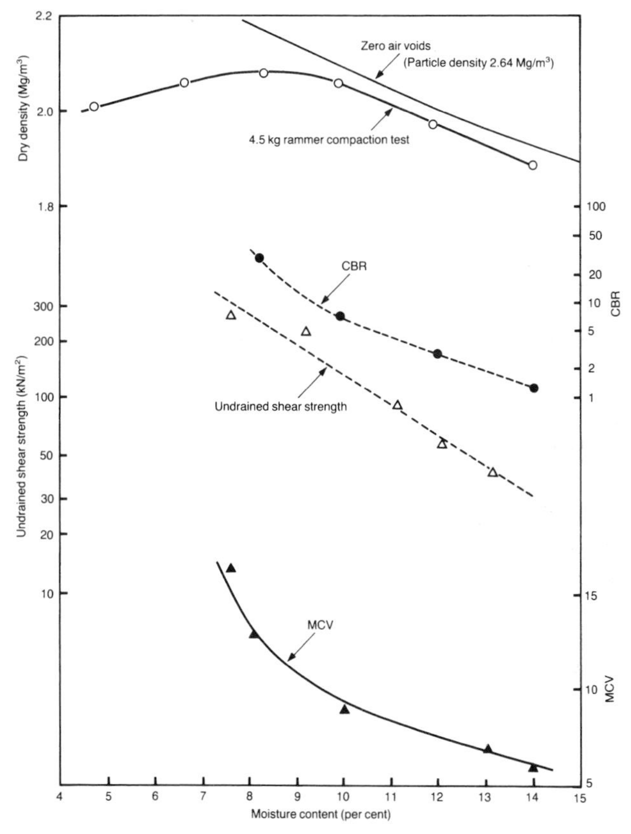

Figure 20.6 Results of laboratory tests on a bulk sample of boulder clay from which limits of acceptability in terms of MCV may be set

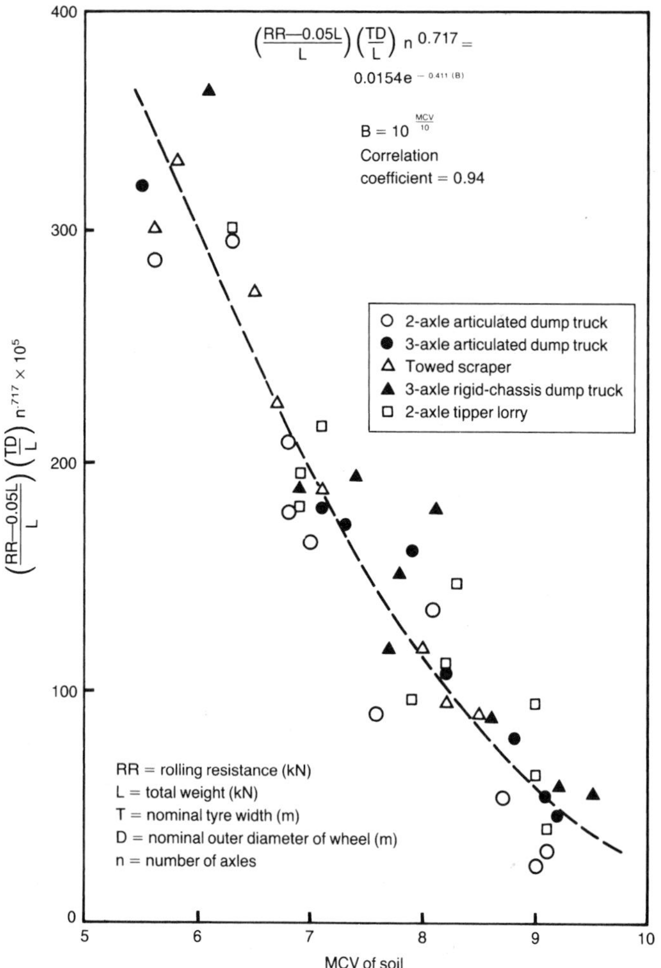

Figure 20.7 General relation for the determination of rolling resistance on compacted sandy clay soil (Parsons and Toombs, 1988)

Tests were made with the soils fully compacted and in 200 mm-thick loose layers. Measurements were made of rolling resistance, wheel slip, depth of wheel rut and soil strength; MCV was used to represent the soil strength. Assessments were made of the trafficability of the machines in the loaded, half-loaded and empty conditions over a range of soil strengths close to the limits of operation of the plant.

20.18 From the results general relations were established between rolling resistance and MCV and between depth of wheel rut and MCV, using total weight, number of axles, nominal tyre width and nominal outer diameter of wheel as variables. The general relation for determining the rolling resistance on compacted sandy clay soil is shown in Figure 20.7. It was found that the effect on rolling resistance of loose soil compared with compacted soil, all other factors being equal, was small at or near the limit of effective operation.

20.19 It was found that the effective rimpull at the onset of wheel slip was about 0.20 of the weight on the driving wheels. Traction force increased with wheel slip, the principal factor influencing the rate of increase being the number of driving wheels.

20.20 From results obtained on sandy clay, two alternative methods were suggested for predicting the maximum drawbar pull of crawler tractors. These were based on either the total weight of the tractor or a combination of the vane shear strength of the soil and the contact area of the tracks; it was suggested that the lower value be applied. These methods were assumed to apply to intermediate and high plasticity clays.

20.21 Charts were produced for predicting the MCV of clay soil under the following conditions:–

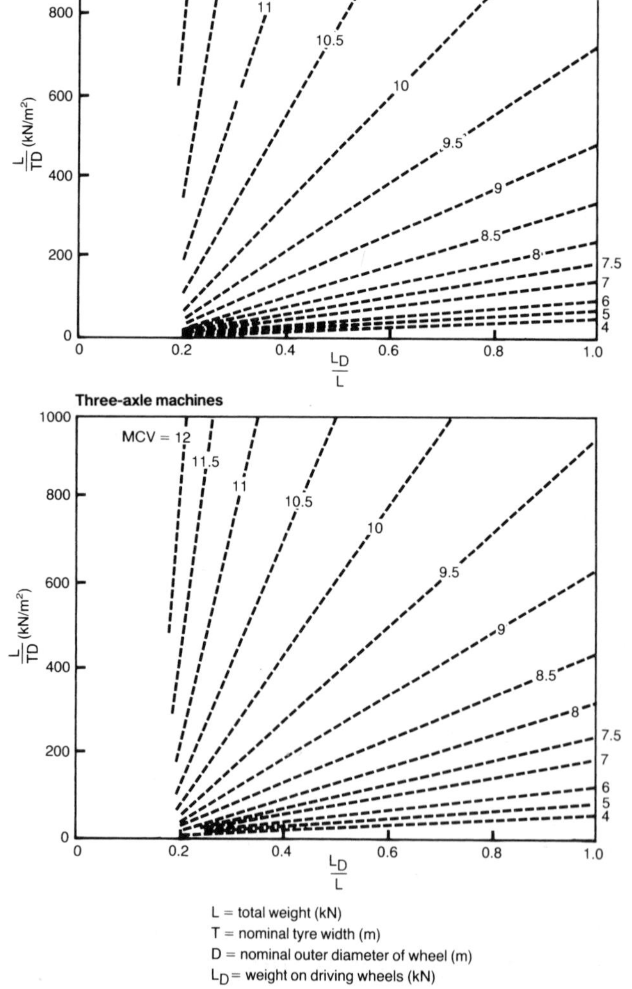

Figure 20.8 Charts for determining the MCV of clay soil at the onset of wheel slip of self-propelled plant (Parsons and Toombs, 1988)

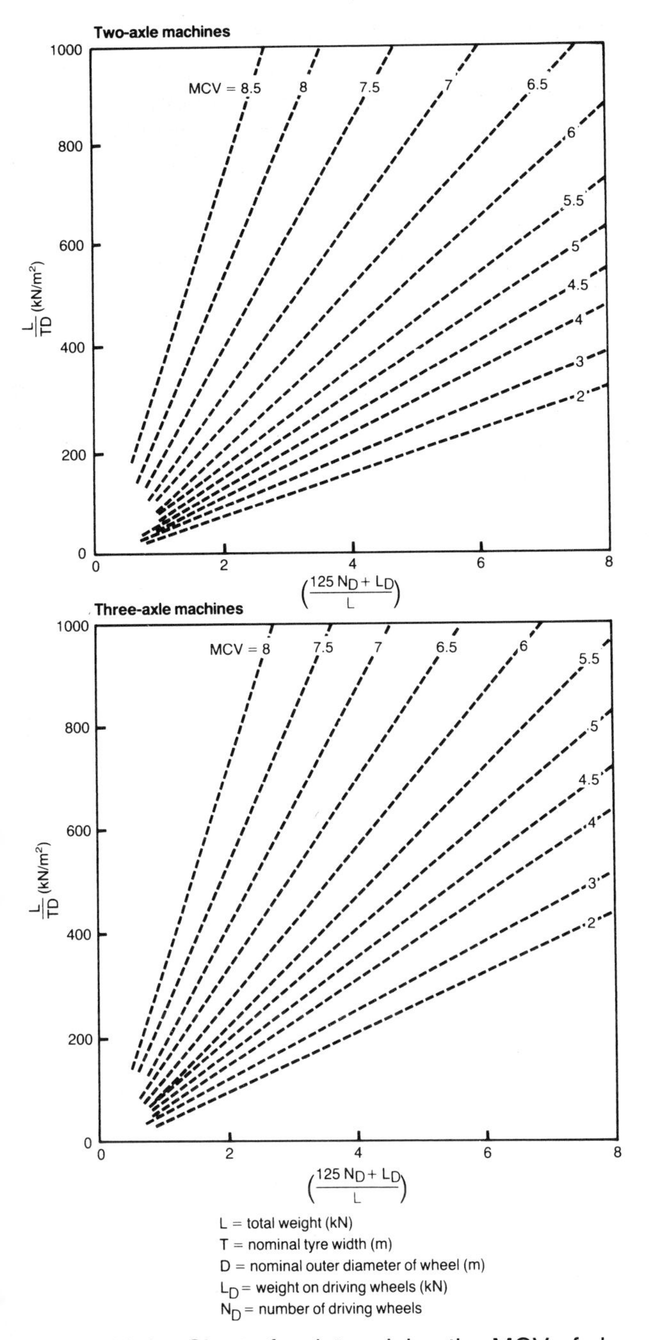

Figure 20.9 Charts for determining the MCV of clay soil at which self-propelled plant become 'bogged down' (Slip = 1.0). Based on results with sandy clay (Parsons and Toombs, 1988)

(i) At the onset of wheel slip of wheeled self-propelled plant (Figure 20.8);

(ii) At total wheel slip of wheeled self-propelled plant (Figure 20.9);

(iii) At the onset of track slip for crawler tractors towing scraper boxes (Figure 20.10);

(iv) At the limit of available rimpull or drawbar pull of all types of machine (Figure 20.11).

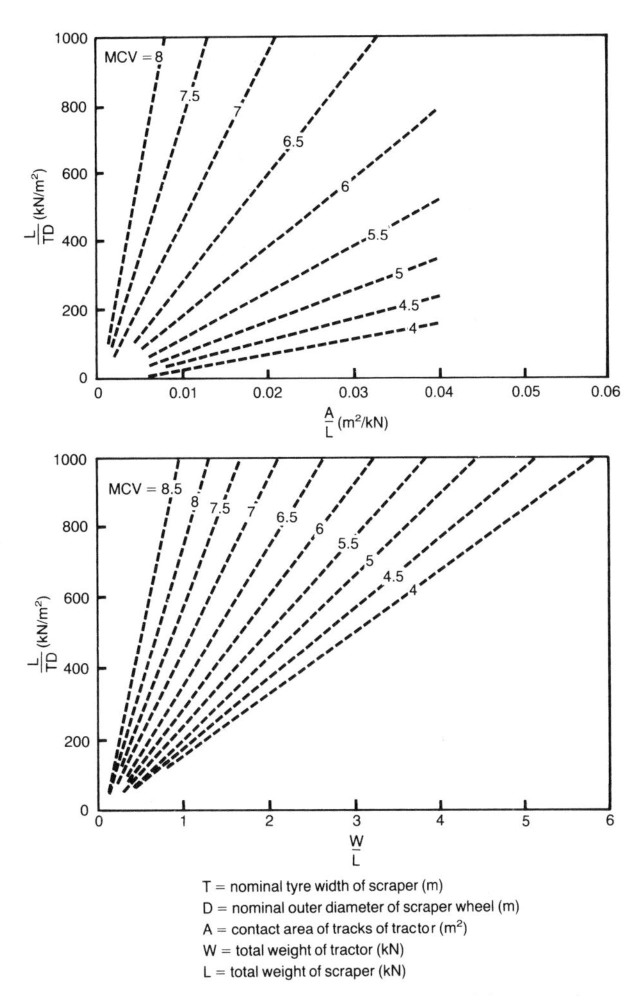

Figure 20.10 Two-axle scrapers towed by crawler tractors: charts for determining the potential limits of operation at onset of track slip with intermediate and high plasticity clays (Parsons and Toombs, 1988)

20.22 With more granular soils, ie soils with less than about 25 per cent passing a 63 μm sieve, the charts given in Figures 20.8 to 20.11 were considered to be conservative (Parsons and Toombs, 1988), so that machines could be expected to operate at lower levels of MCV.

20.23 The results of the pilot-scale studies, in combination with the results of earlier studies on construction sites, were used to produce relations between MCV at the limit of effective operation (wheel slip of 0.25 or onset of track slip) and loaded mass for various types of earthmoving machine available in the United Kingdom. These are shown in Figures 20.12 to 20.16.

20.24 These various Figures provide a comprehensive guide to the possible effects of soil conditions, in terms of MCV, on the effectiveness of the operation of earthmoving vehicles. Where the engineering requirements of

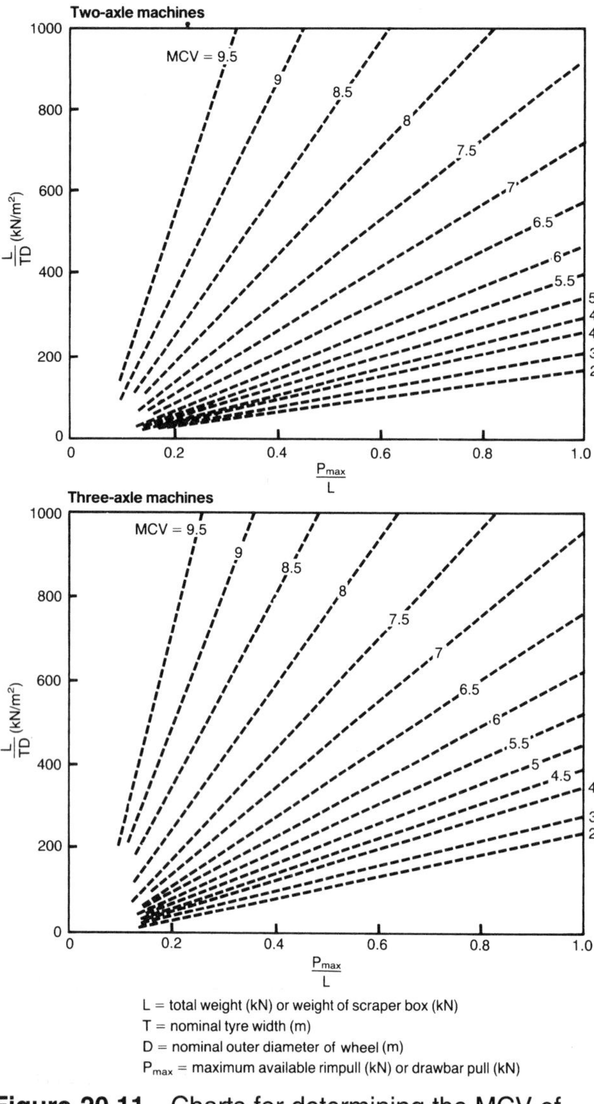

Figure 20.11 Charts for determining the MCV of clay soil below which self-propelled plant have insufficient rimpull or crawler tractors towing scraper boxes have insufficient drawbar pull to overcome rolling resistance (Parsons and Toombs, 1988)

an embankment fill allow the use of comparatively weak material, the information given may be used to select the most effective type of plant; alternatively, an improved limiting soil condition can be set to allow the effective operation of certain types of plant.

Precision of the moisture condition test

20.25 With the increasingly widespread application of the moisture condition test since it was first developed in 1976, the test proved to be very effective on cohesive materials, but on granular, non-cohesive materials the determination of MCV was found to be difficult and liable to error. To obtain information on the precision of the test a study was made with two types of soil, heavy clay and well-graded sand, the characteristics of which are shown in Figure 3.6. Ten independent laboratories co-operated in the study, which has been fully reported by Parsons and Toombs (1987).

20.26 The procedure followed was as detailed in BS 5497 : Part 1 (British Standards Institution, 1979). To produce a range of test values each laboratory was provided with samples at various levels of moisture content. The two soil types were each tested at four levels of MCV, with two replicated tests on each sample. Thus, each laboratory was requested to carry out a total of 16 tests, including with each determination of MCV a measurement of moisture content using the oven-drying method (British Standards Institution, 1990a). Analyses were made of the original results and of the MCVs after correction for moisture content variations at each test level.

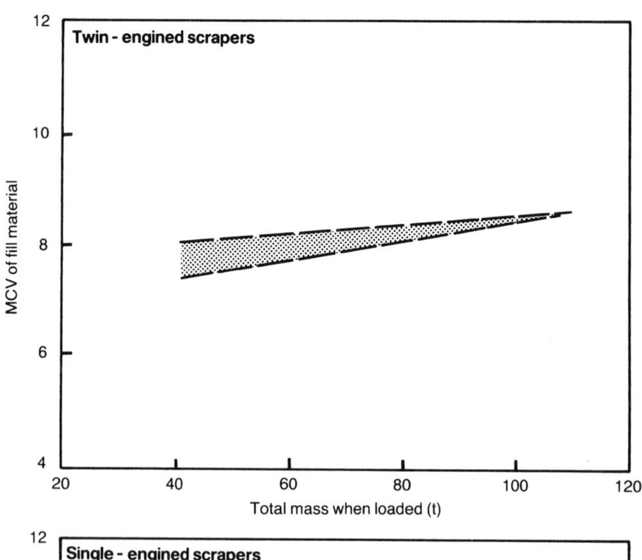

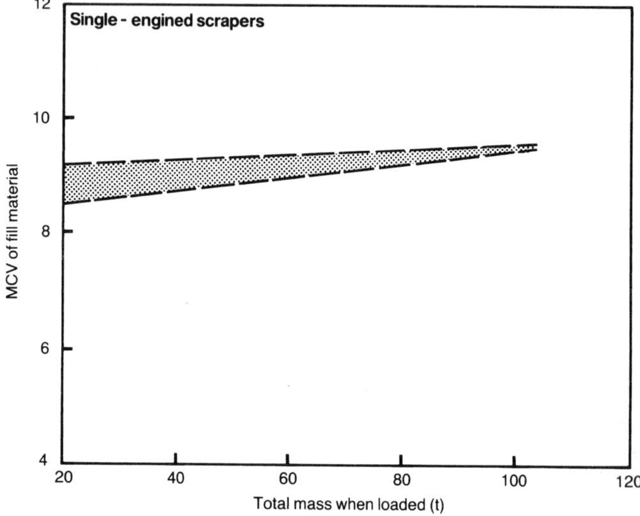

Figure 20.12 Self-propelled scrapers: predicted relations between the MCV of clay soil for a wheel slip of 0.25 and total mass when loaded. All available machines fall within shaded areas (Parsons and Toombs, 1988)

20.27 On collating the results from the ten independent laboratories a number of erratic results were discarded and the reasons for such results subsequently investigated. The number of laboratories contributing to the final determination of precision was reduced, on the heavy clay, to six and on the well-graded sand, to seven. Only for the third and fourth test levels of the heavy clay were the corrected MCVs considered to be more valid than the originals, the differences at these test levels being caused by trends of increasing moisture content throughout the samples in the order in which they were batched. There was no clearly identifiable relation between the values of repeatability and reproducibility and the test level and the mean results, in units of MCV, were as follows:–

Heavy clay	–repeatability	0.8
	reproducibility	1.6
Well-graded sand	–repeatability	1.6
	reproducibility	2.3

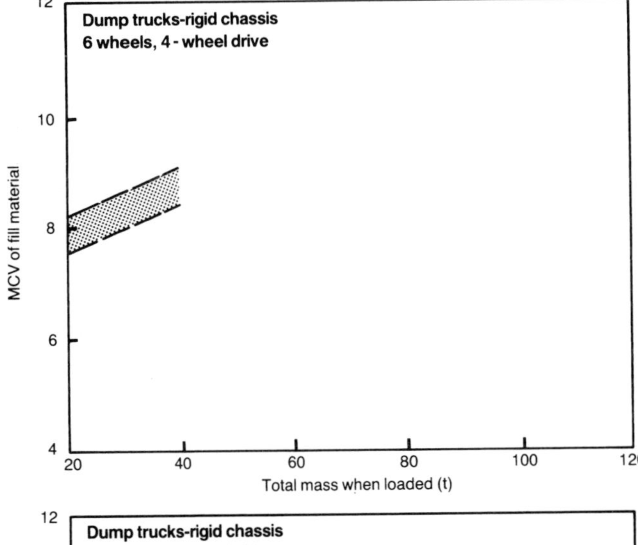

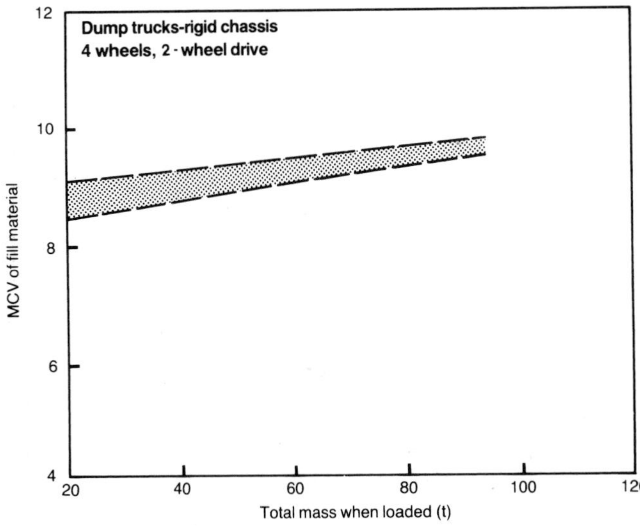

Figure 20.13 Rigid-chassis dump trucks: predicted relations between the MCV of clay soil for a wheel slip of 0.25 and total mass when loaded. All available machines fall within shaded areas (Parsons and Toombs, 1988)

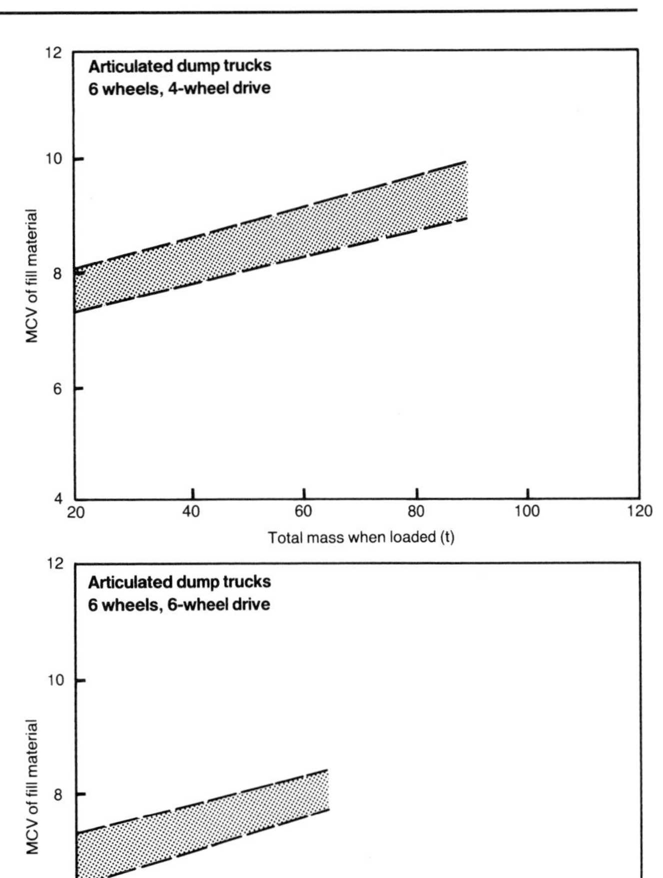

Figure 20.14 Three-axle articulated dump trucks: predicted relations between the MCV of clay soil for a wheel slip of 0.25 and total mass when loaded. All available machines fall within shaded areas (Parsons and Toombs, 1988)

20.28 Typical relations between change in penetration of the rammer and number of blows of the rammer, the latter on a logarithmic scale, as plotted in the determination of MCV (see Paragraphs 20.12 to 20.13), are shown in Figure 20.17. These illustrate the small changes in penetration associated with the well-graded sand in comparison with the heavy clay. Potential errors associated with small inaccuracies in the measurement of the protrusion of the rammer would have had a considerable effect on the derived MCV for the well-graded sand and would have been the principal cause of the relatively poor precision determined for that soil. On the

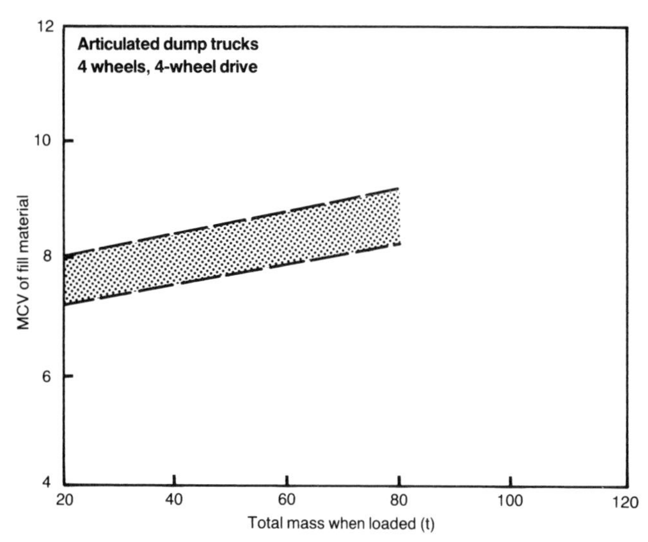

Figure 20.15 Two-axle articulated dump trucks: predicted relation between the MCV of clay soil for a wheel slip of 0.25 and total mass when loaded. All available machines fall within shaded area (Parsons and Toombs, 1988)

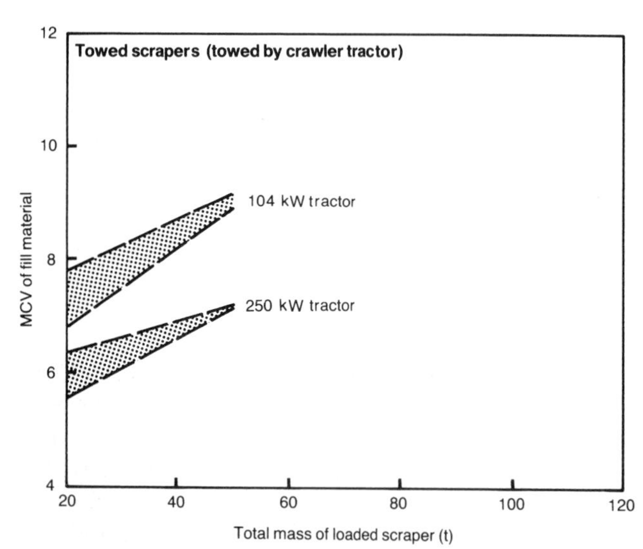

Figure 20.16 Towed scrapers: predicted relations between the MCV of intermediate and high plasticity clays at the onset of track slip of the crawler tractor and total mass of the loaded scraper (see Figure 20.10). All available towed scrapers fall within shaded areas (Parsons and Toombs, 1988)

basis of the potential errors in MCV, it was considered (Parsons and Toombs, 1987) that the values of repeatability and reproducibility for heavy clay, given in Paragraph 20.27, also apply to any soil where the slope of the change in penetration-number of blows relation (Figure 20.17) exceeds 8 mm per logarithmic cycle of blows (1 to 10, 2 to 20, etc). The results given in Paragraph 20.27 for well-graded sand were considered to apply to any soil where the slope of the change in penetration-number of blows relation is in the range of 3 to 8 mm per logarithmic cycle of blows. It can be inferred that if the slope of the relation is less than 3 mm per logarithmic cycle of blows the loss of precision would make the test unworkable as a practical method of controlling the acceptability of earthwork fill material; granular materials with little or no cohesion are most likely to produce relations with such low values of slope.

20.29 An investigation of the reasons for the discarded results showed that in two laboratories procedural errors involved the incorrect definition of the number of blows of the rammer. It was also found that significant, but not necessarily detectable, errors arose because of the loss of free-fall conditions for the rammer. Several clarifying amendments were made to the text of the test procedure to safeguard against these forms of error.

Further applications of the moisture condition test: (a) soil classification

20.30 The potential uses of the moisture condition test to determine the strength of soil, either in terms of undrained shear strength or of CBR, have been discussed in Paragraph 20.15 in the context of setting the limits of acceptability of earthwork material. Its use to predict the performance of earthmoving plant has also been discussed in Paragraphs 20.17 to 20.24.

20.31 Parsons and Boden (1979) suggested that the relation between moisture content and MCV, often a straight line of the form:–

$$w = a - b(MCV),$$

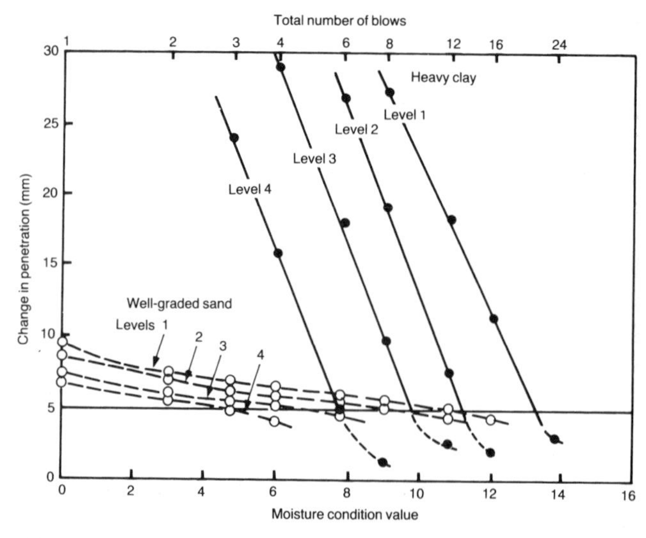

Figure 20.17 Typical relations obtained in the determination of moisture condition value

provided a slope 'b' and an intercept 'a' that were useful indicators of the class of soil. The factor 'b' was considered indicative of the sensitivity of the soil to changes in moisture content whilst 'a' was considered to be an arbitrary low-strength moisture content value which would be used in the same context as the liquid limit of a clay soil. Ewan and West (1983) explored this concept further making direct comparisons between the 'a' – 'b' classification from the MCV calibration and the Casagrande classification based on the liquid limit and plasticity index (Road Research Laboratory, 1952). They concluded that the 'a' – 'b' chart was unlikely to provide a basis for a soil classification system, but with clay soils higher 'a' – 'b' values were indicative of higher plasticity and therefore provided a useful rough guide. Ewan and West (1983) also pointed to the possibility that with clay-type soils plastic limit might be predicted from the moisture content – MCV relation, ie at a unique MCV, although they considered that substantial further testing would be required to verify this.

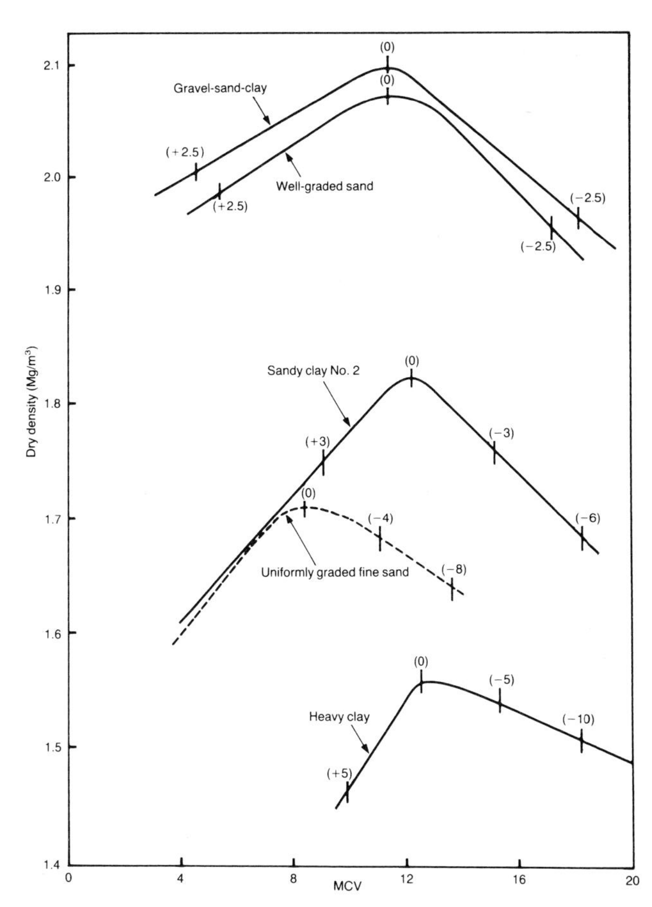

Figure 20.18 Results of 2.5 kg rammer compaction tests in the form of dry density – MCV relations. Values in brackets refer to moisture content relative to the optimum moisture content

Further applications of the moisture condition test : (b) compactability of soil

20.32 The possible refinement of the method specification for the compaction of earthwork materials, whereby the compactive effort would be adjusted according to the MCV of the material, has been discussed in Paragraph 19.66. Variations in compactive effort necessary to yield 95 per cent relative compaction (by comparison with the maximum dry density obtained in the 2.5 kg rammer compaction test) and 10 per cent air voids at different values of moisture content have been given in Tables 15.2 to 15.16. The moisture contents used were related to the optimum moisture content of the 2.5 kg rammer test. Recently obtained results of 2.5 kg rammer compaction tests are given in Figure 20.18 in the form of dry density – MCV relations, ie the moisture content values have been converted to MCVs from the appropriate moisture condition calibrations. Also shown on the curves in the Figure are the points equivalent to the moisture contents used in Tables 15.2 to 15.16. Note that the lowest moisture contents occur to the right of the Figure. Thus, for heavy clay, the points are given on the curves for optimum – 10 per cent, optimum – 5 per cent, optimum, and optimum + 5 per cent; the equivalent MCVs can be derived from the horizontal axis of the Figure.

20.33 For the well-graded granular soils and the cohesive soils the optimum of the 2.5 kg rammer test occurs at an MCV ranging from 11.5 for the gravel-sand-clay and well-graded sand to about 12.5 for the heavy clay. The values of moisture content used in Tables 15.2 to 15.16 and their equivalent MCVs shown in Figure 20.18 are also given in Table 20.1. The equivalent MCVs (Table 20.1) show very similar ranges for the heavy cohesive and light cohesive materials (represented by the heavy clay and sandy clay respectively); each of the two well-graded granular soils (well-graded sand and gravel-sand-clay) has approximately the same range of MCVs, the lowest moisture content (5 per cent below the optimum of the 2.5 kg rammer compaction test) being too dry for the MCV to be measurable).

20.34 The MCV of uniformly graded fine sand at its optimum moisture content, as shown in Figure 20.18, is widely different from the MCVs registered at optimum for the four well-graded soils. With this soil the slope of the change in penetration-number of blows relation plotted in

Table 20.1 Estimated MCVs equivalent to the various moisture contents at which the performances of compaction plant have been calculated in Tables 15.2 to 15.16

Soil type	Moisture content relative to the optimum of the 2.5 kg rammer test (per cent)	Equivalent MCV
Heavy clay	−10	18
	− 5	15.5
	0	12.5
	+ 5	10
Sandy clay No 2	− 6	18
	− 3	15
	0	12
	+ 3	9
Well-graded sand	− 5	–
	− 2.5	17
	0	11.5
	+ 2.5	5.5
Gravel-sand-clay	− 5	–
	− 2.5	18
	0	11.5
	+ 2.5	4.5
Uniformly graded fine sand	− 8	13.5
	− 4	11
	0	8.5

the determination of MCV ranged from 3 to 5 mm per logarithmic cycle of blows; it must be considered, therefore, that the use of the moisture condition test with this soil is only marginally effective (see Paragraph 20.28).

References

British Standards Institution (1979). Precision of test methods : BS 5497 : Part 1 Guide for the determination of repeatability and reproducibility for a standard test method. BSI, London.

British Standards Institution (1990a). British Standard methods of test for soils for civil engineering purposes : BS 1377 : Part 2 Classification tests. BSI, London.

British Standards Institution (1990b). British Standard methods of test for soils for civil engineering purposes : BS 1377 : Part 4 Compaction related tests. BSI, London.

Department of Transport (1986). Specification for highway works. 6th edition. HM Stationery Office, London.

Ewan, V J and West, G (1983). Appraising the moisture condition test for obtaining the Casagrande classification of soils. *Department of the Environment, Department of Transport, TRRL Supplementary Report* 786. Transport and Road Research Laboratory, Crowthorne.

Greenwood, J R (1987). Private communication.

Kennard, M F, Lovenbury, H T, Chartres, F R D and Hoskins, C G (1979). Shear strength specification for clay fills. *Clay Fills.* Institution of Civil Engineers, London, 143–7.

Ministry of Transport (1969). Specification for road and bridge works. 4th edition. HM Stationery Office, London.

Parsons, A W (1965). An investigation of the variability of moisture content and plastic limit of cohesive soils in earthwork construction. *D.S.I.R., Road Research Laboratory Note* No LN/753/AWP (Unpublished).

Parsons, A W (1976). The rapid measurement of the moisture condition of earthwork material. *Department of the Environment, TRRL Laboratory Report* 750. Transport and Road Research Laboratory, Crowthorne.

Parsons, A W and Boden, J B (1979). The moisture condition test and its potential applications. *Department of the Environment, Department of Transport, TRRL Supplementary Report* 522. Transport and Road Research Laboratory, Crowthorne.

Parsons, A W and Toombs, A F (1987). The precision of the moisture condition test. *Department of Transport, TRRL Research Report* 90. Transport and Road Research Laboratory, Crowthorne.

Parsons, A W and Toombs, A F (1988). Pilot-scale studies of the trafficability of soil by earthmoving vehicles. *Department of Transport, TRRL Research Report* 130. Transport and Road Research Laboratory, Crowthorne.

Road Research Laboratory (1952). Soil mechanics for road engineers. HM Stationery Office, London, Chapter 4.

Sherwood, P T (1971). The reproducibility of the results of soil classification and compaction tests. *Ministry of Transport, RRL Report* LR 339. Road Research Laboratory, Crowthorne.

Symons, I F (1979). General report: Performance of clay fills. *Clay Fills.* Institution of Civil Engineers, London, 297–302.

INDEX

Printed in the United Kingdom for HMSO
Dd294752 11/92 C11 G531 10170